World Regional Geography

A Short Introduction

John Rennie Short

University of Maryland, Baltimore County

New York   Oxford
OXFORD UNIVERISTY PRESS

Oxford University Press is a department of the University of Oxford.
It furthers the University's objective of excellence in research, scholarship,
and education by publishing worldwide. Oxford is a registered trademark
of Oxford University Press in the UK and certain other countries.

Published in the United States of America by Oxford University Press,
198 Madison Avenue, New York, NY 10016, United States of America.

For titles covered by Section 112 of the US Higher Education
Opportunity Act, please visit www.oup.com/us/he for the latest
information about pricing and alternate formats.

Library of Congress Cataloging-in-Publication Data

Names: Short, John R., author.
Title: World regional geography : a short introduction / John Rennie Short.
Description: New York : Oxford University Press, [2020] | Includes
    bibliographical references.
Identifiers: LCCN 2019017350 | ISBN 9780190206703 (pbk.)
Subjects: LCSH: Geography—Textbooks.
Classification: LCC G128 .S544 2020 | DDC 910—dc23 LC record available
at https://lccn.loc.gov/2019017350

Printing number: 9 8 7 6 5 4 3 2 1
Printed by Quad/Graphics, Inc.
Mexico

I dedicate this book to the memory of my mother.

In my mind's eye, Netta is forever young, always beautiful. She died when she was only twenty-five and I was only four. This book describes a world she did not get to see but one that her adoring son travels and writes about with her constantly in mind and always in his heart.

Brief Contents

Contents

PART 2   REGIONS OF THE WORLD   39

3   **Central America and the Caribbean   40**

4 South America 66

6 Russia and Its Neighbors   118

7 East Asia   142

⑧ **South East Asia   168**

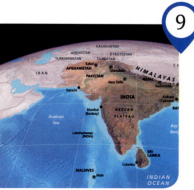

10 The Middle East and North Africa   218

⑬ North America   296

Preface

The primary market for this book is the introductory course in world regional geography. There are a large number of world regional texts, and so in a crowded market, the obvious question is, what makes this book stand out?

This new text allows a fresh departure point from many established world regional texts that over the years and numerous editions have grown to include far too much material, much of it dated and overly congruent between regions. This book introduces students to the character of world regional geography, the distinctiveness of different parts of the world, as well as the linkages and connections. Old established regional texts often have difficulty with the linkages because their focus was originally in demarcation rather than connectedness. This text will start off from the premise of describing both the character of the world's regions but also their interconnections.

This book is concise, but brevity does not compromise depth. The text is designed to appeal to the fundamental excitement of learning about new places. The narrative is crafted to be both interesting and accessible. The text is long enough to cover the main themes but concise enough to maintain interest. Overall, it provides an affordable, accessible, and interesting read that kindles a sense of excitement.

The Need for a New World Regional Geography

The principal market for this book is in the United States. A striking feature of life in the contemporary United States is the enormous ignorance of the outside world. This is not a new state of affairs. The US educational system, especially at the elementary and high school levels, spends little time on the history and geography of other parts of the world. The emphasis is on the here and now of the contemporary United States compared to other parts of the globe and their different histories. The vast bulk of precollege education is concerned with events in the United States. Most educational systems have a national bias; after all, each serves as an element in the creation of national identity and consciousness. But this insularity is reinforced in the United States by three factors. First, as an immigrant society, education has long been dominated by the need to create a national citizenry. The school is one of the principal places of national socialization. Emphasis has long been placed on national history and national identity. Less attention has been spent on teaching students about other places and other times.

Second, in many other countries the simple facts of geography and the disputed facts of history have created a spatial awareness of others. Sharing borders with only two countries, separated by two oceans from Europe and Africa on the one side and Asia from the other, the United States can seem less connected to the rest of the world. The sheer size of the country also means that internal travel provides continental difference. Unlike sun seekers in Northern Europe, people in the United States do not need to leave their own country to experience subtropical climates, see desert landscapes, or go skiing in the high mountains.

Third, the bias of the educational system is reinforced by the military and cultural power of the United States. At a military level, the United States is the undisputed superpower; perhaps the only one with sustained global reach and capabilities. The US projects power onto others, and this brute fact becomes part of a formal and informal understanding of the world. Almost everywhere else in the world the power of others and especially the United States is a fact of both national geography and history. The United States is also a hegemonic cultural power. Symbols and images are projected out from the United States to the rest of the world. There is two-way traffic, but one dominated by the export rather than the import trade. People in other countries learn about the United States. They are made acutely aware every time they hear world news or watch a Hollywood movie that there are other places in the world with a different history.

The rest of the world is not a distant place nor is it a mirror of ourselves. We need to understand the rest of the world as something that is both similar in important ways to us but also different in profound ways. A genuine multiculturalism should be aware of other parts of the world, their historical particularities and geographical differences. An understanding of world regional geography is a good place to start.

Structure of the Book

An introductory chapter presents readers with the opportunities and dangers of using maps, data, and different scales of analysis. Readers are made aware that world regions are useful but a provisional partitioning of the world hides major regional and urban differences, and that similar countries, regions, and neighborhoods are found throughout the world.

Part 2 is the heart of the book and is a concise geography of the major regions of the world:

- Central America and the Caribbean
- South America
- Europe
- Russia and Its Neighbors
- East Asia
- South East Asia
- South Asia
- The Middle East and North Africa
- Sub-Saharan Africa
- Australia and Oceania
- North America

Readers may have their own idea of the ordering of the main regions. Some would start close at home with North America. Others with Europe. A good case could be made for Africa, the original home of humanity and the point of diffusion of humans across the globe. Good justifications can be made for many different starting points. I prefer to start with Central America. Up until 1492, there was a continental divide between the Old and New World. For thousands of years, they remained separate, sometimes with distinctly different flora and fauna and even genetic make-up. After 1492 and the **Columbian Encounter**, the world was seared together in cultural and physical interactions, some accidental, others planned. And many were brutal and exploitative and turned the two separate regions into a global unity. Central America is the scene where a singular world took shape. So I begin this book in the region where the "world" of "world regional geography" was created.

For each region the following topics are covered, although with enough variety to avoid the relentless repetition and tedious march of very similar themes for each region:

- The Environmental Context
- Historical Geographies
- Economic Transformations
- Rural Focus
- Social Geographies
- Urban Trends
- Geopolitics
- Connections
- Subregions
- Focus

In the *Environmental Context*, I will highlight significant environmental challenges. This is not a physical geography text, so the emphasis is on the human–environmental relations and especially on the dominant environmental issues facing the region.

In order to understand the contemporary human geography of a region, it is necessary to have an appreciation of its *Historical Geographies*. In this section I will give a historical dimension to contemporary concerns. An understanding of historical geography is vital to comprehend the contemporary state of affairs.

Economic Transformations looks at economic differences within the region and how the region fits into the global economy. A *Rural Focus* will provide a more granular look at selected rural issues. In *Social Geographies*, I will provide an understanding of the different peoples of the regions, demographic trends, and important issues in the cultural geography of the region. *Urban Trends* looks at patterns of urbanization, the role of cities, and discusses specific cities in a series of focused case studies. In *Geopolitics*, I will discuss current issues that link geography with intra- and international relations. In order to highlight the linkages, *Connections* will focus on some of the flows and transactions between the region and the rest of the world.

In *Subregions*, I will look at the character of different countries within the region in order to provide a finer-grained analysis than the broad regional survey. A series of *Focus* sections will provide more detailed discussions of particular features of the region.

The broad themes are the same for each region, but the specifics are tailored to focus on the distinctiveness of the region. Thus, the *Urban Trends* section of South Asia looks at the dramatic rise of urbanization, while the North America chapter highlights the rise of metropolitan regions, the increasing diversity of big cities, and the urban bias to the rise of the creative-cultural economy. The Australasia region, in contrast, will focus on the domination of just a few very big cities, the suburban spread, and the urban environmental implications of global climate change.

Hopes for the Book

Although a seasoned author and mature academic, I have not lost my child-like sense of awe at being in the world, my love of travel, and my excitement for going to new places and meeting different people. The longer I live, and the more I travel, the more I am aware of the privileged nature of my life and the wondrous nature of our world. I hope that some of this joy and wonder and appreciation find their way into this book.

Acknowledgments

I'd like to thank the many scholars who, during the writing process, dedicated valuable time to reviewing and offering comments on the manuscript. In these busy times, I am so very grateful for their input. Their comments improved the book significantly.

- Angela Antipova, University of Memphis
- Lewis Asimeng-Boahene, Penn State University-Harrisburg
- John T. Bauer, University of Nebraska Kearney
- David Lee Baylis, Delta State University
- Richard W. Benfield, Central Connecticut State University
- Mikhail Blinnikov, St. Cloud State University
- Karl Byrand, University of Wisconsin Colleges
- Craig M. Dalton, Hofstra University
- Neal Devine, Santa Fe College
- Seth Dixon, Rhode Island College
- Catherine Elspeth Doenges, Southern Connecticut State University
- Dawn M. Drake, Missouri Western State University
- Michael Dunbar, Kent State University
- Michael Finewood, Pace University
- F. Tyler Huffman, Eastern Kentucky University
- Injeong Jo, Texas State University
- Juan Miguel Kanai, University of Sheffield
- Yeong-Hyun Kim, Ohio University
- Yong Lao, California State University Monterey Bay
- Max Lu, Kansas State University
- Stephen McFarland, University of Tampa
- Caroline Nagel, University of South Carolina
- Lindsay Naylor, University of Delaware
- Dristi Neog, Westfield State University
- Trushna Parekh, Texas Southern University
- Thomas Pingel, Northern Illinois University
- Adrien M. Ratsimbaharison, Benedict College
- Evelyn Ravuri, Saginaw Valley State University
- Lesli Rawlings, Wayne State College
- Philip D. Roth, Indiana University/Purdue University Indianapolis
- Ginger L. Schmid, Minnesota State University, Mankato
- Andrew Sluyter, Louisiana State University
- Kristin Sorensen, South Plains College
- Jennifer Titanski-Hooper, Wright State University
- Stanley Toops, Miami University
- Annette Watson, College of Charleston
- April Watson, Broward College
- Clayton Whitesides, Coastal Carolina University

I am very lucky to work with a team of great professionals at Oxford University Press. I was fortunate to have not just one but two great editors: Dan Kaveney was involved at the earlier stages, and Daniel Sayre saw the project through to completion. Daniel's enthusiasm for the project is deeply appreciated. I was delighted to work for a second time with OUP's very own Dream Team of Marianne Paul, Production Editor, who turned my words on to the printed page and Michele Laseau, Art Director, who shaped the elegant design of the book. Leslie Anglin copyedited a large and complex manuscript with great care. Thanks to Sarah Goggin for the Learning Objectives. A special thanks to Kevin Lear and his team at International Mapping for the beautiful maps. The book in your hands is the work of these talented professionals. My thanks.

Lisa Benton-Short has travelled with me on many field trips around the world and not only tolerated but also actively encouraged my many solo travels necessary to write this book. Whether at home or abroad, she is the very best of companions.

World Regional Geography

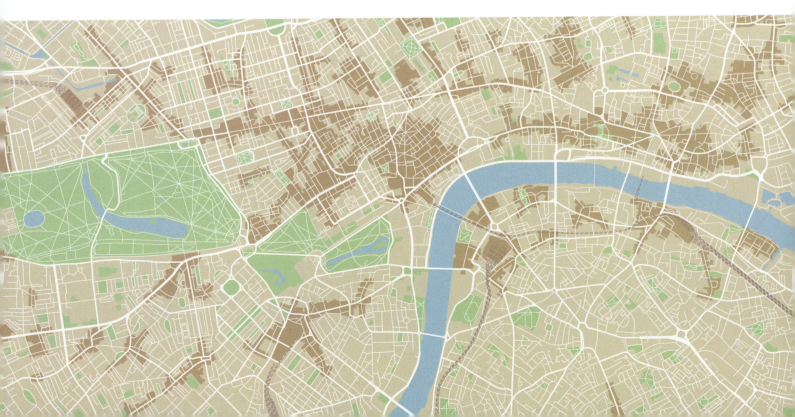

Setting the Scene

Part 1 sets the scene. Chapter 1 is a reminder that the geography of the world is complex. The maps that we use, the terms we employ, and the regional divisions that we create are neither neutral nor fixed. Different maps, other terms, and alternative demarcations would provide a slightly different, if not major difference in perspective. So be wary: this book is just one of many possible regional geographies. And please remember that while the book divides the world into different regions, the geography of the world is also one of flows just as much as territorial borders. Fluidity and movement just as much as fixed territorialities constitute the nature of our world. We inhabit a space of flows.

Chapter 2 considers major themes of global significance, including environmental and geopolitical challenges, socioeconomic change, and demographic transformations. All of the regions face these major challenges.

OUTLINE

1

A World of Difference

LEARNING OBJECTIVES

Describe how the perspectives of early maps are a reflection of human understanding.

Discuss the biases and distortions found in maps, with emphasis on modern versions of mapping.

Recognize the inherent, changing biases in the terminology used to name regions, places, and peoples.

Recall the standard terms that have been used to divide the world and name its regions.

Summarize the inherent flaws of data and scale in categorizing regions of the world.

Discuss the long tradition of writing about the world and the "othering" that is evident in this tradition.

Describe how places have multiple identities and how spaces are composed of ongoing flows.

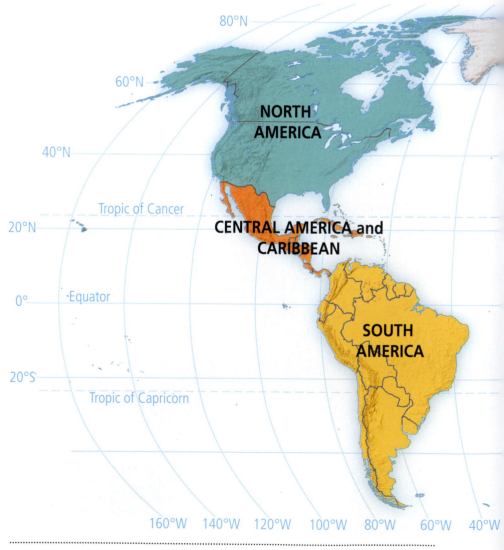

1.1 World map of regions (as used in this book)

This chapter contains words of warning. Studying world regional geography is beset with issues and problems, some more obvious than others. These include the type of maps we use, the terms we employ, the data we analyze, the divisions we create, and the narratives we invoke. World regions are outlined, but you are asked to remember that they are places of incredible internal diversity as well as sites of flows and connection with other parts of the world.

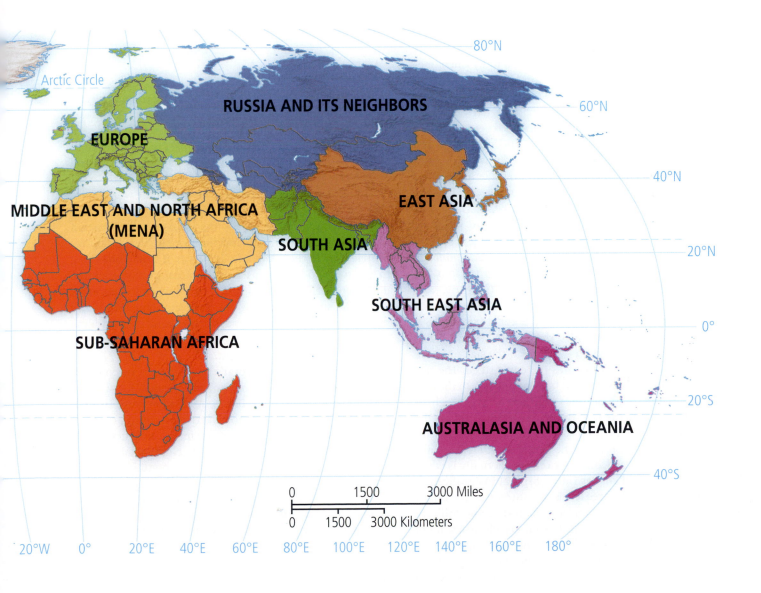

The Maps We Use

It is important to begin a world regional geography with a discussion of how we represent the world. The spatial representation of the world has a long history, and we can uncover important issues if we consider, even in a very brief manner, the long-term evolution of world maps. **T-O maps** of the world were produced in medieval Europe. They were first described in the eighth century and first printed in 1472 (Figure 1.2). In these maps, or series of maps, as a number of variants are recorded, the world is depicted as circular and surrounded by an ocean. The map only represents the view from the northern hemisphere. The land mass is divided into three parts: Africa, Asia, and Europe. The map is orientated (and this word comes from the Latin word *orient* for "East") with East at the top of the map. The name T-O derives from the shape that looks like a T formed inside of an O. The stem of the T is the Mediterranean, and the Nile and Don Rivers form the top bar. Because it is a Christian view of the world, the T is interpreted as a reference to the cross. Jerusalem is at the center of the map, and the three continents are also referred to by the names of Noah's three sons, Sem (Asia), Ham (Africa), and Japhet (Europe). The map was as much an object of faith as a document of knowledge. Other Christian cartographers produced maps that incorporated more recent information from contemporary exploration and discovery. The monk Fra Mauro, for example, produced a world map in the monastery of San Michele di Murano in Venice around 1450 that drew on new geographical knowledge that exceeded the constraints of the traditional T-O design.

We can see a very different world picture when we look at Figure 1.3, which shows one of the oldest Korean maps in existence. Commonly referred to as *Gangnido*, its full title is *Honil kangni yoktae kukto chi to* (*Map of Integrated Lands and Regions of Historical Countries and Capitals*). Its exact date of production is unknown, but estimates place the date between 1479 and 1485, early on in the Joseon dynasty that ruled the country from 1392 to 1910. The map is centered on China, and the size of Korea is exaggerated. China dominates in center stage. The rest of the world forms a distant and hazy periphery. In the left-hand corner, the rough outlines of Africa and the Arabian Peninsula can be made out. India is subsumed under the Chinese continent. The map reflects the dominant Joseon view of the world—China at the center with Japan represented as farther away from Korea than it actually is while the rest of the world shades into marginal insignificance. It is a world map with three levels of accuracy: at the center is an up-to-date picture of the administrative and military geography of Korea; in the next ring are clumsier, more anachronistic depictions of China and Japan; and finally the rest of the world forms a hazy marginal outer ring. While providing a Korean perspective, the map embodies

1.2 T-O map

other influences. The text is in Chinese, but the map draws upon the work of Islamic geographers and map-makers whose maps were brought to China during Mongol rule and then found their way to Korea. There is evidence of cross-cultural encounters as the place names in Africa and Asia derive from Persian and Arabic originals.

The difference between these maps and the world maps that we use today are substantial. The bias of the earlier maps is clearly visible to us—the influence of Christian theology on the T-O map and the Korean bias and Sino-centric view of the *Gangnido*. However, it would be wrong to see our modern maps as value-free, purely scientific documents. The world is a sphere; thus, when we represent it on a flat surface, we need to exaggerate some things and minimize others. Even our maps of the world contain some bias. As one prominent cartographer notes, "Not only is it easy to lie with maps, it's essential."

Figure 1.1 is a contemporary world map. If we compare it to Figures 1.2 and 1.3, it looks more scientific, more objective. But look again. It has biases, some more obvious than others. It is orientated with north at the top giving an inbuilt northern hemisphere bias. There are maps constructed with the South Pole at the top of the map. Notice that the map is centered along the line of 0 degrees longitude. But if we change the center to say the 180-degree longitude, then the world looks more like a giant Pacific base surrounded by land; Australia is now located in a more central position and North West Europe is on the periphery of the world. Centering the map on 90-degree longitude east or west or latitude north and south gives a very different perspective of the world's geography and different takes on what is central and what is peripheral (Figures 1.4 a–d). Moreover, the projection of the maps—these use either a Lambert Azimuthal Equal Area Projection or an Equatorial Orthographic Projection—have their biases, as do all maps that project the three-dimensional spherical earth onto a two-dimensional page.

There are also biases in what a map-maker chooses to present and how. In this case it is a political map of the world depicting countries. It assumes a national homogeneity; each country, after all, is one solid color. It is a political map with a narrow view of politics and no depiction of intra-country differences or intercountry similarities.

So you can see that how we understand the world is a function of how we represent the world. The distortions of the more distant past are easy to identify. It is much more difficult to be aware of the biases in our contemporary, taken-for-granted view of the world. What we take to be given and natural is, on much closer inspection, socially constructed and politically implicated.

1.3 Gangnido map

The Terms We Employ

We not only use maps to depict the world; we use words to describe it. And like world maps, these words are not innocent of wider social implications and deeper political connotations. Let us look at some of the common regional names. Latin America, for example, is often used to describe South America. But the term "Latin" has a Hispanic bias and ignores the pre-Columbian civilizations and indigenous people that are such an important part of the region's history and geography. To use the term "Latin" is to highlight the European dominance and to marginalize the indigenous peoples of the continent. The terms "Middle East" and "Far East," commonly used to describe parts of Asia, assume a description from a place located to the west of the region. In fact, the terms represent the description of the world from people in the European capitals of places like Paris and London, for whom Japan and China were at the other side of the world in the Far East, and the region of present-day Iran and Iraq was halfway there, hence Middle East. We continue to use these terms, a continuing legacy of European imperialism and colonialism, even though global power has shifted from London and Paris to Washington, DC. If the terms reflected newer geopolitical realities, with Washington, DC, at the discursive center, then the previous Far East could be described as the Far West and the old Middle East would become the Far East. An interesting exercise is to imagine new centers for naming the world. From Canberra in Australia, China, Japan, and Korea would be the Far North, and India and Pakistan would be the Northwest. The Americas would be the East, and Europe would be the Far North West.

Consider one region used in this book, *Russia and Its Neighbors*, that includes Kazakhstan, Kyrgyzstan, Tajikistan, Turkmenistan, and Uzbekistan. These former Soviet Republics, now independent countries, are sometimes referred to as Central Asia, a name initially used by the geographer Alexander von Humboldt in 1834 and still in use today. But that depends from where you are writing and for what purpose. For South Asian countries such as Pakistan and India, they are near neighbors that they want to connect with after years of isolation. From Russia it is the South, the Far West from China, and the Greater Middle

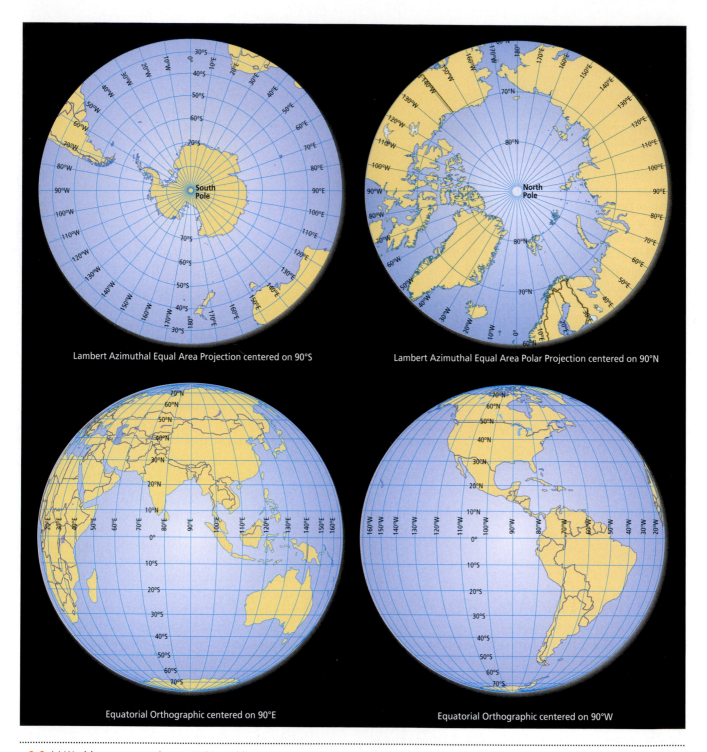

Lambert Azimuthal Equal Area Projection centered on 90°S

Lambert Azimuthal Equal Area Polar Projection centered on 90°N

Equatorial Orthographic centered on 90°E

Equatorial Orthographic centered on 90°W

1.4 (a) World map centered on 90°S (b) world map centered on 90°N (c) world map centered on 90°E and (d) world map centered on 90°W

East for US geopolitical strategy. I have placed these former Soviet Republics in a regional category with Russia because these states owe much of their present human geography to their connection and disconnection from the Russian and Soviet Empires and to their proximity to contemporary Russia.

Names are important because they connote more than just geographical descriptions. When people in the West use the term the "Middle East," very often distinct images, ideas, and reactions are invoked and employed, including the prevalence of terrorism and oppression of women. Others use the term "Middle East" to connote political instability and economic stagnation. The "West," when used by some, also has many negative connotations. For the many critics, the "West" is used as a critical term connected to moral decadence and geopolitical manipulation.

Even the cardinal direction names have deeper meanings. The terms "East" and "West" are loaded when used in certain contexts. For its supporters, the West conjures up images of freedom, individual rights, dynamic economies, and innovative societies. The East, in contrast, is a place of few liberties, lack of rights, stagnant economies, and ossified societies. For its critics, the term "West" often stands for rampant capitalism, mindless consumption, and moral decadence.

Social criticism and political commentary are often encased in geographical descriptions. Simple regional descriptions are enmeshed in wider value judgments, both implied and articulated. So when you see the term "Middle East" from someone writing in the West or the terms "East" and "West" as general cultural and political qualifiers elided into geographical descriptions, then be aware of the underlying implications.

I was reminded of the power of geographical descriptions when I visited what was then Czechoslovakia in 1991. Recently emerging from the grip of Soviet communist control, there was a new self-conscious description of the newly freed country as Central Europe, not East Europe, which was its designation during the Cold War. To be part of East Europe, especially after the fall of the communism, was to be part of a backward, less progressive part of the world. By renaming its wider regional location as Central Europe rather than East Europe, the country was reconnecting with its place in the mainstream of European culture. So the shift from East Europe to Central Europe was not only a geographical renaming; it was a move from the periphery to the center of Europe, a claim to European ideals, and a reconnection with progressive European traditions.

The geographical terms used to describe the world are constantly changing. During the early Cold War, the world was divided into the **First World** of capitalist economies with the United States in a leading role and a **Second World**

of communist societies headed by the USSR. This categorization covered only part of the world. Australia, Britain, France, Iran, Turkey, and Thailand were classified as First World. China, Cuba, Czechoslovakia, and Poland were part of the Second World. But there was a large swath of countries in South America and Asia that were either neutral or nonaligned (see Figure 1.5). The term **Third World** was first used in the 1950s to include this wide range of countries including Argentina, Cameroon, Mexico, India, and Pakistan. This categorization then quickly morphed into a general description of poor countries. The term **Fourth World** was used, not as a grouping of traditional nation-states, but to refer to indigenous people bypassed by economic progress and marginalized in many nation-states around the world. The term is often used to refer to indigenous people in countries long colonized, nations without states, the displaced, and the stateless.

So around 1970 we had four worlds—a First World of rich, affluent, growing economies; a Second World of

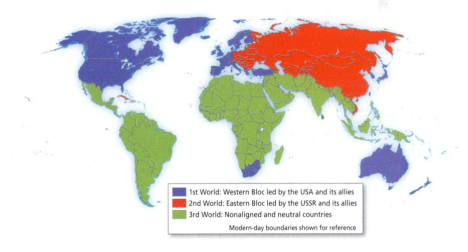

1.5 Three worlds of the Cold War

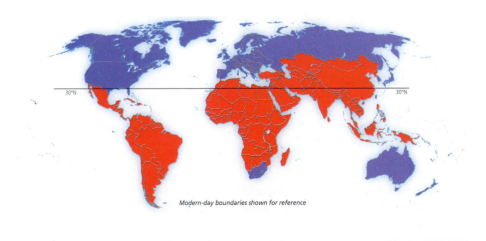

1.6 Two worlds of the global economy

communist countries; a Third World characterized as poor with huge population growth and myriad problems of slums, poor economic performance, corruption, and high mortality; and a Fourth World of the marginalized and the dispossessed.

The ending of the **Cold War** meant the term "Second World" fell out of favor. The term "Third World" was subject to further scrutiny. Industrialization in selected countries, such as initially Japan and South Korea and more recently India and China, created a specific category of **newly industrializing countries** with the membership changing as economies such as Japan and South Korea matured into First World designation. Then the terms "First World" and "Third World" lost their luster and new terms emerged, "developed," "developing," and "underdeveloped." The terms were first used during the Cold War. It slyly indicated some failing—why else a division between developed and underdeveloped? The term "**developing**," however, was sometimes used in a more hopeful sense that progress was in fact taking place. The term suggested a path to progress while the term "**underdeveloped**" or "Third World" had an end state of a more permanent category of poverty. Some scholars pointed out that underdevelopment was created in the process of some economies being developed. It was not the case that developing countries were slower in catching up to the developed but that the wealth of the developed countries was based on the underdevelopment of the so-called developing countries. From this perspective the underdevelopment implied in "developing" was not due to lack of progress but an unfair distribution of the fruits of progress and the transfer of resources from colonies to imperial centers.

The terms "developed" and "developing" are replaced now in the academic literature by the new division into **global North** and **global South** (see Figure 1.6). The term "global South" was first used in a 1980 report by the German Chancellor Willy Brandt to refer to a line roughly about 30 degrees North that divides much of the world into richer and poorer countries. The term is widely used despite the problems with lumping Australia and New Zealand and many South American countries in with some very poor countries.

Currently, we have global North and global South as a standard though flawed categorization of the world divided into rich and poor countries. As the global economy continues to shift and change, investment bankers have come up with even more subcategorizations. In 2001, a report by the global investment firm Goldman Sachs identified a subset of large growing economies that had the long-term potential to provide high returns on investment. The category of Brazil, Russia, India, China, and South Africa was neatly summarized in the acronym **BRICS**. The new classification was a response to the changing geo-economic realities of the world and recognition of rapid growth in these countries. The BRICS as a share of world GDP

TABLE 1.1	World Bank Country Classifications
BY REGION	**BY INCOME**
East Asia and Pacific	Low
Europe and Central Asia	Lower-middle
Latin America and Caribbean	Upper-middle
Middle East and North Africa	High
South Asia	High-income OECD members
Sub-Saharan Africa	

OECD: Organization for Economic Co-operation and Development

increased from 15 percent in 1990 to 26.4 percent in 2010. Even in this small subset of national economies, however, there were differences. Russia's growth was more a function of high prices for its resources of oil and natural gas than the dynamism of economic growth. More recently, Brazil has faltered and falling oil and gas prices crimped Russia's growth, and so the term covers a more mixed bag of countries.

Financial consultants and journalists, in the wake of the wide usage of the term "BRICS," have come up with new sets of countries carefully packaged as slick acronyms. Among the many are the following:

- CIVETS (Colombia, Indonesia, Vietnam, Egypt, Turkey, and South Africa)
- CARBS (Canada, Australia, Russia, Brazil, and South Africa)
- MINTS (Mexico, Indonesia, Nigeria, and Turkey)

These groups were portrayed as good investments because of their natural resources, demographic growth, and projections of high future economic growth rates. In reality, of course, they are more marketing vehicles than categories based on careful geographical and economic analysis. Egypt, Nigeria, and Russia may have some growth potential, but they also have enormous risks that even clever acronyms cannot obscure.

The investment bank categorization reminds us again that how we divide up the world is not innocent of wider economic implications and political readings.

The Data We Analyze

Global **nongovernment organizations** (NGOs), such as the World Bank, International Monetary Fund, and United Nations (UN), also provide categorizations of the world. The World Bank, for example, has a dual classification based on region and income (Table 1.1 and Figure 1.7). The high-income category is further stratified into **Organization for**

Economic Co-operation and Development (OECD) and non-OECD. The OECD is a grouping of the richest thirty-five countries of the world in North America, Western Europe, East Asia, and South America. It was founded in 1961 to promote global trade. Table 1.2 notes some of the basic statistics of the World Bank's four major income categories.

These national economic data are an important way of dividing up the world. However, behind the seeming solidity of the data, there are also two major problems. The first is that the data that are widely used to make country classifications only give us a one-dimensional view. Of course, it is important to identify rich and poor countries, yet this is only one dimension. If we look at other data, then different pictures begin to emerge. The female participation rate measures the proportion of women aged 15 and older that is economically active in the production of goods and services. It does not include non–wage activities such as housework or child care, nor does it include off-the-book transactions as is common in the **informal** and **illegal economies**. Across the world the differences are substantial and do not fit easily into simple income categorization.

Many of the traditional Muslim societies, as we would expect, have low female participation rates, such as Egypt (24 percent) and Jordan (15 percent), but there are also some poor countries with very high rates, including Tanzania (88 percent), Nepal (80 percent), Myanmar (75 percent), Vietnam (73 percent), and Zimbabwe (83 percent). These are substantially higher than many rich countries such as Japan (48 percent) and the United States (57 percent). Female participation rates are in fact highest in central and east Africa. If we were to rank a country by the participation of women in the formal economy, then Tanzania would be a superpower. The measure picks up the large number of women in peasant agriculture, traditionally a female occupation in many parts of the world. So even here the data needs to be treated with caution. But if we use a clearer index, such as the number of women in democratically elected seats in national lower chambers, there is wide variation from the measly figure for some "rich" countries such as Japan (8 percent) and the United States (18 percent) compared to the "poor" countries of Rwanda (64 percent), Uganda (35 percent), and Zimbabwe (32 percent).

TABLE 1.2	World Income Classification		
CATEGORY	**GNI PER CAPITA**	**POPULATION**	**EXAMPLES**
High income	$39,820	1.36 billion	USA, UK, Russia
Upper-middle	$7,598	2.4 billion	Argentina, China, Mexico
Lower-middle	$4,370	2.56 billion	Bolivia, Nigeria, Pakistan
Low income	$709	848 million	Afghanistan, Haiti, Rwanda

GNI = gross national income

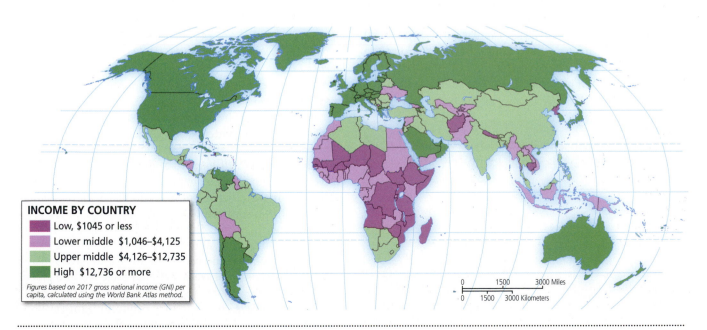

INCOME BY COUNTRY
- Low, $1045 or less
- Lower middle $1,046–$4,125
- Upper middle $4,126–$12,735
- High $12,736 or more

Figures based on 2017 gross national income (GNI) per capita, calculated using the World Bank Atlas method.

0   1500   3000 Miles
0   1500   3000 Kilometers

1.7 World Bank income classification

A crucial indicator is life expectancy: how long people are expected to live. If we look at the figure for men, a so-called rich country such as Russia has a very low life expectancy of only 65 years of age, 6 years less than men in "poor" Vietnam or "impoverished" Sri Lanka. Men in the "rich" United States only live on average 2 years more than men in middle-income China.

Using data is fraught with difficulty. This realization should not mean that we avoid using it, but we should be careful to draw upon a wide range of data and to be very critical in how we use it and interpret the findings.

The second problem is that national data hide substantial regional and urban differences. Consider a crucial development and welfare index, **infant mortality rates**. In 2015 the world average was 31.2 deaths of children aged less than 1 year per 1,000 live births. Behind this figure lies the heart-wrenching tragedy of a life lost too early and families bereft with grief. The United States has a relatively low figure in global terms at 6.0, but when we break the figure down for the individual states, then substantial variations occur (see Table 1.3). Alabama, Louisiana, and Mississippi have figures closer to Sri Lanka and at least double the figure for Cuba. Of course, these countries also have substantial variations by region, but for the moment, their national figures, in contrast with the regional subdivisions of the United States, belie the simple national division into rich and poor countries. Racial and ethnic differences account for some of the difference. In Alabama the infant mortality rate for blacks is twice that of whites in Louisiana and triple what is in Alabama. In the United States, infant mortality rates for non-Hispanic blacks are 13.3 but only 5.6 for non-Hispanic whites. Black infant mortality rates in the United States are closer to the national rates of poor countries such as Albania (13.6) and Jamaica (13.9).

Even within states and cities, there is substantial variation. Consider the case of Washington, DC, where the average family income over the period 2011–2015 was $122,760. But when we break this down into census tracts,

neighborhoods with a population of around 5,000, the differences between the highest and lowest income areas are huge. While some families in the richest parts of the city have incomes of over $349,800, in the poorest part of town, families have to survive on $34,390 for an entire household. There is a fifteen-fold difference between the richest and poorest census tracts of the national capital city. In the center for political democracy, economic inequality is prevalent and marked.

In other words, using aggregate national data, on its own, can skew our understanding and mask deep-seated differences within countries. Readers and writers of a world regional geography should always be acutely aware of this troubling fact. The larger the geographical unit we use—regional bloc or country—the more pronounced the statistical averaging of the diverse experiences at the regional, city, and neighborhood level.

It is very important to bear these comments in mind when looking at national data. In each of the remaining chapters of this book, national data for a number of variables are employed as a way of summarizing trends, placing individual countries in a comparative perspective, and providing a source of information along different dimensions. These are designed as only the starting points for discussion and analysis.

Rich and Poor

The world is composed of countries and world regional groupings of countries. But we should always remember that these larger units are composed of the finer grained levels of region, city, and neighborhood. Poor people live in rich countries, and rich people live in poor countries. Individual lives, neighborhood experiences, and city economies can be similar in very different parts of the world. Let us consider some examples by looking at the very rich and the very poor.

The super-rich inhabit multiple spaces. They occupy the fast lane of the flattened world: not simply flying at the front of commercial aircraft but owning their own planes. Moving through VIP lounges and residing in luxury enclaves, they inhabit transnational spaces that some have called Richistan. This is not a single place, or individual country; it is a separate universe of privileged sites scattered around the world. The very rich have the lightness of a smooth mobility that allows them to crisscross the globe, but enough political and economic weight to have a heavy influence on their local areas, national societies, and global discourses.

Sentosa Cove is one—a luxury condominium development on a private island enclave just off the coast of Singapore. More than 2,500 units are located on a 290-acre

TABLE 1.3 Infant Mortality Rates, 2015	
COUNTRY	**INFANT MORTALITY RATE**
United States	6.0
Alabama	8.3
Mississippi	11.2
Louisiana	8.1
Cuba	4.0
Sri Lanka	8.0
World	31.2

Infant mortality rate is the number of deaths of infants under 1 year old per 1,000 live births.

waterfront gated community. In June 2012, a Chinese billionaire purchased a bungalow for S$36 million. Two Australian billionaires, Gina Rinehart and Nathan Tinkler, have residences there. Wealthy foreigners purchasing 99-year leases are given an initial three-residence pass that is renewable every 5 years. A survey of a sample of residents in the exclusive resort by the urban geographer Choon-Piew Pow identified people from Australia, Britain, China, France, India, Indonesia, New Zealand, South Korea, and Taiwan with occupations such as owner of a shipping company, a commodity trader, a former investment banker, and owners of various businesses, including garment factories and a global logistics firm.

Enclaves of the very rich can be found all over the world. Palo Alto in California is where many of the rich people who work in Silicon Valley now live. In a city with a population of 60,000, more than four out of every five people has at least a bachelor's degree. The median value of dwellings is $1 million (see Figure 1.8). Rublyovka is a very rich suburb in the Russian capital of Moscow. Large houses sit behind high, protected walls in a sylvan neighborhood of dense fir forest. Once the home of the Soviet political elite, it is now the fashionable place for very rich Russians to have a home. Presidents, former and current, of the Russian Federation have residences. Nearby are the services for the wealthy consumer, including a Gucci store, European sports car importers, and a helicopter dealership. In central London, One Hyde Park is the most expensive residential building in the city with apartments selling for $214 million. There is a connection between the two places as some of the Russian rich are thought to have apartments in London residences as ways to move some of their assets offshore to the safer confines of the United Kingdom.

In 2012 the pop record *Gangnam Style* by South Korean PSY and accompanying video became a global sensation. It achieved over 1 billion hits on *YouTube* in less than 5 months and spawned numerous parodies; my favorite is by the Chinese dissident artist Ai Weiwei. The song was a gentle satire on the rich of the Gangnam district in Seoul. Gangnam houses many of the headquarters of major Korean companies such as Samsung and Hankook. There are consumption palaces for the rich with flagship stores of global fashion brands, cosmetic surgeries, and English language schools. At night the place becomes party central for Seoul's rich young elite. It is to South Korea what Beverly Hills is to the United States.

1.8 Richistan; Palo Alto, California (photo: John Rennie Short)

In some cases the rich partition themselves off from the surrounding poor. Exclusive gated communities are common in Latin America; Maria Jose Álvarez-Rivadulla, for example, describes the "golden ghettoes" of Montevideo, Uruguay, where the very rich live in order to distance themselves from poorer people and to live among people of their own class (see Figures 1.9 and 1.10).

Montevideo also houses a very large number of poor people, as well as a small number of rich people. At the same time as the wealthy inhabit exclusive neighborhoods, the slum settlements expand as the poor make their own dwellings in marginal sites. The older slums in the city center, the slums known as *cantegriles*, house migrants from rural areas who moved in the 1950s and 1960s. More

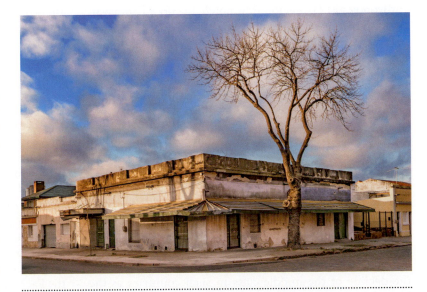

1.9 Poor city: slum in Montevideo, Uruguay (Daniel Ferreira-Leites Ciccarino/123RF)

1.10 Affluent neighborhood in Montevideo, Uruguay (photo: John Rennie Short)

recently, more people have been forced into the newer slums on the city's edge, known as *asentamientos*. As the anthropologist Daniel Renfrew noted, the poor communities suffer from an epidemic of lead pollution.

Comparing the urban worlds described by the two authors, Maria Jose Álvarez-Rivadulla and Daniel Renfrew, reveals the very differing worlds of people living in the same city. Across the globe there are very different life experiences in the same country and city.

The Regional Divisions We Create

The categorization of the world is not easy. The world is constantly changing. The global shift of manufacturing, the decline of formerly traditional manufacturing areas, the decline of the middle class in such countries as the United States and its rise in countries such as China and India all make even the simple national division of rich and poor countries very problematic.

In many different countries, there are similar processes. The loss of manufacturing jobs can be identified from Detroit and Birmingham, United States; to Birmingham, United Kingdom; to Seoul; and even more recently to Shanghai. Centers of high-tech industries can be found in Cambridge, Massachusetts, and Cambridge, England. Silicon Valley in California is cloned in places such as Bangalore in India. The similarities come from shared processes of deindustrialization, economic growth, and high-tech concentration.

Some countries move quickly from low income to middle income. Others decline. There is no easy way to demarcate a rapidly changing world. This does not nullify the need for categorization but reinforces a major point of this introductory chapter: categorizations are always made from some places at certain times with underlying social implications. The world can be divided up into regional subdivisions, but this demarcation is always provisional.

There are a number of different ways to approach a world regional geography. We could, for example, have a discussion based on a division of high-, middle-, and low-income countries. This would, for example, focus attention of the shared problems of poverty for the low-income countries, whether it is Dominica, Nepal, or Central African Republic. We could also have a more people-orientated approach that focuses on the high-, middle-, and low-income people around the world. This would link the marginalized of New York City with the poor of Lagos and the billionaires of China with the Internet tycoons of Silicon Valley.

In this book I will adopt the standard approach of dividing up the world into standard regional divisions. However, I will draw upon the perspective of these two other approaches to enliven the text. This means that the regional breakdowns are only a first cut at understanding the world. The regional breakdown is best seen as one of many possible and useful ways to deal with the complexity of the world rather than the primary source of difference in the world. I will spend time looking at the connection between people and places around the world.

There are existing divisions of the world. Some are the basis for data collection and presentation. The UN *Demographic Yearbook* uses the continental groupings of Africa, North America, South America, Asia, Europe, and Oceania, an obvious and seemingly clear division. But what about Egypt? Most of us tend to think of it as more of a Middle East rather than as an African country. All of Central America and the Caribbean is included in North America. The division between Europe and Asia is not clear-cut. The category of Asia includes Cyprus, an island firmly located in the history and geography of the Mediterranean. Russia, with a huge land mass that stretches to the Pacific and has a longer border with Asia than Europe, is classified as part of Europe while adjoining Georgia and Azerbaijan, with more territories farther west than Russia, are classified as part of Asia.

The World Bank at its country site (http://www .worldbank.org./en/country) uses a number of classifications, but the main geographical divisions are Africa, East Asia and Pacific, Europe and Central Asia, Latin America and the Caribbean, Middle East and North Africa, and South Asia. This is a finer scaled division than the UN, but differences still arise. Turkey is considered part of Europe and Central Asia, where a case could also be made for it to be part of the Middle East and North Africa. Poland is considered part of Europe and Central Asia and not in the same category as France or Germany despite being a member of the North Atlantic Treaty Organization (NATO) and the European Union (EU). Myanmar is classified as part of East Asia and Pacific despite having a shared colonial experience with Pakistan and India.

These comments are not meant as cheap shots but to make the point that all regional divisions have a strong core with more debatable outer boundaries. Most people would agree that Brazil is part of South America, but what about Panama? Is it in Central or South America? And is Cyprus really part of Asia? And under what criteria is Indonesia always described as part of Asia, or South East Asia, but not Australia despite the two countries being very close neighbors? Only 125 miles separate the two countries. Australia is often lumped together with New Zealand, although they are 2,500 miles apart.

In this book I will use the following geographic divisions:

- Central America and the Caribbean
- South America
- Europe
- Russia and Its Neighbors
- East Asia
- South East Asia
- South Asia
- Middle East and North Africa
- Sub-Saharan Africa
- Australia and Oceania
- North America

The division has the advantages of being well known and often used. An important point to note is that the classification reduces the enormous diversity of the world into manageable chunks that have some form of geographical cohesion, but it does not solve all the issues of dividing up what is essentially and increasingly an interlinked world into different parts. Cyprus will be included as part of Europe. Australia will be in a different classification from Indonesia despite their geographic intimacy. In terms of geographic proximity, Australia should rightly be part of South East Asia. But differences are profound enough to override this geographic fact. Although it is important to note that in a globalizing world, Australia has increasing

and deepening trade ties, and migration patterns and political organizations, with Asian countries.

The regional divisions act as a convenient framework, but I will also employ a section entitled "Connections" to consider many of the linkages that connect different parts of the world.

The Narratives We Invoke

To write and read a regional geography is to take part in a long tradition of writing and reading about the world. It is a tradition that stretches across space and time. The Chinese geographer Xu Xiake (1587–1641) wrote during the time of the Ming Dynasty. Herodotus (c. 484–425 BCE) wrote about the Greco-Persian wars and provides a description of the world, as he knew it. He claimed to have travelled widely and drew upon the information provided by other scholars and local informants. It is an early form of world regional geography, though a world centered on ancient Greece and extending only as far east as the Indus, west to Spain, south to lower Egypt, and north to the Caucasus.

Herodotus's geography shows in most stark form the characteristics of world regional geographies. First, they are always written from a center somewhere in the world that guides perception of what is central and what is peripheral. Second, while the area of centrality is well known, accurate information declines with distance, which leads to the peripheral area being filled with strange exotica, such as ants the size of foxes, according to Herodotus, that could chase camels. The more distant and less well known, compared to the better known close at hand, becomes an empty narrative space that can be easily filled with all kinds of fanciful imaginings. We can see this in the work of the Greek geographer Strabo (64 BCE–24 CE) whose *Geographica* was centered in Greece and the Aegean, a region given five books of description, compared to only one book for Britain, Gaul, and the Alps in his fifteen books of world regional geography. Pliny the Elder's *Natural History* was published around 77–79 CE. It is an encyclopedic coverage of knowledge at the time when more areas of the world were coming under Roman influence and scrutiny. For many of these new areas, Pliny recounts some of Herodotus's tales of men with dog's heads, men with one giant leg and a monstrous foot so large that it acted as sun shade, and a race of people who had no mouth and lived on the scent of food. The distortions are obvious and laughable to modern-day readers, but they record a deeper truth; the new and the foreign are always subject to more imaginative readings than the strictly empirical accorded to the well-known and the nearby. This is the first law of regional geography. When places and people are far away, they are more subject to "othering" that can involve a description that makes them seem so much more different. We no longer believe that there is a race of people with a giant sun-shading foot, but we can still slip into seeing the distant as different from the near.

Sebastian Munster was born near Mainz. In 1505 he went to Heidelberg to enter a Franciscan order. Two years later he went to Louvain to study mathematics, geography, and astronomy. In 1524 he was appointed professor of Hebrew at Heidelberg University, where he also lectured on mathematics and cosmography. Munster's major work, *Cosmography*, was first published in Geneva in 1544. It was a massive work. The first edition had 659 pages with 520 woodcut maps and illustrations. Subsequent editions increased in size. By the 1550 version, the work had reached gargantuan proportions of 1,233 pages and 910 woodcuts. An eclectic collection of material, some old, some new, part old myth, part new fact, the book contains material on surveying and mining techniques as well as discussions of the phoenix, goblins, and spirits; it includes material detailing the reasons for sugar growing in parts of Italy as well as a discourse on the one-eyed and large-eared people who were supposed to inhabit parts of India. The success of Munster's book was in no small part due to its ambitious comprehensive coverage: the world was being ordered, comprehended in one narrative sweep, encompassed in a single text.

Today the emphasis is on empirical rationality. However, we should be aware that our narratives of other places are still filled with ideologies and imagining as well as measurements and arguments. By way of an example, consider the work of two US demographers, Anlsey Coale and Edgar M. Hoover. They looked at the population data from a number of developing countries, and especially the results of the 1951 Census of India. They came to the conclusion that high population growth hampered development. This became an influential theory and one that guided US aid programs that promoted birth control policies. It also entered the popular imagination as ideas about Third World populations overwhelming the carrying capacity of the world. The response was for the control and management of Third World populations. No mention that First World populations have the largest environmental impact per capita. If environmental impact was the real goal, then the answer was to limit First World populations. This idea was never seriously considered; instead, the "blame," as it often is, was apportioned to the poorest and weakest. The work was based on incomplete and wildly inaccurate data extrapolated into a general conclusion that developing countries had unsustainable population growth rates. This is debatable on a number of levels. Many countries, including the United States, go through a period of rapid economic growth during the process of **demographic transition** when the proportion of the working population increases in both absolute and relative terms. Large

populations can in fact create a **demographic dividend**. What was at work here was less a sober assessment of the facts and more an ideological positioning of the developing world's population different from the populations of Europe and North America, a source of danger to be controlled and managed. Similar to the people with no mouths of Herodotus and Pliny.

Another example: the human geographer Thomas Klak compared how Havana, Cuba, and Kingston, Jamaica, were represented in major US newspapers. Both cities face problems of poverty. However while Kingston was shown to have problems typical of Caribbean islands, the poverty of Havana was depicted as a failure of the communist system. Jamaica's poverty was endemic, unrelated to the politico-economic system, while Havana's was a function of its political system. One was part of a discussion of endemic Caribbean poverty, while another was shown as an example of the failure of communism.

Fixity and Flow

To demarcate regions of the world is to concentrate on fixed categories, seemingly stable units separated from one another. But much of the contemporary world is composed of the flows and connections as people, goods, and capital as well as ideas and practice circulate around the globe. Space is best theorized not just as different places but also as sites of flows and connections—spaces of flows as well as spatial demarcations.

Conclusions by Way of Warnings

The goal of this book is to give a greater sense of both the unity and the diversity of the world. It seeks to find similarity in a world of geographical difference and difference in the world of similarity.

Let us summarize what we have learned about the enterprise of a world regional geography:

1. The maps we use are not neutral depictions of the world but social documents with biases and implications.

2. The very terms we use to describe and demarcate the world are also not simply geographical descriptions but terms laden and freighted with social implication and political ramifications.

3. The data we rely upon provide a useful but always partial understanding of the world.

4. The scale of our analysis is crucially important. National data are crude aggregates of regional and local differences. Child mortality rates, for example, show how national figures conceal enormous regional and racial differences.

5. The terms "rich country" and "poor city" hide the fact that there are poor people in rich countries and rich people in poor cities.

6. Regional geographies employ narratives that contain, sometimes obvious and sometimes more subtle, messages about similarity and difference, power and process. For example, if we describe a Third World as a place of natural disasters, might we imply that the poverty in these countries is the result of "nature"? And hence that poverty is more natural than social and more easily justified than challenged?

7. Regions, and in fact all places, have multiple identities. We should be careful in ascribing only one meaning. Spatial divisions are ongoing processes rather than fixed categories; they are not enclosed, but sites of flows with other places.

These points suggest the need for a thoughtful approach. A world regional geography needs to be approached with eyes wide open to the pitfalls and difficulties of the enterprise.

The world is a mosaic of complexity and variety. We live in a world of connecting similarities and wrenching differences. The creation, maintenance, and spatial transformations of this complex blend of the shared and the different are the subject of this book.

Focus: The World as Space-Time: The Geography of World Time Zones

The world is a clock. It moves through time as well as space. Indeed, its movement creates time as the journey around the sun creates the seasons and the spin around its own axis is the basis for our 24-hour day. Not so long ago, just over 130 years ago, local places kept their own time. We still have legacies; when it is 9 p.m. across the United Kingdom, Great Tom, the bell of Christ College, Oxford still rings out at 9:05, reminding us of when Oxford, barely more than 70 miles west of London, was on a different time from the nation's capital.

When people walked, rode, or took a stagecoach between Oxford and London, the time difference was barely noticeable. But with the coming of the railways in the nineteenth century, space and time collapsed; the railway travelled between Oxford and London in just under an hour. Trains covered space too quickly for local times to be effective. In 1840 the Great Western Railway adopted London time. When it was 9 p.m. in London, it would be 9 p.m. in Oxford, no matter when the bell rang at Christ Church. By 1847 all the railway companies in Britain used London time, and by 1855 all public clocks chimed at noon in unison. It was a relatively easy thing to do. London was by far the dominant city, and the country was relatively narrow. North America, in contrast, was a giant continent. In the middle of the nineteenth century, there were 144 official times in North America. With the coming of the railway, it became important to have more precise and standardized time keeping. Initially the railway companies kept their own time. The Pennsylvania railroad maintained Philadelphia time along its entire network. New York Central used the time at Grand Central Station along all of its routes. In St. Louis, a major railway junction, there were six official different railroad times. In 1883 the managers of the fifty largest railway companies decided to reduce the now fifty different times zones to just four: Eastern, Central, Mountain, and Pacific. The new time zones came into effect on Sunday morning, November 18, 1883. In 1918, Congress ratified the arrangement.

The move toward standardized time raises two more issues: Where would the starting point be? And how could we negotiate the difference between the rigid geometry of longitude-based time zones—a 15-degree arc of longitude equals 1 hour time difference—with the messy geography of the world's political geography?

Agreeing upon a universal **prime meridian**, zero degrees longitude, was a difficult though necessary first step. Previously, states used their own capitals as their prime meridian. The French used Paris, the British used Greenwich in London, the Dutch Amsterdam, the Belgians Brussels, and the Portuguese Lisbon. After the American Revolution, map-makers in the United States changed the prime meridian from London to Philadelphia (the capital from 1790 to 1800) and then to Washington. Throughout the nineteenth century, many US maps would use a double system on the same page, with longitude from Washington on the bottom and Greenwich at the top. At the International Geographical Conference in Rome in 1875, it was agreed to use Greenwich as the prime meridian on land maps, although the conference hoped that this honor would make Britain adopt the metric system. Britain proved recalcitrant on this matter, and feet and yards, stones and pounds, pints and quarts were to last well into the next century. The delegates at the 1884 International Meridian conference meeting in Washington, DC, agreed Greenwich would become the prime meridian. The world now had a global metric; local time had been replaced by a standard time centered on Greenwich. Space had triumphed over place. Longitude would be measured east

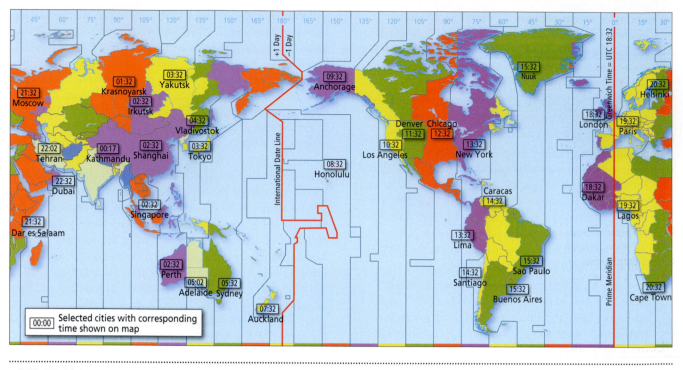

1.11 World time zones

and west of the Greenwich line and continuing the line on the other side of the world yielded the **International Date Line**, where a new day began when it was noon in Greenwich.

The conference also decided upon a 24-hour clock and the resultant division of the world into twenty-four separate time zones. The initial idea was to have a perfectly even distribution of the earth's surface into equal one-hour units. In reality, as any glance at current maps of time zones reveals, politics bends the straight lines (Figure 1.11). The International Date Line, for example, is bent so that far eastern Russia is not placed in the same day as Alaska. Then there are the anomalies. Australia is divided evenly into three time zones across the country, but central Australia is not as the name suggests in the middle of the time zones, but eccentrically located, ½-hour difference from the east coast zone, and 1½-hour difference from the west coast zone. Just to add to the chronological confusion, one of the states in the eastern zone, Queensland, does not operate daylight savings during the summer months, while the others do. In the central zone, one state uses daylight savings, South Australia, and one does not, Northern Territory. On its own in the western time zone, Western Australia felt free to operate Summer Time, but only from 2006 to 2009. A time zone is also bent so that Iceland is in the same time zone as Greenwich despite being well to the west.

Sometimes political differences are expressed in time zone designations. North and South Korea used to be in the same time zone. In 2015 North Korea turned its clocks back 30 minutes to establish a new standard time, half an hour different from South Korea.

Time zones are not only bent; they are also erased. China, a vast country as wide as the four–time zone United States, shares just one time zone. When it is noon in Beijing, it is also noon in Kasha almost three thousand miles to the west, over 3½ hours later as the sun moves. A Chinese national standard time rules over both local time and world standard time.

Look at the accompanying map. Where else is the elegant geometry distorted by the eccentric geography of political realities?

Select Bibliography

Álvarez-Rivadulla, M. J. 2007. "Golden Ghettos: Gated Communities and Class Residential Segregation in Montevideo, Uruguay." *Environment and Planning A* 39:47–63.

Klak, T. 1994. "Havana and Kingston: Mass Media Images and Empirical Observations of Two Caribbean Cities in Crisis." *Urban Geography* 15:318–344.

Marozzi, J. 2008. *The Way of Herodotus*. Cambridge, MA: Da Capo Press.

Pow, C-P. 2011. "Living It Up—Super-Rich Enclave and Transnational Elite Urbanism in Singapore." *Geoforum* 42:383–393.

Renfrew, D. 2013. "We Are Not Marginals": The Cultural Politics of Lead Poisoning in Montevideo, Uruguay. *Latin American Perspectives* 40:202–217.

Short, J. R. 2004. *Making Space: Revisioning the World, 1475–1600*. Syracuse, NY: Syracuse University Press.

Websites

On map projections:
   http://www.radicalcartography.net/index.html?projectionref
Global databases:
   http://data.worldbank.org/products/data-visualization-tools/eatlas

http://data.un.org
World Bank data site:
   http://data.worldbank.org

Learning Outcomes

Understanding how we view and represent the world through the evolution of maps.

Early maps centered geographic areas based on the location and perspective of the creators of those maps, such as was found in T-O maps that centered on the northern hemisphere or the Gangnido, which placed China at its center.

Although modern maps tend to contain fewer obvious biases than earlier versions, modern maps still contain both obvious and less obvious biases.

The terminology used to describe places also contains inherent biases that emphasize and deemphasize the dominance of certain peoples and cultures, and are constantly changing.

Although many terms have been used in an attempt to place countries into specific categories, currently "global North" and "global South" are the standard, yet imperfect designations used to divide the world into rich and poor countries.

Although national economic data are an important way of dividing up the world, there are two major problems with this method. First, the data that are widely used to make country classifications only give us a one-dimensional view. The second problem is that national data hide substantial regional differences.

World regional groupings of countries are composed of the finer grained levels of region, city, and neighborhoods.

Categorizing the world into rich and poor areas, or into other categories, is not an easy task as the world's constant changes make even these seemingly simple national divisions very problematic.

There are many different ways of dividing the countries of the world into distinct groupings. This book employs the following geographic divisions: North America, Central America and the Caribbean, South America, Europe, Russia and Its Neighbors, East Asia, South East Asia, South Asia, Middle East and North Africa, Sub-Saharan Africa, and Australia and Oceania.

The first law of regional geography is that the new and the foreign are always subject to more imaginative readings than the strictly empirical accorded to the well-known and the nearby.

Places have multiple identities. We should be careful in ascribing only one meaning.

Spatial divisions are ongoing processes rather than fixed categories; they are not enclosed, but sites of flows with other places.

A Global Context

LEARNING OBJECTIVES

Explain the patterns of world population and environmental impact, focusing on sustainable goals and global climate change.

Identify the phases and forms of globalization and evaluate the backlash against globalization.

Summarize the three distinct global economic trends that are shaping today's world.

Discuss global demographic shifts and trends in cultural differences around the world.

Describe the historical and contemporary growth of cities and relate urbanization to rural landscapes and global climate change.

Recognize the geopolitical organization of the world and outline the three trends of the new world order.

Identify the positive trends that are evident across world regions today.

2.1 World map of countries (national boundaries)

This chapter examines the major trends that we will explore in regional detail in later chapters: environmental challenges, globalization and its discontents, economic transformations, demographic shifts, cultural geographies in a globalizing world, urban growth, and geopolitical conflict. Consider it as a metageography, a global framework for the regional chapters that follow.

Environmental Challenges

Environmental challenges take a number of forms. They include increased pressure on our life-supporting ecological systems and the emergence of new environmental threats, as well as those of long standing.

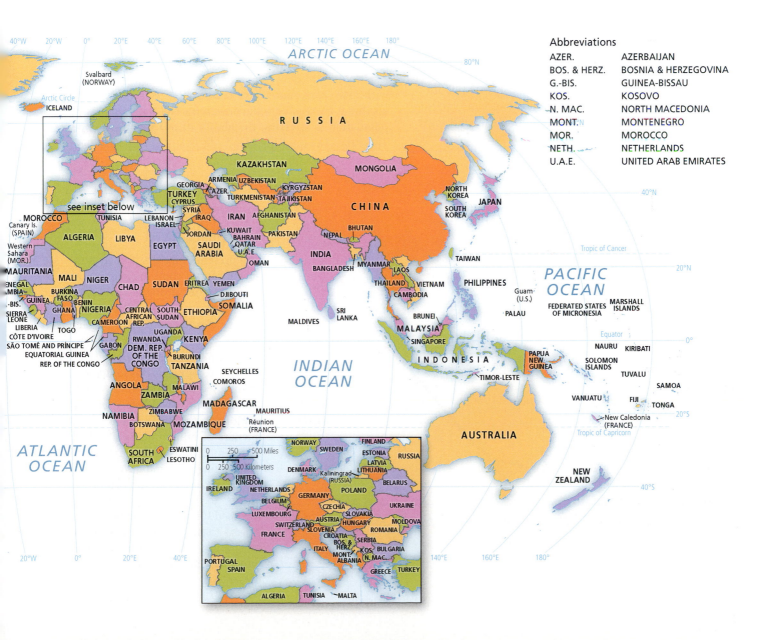

Abbreviations	
AZER.	AZERBAIJAN
BOS. & HERZ.	BOSNIA & HERZEGOVINA
G.-BIS.	GUINEA-BISSAU
KOS.	KOSOVO
N. MAC.	NORTH MACEDONIA
MONT.	MONTENEGRO
MOR.	MOROCCO
NETH.	NETHERLANDS
U.A.E.	UNITED ARAB EMIRATES

The Weight of Population

The most obvious environmental challenge is the relentless increase in the amount of people now living in the world and their growing consumption that increases environmental demands.

At the beginning of the last century, in 1900, the world's population was under 2 billion. It is now fast approaching 7.5 billion. Figure 2.2 shows the steady increase and indicates possible future scenarios for 2100: from a high estimate of 16 billion to a low of 6 billion. This wide disparity is because past patterns may not be a secure guide to the future. If affluence spreads more quickly, widely, and deeply across the world, then population growth will slow down and may even cease. In Japan, for example, the birth rate is 1.4, less than the replacement rate. The more widespread the affluence, the lower the birth rate. The higher estimates of growth assume that if affluence does not spread so widely, then birth rates may remain high in much of the world, especially in Asia and sub-Saharan Africa.

It is not just that more people now live in the world; there are many more affluent people demanding more. Since humans consume resources and generate pollution, this extra weight is taxing the resilience of vital environmental systems of air, land, and water. These vital ingredients of our life on earth are being so stressed that they may no longer provide the basis of the long-term livability of our planet.

This environmental pressure takes different forms around the world: from issues of water supply in the Middle East to increasing air pollution in China. They share a similar cause: more people demanding more from the environment.

We can get a sense of the differential impacts by looking at the ecological footprints of cities in different parts of the world. The **ecological footprint** measures how much land and water area a city requires to produce the resources it consumes and to absorb its wastes. It is measured in global hectares (gha) per capita. The global average is around 2.6. The footprint of London, for example, was measured at 4.5. In San Francisco the value was calculated at 7.1 gha, while in Calgary, Canada, the value was 9.8 gha. Winters are cold in Calgary, and most people use cars to get around. While the global average is 2.6, these three cities in the richer parts of the world, in this case Canada, the United Kingdom, and the United States, exert a much heavier footprint.

The more affluent the people, the heavier the footprint. As societies become more affluent, there are more people with more cars, more fresh fruit, more meat, more of everything. Increasing affluence combined with increasing population is stressing the ability of our planet to cope. This raises the question of long-term sustainability.

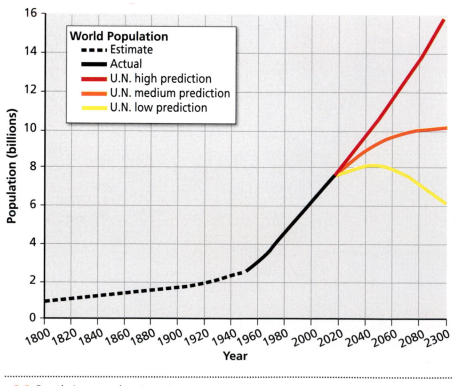

2.2 Population growth trajectories

Sustainable Development

Our present trajectory is unsustainable; it cannot continue in the same way, with consuming resources and generating waste. **Sustainability** is now a major theme and is promoted as the way forward for human–environmental relations.

This recognition has long roots. In his 1864 book *Man and Nature*, George Perkins Marsh drew attention to the unforeseen consequences of human actions such as floods and landslides as a result of overgrazing. Marsh emphasized the food chain, the importance of forests to soil conservation, and the need for sound ecological principles. These are now such taken-for-granted notions that we often forget they had to be developed and presented.

Gifford Pinchot (1865–1946) introduced the term "sustained yield" as a way to practice wise use resource management based on the maximization of benefit for the greatest number of people rather than for the profit of a few. He promoted long-term interest over short-term concerns.

During the 1960s and early 1970s, numerous writers and thinkers expanded discussions about nature–society relations. In 1962, Rachel Carson published *Silent Spring*. It begins with the image of a birdless spring, a result of pesticides such as DDT destroying birdlife. Her book chronicled the growing use of chemicals like DDT, deildrin, endrin, and parathion, detailing their deleterious effects on humans, plants, and animals. The book drew upon a wide body of scientific writing; her bibliography ran over fifty pages and included reports and papers in science journals. Today her message seems eminently sensible. At the time, however, there was great controversy. Chemical manufacturers mounted a vicious campaign against her and tried to block publication of the book. *Silent Spring* marked a new shift in environmental awareness. The earlier conservation and preservation movements had focused on the danger of overexploitation of resources. Carson shifted concern to the threat of human and animal extinction.

In 1987, the United Nations World Commission on Environment and Development issued a report called *Our Common Future*. The report is often referred to as the *Brundtland Report*, after Gro Harlem Brundtland, chair of the Commission. The report defined sustainable development as "development that meets the needs of the present without compromising the ability of future generations to meet their own needs." Sustainability was seen as the guiding principle for long-term global development. It consists of three pillars: economic development, social development, and environmental protection. These are often referred to as the three E's—economy, ecology, and equity.

International debates continued with Agenda 21, the primary outcome of the 1992 Rio Earth Summit, where 178 governments voted to adopt the program. The "21" refers to an agenda for the twenty-first century. The preamble to Agenda 21 called for a global partnership for sustainable development. In 2012, Rio +20 was held and participants reaffirmed their commitment to Agenda 21 in their outcome document, "The Future We Want."

Sustainability is now a guiding agenda. It is a global discourse that is acknowledged and addressed with varying degrees of success. Its popularity reflects not only its wide adoption but also an articulation of a deeply felt need. Sustainability focuses attention on our current and long-term human–environment relations.

In 2015, the United Nations widened the notion of sustainable development. The seventeen Sustainable Development Goals (SDGs) goals are outlined in Table 2.1. Each of the

TABLE 2.1 Sustainable Development Goals
Goal 1. End poverty in all its forms everywhere
Goal 2. End hunger, achieve food security and improved nutrition, and promote sustainable agriculture
Goal 3. Ensure healthy lives and promote well-being for all at all ages
Goal 4. Ensure inclusive and equitable quality education and promote lifelong learning opportunities for all
Goal 5. Achieve gender equality and empower all women and girls
Goal 6. Ensure availability and sustainable management of water and sanitation for all
Goal 7. Ensure access to affordable, reliable, sustainable, and modern energy for all
Goal 8. Promote sustained, inclusive, and sustainable economic growth, full and productive employment, and decent work for all
Goal 9. Build resilient infrastructure, promote inclusive and sustainable industrialization, and foster innovation
Goal 10. Reduce inequality within and among countries
Goal 11. Make cities and human settlements inclusive, safe, resilient, and sustainable
Goal 12. Ensure sustainable consumption and production patterns
Goal 13. Take urgent action to combat climate change and its impacts
Goal 14. Conserve and sustainably use the oceans, seas, and marine resources for sustainable development
Goal 15. Protect, restore, and promote sustainable use of terrestrial ecosystems; sustainably manage forests; combat desertification; halt and reverse land degradation; and halt biodiversity loss
Goal 16. Promote peaceful and inclusive societies for sustainable development; provide access to justice for all; and build effective, accountable, and inclusive institutions at all levels
Goal 17. Strengthen the means of implementation and revitalize the Global Partnership for Sustainable Development

goals also had specific targets. For example, Goal 12, *Ensure sustainable consumption and production patterns*, contains the following targets for 2030:

- Achieve the sustainable management and efficient use of natural resources.
- Halve per-capita global food waste at the retail and consumer levels and reduce food losses along production and supply chains, including postharvest losses.
- Substantially reduce waste generation through prevention, reduction, recycling, and reuse.
- Encourage companies, especially large and transnational companies, to adopt sustainable practices and to integrate sustainability information into their reporting cycle.

- Promote public procurement practices that are sustainable, in accordance with national policies and priorities.

- Ensure that people everywhere have the relevant information and awareness for sustainable development and lifestyles in harmony with nature.

These are ambitious goals with recognizable benchmarks to assess progress. They provide a metric for assessing the global shift toward more sustainable development.

Their achievement will be difficult, but they do indicate a global commitment to rebalancing human–environment relations in order to meet our greatest environmental threat: the continuing livability of our planet.

Global Climate Change

Global climate change is one of our most severe specific environmental challenges. Global climate change is not a new phenomenon. We are in a postglacial period after millions of years of an Ice Age. And there is variability as weather patterns change day to day and season to season. But the term "global climate change" refers to the impacts of an increase in land and sea temperatures across the globe in the last 150 years due to the steady increase in the emission of fossil fuels. Close to 40 billion tons of carbon dioxide are released each year into the atmosphere.

In the past 100 years, our planet has heated up by 1.5°F (0.8°C) and is projected to rise even more in the next 100 years due to greenhouse gas emissions, especially **carbon**

dioxide, which warms the atmosphere. The long-term trend is for a steady increase in sea and land temperatures.

The **Intergovernmental Panel on Climate Change (IPCC)** is the single best source for documentation of climate change. The IPCC had produced five reports, at the time of writing, that provide a wealth of data. In the fifth report produced in 2014, they highlighted a number of major changes caused by a warming of the planet because of greenhouse gas emissions:

- Melting snow and ice, including permafrost thawing
- Sea level rise
- Loss of unique ecosystems, such as coral reefs
- Plant and animal species shifting their geographic range due to climate change
- Mainly negative impacts on crop yields
- More extreme weather event such as heat waves, droughts, floods, cyclones, and wildfires
- The greatest impact is on the more vulnerable population in the world, especially the poor.

Figure 2.3 depicts some of the main changes, and Table 2.2 lists only some of the main risks in the major regions of the world.

Global climate change takes different forms across the globe. In the Southwest and West of the United States, it is evidenced in warmer drier conditions; in far northern Canada, it is witnessed in the melting of the permafrost;

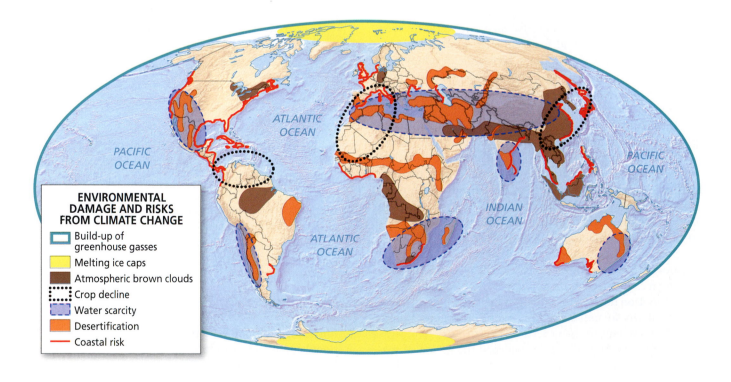

2.3 Global climate change

TABLE 2.2 Regional Risk of Climate Change

REGION	KEY RISKS
Africa	Water resources, decreased crop productivity, more diseases
Europe	Coastal flooding and sea level rise, water restrictions, extreme heat events
Asia	Flooding, extreme heat events, drought
Australasia	Coral reef destruction, sea level rise
North America	Wildfires, extreme heat events, flooding, heavier storms
Central-South America	Water availability, decreased food production, spread of disease
Polar regions	Ecosystem destruction
Small islands	Sea level rise
Oceans	Change in fish distributions, reduced biodiversity, coral bleaching

and it threatens to flood small Pacific island nations and to cause more wildfires in Alaska, more typhoons in the Philippines, more flooding in Bangladesh, and increased aridity in the Sahel and in Australia.

As a snapshot, consider two different parts of the world. In August 2016, more than 82,000 people in the western United States were forced to evacuate their homes due to wildfire. Global warming in this part of the United States is producing higher temperatures and more arid conditions. More trees are dying, which provides more fuel for wildfires. More than 66 million trees in the region are dead, and 30 million more are water stressed. And more people are moving into the urban wilderness interface, creating greater risk and vulnerability to fire. Global warming has heightened the drought and made wildfires more likely, increasing the length of the fire season and making the fires even larger than usual.

On the other side of the world, the Chitral Valley in Pakistan sits at the foot of Awi glacier. From 1960 to 2010, the average temperature increased by 2.1°F (1.2°C). The snow line is creeping up the mountainside, and the local glaciers are melting, causing severe flooding. The situation, as with people moving closer to the forest interface in California, is heightened by social activity. More people are moving into the valley; more trees are being cut down, which exacerbates the risk of flooding. The snow pack is now exploited by more people to provide refrigeration.

Climate change, in association with social activity, can create problems from wildfires in suburban California to flooding in valleys in Pakistan.

The earth's warming produces more precipitation in some locations, less rain in others; here more floods, there more hurricanes. As the oceans warm and become more acidic, they destroy the coral reefs that fringe the tropical seas in necklaces of indescribable beauty.

Climate change is exposing more people and the ecosystems they rely on to the greater risk of environmental hazards as the formerly 100-year flood occurs every 20 years, and the 25-year storm becomes a more regular event. More extreme weather is occurring on a much more regular basis. The vulnerability to hazards varies. Poor people in the poorer countries of the world have the greatest vulnerability. Figure 2.4 shows the global distribution of vulnerability to climate change.

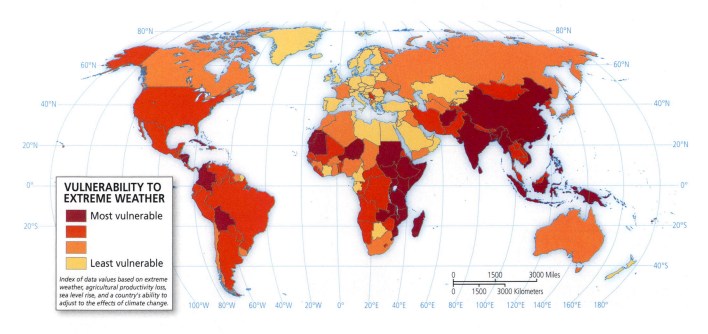

2.4 Vulnerability to climate change

Shaky Ground

We sit on a thin layer of rock, the lithosphere that floats on an ocean of molten magma. The earth is a sphere, so the lithosphere, no more than 125 miles (200 km) thick, is broken up into plates that essentially float on the magma. These plates are on the move; their annual rate is about the same as the growth rate of our nails or that of our hair. The edges of the plates where they collide and bump against each other or spread apart are places of intense tectonic activity, creating volcanoes and earthquakes. When the plates move apart, magma rises, cools, and solidifies to create new igneous rock. This sea floor spreading occurs at the plate boundary in the middle of the Atlantic Ocean. Mountain building occurs where plates collide, as in the case of the Andes and Himalayas. Plates also slip and slide past each other. Two plates slid past each other off the coast of South East Asia in 2004. The tectonic activity displaced a huge quantity of water, which turned into a tsunami that swept across the Indian Ocean. More than a quarter of a million people were killed.

Earthquakes and volcanic eruptions are major environmental hazards along the edges of the plates. While some environmental hazards such as hurricanes are now well-modeled—our predictions of hurricane intensity and tracks are now remarkably accurate, allowing people to evacuate—tectonic activity is nowhere near as predictable. We know that slippage may occur along a plate boundary, but we do not know exactly when or how powerful it will be. Predicting earthquakes is still at an elementary stage.

Another problem is that people continue to build and live in vulnerable areas. The threat of earthquake along the San Andreas Fault has not stopped the building up of Los Angeles or San Francisco, despite the certainty that an earthquake will impact both of these cities sometime in the future (Figure 2.5). Settlement and building continue along the coastal areas of the world, that are vulnerable to tsunamis.

Environmental hazards are also made worse by social factors. Around 4:00 p.m., on January 12, 2010, an earthquake measuring a devastating 7 on the Richter scale occurred in Haiti. Over 300,000 were killed, a similar number were injured, almost a million people were made homeless, and much of the capital city of Port au Prince was reduced to rubble. Despite international relief efforts, even by 2012, half a million people still lived in temporary shelter, often in appalling conditions. The environmental disaster revealed not only the unstable nature of the underlying geology but also the fractures in society: the dysfunctional nature of the Haitian state, the shoddy building in the nation's capital, the crippling poverty of most of its citizens, and the marked inequality that allowed a tiny elite to remained unaffected while the poorest people were most negatively impacted.

On March 11, 2011, an earthquake on the ocean floor 230 miles northeast of Tokyo, Japan, registered 9 on the Richter scale. The resulting tsunami waves lashed over the coastline, eradicating towns and villages. More than 20,000 people were killed, and the total damage is estimated at $300 billion. This natural hazard then turned into a nuclear disaster. On the day of the earthquake, at the Fukushima Daiichi nuclear power plant, 150 miles north of Tokyo on the Pacific Coast, three of the six power reactors were already closed for routine maintenance. The other three shut down automatically in response to the first seismic waves that hit at 2:46 p.m. Forty-one minutes later the first tsunami wave hit the area but did not reach over the 33-foot seawall. Less than 10 minutes later, a second giant wave, 77 feet high and traveling at around 500 mph, easily breached the seawall defenses, crashed into the nuclear plant, and destroyed the power and the emergency diesel generators that were used to cool the nuclear fuel rods. The rods produce tremendous heat to generate steam to drive turbines that generate electricity. The nuclear reactors were now generating vast amounts of heat with no sources of power to cool them down. That night a decision was made to release contaminated steam in order to avoid a giant explosion. Subsequent explosions also resulted in radiation leakage that spread across 700 square miles and forced the evacuation of 100,000 people.

Location along an unstable plate boundary and resultant vulnerability to earthquakes has not halted development. On the morning of April 18, 1906, San Francisco was shaken by an earthquake that registered 8.25 on the Richter

2.5 Los Angeles; sitting on a major fault line (photo: John Rennie Short)

scale and destroyed 25,000 buildings. Gas mains broke open, and power lines were downed. Fires raged through the city for 3 days. More than seven hundred people were killed, and a quarter of a million were made homeless. The city rose from the destruction relatively quickly, rebuilding within a few years. However, the threat of earthquakes remains as the city sits on the San Andreas Fault, an unstable plate boundary. In 2006, to coincide with the 100-year anniversary, researchers simulated the effects of a similar strength quake. At worst, the scenario predicted 3,500 people killed, 130,000 buildings damaged, and 700,000 people made homeless. The 1906 earthquake affected a city of 400,000; today there are more than 7 million people in the entire San Francisco Bay area. The scary truth is that the issue is not *if* the big one happens but *when*.

Globalization and Its Discontents

A Globalizing World

We can identify three different phases when the world was pulled together and zipped up into a more cohesive unit.

After 1492, the Columbian Encounter bridged the hemispheric divide in a series of transactions and exchanges of people, plants, viruses, and animals that created a global world. We live in a post-Columbian global society.

Through this new globalization, local places were incorporated into the space of a global economy, and the global economy was articulated through a series of connected places. Local ways and practices were transformed by the new connections to distant markets and foreign influences. One example: Native Americans in the seventeenth century on the banks of the Hudson, at present-day Albany, traded beavers with Europeans for alcohol, guns, and iron tools. It was more than just an exchange of goods; it was the transformation of place into space.

Globalization in this first phase was the enforced incorporation of the world into European control. The story is one of increasing territorial annexation, a widening of the flows of the globalization, and a deepening as the economies and fortunes of Europe and the rest of the world became more intimately linked.

In the second phase, from 1865 to 1914, globalization became more pronounced. It is a time of low tariffs, an international labor market, and relatively free capital mobility. This regime of free trade did not reign everywhere; critics of the time called it a British internationalism. British overseas investments were twice those of France and five times those of the United States. It was an economic globalization firmly centered in London. Other countries, for example, France, Germany, and the United States, were also involved, albeit in their smaller trading empires.

Economic integration was reinforced by changes in transportation that compressed space and time. Railways had been introduced before 1865 but were limited to partial networks in only a few countries. After 1865 there was a widening and deepening of the railway network around the world. In 1870, there were only 125,000 miles of track in the world; by 1911, this had increased to 657,000 miles. Places were pulled closer together as it took less time to cover distance. The railways were a form of **space-time compression** that brought places closer together and made the world a smaller place. Far away from the railroad track, life may have been unchanged; but the whistle of the train speeded up the world.

There was also an expansion in shipping. In 1865, sailing ships carried most of the world's trade. By 1914, the more reliable steam engine replaced the vagaries of the wind. It became easier for steamers to take emigrants across the Atlantic, transport beef from Latin America, ship tea from China, and export iron railway lines to India. The Panama Canal (1914) and Suez Canal (1869) cut shipping durations even further (see Figure 2.6).

The period from 1865 to 1914 marks a widening and deepening of economic integration around the world, with London as the pivot of the international trading system, the sun of the trading universe. The formation of a world economy involved the incorporation of larger areas of

2.6 The Panama Canal; opened in 1914 and still used as a major shipping route (photo: John Rennie Short)

the world. Large parts of Africa, Asia, and even faraway islands in the South Pacific were annexed to a global economic order. Overseas territories provided cheap raw materials and secure markets for manufactured goods—they were sources of national prestige and pawns in the great game of global geopolitics.

A flattening of the world in terms of ease and cost of transporting people and goods marks the third wave that began in earnest after 1989. This current phase is also characterized by economic, political, and cultural globalization. **Economic globalization** involves the emergence of a global economy and the increasing and tightening linkages between different national and regional economies. Easier and cheaper transportation allows long production chains to snake their way across the globe in search of cheap labor areas. Manufacturing, which hitherto had been the preserve of Europe and North America, shifted because of easier transport and cheaper labor to places such as Japan and South Korea and then, when their labor costs increased, to China.

Political globalization is the shift toward more global governance concerned with regulation and control. In the post-1944 era and particularly since 1989, the global economy has been governed by a series of institutions and rules. Global governance has been most successful in the trend toward economic integration through such institutions as the International Monetary Fund (IMF), the World Bank, and the World Trade Organization (WTO). The IMF currently has 182 members, including former Eastern bloc countries. Each member country contributes a certain sum of money as a credit deposit. The IMF wields its power through its surveillance system and the strings it attaches to lending. "Surveillance," the term used by the IMF, involves monitoring of a member's economic policies and evaluation of its economy. The World Bank provides market-based loans to middle-income countries, gives interest-free loans to low-income countries, provides loans to private investors setting up business ventures in developing countries, and underwrites private investment in developing countries. The WTO establishes the global rules of trade between nations. It focuses on reducing tariff barriers, free trade, and open markets.

Recent globalization is dominated by three things: the power of the rich countries, such as the United States, to set the agenda of global governance; the power of international bureaucracies to operate global governance without democratic consultation; and the power of multinationals, banks, and financial institutions to dominate the pace, direction, and consequences of globalization.

Cultural globalization occurs through the increasing and deepening flows of goods, people, capital, ideas, and information across national boundaries. While some argue that a cultural homogeneity is the result, an alternative thesis argues that, while particular television programs, sport spectacles, network news, advertisements, and films may rapidly encircle the globe, this does not mean that the responses of those viewing and listening will be uniform. Goods, ideas, and symbols may be diffused globally, but they are consumed within national and local cultures. Ideas, symbols, and goods that circulate around the world are consumed in national contexts and in local circumstances.

And Its Discontents

There is a backlash to globalization. There is indignation against the seeming unfairness of the globalization: it allows the wealthy to become even wealthier, and the powerful to become even more powerful.

There is also growing resentment. Political and economic elites in the West argued that free trade, global markets, and production chains that snaked across national borders would eventually raise all living standards. But as no alternative vision was offered, a chasm grew between these elites and the mass of blue-collar workers who saw little increase in wages. The backlash against economic globalization is most marked in those countries such as the United States, where economic dislocation unfolds with weak safety nets and limited government investment in job retraining or continuing and lifetime education.

Many of those who fear globalization rightly point to the fact that unelected, undemocratic, regulatory bodies now set the framework for global trade and interaction.

There is also a genuine fear that globalization is eroding cultural identity. The connection between globalization and nationalism is not clear and simple. Nationalist identity can be reinforced, represented, and recreated when a country is undergoing rapid integration into the global economy or experiencing quick immersions into global cultural flows and political globalization. In those countries where the connection with the global economy is experienced as a loss of jobs or large-scale immigration of "foreigners," cultural politics can take a particularly nasty turn. The rise of racist parties and racist rhetoric, even if they are not electorally successful, can wield power by the calculated response of more mainstream parties who adopt an anti-immigrant stance, in order to cut off the support to the more racist parties. While states want an integrated economy, they may also want to strengthen the barriers to immigration. The global economy is an abstract idea; the local immigrant community is a more obvious target of resentment and disquiet.

Globalization is often the name given to the uncertainty and fragmentation of a world of rapid economic changes.

The Geography of Globalization

There is a spatial dialectic to globalization. On the one hand, some places have moved closer together in relative space. The trajectories of national, regional, and local economies have become even more enmeshed within a network of global financial flows and transactions. But on the other hand, some places have moved farther apart in relative space, as they have been subject to a process of financial exclusion, and this has led to a widening of economic and social spaces between places of exclusion and inclusion. Some places are marginalized, while others are heavily interrelated in the global economic system. Some of the most global cities have low-income areas that are starved of resources and are disconnected from the circuits of globalization. There is a new geography of global centrality and peripheralization cutting across national boundaries.

With increasing economic competition and capital mobility, the outcome is often increased uneven development and spatial differentiation rather than homogenization. The world is becoming more interconnected, but the world is not necessarily becoming more of the same.

The rapid flows of capital, the decline of transport costs, and the rise of electronic communication have prompted some analysts to write of the end of geography. But in a competitive world, the small differences in relative space become even more important. Whether it is in locations relative to markets, perceived quality of life, quality of environment, variations in wage rates, systems of regulation, or local business culture, characteristics of place take on crucial significance. The friction of distance has not yet become the fiction of distance. Against a background of a shrinking world, geography becomes more important, not less.

Economic Transformations

There are three distinct global economic trends.

Global Shift

The first is referred to as **global shift**. This refers to the shift in manufacturing from the developed to the developing world. The industrial base has shifted from the high-wage areas of North America and Western Europe to the cheaper wage areas of East Asia: first Japan, then South Korea, and more recently China. As a result, there was a global redistribution of wealth. In the West, factories shuttered, mechanized, or moved overseas, while in China

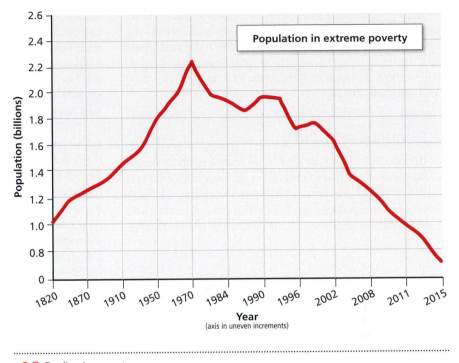

2.7 Decline in poverty

prosperity grew, with the poverty rate falling from 84 percent in 1981 to only 12 percent by 2010.

Global shift implies the decline of the blue-collar workers in North America and Europe and an increase in the manufacturing employment in South and East Asia. There are associated socioeconomic impacts with a severe decline in the purchasing power of the industrial workers in the rich world and an increase in the developing world. There are, of course, regional differences. Take the case of East Asia. Japan and South Korea saw a decline in manufacturing, while China experienced marked increase in the past 30 years.

Decline in Poverty

A second trend: A billion people have been lifted out of poverty in the last 40 years. The rapid economic growth of China and India has played an enormous role; it is responsible for close to 70 percent of this dramatic improvement. Changing status from poor to low income may not seem like much, but it represents a huge improvement in peoples' lives.

There is also a marked and rapid decline in the numbers of those living in extreme poverty, now defined as those living on $1.90 or less a day. In 1970, at its peak, the number of people in absolute poverty was 2.22 billion, roughly 60 percent of the world population. But by 2015, the number was 702 million, or 9.6 percent, of the global population (see Figure 2.7). Economic growth and investments in education and social safety nets have all made a difference.

There is a distinct geography to global poverty. While poor people live in all societies, even rich countries, the vast majority of global poverty is in South Asia, South East Asia, and sub-Saharan Africa. In East Asia, it is 4.1 percent, but in sub-Saharan Africa it is 35.2 percent, down by 7 percent from 2012.

Growing Inequality

Around the world, inequality has increased as more wealth is concentrated at the top end of the socioeconomic hierarchy. Some wealth has percolated down to raise people from poverty, but inequality, especially when it is entrenched and growing, creates conditions for political and social stability.

Demographic Shifts

The Demographic Transition

The **demographic transition model** posits a number of stages (Figure 2.8). In stage 1, birth and death rates are high, and population levels are stable. Then in stage 2, as public health improves, death rates fall rapidly. Because birth rates remain high, there is a very rapid increase in population. In stage 3, as death rates continue to fall, birth rates begin to fall as great affluence means that parents no longer need to have as many children. More resources can be allocated to each child in order to equip them for a more

complex economy. Thus, we shift from needing as many hands as possible to work in the fields to having fewer children but spending more on their education. Population increases only slowly, in stage 4, when births and death rates are low and population is either stable or only increases slightly. In stage 5, birth rates may drop even further so that population may even show a decrease.

Different regions and countries in the world are at different stages of the demographic transition. Figure 2.9 shows the variation in birth rates around the world. Notice the high figures for a band of countries in sub-Saharan Africa and the Middle East. The distribution on this map is repeated when we consider population growth and life expectancy. Examples include Afghanistan, where on average each woman has over five children, the population growth rate is 2.4 percent, and average life expectancy is only 52 years. In the United States, the comparable figures are two children, 0.8 percent, and 79 years. High-fertility-rate countries are also associated with rapid population growth and lower life expectancy.

Between stages 1 and 5 are countries such as Brazil, where fertility rates have declined from 6.3 in 1960 to 1.9 by 2015. Others are in stage 5. Japan, for example, is at stage 5 with very low birth rates, where the average woman has only 1.34 children. The replacement rate is just over 2; so Japan is experiencing a decline in population. From 2010 to 2015, the population shrunk by over 1 million. In the rural areas the depopulation is obvious as the young move to the city, leaving a behind a shrinking and aging population. By the end of this century, on current

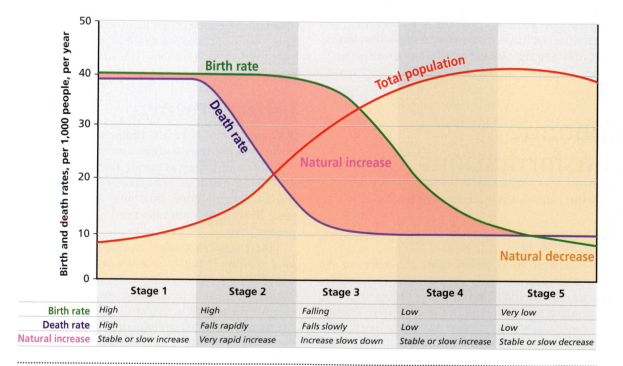

	Stage 1	Stage 2	Stage 3	Stage 4	Stage 5
Birth rate	High	High	Falling	Low	Very low
Death rate	High	Falls rapidly	Falls slowly	Low	Low
Natural increase	Stable or slow increase	Very rapid increase	Increase slows down	Stable or slow increase	Stable or slow decrease

2.8 Demographic transition model

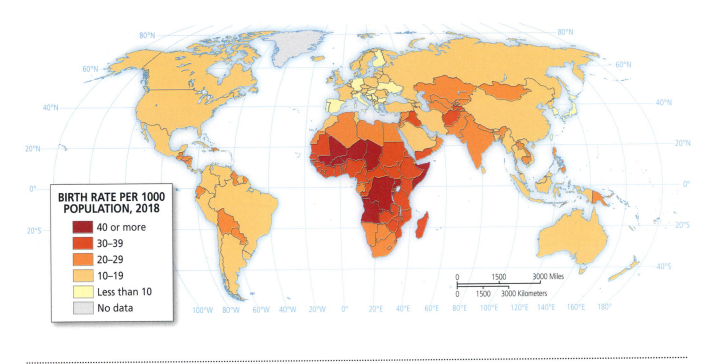

2.9 Birth rates

trends, the population of Japan will lose a third of its current population.

In some countries, declining birth rates can be offset by immigration from other countries. Immigration can be used to replace an aging workforce, as is the case in much of Western Europe. While this halts population decline, it can raise social tensions, especially when immigration levels remain high and economic opportunities shrink during short- and long-term recessions.

Countries at the early stage of the transition experience rapid population growth. Take the case of Nigeria, where the population has increased from 45 million in 1960 to 182 million in 2015. In 1960, Nigeria had only 25 percent of the population of the United States; by 2015, it had increased to 57 percent.

Rapid growth comes with problems. The rate of growth may overwhelm the ability of the economy to provide jobs or the government to provide services. Across sub-Saharan Africa and other regions of rapid growth, there are high levels of unemployment and slum formation.

There is also an economic drag in the short to medium term as the large, both absolute and relative, numbers of very young children shift the balance from the working to the dependent population. When more of the population is aged under 15 years, then more resources have to be extracted from the working population to support the young.

The Youth Bulge

A **youth bulge** occurs in stages 2 and 3 when an increasing share of the population is composed of children and young adults. There is a bulge in the youth population compared to other generations. In Africa, 40 percent of the population is aged less than 15 years, and 70 percent is aged less than 30 years.

The youth bulge can be a demographic time bomb if it is not associated with employment opportunities and social openings. High rates of youth unemployment are often associated with political instability: because young people have less to lose and less access to opportunities, they are more willing to take their complaints to the streets. Youth unemployment is a problem in much of Africa, the Middle East, and large parts of Europe. Very high levels of youth unemployment did not in themselves cause the Arab Spring, but they were an important element as young people, disenfranchised from power and economically marginalized, protested and overturned decades-old regimes.

The Demographic Dividend

The youth bulge can turn into the **demographic dividend**. As the very young age, they become the economically active. This demographic sweet spot is referred to as the demographic dividend, roughly a 20- to 30-year period when the very young move into the productive stage of their lives and before the population ages. During this period more of the population is aged between 6 and 65 years. In effect, more of the total population has the potential to become economically active. Less money needs to be spent on the needs of the very young and the old, so more can be invested directly into improving economic productivity. Brazil, China, and India are all experiencing a demographic dividend.

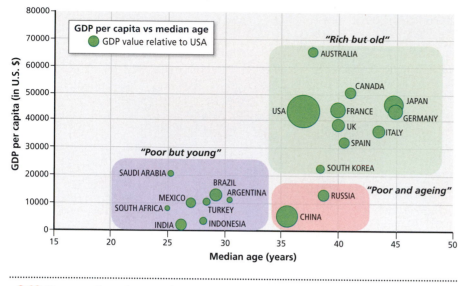

2.10 Demography and economics

In China, for example, there are twice as many people of working age, generally taken to include those aged over 15 and less than 65 years, compared to the rest. The **dependency ratio** measures the population aged 65 years and over as per 100 of those aged between 15 and 64 years. While the figure is 65 years for Japan, it is only 16 years for India.

The Geography of Aging

At stages 4 and 5 of the demographic transition, population growth stabilizes and may even decline. The composition of the population also changes because as birth rates plummet and death rates decline, the population ages. The late stages of the transition are associated with the graying of the population. The dependency ratio increases. In Japan it increased from 17 years in 1960 to 65 years in 2015.

The aging of the population is not in itself a problem as there are more people living longer and more productive lives, and more able to use their life experience and pass on their skills and knowledge. However, there are problems if there are intergenerational transfers of income. Take the case of Social Security and Medicare in the United States. In 1950, as the United States was experiencing a demographic dividend, 48.2 million workers supported 2.9 million beneficiaries; that is, 16 workers for every beneficiary. Seventy years later, only 156 million workers were supporting 53 million beneficiaries; that is, barely 3 workers for every beneficiary. As more people are eligible for Medicare, and as people live longer and medical costs in the United States seem to defy the laws of gravity, the working population has to support an increasingly larger number of old people.

We already know that where you are born has a huge influence on the quality of your life. We also need to add the when. Those born in good times get advantages over those born in bad times. Assume the (fictional) average

American. Born in 1900, you experienced the Great Depression and World War II. It was only in your fifties that things began to turn around. Born in 1940, in contrast, you were carried along on the great postwar expansion of economic growth, rising incomes, and new and extended benefits. If you were white, it was easy to get a job and to do well. Born in 2000, you are coming into a job market in the aftermath of the Great Recession, with well over a generation of stagnant incomes and increasing costs. To add insult to injury, you have to work to keep your elders in a state of privileged health care and generous social security payouts that you are unlikely to see for yourself. Sandwiched between tough times, those born between 1935 and 1970 are the lucky generations.

There are lucky and unlucky generations. Generational inequity really kicks in if those paying for the elderly are unlikely to see the same benefits. Older adults are advantaged because they have publicly provided pensions, health care, housing subsidies, tax breaks, and other benefits that younger age groups do not. What is more, the younger groups, for a variety of reasons, including the relative decline in US wages and incomes due to globalization, are unlikely to receive the same level of benefits. Class, race, and gender have long been identified as sources of difference, advantage and disadvantage. We also need to add age to that list.

In summary, we can identify a continuum of countries from those with rapid population and limited GDP per capita, such as many countries in central Africa, to those such as Japan with declining populations yet high GDP per capita. In between are those undergoing the demographic dividend, such as Brazil and China. We can identify "young" and "old" countries and rich and poor countries. Some of the combinations are shown in Figure 2.10.

Cultural Differences

The Empowerment of Women

Changes in reproductive technologies in association with changing social attitudes have resulted in a greater empowerment of women in many countries in the world. In the last 100 years, more women are able to become recognized as active members in the political and economic life of societies. In many countries, gender demarcation still restricts women to the home and the domestic sphere. Female

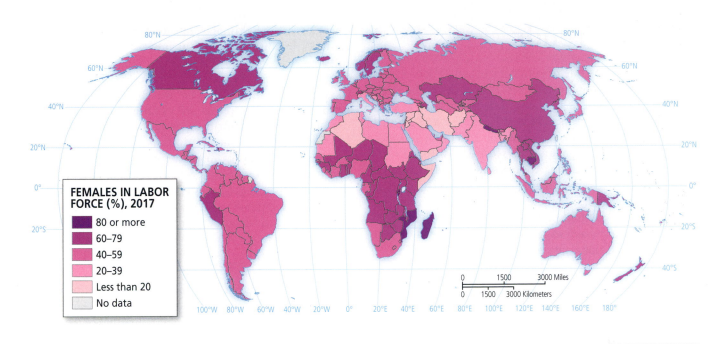

2.11 Female participation rates

participation rates are the number of women aged 15 years and over working outside the home. The rate varies around the world (Figure 2.11). The highest participation rates occur in non-Muslim low-income countries, where many women work in the agricultural sectors. Some of the poorer, more rural societies have the highest rates, such as Cambodia with 76 percent participation rates and Uganda with 82 percent. The Arab world has only a 23 percent participation rate.

In some places law and custom tightly circumscribe the role of women. Elsewhere, women are becoming more empowered, though even in a rich country such as the United States, there is still a difference in political representation and economic earning power. In 2017, of the one hundred US senators, only twenty-one were women. In the same year, on average, a woman doing the same job earned only 79 cents for every dollar a man earned. The wage gap has narrowed, but there is still a long way to go. And many women are discriminated against in the workforce and in the public sphere.

Sources of Difference

Populations are not homogenous. We have already made a distinction between men and women. We can also draw attention to differences in religious persuasion, sexual orientation, and ethnicity. In the previous section, for example, we highlighted the differences between men and women's pay in the United States. We can break this down even further. Figure 2.12 shows that there are substantial differences among Hispanic, Asian, and white women.

Ethnicity is a social category, not a biological one. Ethnicity is imposed as well as adopted, celebrated as well as undermined. In the regional chapters to follow, we will highlight the plasticity of ethnicity, how it changes over time, but also how it is maintained and reinforced. We will also explore other cultural differences such as religious identification.

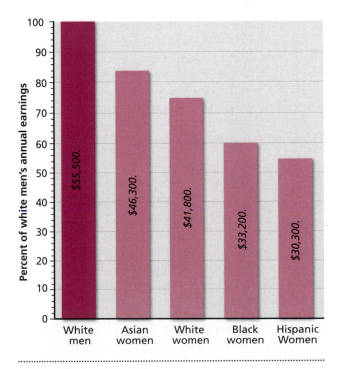

2.12 Differences in earnings

Urban Trends

We are living in a time of planetary urbanization. A majority of people now live in cities. Across the globe there is an urban growth change and resurgence. It is of such major and global significance that I describe it as a Third Urban Revolution. The first saw the invention of cities around 5,000 years ago, and the second was linked with the Industrial Revolution. The present one is associated with three trends: rapid urbanization across the globe; the growth of large cities, as there are now over four hundred cities with a population of over a million; and the widening metropolitan reach as giant urban regions extend well beyond the traditional city boundary.

Urban Growth

Urban growth is apparent across much of the global South. The rates of growth are staggering. In 1950, Dhaka had a population of 336,000. By 2015, it was 15.6 million and estimated to rise to 20 million by 2025. The swell of population has often overwhelmed the market's ability to cope or the local government's ability to organize. And the result is formal economies and housing markets that are simply inadequate to deal with the demand for jobs and the need for housing.

Cities are, of course, sites of problems. In much of the global urban South, the expansion of informal employment and housing results from the inability of the formal sector to provide jobs or housing. But cities are also places for innovative solutions of civil society. In the past 60 years, between 1 and 2 billion people have built their own homes and communities in cities all over the world.

Some cities are witnessing new influxes of people, quite literally revitalizing the urban experience. The population of New York City declined steadily from 1950, bottomed out in 1980, and has now bounced back. It is not just a US phenomenon: London saw a steady population decline from 1951 until 1991 when population growth began to surge past 1951 figures. A similar trend is apparent in Paris and Berlin.

Urban Decline

But not all cities are experiencing this resurgence. Former industrial cities, such as Baltimore or Detroit, that are unable to replace the lost industrial jobs continue to lose population and fail to attract investment. In 1950, Baltimore had a population of 950,000 and, like many cities in the United States, a vibrant manufacturing base that provided jobs and economic security. The magnet of jobs attracted black migrants from the South. Since the mid-1970s, though, there has been a steady loss of manufacturing jobs due to offshoring, relocation to suburbs in nonunion areas of the United States, and increased productivity. By 2016, Baltimore's population had declined to just over 614,000.

There are other Baltimores outside of Maryland. They include Akron, Birmingham, Cincinnati, Cleveland, Detroit, Pittsburgh, and Toledo. It is not just an inner-city problem. There is an inner ring of suburbs in crisis. There are also the bleak areas in the cracks of the metropolis: the trailer parks and suburban rental units that house those pushed out of the city by gentrification and redevelopment. Baltimores of economic neglect, massive job loss, aggressive policing, and multiple deprivations are found throughout metropolitan regions across the country. They are the places of despair that house the voiceless of the US political system; the marginalized of the US economy; and those left behind in the commodification of US society.

Cities and Climate Change

The high concentration of populations and investment puts cities at the very heart of climate change issues. Many of the world's cities are close to the sea, and many of the most vulnerable ones are those in coastal locations. Cities in the developing world, in particular, are often more vulnerable to natural disasters but are less able to spend billions of dollars to upgrade their infrastructure to better withstand flooding or to take similar measures. Cities such as Dhaka, Mumbai, Bangkok, Manila, and Ho Chi Minh City are already in low-lying areas under threat of increased flooding from extreme weather. The city of Jakarta in Indonesia, for example, is challenged by the flooding that accompanies the yearly monsoons. But land subsidence due to soil compaction from new skyscrapers, and increased groundwater extraction for a growing population, has caused the city to sink 10 times faster than the Java Sea is rising because of climate change.

Figure 2.13 shows when different cities around the world will experience major climate shift. Notice how many of the poorer cities in the tropics will feel the change much sooner than the more affluent cities in the more temperate regions. Climate change varies in its temporal impact across the globe with the poorest feeling the heat before the wealthiest. There is unevenness to those most at risk; the poor, infants, and elderly are most vulnerable.

People in cities are responding with both **climate change mitigation** and **climate change adaption**. Mitigation focuses on reducing the concentrations of greenhouse gases by using alternative energy sources, encouraging greater energy efficiency and conservation, and through the promotion of carbon sinks by planting trees. Curitiba in Brazil is the showcase for many successful policies, including the integration of green spaces within the city, a widely used public transportation system, and policies for the reduction of waste.

Cities are also adapting to the effects of climate change. Chicago, for example, has developed policies anticipating a hotter and wetter climate by repaving its roads with permeable materials, planting more trees, and offering tax incentives to encourage green office roofs.

2.13 Cities and climate change

Geopolitics

Global geopolitics is dominated by three distinct trends.

From Bipolar to Multipolar

The first is the decline of a bipolar world that lasted from 1946 to 1989 separated into the communist and capitalist blocs led, respectively, by the USSR and the United States. The communist bloc collapsed in 1989 and signaled the end of a world riven by a Cold War and an Iron Curtain that stretched across Europe.

The fall of the Soviet Union in 1991 and the end of the Soviet bloc transformed the world. New states emerged in the break-up of the USSR and Yugoslavia. East Germany rejoined West Germany. Across the globe, former communist economies become more integrated into the global economy. There was democratization in some but a quick return to authoritarian governments in many others.

The Cold War era was marked by a simple polarity: East and West. There were some nonaligned countries such as India, but by and large most of the world was enmeshed in the grid of superpower relations. Today, however, there are other sources of power. The rise of the European Union as an economic entity now rivals the US economy while Russia and China are major regional powers that exercise power in their respective neighborhoods.

To be sure, tensions remain. Russia is pursuing a more aggressive foreign policy after decades on the mat and continues to exert control over adjoining countries and casts a shadow over former Soviet republics such as Georgia and Ukraine. The Russian annexation of Crimea from Ukraine in 2014 reminds us that Russian power in the region is still strong.

The past 20 years have also seen the rise of China as a powerful economic agent across the world as it imports commodities, exports goods, and invests in infrastructure. China's economic growth allows it a more imposing military posture, evident as it flexes its maritime muscles in the South China Seas.

Issues are no longer so susceptible to an East-West perspective. Ethnic genocide in the Balkans or in Rwanda, social revolution in Syria, or state failures in the Horn of Africa are the result of complex factors no longer so amenable to superpower interference. In order to get things done, multilateral arrangements are now the order of the day as in the case of the alliance against ISIS. A war on terrorism requires multinational responses and multilateral organizations.

The United States remains the only global superpower able to exert its influence around the world. But this power is hemmed in by multinational agreements and, in the wake of the fiascos of the invasion of Iraq, a popular resistance to foreign interventions. While some write about the decline of the United States in terms of military reach and economic power, the United States retains its prime position in a multipolar world. But it faces daunting challenges from how to negotiate with North Korea, to dealing with an emerging China superpower and an empowered Russia, and how to respond to dysfunctional states that act as training grounds of terrorist groups. The United States has a difficult role as the world's superpower at a time of new rising powers, continuing regional conflicts, and new threats to global security.

Conflict and Cooperation

Reading the news headlines can sometimes be depressing. But one glimmer of hope is an overall decline in the level of political violence. The number of civil wars, after peaking in 1992, has declined. Invasions are rare, with the US invasion of Iraq a tragic exception.

Reasons for this decline include a long peace between most of the nations in the world, a reduction in civil unrest, and greater global scrutiny of state violence. It is still a violent world where violence and civil war persist. Bosnia, Myanmar, Rwanda, and Syria attest to the continuing nature of state-sanctioned violence against social groups. Genocide persists.

But across much of sub-Saharan Africa and Central and South America, civil wars that caused such havoc in the 1990s are now at an end. Since 1946 the absolute number of wars has declined. We have no contemporary equivalent of major conflicts such as the Korean and Vietnam Wars. Although perhaps hard to comprehend, the world is becoming a much less violent place. In fact, the world is a much less violent place than it was just 50 years ago. Peace has broken out across much of the globe, but that makes the remaining centers of violence and unrest all the more disturbing.

There are still major areas of conflict. In 2015, between 15,000 and 37,000 people died in Afghanistan, 21,000 in Iraq, 112,000 in Africa at the hands of Boko Haram, and 55,000 died in the Syrian civil war. But what is most marked is the decline in the violence between the formal armies of rival states. Much more common are civil wars and insurgencies.

Civil wars are more frequent in countries where economic and political inequality is mapped onto racial, ethnic, or religious differences; there is not a long history of national cohesion; the central authorities are seen as not meeting populist demands; or the authorities are very weak or dysfunctional.

Non-State Actors and Fragile States

Violence which for centuries was dominated by interstate rivalry is being displaced by the violence of non-state actors. We have the rise of non-state actors and especially the rise of militant organizations such as Al Qaeda, ISIS, and Boko Haram. These organizations are networks rather than territorially organized, although ISIS did try to establish the territorial caliphate. The bombing of the Twin Towers on 9/11 revealed the enormous damage that small organizations could do. There is an asymmetry between the power of the big state such as the United States and non-state actors. At the start of the long war on terror, the state actors had to catch up to small, mobile, spectral enemies. The surveillance of the Web and the rise of drone warfare are just some of the ways that states are dealing with the non-state actors.

Figure 2.14 shows the distribution of fragile states that are more vulnerable to instability and disruption and give more space for non-state actors such as armed secessionist

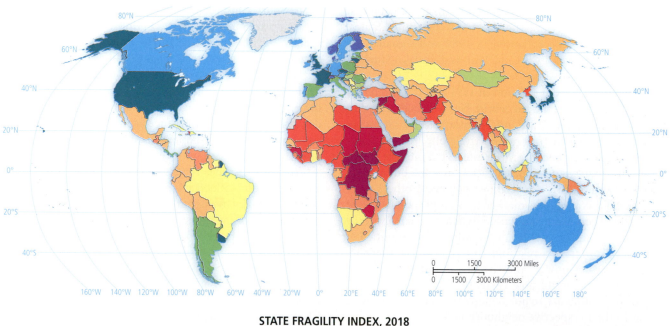

STATE FRAGILITY INDEX, 2018

Sustainable — Stable — Warning — Alert

2.14 State fragility

movements, terrorist groups, and endemic civil unrest. Notice the heavy concentration of state fragility in Africa and the Middle East.

Non-state actors look for weak or dysfunctional states in order to expand their operation. Al Qaeda operated out of Afghanistan, finding a safe haven under the Taliban, and used the liminal border zone between Pakistan and Afghanistan. Boko Haram operates in the border zones of Nigeria and its neighbors. Weak, collapsing, and dysfunctional states provide the space for non-state terrorist groups to flourish.

Sources of Hope

It is relatively easy to be disheartened. The news, especially international news, tends to cover tragedies and miseries: the devastating typhoon, the terrorist attack, or the outbreak of a fatal disease. We should not turn our eyes away from these events or lose our moral outrage; but neither should we forget the long-term, often rarely reported positive trends. In this chapter we have already noted that across the globe, child mortality, political violence, and absolute poverty are all declining. People are living longer. In 1950, the average life expectancy across the world was 48 years. By 2015, it had increased to 71 years. Across the world the overall story is positive with a marked decline in child mortality and an increase in the length and quality of life. The number of undernourished also continues to decline at a steady rate. We only have data from 1991; but since then, when 18.8 percent of the world population was considered undernourished, by 2015 it had almost halved to 10.8 percent. There are still too many people living in difficult conditions of war, disease, and hunger. But over the last decades their numbers have declined.

Select Websites

The global trends discussed in this chapter are constantly changing. Here is a range of reliable and constantly updated websites that give access to global data:

BBC Country Profiles:
http://news.bbc.co.uk/2/hi/country_profiles/default.stm

CIA World Factbook:
https://www.cia.gov/library/publications/the-world-factbook/
Our World in Data:
https://ourworldindata.org
UN Statistics:
https://unstats.un.org/unsd/databases.htmWorld Bank

Learning Outcomes

The world population is increasing in both number and environmental impact.

Global climate change is disrupting traditional weather patterns across the globe with differential impacts that include higher sea levels, warming oceans, longer dry spells in some areas, and more intense storms in others.

The world is pulled closer together through economic, cultural, and political forms of globalization. In some places, there is backlash to globalization.

There has been a global shift in manufacturing away from mature toward developing economies.

There has been a dramatic decline in the number of people living in absolute poverty.

People are living longer and healthier lives. Although there are still profound differences in fertility, child mortality and life expectancy have improved.

Gender, ethnicity, and religious identity continue to be major forms of social difference.

We are becoming a more urban society as more people live in cities and cities get larger.

We have moved from a bipolar to a multipolar world. The United States remains the preeminent economic and political power in the world, but the economic rise of a united Europe, the military power of Russia, the rise to superpower status of China, and a range of non-state actors are all sources of tension in this new world order.

Despite the many crises and problems that continue to create human suffering, the past 30 years have seen a marked improvement in peoples' lives across much of the world.

PART 2

Regions of the World

The following chapters describe the main characteristics of important world regions. We will look at their environmental challenges, economic transformations, demographic shifts, geopolitics, and urban and cultural geographies. At the margins of each region, some countries are in a liminal position. Belarus straddles the chapters devoted to *Europe* and *Russia and Its Neighbors*, while Myanmar could be considered as much part of *South East Asia* as *South Asia*, and the line between *MENA* (Middle East and North Africa) and *Sub-Saharan Africa* divides such interconnected countries as Somalia and Kenya. The regional divisions are a useful starting point, not a final destination for understanding the world.

3

Central America and the Caribbean

The region of Central America and the Caribbean consists of the narrow isthmus between North and South America, sometimes referred to as Meso-america, and the island nations in the Caribbean Sea. Once the setting for imperial rivalries, exploitative plantation economies, competing Cold War ideologies, and violent political struggles, the region is undergoing the difficult transition to a more democratically based politics and a more postcolonial economy. Mexico is by far the largest country in a region of small states often hampered by social conflict. It was long under the influence of Spain and then the United States, which continues to play a defining role in the region's economy and politics.

LEARNING OBJECTIVES

Summarize the region's geologic hazards and atmospheric situation; then relate these to its human activities.

Discuss the arrival of Europeans to the region and survey the impacts of the Columbian Encounter.

Explain the region's postcolonial economic development and distinguish its major economic sectors.

Relate the region's growth to the demographic transition and discuss cultural shifts in its language and religion.

Define land grab as a process; then connect it to the region's experience with rural peasant activism.

Identify drivers of urban growth and describe the region's informal urbanism and urban primacy.

Outline instances of colonial and US influence in the region; then examine interstate relations over border disputes.

The Environmental Context

Restless Geology, Social Instability

This is an area of intense geological activity: volcanoes and earthquakes constitute major environmental hazards. Eruptions and quakes not only shift the ground; they also shake the social foundations.

The Caribbean Plate is a relatively small plate, 1.2 million square miles, with active boundaries with at least three other plates, the Cocos, the South American, and the North American (Figure 3.2). As the Caribbean Plate moves eastward, the North and South American Plates are both pushed under in the process known as **subduction**, that creates the islands of the Lesser Antilles in two arcs. An older outer arc from Anguilla to Guadeloupe consists of long-extinct volcanoes that have weathered to form low-lying islands. An inner, younger arc stretches from Saba to Grenada. There are at least seventeen active volcanoes along this stretch of the earth's torn fabric. The edge of the Cocos and Caribbean plate form the 930-mile Central American Volcanic Arc from Guatemala to Panama.

The history of the region is shaped by volcanic events along this tear line as earthquakes create the setting for social change and political upheaval. Consider the case of the Spanish administrative center in Antigua, in what

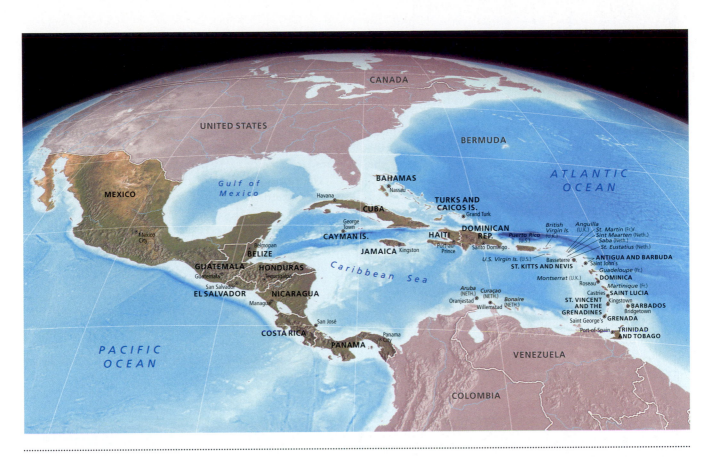

3.1 Regional map

is now Guatemala. An earthquake in 1771 destroyed over three thousand buildings, and in 1774 a large earthquake—estimated at 7.4 on the **Richter scale**—destroyed much of the city's fabric. The Spaniards, probably fed up with the constant destruction, then moved the capital only 30 miles away to the site of present-day Guatemala City. The new capital is still in a vulnerable position. In 1976, over 23,000 people were killed and 1 million rendered homeless when an earthquake struck. The epicenter was around 75 miles (120 km) from the city, but the shock waves still caused massive devastation. The social dislocation in the wake of the earthquake also led to a major population shift in the surrounding countryside as internal refugees, made homeless by the disaster, moved to the city. More than a quarter of a million Mayan people

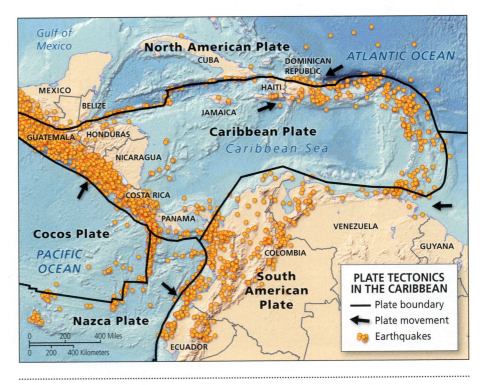

3.2 Plate tectonics in the Caribbean

moved to the city from their devastated region. From 1976 to 1987, the city's population exploded, mostly in the slum neighborhoods, from 675,000 to over 2.3 million.

In 1972, an earthquake devastated the capital city of Nicaragua, Managua. Buildings were demolished in the city center, over 6,000 were killed, 20,000 were injured, a quarter of a million were made homeless, and two-thirds of the 1 million inhabitants were displaced. The earthquake also had political shock waves. Criticisms of the ruling regime, their inadequate response to the crisis, and rumors that the political elite had stolen foreign aid helped to crystallize and fuel a popular resistance movement that ultimately led to the Nicaraguan Revolution that overthrew the dictator Somoza in 1979. Because of the long civil war, the rebuilding of the city was delayed almost until the end of the century. The earthquake and the long legacy of damaged buildings led to an outward movement from the city center. Many people, especially the rich, moved out of the city into the new suburban areas. Temporary squatter camps in the city became permanent slums.

In Chapter 2 we already noted the devastating impact of the 2010 earthquake in Haiti. Over 300,000 were killed and a million people were made homeless, and much of the capital city of Port au Prince was reduced to rubble (Figure 3.3). The rich elite, who lived in the green suburb of Petionville, up in the hills above the city, escaped the plight of the urban poor trapped in the devastated city below. In Petionville, few homes were destroyed and police were quickly mobilized to protect the residents and their property. Haiti's elite, because they were rich enough to live in Petionville, were spared from much of the devastation of the earthquake.

The restless geology of the region generates environmental hazards that have differential impact on the rich and the poor. Geologic activity does not create poverty and inequality; but it can exacerbate them.

Living in a Hurricane Zone

This region is located in the middle of a hurricane zone (Figure 3.4). Hurricanes emerging in the eastern Atlantic are driven west by the tropical **easterlies**, first hitting the Caribbean islands or the coast of Florida and then, around 30 degrees north of the equator, are driven north and east by the prevailing **westerlies**.

3.3 The 2010 earthquake destroys a poor neighborhood in Port au Prince, Haiti
(UN Photo/Logan Abassi, United Nations Development Programme via Wikimedia Commons)

Since the mid-1970s, tropical cyclones have been increasing in spatial range, storm lifetime, and intensity. The increase is correlated with the rise in tropical **sea surface temperatures** (SST), one of the consequences of global climate change. As the ocean water stays warmer longer, the hurricane season is extended and more and larger hurricanes are created. Average SST in the Caribbean is now averaging 2.5°F degrees (1.41°C) above historic monthly maximums. The water is warmer and stays warmer for longer, thus effectively increasing the amount of fuel for hurricanes, the time available for hurricane formation, and the spatial range of possible hurricane activity.

A hurricane can devastate a town in the United States, but the same storm can destroy the capital city and the entire national economy of a small island nation. In 2017, in one of the most active seasons to date, a series of punishing hurricanes pummeled the Caribbean and parts of the mainland United States. Hurricane Irma was a Category 5 storm that in the first 2 weeks of September devastated the Leeward Islands, Greater Antilles, Bahamas, and parts of the United States. Barbuda was declared uninhabitable. The tiny island nation felt the full force of the storm as the strong winds destroyed most buildings on the island. Throughout the region, the storm caused damage of over $62 billion and 134 deaths (Figure 3.5). Two weeks later in the middle of September, Hurricane Maria battered Dominica, Guadeloupe, Puerto Rico, among other islands, and Florida and caused $51 billion in damages and 66 direct deaths. Most of Puerto Rico lost power and running water.

Historical Geographies

Wind and Weather Patterns Impact Trade and Colonization

For centuries before the coming of the steamships in the nineteenth century, maritime trade and exploration were by sail and totally dependent on wind. Wind patterns made some routes easier than others. The **Azores-Bermuda High Pressure Zone** that spins winds clockwise around an area in the Azores played a huge part in the connection of this region with Europe. Ships could sail from Europe by harnessing the power of the trade winds to sail into the Caribbean. Then sailing north and east, they could then use the westerlies to sail back home. Note the name for winds refers to their origin, not their destination, so the westerlies blow from west to east. This wind pattern is an important background factor in the early European colonization of the region, the rise of the North Atlantic **triangular trade route,** and the creation of an Atlantic economy that linked Europe, Africa, Central America/Caribbean, and North America and the rise of port cities such as Seville, Nantes, Liverpool, and Amsterdam.

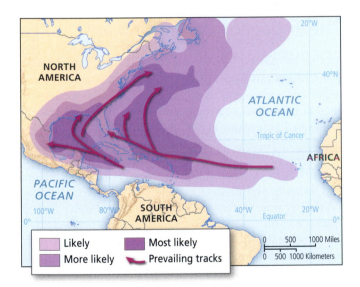

3.4 Likely track of Hurricanes in September

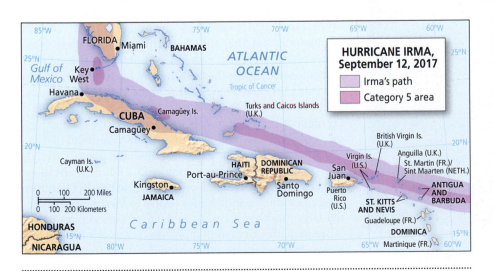

3.5 Hurricane Irma

The Lure of the Tropics

Most of the region is firmly centered in the **tropical zone**, defined as the area between the Tropic of Cancer (23.43 degrees North) and the Tropic of Capricorn (23.43 degrees South). This location provides the basis for a distinctive flora and fauna. Much of the region was initially covered in tropical rainforests, although altitude and rainfall differences create a mosaic of microclimates. In Puerto Rico, for example, there is a tropical rainforest known as El Yunque in the eastern part of the island. The slopes carry lush vegetation (Figure 3.6). To the west and south, there is a state forest, Bosque Seco (Dry Forest), where the low rainfall, due to being on the leeside of rain-bearing winds, results in a dry tropical forest of small stunted trees and bushes, almost alpine in their delicate size. Both forests were spared the common fate of deforestation in the region through the protective designation as national and state forests, a relatively rare event in this part of the world. Alas, many trees were blown over by Hurricane Maria in 2017. Altitude impacts vegetation: at higher altitude tropical rainforests give way to dry forest and, in the case of Mexico, high deserts.

The tropical environment of the region was a major attraction for Europeans, who were drawn by the opportunities to exploit an environment so very different from their own. Crops such as sugar that were not easily or cheaply available in Europe could be grown in this environment. There was also a geographical imagination in which the tropics were depicted as a place of wonder, the lush vegetation a metaphor for nature's bounty and an opportunity for easy wealth creation. The rich gold and silver mines and the ostentatious wealth of the Incas and Aztecs rightly provided further evidence that this was a land of plenty. There was also the racist ideology that promoted the idea of "superior" Europeans to "inferior" natives. The idea of a tropical bounty readily available and a "backward" pliant population fueled both the imagination of the writers as well as the hard-headed calculation of the investors. **Geographical imaginations** play a role in historical geography as much as geographical realities. Because most of the early European colonists came from temperate environments, the tropics became as much a place of imaginings as a place of rational calculation.

The Impact of the Columbian Encounter

The **Columbian Encounter** of 1492 was a pivot point for major environmental, social, and political change. The arrival of Columbus and the beginning of European colonization had four major impacts.

THE COLUMBIAN EXCHANGE

The first was the **Columbian Exchange** in which plants and animals unique to either the Old World or the New World became part of a global exchange. From Central America came the chocolate and maize that were created from, respectively, cacao seeds and experimenting with native grasses by the ingenious Mayans. Chocolate lovers everywhere owe a debt to the early Mayan.

EUROPEAN CONTROL

The second was the imposition of Spanish and European control. The Spaniards destroyed the existing political orders and then established towns as political control points, often on the site of existing urban centers. Vast tracts of land were taken and doled out to Spanish soldiers and colonists, laying the basis for the massive inequalities of wealth between a landed elite and a landless peasantry. These inequalities exist to this day. The Spaniards left a permanent legacy of grid-plated towns, the Catholic religions, the Spanish language, and, on the mainland of Central America, Hispanic elites who retained power through the centuries. There is a long and enduring division, marked by political and economic inequality, between the Hispanics and the indigenous people of Mesoamerica, particularly marked in countries such as Guatemala, where Hispanic elites have long dominated over the indigenous Maya. *Mestizo* is also used throughout the region to refer to people of mixed European-indigenous origin; in reality this includes the vast majority of the people of Central

3.6 El Yunque upland tropical forest. Throughout the region, much of the original rainforest is gone. Park designation saved this ecosystem in Puerto Rico. (photo: John Rennie Short)

America. Political elites throughout Mesoamerica, and indeed Latin America as a whole, still tend to look and live in ways that are more Hispanic than indigenous.

While Spain gained control over much of Mesoamerica and the large islands of the Caribbean, other European powers, especially the English/British, French, and Dutch, nibbled at the edges of Spanish power and after 1600 began to annex islands. The huge demand for sugar made the control of sugar islands an important commercial and geopolitical goal. Soon every island in the Caribbean was incorporated into the sphere of influence of one of the major European powers.

ENVIRONMENTAL TRANSFORMATIONS

Third, there was also an environmental transformation as natural forests were cleared for plantations and soil erosion washed away many of the nutrients. The Europeans depended on the crops developed by the Amerindians such as cassava, corn, and sweet potatoes to survive, but they developed cash plantation crops such as cotton, sugar, tobacco, ginger, and indigo. The get-rich-quick exploitative form of plantation agriculture led to deforestation, soil erosion, and loss of natural habitats. When the plantations were abandoned, they were replaced by low scrub woodland.

DEMOGRAPHIC HOLOCAUST

Fourth, the coming of the Europeans also caused what is referred to as the **demographic holocaust** as the indigenous populations, susceptive to everyday diseases that were not fatal to the Europeans such as measles and influenza, died by the millions. The indigenous Caribs almost disappeared and the population of Hispaniola (Haiti and Dominican Republic) declined from around 4 million in 1492 to only 60,000 in 1518 (Figure 3.7). The population of the entire region declined by around 90 percent between 1500 and 1700.

The loss of population meant that land and mine owners needed a new source of labor. Indentured labor from Europe was initially used, but the death rate was high and the system did not provide a steady enough supply of workers. The slave trade was the response. Almost 12 million people were shipped from Africa to the New World. Slaves were first sent to the Spanish and Portuguese colonies and then later to the colonies of the British, French, and Dutch. More than half died during the sea crossing, while 20 percent died within the first year of work. Working under brutal conditions, the life expectancy of slaves was short. Conditions were so bad and life expectancy so short that slaves felt they could lose little and gain a lot by resisting and revolting. The fear and reality of slave revolts became a constant threat to the white elites in the region.

Enslaved Africans also brought their knowledge of farming techniques. Forms of cattle ranching, horticulture, and rice growing were developed in the New World on the basis of African agricultural knowledge.

3.7 I met this lady while travelling by a public bus in Dominica. She told me she was a Carib and asked for her picture to be taken to let people know that the Caribs still survive. Her name is Janet. (photo: John Rennie Short)

The demography of the region, and especially the Caribbean islands, was transformed by the slave trade. In 1700, the white-to-black ratio on the island of Grenada was 1:2; by 1783 the ratio was 1:25. In Antigua in 1666, there were 300 whites and around 500 blacks. By 1734, there were 3,700 whites and 24,500 blacks. A small, white elite was serviced by a large, black underclass of servants and slaves.

Sugar, Slaves, and the Atlantic Economy

An important element in the historical geography of the region is the sugar trade. The Spaniards introduced sugar, and by 1650 it became the principal cash crop for the entire Caribbean region. The sugar boom began around 1700. Sugar cultivation required heavy and constant work and a huge demand for slave labor. Over half a million slaves were imported in Jamaica between 1700 and 1786. Sugar, like oil today, became such a prized commodity that it provoked wars, national rivalries, and imperial clashes. Sugar islands were fought over between the competing European powers. When British negotiators received Canada from France in return for the two sugar islands of Martinique and Guadeloupe after the war of 1756–1763, they were widely condemned in the London press for giving up a hugely valuable resource in return for what was dismissed as "tundra." Sugar islands were treasures, to be guarded and appropriated, and were perceived as more valuable than the vast, cold, empty spaces of Canada.

Sugar plantations can still be found in the region, but the commodity is no longer so prized that countries would go to war.

Economic Transformations

From Colonial to Postcolonial Economies

The economy of the region was transformed after the Columbian Encounter so as to provide tropical resources and cheap commodities for European markets. The colonial system was an uneven exchange as the region exported cheap **primary commodities** and imported the more expensive (because it was more value added) manufactured items. While the primary products became the basis for processing industries in Baltimore, Liverpool, and Bristol, they were not the platform for economic development in the region. Without a large and dynamic secondary and tertiary economy, the region was, and still remains, handicapped by limited employment opportunities. Unable to break out entirely of the colonial legacy, many of the region's population survive on low to modest incomes. Because of the limited economic opportunities in the formal economy, there is a large **informal sector** as people try to make a living as best they can. Enormous ingenuity and energy are spent trying to make a living in economies that provide few opportunities and limited wage-paying jobs.

Primary commodity production is still an important part of the economy of the region (Figure 3.8). Coffee, bananas, beef, sugar cane, and cotton are still grown and cultivated for export markets. The Guatemalan port of Puerto Barrios has a terminal entirely devoted to shipping containers for Dole, Del Monte, and Chiquita to take tropical produce from the region to North America.

In some of the large plantations such as the coastal sugar estates in Guatemala, the overuse of pesticides has eventually led to the decline of production levels. A flood of cheap maize from the United States has also undercut local maize growers, especially the small-scale farmers.

Economies that rely too much on primary commodities are subject to fluctuating prices that can swing from boom to bust. Much of the region has been unable to move up the value-added chain to an economy more based on manufacturing and services.

Economic development was further hindered by the nature of landholdings in Central America, where vast estates provided wealth and power for an elite but kept the majority as rural poor. In turn, this led to political instability that made it more difficult to attract the foreign investment necessary to promote large-scale manufacturing and a more service-orientated economy.

In recent years there have been three developments.

The Emergence of Manufacturing

The first is the emergence, on a limited scale, of a manufacturing sector. This is more prevalent in Mexico than in the other Central American economies and not at all well developed in the small Caribbean island nations where problems of transport, infrastructure, and lack of skilled labor make an industrial transformation difficult. Yet even here there is some growth in manufacturing employment. The Dominican Republic established **free trade zones (FTZs)** where manufacturers were tempted by low-cost labor and tax concessions. By 2000, forty-six FTZs employed 145,000 workers on the island making pants and shirts for the US export market. However, by 2009, the number of workers fell to 41,000 as competition from China and other Asian

3.8 Organic banana plantation on the east coast of Costa Rica (Photo: John Rennie Short)

countries, including Bangladesh and Vietnam, undercut even the cheap cost of clothes assembled in the Dominican Republic. In a globalized economy, offering cheap labor is not a surefire way to succeed, especially if energy and transport costs are high and there are other parts of the world with even cheaper labor costs, less environmental protection, and lower standards for worker safety.

Mexico was more successful in establishing a manufacturing sector. The country is now the world's seventh largest auto manufacturer as foreign companies such as Ford, Toyota, and Mazda continue to build factories. Over $7 billion was invested in the auto sector from 2012 to 2015. In 2014 alone, eighteen factories produced 3.2 million cars. This sector provides above-average wages, on average $8 an hour in wages and benefits, although this is about a quarter of what they would be in the United States. The cheaper labor is the main reason that the car companies make cars in Mexico.

Tourism

Tourism is a significant economic sector. And here we can recall the comment made earlier about the tropical imagination. For many tourists, the Caribbean conjures up images of sun-kissed, palm-fringed beaches, translucent seas, and plentiful supplies of rum (Figure 3.9). Mass tourism in the Caribbean was long hindered by inaccessibility. In the first two-thirds of the twentieth century, only a small elite group composed of rich Europeans and North Americans could afford to vacation in the pleasant warm winters. Millionaire Laurence Rockefeller purchased much of Saint John in the US Virgin Islands. In 1956, he donated the land to the US National Park Service on condition that it was left largely unexploited. Today three-quarters of the island is a national park. It is an early example of ecotourism in the region. The French island of St Barts maintains its exclusivity through the powerful elite control of land construction and services. To maintain their special quality, cruise ships are severely restricted, unlike in most other islands where cruise ships are welcomed.

The Caribbean is now a tourist destination site for a variety of income groups from the select and expensive resorts such as St Barts and sailors in luxury private yachts to more mass tourism and cheaper cruises. Tourist towns such as Cabo San Lucas, Cancun, and Cozumel in Mexico form a distinct enclave, sometimes referred to as gringoland—the resort equivalent of gated communities, which are located in Mexico but very much designed and maintained for North American and European tourists. They are in Mexico but not of Mexico.

The cruise ship season in the Caribbean runs from December through March/April when the winter warmth attracts vacationers from colder climes and before the hurricane season makes smooth sailing more problematic. Cruise ships provide a steady supply of consumers to island and resort economies. Elsewhere, gated resorts on the coast provide a safe environment for rich tourists. While the economic benefits are unequally shared with most of the profits siphoned off from the region, the industry does provide much needed capital investment and employment opportunities in a region long starved of both. There is also the argument that as more tourists look for more than just sea, sand, and rum, there are opportunities for ecological tourism that can provide an economic rationale for habitat diversity, ecological diversity, and greater environmental protection. When people can go zip lining through the rainforest, then the rainforest is more likely to be preserved than cut down for timber. In some countries such as Costa Rica, an ecological tourism is actively encouraged and promoted.

Toward a More Service-Based Economy

Finally, a selected part of the region is successful in moving to a more service-based economy. More than three-quarters of Panama's GDP is in the services sector. Ship insurance, marine logistics, and a large banking industry in Panama City have lifted the country into upper-middle-income status. Costa Rica's economy still has primary

3.9 Archetypal Caribbean beach view: Bahamas (photo: John Rennie Short)

production—coffee is still important in the highland regions and sugar in the lowlands—but the capital city of San Jose has a range of secondary and tertiary economic sectors. The relative political stability of both countries attracts foreign investment more easily than the political instability of the other Central American republics.

Social Geographies

Demographics

In the last 50 years there was explosive population growth in the region with a marked concentration in cities. The population of Haiti, for example, more than doubled from 3.8 million in 1960 to over 10 million in 2012. The population of Guatemala increased in the same period from 4.1 million to 15 million. The rapid growth was a result of improved medical care that reduced infant mortality. The region experienced the second stage of the demographic transition in the 1960s and 1970s, when birth rates remained high as death rates fell, leading to rapid demographic growth. The result was increased pressure on resources; in some cases, this led peasant farmers to move into marginal areas, and in others, for more people to move to cities in search of employment opportunities.

Most of the region has now passed through the first stage of the demographic transition of high birth and death rates and is now in the second and third stages of decreasing birth rates and the extension of life expectancy. Many countries are at the peak of the **demographic dividend** with the number of people in the working age group from 14 to 64 years at its all-time high.

A Melting Pot

The melding of indigenous, European, and African legacies generates a vitality and enormous diversity to the region. Many different types of people shape the culture of this region. We can consider just two examples of this cultural exchange and creation of hybrid identities.

THE LANGUAGE OF ARUBA

We can begin with the language of the tiny island of Aruba. The official languages are Dutch and Papiamento, a creole that draws on the Amerindian, African, Portuguese, Dutch, Spanish, and English languages. Creole languages develop from a form of pidgin, a simplified language for communication between two different languages. The pidgin stock of Papiamento is the simplified language that developed in the interaction between Portuguese slavers and African slaves. At its base, then, Papiamento is built on the slave trade that played such an important part of the region's economic history. Other additions to the language include

a pre-Columbian linguistic heritage, Spanish control of the island for a hundred years, and, since 1636, Dutch control. The island is still considered part of the Netherlands. With its rich mix of languages from three continents, Papiamento embodies the complex history of the region.

GARIFUNA

The Garifuna is a distinct ethnic and language group whose origins give us a reminder of resistances to slavery and colonialism. The Carib people of the Lesser Antilles fiercely resisted French and English (then British) domination. They refused to work as laborers in the French plantation of Martinique. When French authorities defeated the Carib in 1660, they expelled them from the island. An early example of **ethnic cleansing**, many ended up in Dominica and St. Vincent. Runaway black slaves set up shared households with the Carib. In 1763, the British gained control over St. Vincent and again the Carib resisted. Defeated in 1796, the British deported around 5,000 people. They selected the more African-looking Carib from the Amerindian-looking Caribs and banished them to an island off the coast of South America. Only around 2,500 people survived the voyage. The people soon moved off the island and settled along the Caribbean coast. They are known as Garifuna and now number close to 600,000. They are found in Honduras and Guatemala, and with subsequent migrations also in New York City and Los Angeles. They are located along the coast in small towns like Trujillo on the coast of Honduras. Their language derives from the Amerindian languages of Arawak and Carib as well as English and French. The Garifuna, with their distinctive cuisine, language, dance, and music, are only a small part of the rich multicultural tapestry of the region.

Religion

Religion plays an important part in the life of people in the region. In much of Central America the Spanish left a legacy of Catholicism. In some regions with a large indigenous population, such as the highland of Guatemala, Catholic rituals were grafted into pre-Columbian religious beliefs and practices. Elsewhere, **syncretic religions** emerged.

Vodou, also written as voodoo, is an example of a syncretic religion, a faith that draws upon a variety of religious traditions. Vodou emerged in Haiti from the experience of slavery. Slaves brought from Africa were not allowed to practice their traditional religions of West Africa and were forced into Catholicism. Yet they maintained their traditional beliefs by tweaking their religious ceremonies to appease the Christian overseers. Traditional African spirits were disguised as Catholic saints, and Catholic rites were superimposed on traditional practices. They even drew upon the beliefs and rituals of the Amerindians. Vodou developed into a distinctive religious practice, a New

World creation from Old World roots. Today, almost half the population of Haiti practice some form of Vodou.

Vodou has a bad reputation as a form of devil worship. But if you think about it, you can see why the colonial powers would label it as devil worship as it allowed the practitioners to be seen as less than civilized. When Haiti became the first slave colony to throw off its colonial yoke, the depiction of Vodou became part of the way that white supremacists delegitimized the slaves' claim to freedom. Vodou is a common religious practice in Haiti as well as throughout the Haitian diaspora, especially in New York City.

Rural Focus

Land Grabs

Across the globe the land of indigenous peoples and poor peasants is being appropriated as companies and governments enclose commons and dispossess the peasantry in a global land grab. The process is fueled by the commercialization of agriculture as food production in many regions becomes more linked to global markets and the growing demand for biofuels. When agricultural land becomes more valuable as a commodity, then rich powerful interests seek to possess the land. This often involves the dispossession of local communities with long-established links to particular parcels of land. Peasant societies struggle against well-connected corporations and powerful state bureaucracies. The Bajo Aguan region in Honduras has been a scene of intense conflict over the last two decades. In the 1990s, large landowners purchased land from farmer cooperatives to harvest palm oil for export. Local activists complained that the deals were unfair and people were not made aware that they were signing away land rights. One company owned by a member of the very wealthy Honduran elite amassed one-fifth of all the land in the region. In 2009, peasants invaded and occupied company land. Troops were sent in, people were evicted, and more than forty people were killed. The peasants resisted, and the bad publicity led to one German bank withdrawing loans to some of the large companies. In 2011, the Honduran government passed a law that allowed peasant farmers in Bajo Aguan to purchase 4,000 hectares at favorable interest rates. The peasants resisted and forced a change in the government attitude.

Urban Trends

City populations in the region increased through natural growth as well as massive emigration from the countryside as rural impoverishment restricted opportunities for an expanding population. In 1960, the population of Kingston,

TABLE 3.1 Metropolitan Areas in the Region		
METRO AREA	**POPULATION (MILLIONS)**	**COUNTRY**
Mexico City	21.2	Mexico
Guatemala City	4.1	Guatemala
Santo Domingo	3.7	Dominican Republic
San Salvador	2.4	El Salvador
San Juan	2.4	Puerto Rico
Port au Prince	2.4	Haiti
Havana	2.1	Cuba
San Jose	1.7	Costa Rica
Tegucigalpa	1.3	Honduras
Managua	1.3	Nicaragua
Panama City	1.2	Panama

Jamaica, the largest English-speaking city in the entire region, was close to 340,000, but by 2012, it was almost 940,000. As a percentage of the island's total population Kingston's share increased from 21 percent to 34 percent. There was an urban explosion across the region. Table 3.1 lists some of the largest cities in the region. Mexico City is by far the largest city in the region.

Informal Urbanism

The cities are characterized by informal housing and an informal economy. Informal housing has various names, including marginal housing and slums. Built on the most marginal land, prone to landslips, flooding, and other hazards, and with uncertain legal status, over half of the city's population lives in self-built accommodation (Figure 3.10). Over the years residents may be able to organize and obtain services such as water and power. All the major cities in the region have substantial numbers of slum neighborhoods.

Because of limited formal employment, between 40 and 50 percent of the people work in the informal economy recycling trash, buying and selling, and providing goods and services in an economy that is neither recorded nor taxed.

The cities are sites of marked inequality between the gated communities of the rich minority and the slums dwellers living on the margins. These economic divisions are often overlain with racial and ethnic differences—between indigenous and Latino in Central America and between shades of black and white in the Caribbean. Many of the slum dwellers in Guatemala City, for example, are Mayan.

Urban Primacy

Urban patterns in the region are distinguished by an **urban primacy** in which one city tends to dominate the national urban system. This is a region with a long urban

3.10 Informal housing in Dominica (photo: John Rennie Short)

history: many of the cities acted as control centers for pre-Columbian and then colonial powers and continued to attract further growth and development. Established as ports and as control sites of colonial power, these centers are often the largest in the country. The sheer size of the city attracts further investment, as multinational companies as well as government offices are located in the major capital cities, attracting other sectors of the economy and a steady stream of migrants. There is an urban bias to economic development as cities continue to attract investment and migrants and become the economic focus and political center of the entire nation.

City Focus: Mexico City

When the conquistadores arrived in the area where Mexico City now stands, they found a large city, the capital of the Aztec Empire. The Aztecs founded Tenochtitlan in the early fourteenth century. The city was built on an island in the middle of a shallow lake. At its center was the Great Temple precinct with a double pyramid, edifices built to honor the rain god. At a rededication of the great pyramid, 20,000 men were sacrificed in 4 days. Around the city's center lived the high nobles and senior officials.

Commoners lived outside the island city. The Spaniards eventually leveled much of this beautiful city.

The city of the Aztecs became the center of Spanish power and eventually the capital of independent Mexico. Under the authoritarian rule of President Diaz (1876–1919), the city's infrastructure was expanded. It became the most developed Mexican city, primed for industrial takeoff. Under the authoritarian regime of the Institutional Revolutionary Party (known by its acronym, the PRI), from 1929 to the early 1990s, the city benefitted from the high tariffs imposed on a range of manufactured imports that encouraged domestic industrial production in Mexico City. The city expanded and grew as it attracted more industry and workers. The population grew from 1.6 million in 1940 to 5.4 million in 1960. It is now one of the largest cities in the world. The city's population is 8.8 million. The population of the entire metropolitan region is now 21.2 million (Figure 3.11).

Mexico City like many capital cities in Central and South America is a **primate city**: that means it is by far the largest city in the country and the center of economic and political life. While the industrialization of the northern border region has shifted some of the economic center of gravity and manufacturing farther north, Mexico City remains the unchallenged city in the country, home to the political elites and the location of major companies and foreign corporations. The second largest city, essentially a suburb of the metro Mexico City, has a population of 1.6 million. The largest independent city is Guadalajara at 1.5 million.

The city's population doubled from 1960 to 1980 as rural migrants flocked there. Slums emerged all around the city. The rapid growth strained the environmental capacity as air and water pollution posed significant health risks. The city sits in a high plateau surrounded by mountains that traps in the polluted air. The city is also vulnerable to earthquakes.

Mexico City has numerous environmental problems, such as illegal dumping of trash, subsidence as the city exploits the underground aquifers beyond replacement levels, and ongoing environmental problems with 30 percent of households lacking access to toilets and most wastewater discharged without treatment. But the problems are also the opportunity for solutions. Recycling is encouraged, public transport is promoted, and air quality levels have improved

3.11 Map of Mexico City

dramatically. Since 1990, lead in the air has been reduced by 90 percent and ozone levels have dropped by 75 percent, the result of the promotion of public transport, the relocation of polluting industries, and the improvement of automobile exhausts. There are also imaginative designs such as three giant vertical gardens of 50,000 plants that absorb pollution. Mexico City is a source of problems but also a site of solutions as its citizens strive to create a more livable and sustainable city.

Geopolitics

Colonial Legacy

The colonial legacy lives on in the language of Spanish-speaking republics, cricket playing in Barbados, or the French-influenced cuisine of Martinique. There are also fragments of the British Empire that still remain. These include the British territories of the British Virgin Islands, the Cayman Islands, Montserrat, and the Turks and Caicos

Islands. The Cayman Islands are now a major site of off-shore banking tax avoidance and money washing because it is linked to the global financial system but has the political stability of a British territory and the advantages of bank secrecy. The French still retain control over Guadeloupe, Martinique, and St Barts. The Netherlands has title over the islands of Aruba, Bonaire, and Curacao that sit just offshore from Venezuela. And in a reminder of former rivalries, the tiny island of St Martin/Maarten, only 33.5 square miles, was divided in 1648 between the Netherlands (where it is known as Sint Maarten) and the French-controlled Saint Martin. The boundary crossing between the two different national areas is one of the easiest in the world with no passports or visas necessary and no border guards visible. Empire lives on in these islands of the Caribbean.

The Role of the United States

The latest outside power to enter the area was the United States, which aggressively asserted its role in the **Monroe Doctrine** that was proclaimed in 1823 and laid claim to geopolitical influence in the region. It stated that while existing boundaries would be honored, the United States would not allow any further excursion by European powers into the region. It only came into effect after 1850 when the United Kingdom, the unrivalled global superpower of the time, agreed. Issued like a blank check at a time when the military power of the United States was severely limited, it has continually been cashed in as the United States has grown in military dominance.

The United States has exerted a huge influence in the region. Cuba was invaded, and Puerto Rico was permanently annexed. The United States purchased what is now the US Virgin Islands from Denmark in 1916 and controlled the Panama Canal Zone from 1904 to 2000. The Monroe Doctrine was cited as a rationale for numerous interventions over the years. In some cases the military intervened as in Honduras a number of times from 1903 to 1925. In other cases there was outright occupation, including Nicaragua from 1912 to 1933 and Haiti from 1915 to 1934. There were also more covert actions of regime change and attempted regime change, including Guatemala in 1954, Cuba in 1959, and the Dominican Republic in 1961. Table 3.2 lists only some of the better-recorded direct and indirect interventions. Most of these interventions deposed leftist governments and regimes and supported the landowning class and US corporate interests.

Boundary Disputes

Boundary disputes between nation-states sometimes reflected the uncertain nature of territorial formation in the wake of Spain's withdrawal. A brief war between El Salvador and Honduras erupted in 1969 after a football (soccer) match between the two countries. It is known as the Football War (La Guerra del Futbol), but its real origins lie in the border disputes and tension caused by immigration from El Salvador to Honduras. The war lasted less than

100 hours. The two countries signed a peace treaty, and the International Court of Justice demarcated the border, with most of the disputed territory going to Honduras.

While many of the land border disputes are either settled or in the process of adjudication, maritime boundaries are an emerging site of interstate relations in the region (Figure 3.12). The Caribbean is a relatively small body of water where thirty-nine states (twenty-two independent countries and seventeen territories) lay claim to what is essentially a closed sea. The United Nations Convention of the Law of the Sea, ratified in 1994, allows states to claim up to 200 nautical miles of their coast as an exclusive economic zone. This extends the reach of small island states such as Barbados that claim up to ten times more maritime space than terrestrial space. These claims are becoming more important as oil, gas, and valuable mineral deposits are found offshore. In 2001, Nicaragua took Colombia to the International Court of Justice over a disputed maritime boundary. In 2012, the Court granted Nicaragua an **Exclusive Economic Zone** (EEZ) that shifted the previous maritime boundary eastward, roughly along the 82-degree meridian that effectively meant a transfer of 30,000 square miles from Colombia to Nicaragua.

While the presence of valuable resources may heighten tensions, the need for cooperation both to exploit resources that are found in adjacent EEZs and to manage conservation and marine protection policies provides a basis for greater cooperation between the states in the region.

Connections

Since the fifteenth century, this region has been an important part of the global economy. In this section we can consider two types of more contemporary economic connections.

The Panama Canal

The narrow land isthmus between the Pacific Ocean and the Caribbean has long attracted attention. For centuries, goods were shipped across the narrow divide between Panama City and Portobelo. In 1689, Scots investors put together the **Darien Scheme** to establish a colony of Scottish settlers across the narrow isthmus. The scheme was a disaster as colonists died from disease and were eventually evicted by the Spanish. The disaster—almost a quarter of all money circulating in Scotland at the time was invested in the failed enterprise—led many to support the Act of Union in England in 1707 as a way to regroup and cash in on the more successful English colonial schemes. The linkage between Scotland and England is, in part, a result of the Darien disaster.

The isthmus between the Atlantic and Pacific is so narrow in the region of modern-day Panama that it long attracted interest. Pack animals moved goods between the

TABLE 3.2 US Interventions in the Region	
COUNTRY	**DATES OF INTERVENTIONS**
Cuba	1898, 1906–1909, 1959
Dominican Republic	1905, 1916–1924, 1965
El Salvador	1979–1982
Grenada	1983
Guatemala	1954
Haiti	1915–1934, 1994–1995
Honduras	1903–1925
Mexico	1846–1848, 1914–1917
Nicaragua	1909, 1923–1928, 1981–1990
Panama	1989

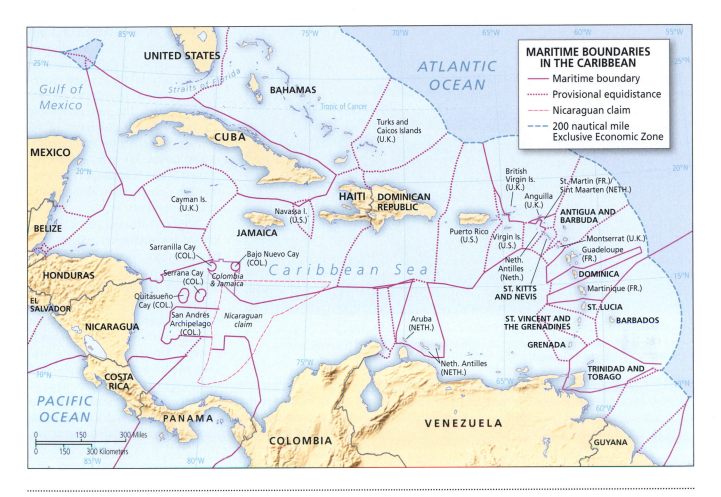

3.12 Maritime boundaries in the Caribbean

two oceans, but the journey was hazardous and trade was limited. A canal would reduce the long and hazardous journey around Cape Horn. The French, fresh off their success from building the Suez Canal, started a cut in 1879. But the Suez was a giant trench cut through hard ground. This was a more difficult proposition. Unstable soils made a simple trench impossible. Disease also created high mortality rates; more than 22,000 workers died, and the scheme was abandoned in 1893. The United States then entered the scene as a global power with a strong navy eager to establish its role in world affairs. The United States backed a revolution that broke off Panama from Colombia and, in exchange for $10 million, the United States obtained from the new Panamanian government control over a 5-mile-wide Canal Zone that ran through the middle of the new country.

The Canal that the US Army engineers designed and Caribbean labor built was a lock, not a trench system. Nine miles were cut through high ground, and Gatun Lake was flooded to provide water for a lock system that raised ships up and down over 48 miles using a gravity system. It was a simple but elegant system built ahead of schedule and under budget. It opened in August 1914. Ships sailing along the Canal were in what was essentially US territory.

A treaty signed in 1977 returned the Canal and the zone from the United States to Panama in 2000, although the United States remains in charge of protecting its neutrality. Today, around 14,000 vessels pass through the canal each year carrying a total of 300 million tons of cargo. The transit rates vary but approximate at around $3–$4 a ton. Built in the early twentieth century, by the early twenty-first century the canal was losing traffic because of its restrictive size. In order to accommodate the large container ships, wider locks were built that increased the capacity from 1,050 feet long, 115 feet wide, and 41 feet deep to 1,400 feet long, 180 feet wide, and 60 feet deep (see Figure 3.13). The enlarged canal opened in 2015 at a cost of $5 billion. The expansion allows an increase of annual tonnage from 300 million to 600 million.

Migration and Remittances

In this region there are more people than employment opportunities. Economic growth has not kept pace with population growth, and unemployment is high. One response is for people to leave in search of employment. The United States, with a huge economy and lots of opportunities,

3.13 A giant container ship in Gatun Lake about to enter the locks of the new Panama Canal (photo: John Rennie Short)

especially for semiskilled and unskilled workers, acts as an alluring magnet. Table 3.3 lists the number of migrants to the United States from selected countries in the region in 2015. The percent of these US migrants as a percent of all overseas migrants for that country is also noted. We can see that the United States is an important destination point both in absolute and relative numbers.

Migrants also send back money, known as **remittances**, to the country of origin, helping to provide support for friends and families (see Table 3.4). The flow of remittances is now much larger because of the ease of money transactions even of small amounts. The United States is a major source of remittance, and in cases such as Mexico the absolute sums are substantial. When expressed as a percentage of GDP, we can see that remittances play an enormous role in the national economy of smaller nations such as Honduras, Haiti, and Jamaica. In many villages and small towns across the region, many of the younger and more educated people work overseas, sending money to families back home.

Subregions

The region can be divided into three distinct groups of countries: Mexico, Central American republics, and the small island nations of the Caribbean.

TABLE 3.3 Immigration to the United States		
COUNTRY	**NO. OF MIGRANTS TO US**	**% OF ALL EMIGRANTS FROM COUNTRY**
Dominican Republic	787,015	75.9
Guatemala	753,720	86.4
Haiti	545,437	54.0
Honduras	469,202	82.3
Jamaica	649,046	65.8
Mexico	11,635,996	98.1
Nicaragua	242,886	33.3
Grenada	33,568	49.1

TABLE 3.4 Remittances from the United States			
COUNTRY	**TOTAL REMITTANCES FROM US ($MILLION)**	**AS % ALL REMITTANCES**	**AS % OF GDP**
Dominican Republic	2,732	77.9	7.3
Guatemala	4,400	89.3	9.8
Haiti	1,088	66.9	15.4
Honduras	2,579	86.8	19.3
Jamaica	1,465	67.8	13.8
Mexico	22,811	98.2	2.5
Nicaragua	430	42.5	10.3
Grenada	30	50.8	8.7

The tables were calculated from World Bank data on migration and remittances.

Mexico: The Giant of Central America

Mexico is the largest country by far, both in terms of area and population, in the region (Figure 3.14). With a population of close to 115 million, it is the most populous Spanish-speaking country in the world. The population of the capital city alone is greater than any other country in the region. Mexico now has the largest per-capita GDP in the region.

It also has a distinctive location as it shares a long border with the United States, which has shaped its migration patterns, trade flows, and economic interactions, especially in the wake of **North American Free Trade Association** (NAFTA).

There are echoes of pre-Columbian times in the ancient ruins of old cities and in the living presence of indigenous people who make up around 30 percent of the population. They are concentrated in the south of the country in regions such as Chiapas, Oaxaca, and the Yucatan and include Mayans, Mixtecs, and Zapotecs. They sit at the bottom of the socioeconomic hierarchy. Their world is often described as "deep Mexico," and their traditional landholdings are under threat from privatization. Mixed race or mestizos make up 60 percent.

The country bears the mark of Spain in its language and religion and towns. Town buildings are clustered in a grid around a central plaza where often the church and municipal office sit. The country has long been ruled by Hispanic elites. The country was ruled by a single party throughout most of the twentieth century, but since 2001, a more democratic system has taken hold.

In terms of the economic development and the demographic transition, Mexico is between the wealthy United States and the poorer nations of the Caribbean. An old Spanish phrase translates into "Pity poor Mexico, so far from God, so close to the United States." The phrase recalls the US invasions of Mexico, but closeness to the United States has also benefitted the country in a number of ways. The United States is a huge market for agricultural produce and for manufactured goods. Free trade zones along the border have attracted manufacturing operations, referred to as maquilas or **maquiladoras.** Along the border, these manufacturing and assembly factories now employ around 1.3 million workers, many of them young women, and constitute around half of all of Mexico's exports. Trade increased in both directions after the signing of NAFTA. Mexican imports from the United States rose from 6.1 percent in 1990 to 12.3 percent in 2012. Criticisms of the system cite low wages, difficult working conditions, and poor environmental standards. However, the northern Mexican states closer to the United States have higher incomes than states in the center and south. The poorest part of the country is along the southern Pacific coast.

The population of Mexico is unevenly distributed. The development of the maquiladoras shifted population toward

3.14 Map of Mexico

the cities of the northern border. There is still a major concentration around Mexico City and the central plateau with pockets of concentration in selected coastal sites.

There are broad regional divisions within the country. The Baja peninsula is the thin sliver of land in the northwest of the country. The border town of Tijuana now tops 2 million people, its growth a function of its border location with the United States and the expansion of maquiladoras along the border. Further along the coast, a number of resorts cater to wealthy North Americans and Europeans. Cabo San Lucas was a tiny fishing village, at the end of the long, dry peninsula where the Pacific and the Sea of Cortez meet. Since the 1980s it has been transformed into an enclave for expatriates and foreign tourists, filled with nightclubs and million-dollar homes.

The north of the country consists of harsh rugged terrain in the states of Chihuahua, Coahuila, Nuevo Leon, Sinaloa, Sonora, and Tamaulipas. Along the border with the United States, towns such as Ciudad Juarez, Matamoros, Nogales, and Reynosa have burgeoned with cross-border trade. The largest town in the region, Monterrey, with a population of 4 million in the metro region, has experienced deindustrialization of heavy industry. The large steel works closed in 1986. The city has also experienced calamities, both environmental and social. In 2010, Hurricane Alex hit the city hard. Cartel violence, which erupted in 2011, has also cast a pall over the city.

In Lower Mexico, also known as the Bajio, consisting of states such as Aguascalientes, Jalisco, and Zacatecas, the more fertile valleys support a denser population than North Mexico. There is still a rich legacy of Spanish architecture in the town and cities.

Mexico City and the surrounding capital region are the heart of Mexico. This is the largest metropolitan region in Mexico and Central America combined with more than 21 million people. Mexico has a primate city distribution because of the concentration of political, economic, and social power in its capital.

Moving southward from this region is to shift from the Hispanic to the more rural and indigenous. Campeche, Chiapas, Oaxaca, and the Yucatan have significant proportions of indigenous peoples, around 50 percent in Yucatan and Oaxaca, 27 percent in the Chiapas. This is what is referred to as deep Mexico; rural, indigenous, where pre-Columbian traditions and beliefs persist alongside the Hispanic Catholic order imposed by the Spaniards. In selected coastal sites such as Cozumel and Cancun in the Yucatan and Huatulco in Oaxaca, enclaves have been created for tourists from North America and Europe who come for the beaches and to visit the ruins of pre-Columbian cities in the Yucatan jungle. The tourist island of Cozumel was badly hit by the destructive power of Hurricane Wilma in 2005 when winds reached 183 mph.

Mexico is designated as an upper-middle-income country, and growth has resulted in the enlargement of what was traditionally a very small middle class. Income inequalities persist and poverty remains a stubborn problem with more than one in five below the official poverty line and close to 50 percent according to other estimates.

The population has also moved into the later stages of the demographic transition. In the 1960s the average family consisted of seven children; now it is closer to two. Life expectancy has increased to 77 years, close to the figure for the United States. Mexico's demographic dividend will only pay off if more money is spent on providing education and training.

Central American Republics

The small Central American republics of this region include Belize, Costa Rica, El Salvador, Guatemala, Honduras, Nicaragua, and Panama (Figure 3.15). Close to the equator, they have a hot and tropical climate, ideal for the growing of sugar and bananas and the exploitation of tropical hardwoods. There is a small manufacturing sector, but tourism generates a large part of foreign exchange. Remittances are an important source of national income.

Only Belize was not part of the Spanish Empire. From 1862 until independence in 1981, it was part of the British Empire and known as British Honduras. Guatemala had laid claims on the territory and only recognized the country in the 1990s. Almost 15 percent of citizens live abroad, and an equal number have moved into the country from neighboring Central American countries, making Belize more like the rest of Central America. There is a stream of retirees attracted by the all-year warmth, generous tax arrangements, and, for elderly US citizens, cheaper health care.

TROUBLED NATIONS

El Salvador, Guatemala, Honduras, and Nicaragua form a bloc of Spanish-speaking republics that obtained independence from Spain in the early nineteenth century. They are lower-middle-income countries with economies based on primary production, marked income inequality, and a political structure of a vast majority of poor ruled by a small, very rich elite. The derisive term of "**banana republics**" was used to describe them because their economies were based on tropical primary commodities such as bananas, but the term also entails the notion of political corruption and instability. The term had other more specific connotations because the US United Fruit Company played an outsized role in the economy and policies of the region. For many years they were republics but certainly not democracies as the small elite landowning Hispanic groups and later an army officer corps dominated.

During the Cold War these countries were sites of major social conflicts. In El Salvador a civil war, fought from 1979 to 1992 between leftists and US-backed government forces, devastated the country. Almost 75,000 people died. Many people fled and now one in five citizens lives

3.15 Map of Central American republics

abroad and remittances are a major source of national income. In Guatemala, a 36-year civil war only ended in 1996. More than 200,000 people died and 1 million were displaced, especially in the Mayan region of the country where government forces burned villages and removed people to punish the insurgency. More than one in three of the population is of Mayan origin. In 1979, in Nicaragua, leftists overthrew a dictatorship. The United States then funded a guerrilla movement for over a decade, and the ongoing conflict weakened the economy. Before leaving office in 1990, leaders of the leftist government appropriated property and millions of dollars from the state treasury. Honduras, unlike the other republics, did not have a civil war but was a haven for military forces, guerrillas, and insurgents fighting in Nicaragua and El Salvador. The legacy of conflict still lingers, as the republics inch toward a more normal form of political debate and negotiation.

These four countries are some of the poorest in the continent with endemic violence. Poverty and growing violence in Guatemala, Honduras, and El Salvador trigger migration to the United States. One heart-rending feature in the past few years is the growing number of women and young children escaping crushing poverty and threats of gang and drug cartel violence. Two gangs in El Salvador, Barrio 18 and Mara Salvatrucha gang, have more than 70,000 members combined. In the first half of 2014, more than 75,000 unaccompanied minors traveled north to try to gain entry into the United States. More than two-thirds of the children had suffered or been threatened with serious harm in their home regions.

A NEW CANAL

In one of the more ambitious attempts to generate economic growth in the region, the government of Nicaragua has plans for a canal to rival the Panama Canal. If you look at the map, you can see that southern Nicaragua is only a narrow band of land between a large lake and a land connection between the Atlantic and the Pacific. Numerous schemes were planned over the years. During the San Francisco Gold Rush, migrants crossed the land by stagecoach and the lake by steamer. The US government considered possible canal routes in both Nicaragua and Panama before finally choosing Panama, in part because of the vulnerability to earthquakes in Nicaragua. The

United States also worked to reduce Nicaragua's ability to compete with the Panama Canal. Under a new plan, Lake Nicaragua will be dredged, existing rivers widened, and locks built to enable huge ships to move from the Atlantic to the Pacific. It is a massive undertaking—one of the largest construction projects in the world, involving the excavation of 175 billion cubic feet of earth and the building of roads, towns, ports, and airports. The canal could serve almost 5 percent of the world shipping traffic and especially the biggest container ships. The estimated cost is $40 billion. The project is to be financed and built by the China Railroad Construction Corporation who will give the government $10 million a year for the first decade of operations and then a share of the profits. It is one element in the emerging economic and geopolitical power of China in the region. Almost 50,000 jobs would be created in the construction, but most of them are to be allocated to Chinese workers. The project would involve the displacement of thousands of small farmers and construction in the Indio Maiz ecological reserve.

STABLE NATIONS

It is no accident that the only countries in Central America apart from Mexico to make it to upper-middle-income status are countries with a more peaceful recent past, Costa Rica and Panama. Costa Rica is one of the richest countries in Central America, in part because of a long period of stable democracy and, since 1948, the abolition of armed forces. Some of the stability is in part to the landholding system that is slightly different from the other republics; less common are the huge estates of the very wealthy and more common are the small holdings especially in the upland coffee growing district. Political stability and economic growth feed off each other as the country can afford to create a more educated workforce that in turn attracts companies seeking more skilled labor. Manufacturing, services, and tourism have created more job opportunities that in turn attract migrants, especially from Nicaragua.

Panama is also relatively peaceful, apart from the invasion by US forces in 1989. A source of direct and indirect income is the Panama Canal that came under Panamanian control in 1999. The economy is more diversified than other Central American countries as it relies more on services, some of them connected to the presence of the canal and the busy maritime trade such as registry of ships, maritime logistics, and container port activity. Panama City is now a major site of banking and commerce. The construction and operation of the Panama Canal was done in US dollars, and Panama City developed as a dollar-friendly banking site. Shops passing through the Canal paid their fees in US dollars. The currency of Panama is the US dollar. For years its banking secrecy and lack of transparency made it a haven for offshore investment (Figure 3.16). Wealthy individuals and corporations placed their liquid assets in Panamanian banks. Since 2010, the country has signed onto an international effort to improve bank transparency and catch tax evaders. Still, neighboring countries such as Colombia placed Panama on a black list as a place for Colombian tax evaders to invest their money. More than a third of Colombian capital shipped abroad ends up in banks in Panama. With low taxes, a corporate-friendly government, and financial secrecy, Panama City is now a global hub for finance, second only to Hong Kong in the number of foreign companies registered in the city. Wealth is not widely distributed. Panama has one of the more unequal distributions of income and wealth on the continent.

Island Nations of the Caribbean

The island nations of the Caribbean include the more populous states of the Greater Antilles, including Cuba (11 million), Haiti (10 million), and the Dominican Republic (10.3 million) (Figure 3.17). Although close together and with a shared colonial history, they provide a rich contrast (see Table 3.5).

CUBA

The one major exception to US dominance in the Caribbean is Cuba. Initially a Spanish colony, it became independent in 1902 but with a constitution that gave the United States the right to intervene and run its national finances and foreign relations. A permanent US naval station was also established at Guantanamo, and a US governor ruled the country from 1899 to 1906.

3.16 Skyline of Panama City (photo: John Rennie Short)

3.17 Map of Caribbean islands

TABLE 3.5	Contrasts in the Greater Antilles		
	POPULATION (MILL.)	**GNI PER CAPITA**	**LIFE EXPECTANCY**
Cuba	11.2	$5,890	79
Dominican Republic	10.4	$5,770	73
Haiti	10.3	$810	63

GNI = gross national income

The distance from Key West in the United States to Havana, Cuba, is little more than 100 miles. The US influence on the economic life and politics of the country was profound and pervasive. When Fidel Castro came to power in 1959 and redistributed all landholdings over 1,000 acres, the United States saw Castro as a left-wing leader in a US-dominated region. When Cuba signed an agreement with USSR that essentially guaranteed a steady market for the Cuban sugar crop, Cold War ideology now saw Cuba as a danger to US interest. The United States imposed sanctions in an attempt to undermine the regime. The United States' worst fears were confirmed when the USSR exported missiles to the islands that led to the Cuban missile crisis of 1962. Cuban exiles in Florida played a major part in generating anti-Castro sentiment in the United States, and the political importance of Florida in presidential elections assured them a receptive audience at the highest levels of government. US sanctions against Cuba persisted, but there is recent evidence of a thawing of the relationship with a new generation of Cuban Americans softening in their attitudes, especially as the Castro regime looks close to the end. Most other countries now trade with Cuba, and more than 2.6 million tourists visit the island mainly from Canada and the European Union. With moves toward marketization and democratization and a thawing of US attitudes, Cuba is once again becoming part of the global economy.

Cuba under Castro pursued a route of centralized political and economic control. The system was successful in maintaining political control, but economic growth was sluggish. The system gave few political freedoms but ensured a basic standard of living. Life expectancy is one of the highest in the region, and income inequality is relatively small. After years of trade embargo by the United States, trade relations are slowly normalizing.

THE CONTRAST BETWEEN THE DOMINICAN REPUBLIC AND HAITI

The Dominican Republic and Haiti did not follow the Cuban model, but they do provide an incredible contrast for countries sharing one island. The Dominican Republic is an upper-middle-income country. Haiti, in contrast, is one of the poorest countries in the region with a high poverty level of close to 80 percent and a life expectancy of only 63 years. Why is Haiti so poor, especially when the country right next door, with similar environmental conditions and colonial history, is so much richer? Although the Dominican Republic was under the control of the Spanish while Haiti was part of the French empire, they share a similar history of slave-based economies based on primary production, especially sugar plantations. Haiti gained independence from France in 1804. In newly independent Haiti there was factional infighting especially between blacks and mulattos—a divide that persists. The political instability created a condition of **predatory elites** in which any group that came to power sought not to develop the country but to enrich themselves and their supporters. The result was a slow steady spiral downward. The United States occupied the island from 1915 to 1934 in order to protect US economic interests.

The political dictatorship of the Duvalier family, from 1971 to 1986, did little to aid the country and in fact led to further impoverishment through outright corruption and gross mismanagement. The treasury was looted and little investment was made in infrastructure, education, or economic development. Rapid population growth in the countryside led to massive deforestation as poor farmers cut down trees to sell charcoal. The mass of people are poor and illiterate. The poverty rate is close to 80 percent, one of the highest in the world. With few economic opportunities, almost 84 percent of those with tertiary education leave the country in search of employment. There are remittance corridors that link New York City and Haiti. Haiti remains one of the poorest countries, an example of what damage generations of predatory elites can wreck. The 2010 earthquake was a catastrophe for an already poor nation. Even as late as 2018, 55,000 people still lived in campsites, over 800,000 were infected with cholera, and more than 2.5 million relied on humanitarian aid.

THE LESSER ANTILLES

The Lesser Antilles is composed of a number of relatively small independent nations. They face not only the problems of economies still heavily dependent on just a few commodities and industries, such as tourism, but the environmental hazards of volcanoes and hurricanes. The small size of the island nations makes them more vulnerable. One major hurricane can impact the entire national economy such as when Hurricane Ivan devastated Grenada in 2004 or when in 2017 Hurricane Irma destroyed most of Barbuda. One journalist referred to the storm as "the night Barbuda died." A similar hurricane in the United States would impact a major city or region but not the entire nation. Powerful hurricanes in small Caribbean nations can devastate the entire national economy.

Smaller island states have difficulty in generating enough revenue in order to pursue ambitious plans for infrastructure or social development. As a result, educational and employment opportunities are limited, and many people leave the island to search for jobs in Britain, Canada, and the United States (Figure 3.18).

Throughout much of the Caribbean, the small island nations have problems of huge government debt that puts limits on social investment, endemic crime, and a middle-class brain drain as the more talented and educated leave the restricted opportunities of the islands for the greater opportunities in North America and Europe. More than 70 percent of the people with tertiary education leave the islands. This drain of the better educated and more often more politically active citizens creates the conditions for political apathy as the brightest and best leave.

The small island states have the trappings of statehood without the fiscal ability to meet all the standards of modern states. Martinique and Guadeloupe, in contrast, are part of France and thus able to access the resources of the large and wealthy French state. The islands are considered part of France. Elsewhere, the fragmentation of sovereignty into small island nations makes it difficult for governments hamstrung by lack of revenue to respond to economic recession, business downturns, and recurring environmental hazards. While big rich countries can borrow more money, small states have limited credit. The one exception to the economic difficulties of small island nations is Trinidad and Tobago that have the advantages of oil deposits. Oil and gas account for 80 percent of exports and provide enough revenue to support high levels of public investment and one of the largest average incomes in the entire region. The tax haven of the Cayman Islands is also one of the richer parts in the region.

There is an asymmetry between small island governments, on the one hand, and powerful outside interests, on the other. Major corporations have greater bargaining power than the state, while criminal elements such as gangs and drug cartels can often wield greater muscle than government forces. Small island governments are vulnerable in a world of mobile giant corporations and powerful and ruthless cartels.

3.18 Young citizens of St Vincent and the Grenadines. They will probably have to leave their island home to find good-paying jobs. (Photo: John Rennie Short)

Focus: Cocaine Capitalism, the Narco Economy, and the Narco State

This region is an important transmission point for the supply of drugs to the huge US market. Hard data are difficult to come by for US illegal drug consumption, but estimates are in the region of $80 million spent on illegal drugs each and every day. To supply this huge demand, there is an international supply chain. Coca grown in Peru and Bolivia is turned into cocaine in Colombia and shipped through Central America into the United States. The huge profits fuel the trade and pay for the bribes, the arms and muscle that characterize drug trading organizations (DTOs). **Cocaine capitalism** is an important part

of the connections between this region and the rest of the world and especially the United States.

A combination of huge demand, weak and corrupt states, and porous borders creates the condition for a **narco economy,** in which the drug trade dominates economic activities, directly or indirectly. A criminal state is when the political leadership is integrated with criminal activities. A narco state is defined as one controlled and corrupted by powerful drug cartels. Numerous national and local politicians in the region have worked and continue to work closely with DTOs.

Consider the case of Mexico and Honduras. Mexico is a transition point for drugs entering the United States as well as a production site for heroin and methamphetamines. When the DTOs in Colombia were dismantled in the 1990s, Mexican cartels emerged as the distribution agents for cocaine made in South America and consumed in North America (Figure 3.19). More money is now made from control of distribution rather than production. Cartels emerged in different parts of the country. La Familia

MEXICAN CARTELS
- Sinaloa cartel
- Gulf cartel
- Los Zetas
- Los Caballeros Templarios
- Cártel Jalisco Nueva Generación
- Juárez Cartel
- Beltrán-Leyva Organization
- La Familia Michoacana

3.19 Drug cartels in Mexico

controls the state of Michocana, Los Zetas controls much of the east coast, the Gulf cartel the border area in the northeast, and the Sinaloa cartel much of the central and western regions.

In 2006, the Mexican government launched a war on organized crime and the drug trade. The war on drugs militarized the nation. Military and federal police were dispatched to states around the country, an acknowledgment of the endemic corruption of local police by the cartels. Over the next 10 years, at least 45,000 troops were deployed. The cartels easily outmatch the limited resources of municipal government and local police. It is not only that crime creates corruption but also that the corruption of public institutions facilitates crime. The convicted drug lord of the Sinaloa cartel Joaquin Guzman, also known as El Chapo Guzman, simply walked out of a supposedly high-security jail and lived in the state of Sinaloa until his recapture in 2014. He managed to escape in 2015 but was recaptured in January 2016.

In Mexico as a whole, the annual homicide rate increased from 2,120 in 2006 to 12,000 in 2011. The war on drugs was bloody: from 2006 to 2016, at least 100,000 people were killed and 30,000 people "disappeared." Homicide rates rose dramatically because of the government's "war" on the cartels and also because of intense competition between the cartels for monopoly power. There was also intracartel violence as leadership struggles turned bloody. The violence became more gruesome with each cycle becoming more savage than the last in an

escalation of savagery. Violence would visit a city when rival cartels struggled for power. The town of Tampico on the Caribbean coast was a relatively peaceful oil town before competition between the incoming Los Zetas and the long-established Gulf cartel escalated into a rising body count on the streets of the city.

The wealth made from the drug trade is invested in legitimate organizations and businesses such as restaurants, farms, and day care centers. In some Mexican states such as Sinaloa more than two-thirds of the state's economy is intertwined with drug money. In some cases, the cartels provide a range of public services that the state is unable to provide, binding local institutions to the cartel. DTOs are bound up with local economies, police forces, and political leadership.

Ninety percent of the cocaine refined in Colombia for the US market passes through Honduras. It is part of the Northern Triangle of El Salvador and Guatemala, where drugs are trafficked into Mexico. The region's coastline and national borders are lightly policed, and the resources of the small state have difficulty controlling powerful rich cartels. Planes filled with drugs land in custom-built airstrips along the Atlantic coast of Honduras. Smugglers take the drugs to the United States or to the United States through Mexico. Mexican drug cartels, and especially the Sinaloa cartel, have now penetrated the political and economic life of the country. The lack of democratic accountability and endemic corruption make the country particularly vulnerable to drug cartels. The cartels also provide public service, becoming in effect a shadow government more powerful and sometimes more sensitive and responsive to local needs than the official state. Crime rates have increased in Honduras; the country now has one the highest murder rates in the world, because of competition between rival cartels. Violence pervades the country. In 2013, 1,013 people aged under 23 years were killed. In the first 6 months of 2014, 409 children aged under 18 years were killed. The violence fueled the movement of young people and children northward to the United States. The local people of the region, the Miskitu, are often caught between the US-Honduran military forces and the drug cartels. The Mosquito Coast is a battle zone between competing militaries with local people caught in the awkward middle.

Focus: The Real Pirates of the Caribbean

Piracy. It is now an integral part of the marketing, selling, and representation of the Caribbean to a wider world. Piracy has long moved from fact to a commodity in popular culture. But who were the real pirates of the Caribbean, as opposed to the Johnny Depp characterizations?

The wider historical context is the imperial competition between the dominant European powers in the region—Spain and upstarts into the region, the English, French, and Dutch. Spanish treasure ships transferred gold and silver from Central and South America. There was a transfer of precious cargo on well-known routes at well-known times. The treasure ships were thus vulnerable as they sailed through dense shipping lanes relatively close to the coast via sea lanes dotted with small islands that provided escape and refuges for desperate men. Runaways, indentured sailors, servants, and slaves provided the pool of pirate conscripts.

The main era of piracy was from 1560 to 1750 with the golden age—the era depicted in novels and movies—being a relatively short period from the 1640s to the 1680s. At its height piracy was also a form of state-sponsored terrorism. The English and French governments encouraged piracy as a way to undermine the power of Spain without spending public money. They offloaded the war onto privateers. Governments retained plausible deniability, reduced war cost, and shared profits. After 1700, geopolitics changed as the British gained dominance and formed an alliance with Spain and the British Navy began to police the seas from its base in Bermuda and other naval stations in the Caribbean. The Caribbean became less a war-torn space filled with desperate pirates than a policed space for the regular international trade. There were also fewer treasure ships as trade changed from the export of precious minerals to the export of primary produce such as sugar and rum. (There is less immediate cash value to bananas and sugar than gold and silver.)

The better known pirates include William Kidd (1645–1701), Edward Teach aka Blackbeard (1680–1718), Black Bart Roberts (1682–1722), and, just to show that piracy was not an all-male profession, Anne Bonney (1700–?). Their lives are part fact, part myth, so for greater historical accuracy we should consider the more documented figures. Three pirates embody the rise and fall of Caribbean piracy. Sir Francis Drake (1540–1596) was born in England's West Country, a region with a long maritime tradition. He was involved in the slave trade, and from 1577 to 1581 circumnavigated the world, amassing stolen treasure on the way and claiming land for England. Queen Elizabeth knighted him. On his return he fought against the Spanish and sailed in the Caribbean as a privateer. He died of dysentery while attacking San Juan in Puerto Rico (Figure 3.17). Sir Henry Morgan (1635–1688) also attacked Spanish ships and ports in Cuba, Panama, and Venezuela. He engaged in ransom, theft, and extortion, but was knighted in 1674. William Dampier (1651–1715) arrived on the scene at the tail end of piracy in the region. In 1680, he joined a pirate fleet of twelve ships led by shipmates of Morgan. The pirate fleet attacked Spanish ships and ports throughout the region. Dampier gave up his pirate ways to be a captain in the Royal Navy. He sailed to the Far East and around Australia. He wrote regional geographies on his travels, so his full accomplishments range from pirate, explorer, and author to naturalist, geographer, and scientist; he was one of the first to create a genre of geographic fieldwork. Not quite your Johnny Depp character of charming criminal charisma but an interesting character nevertheless, who straddles an unlikely pair of pursuits: piracy and geography.

Select Bibliography

Anderson, T. D. 1984. *Geopolitics of the Caribbean*. New York: Praeger.

Ashcroft, B., G. Griffiths, H. Tiffin, and B. Ashcroft 2007. *Postcolonial Studies: The Key Concepts*. Hoboken, NJ: Taylor & Francis.

Barker, D. 2012. "Caribbean Agriculture in a Period of Global Change: Vulnerabilities and Opportunities." *Caribbean Studies* 40:41–61.

Bishop, M. L. 2013. *The Political Economy of Caribbean Development*. New York: Palgrave Macmillan.

Boskin, M. J. 2014. *NAFTA at 20: The North American Free Trade Agreement's Achievements and Challenges*. Stanford, CA: Hoover Institution Press.

Brothers, T., O. Dwyer, and J. Wilson. 2008. *Caribbean Landscapes: An Interpretive Atlas*. Coconut Creek, Florida: Caribbean Studies Press.

Burkholder, M., and L. Johnson. 2012. *Colonial Latin America*. 8th ed. Oxford: Oxford University Press.

Carmack, R. M., J. Gasco, and G. H. Gossen. 2007. *The Legacy of Mesoamerica: History and Culture of a Native American Civilization*. Upper Saddle River, NJ: Pearson/Prentice Hall.

Clawson, D. L. 2012. *Latin America and the Caribbean*. 5th ed. Oxford: Oxford University Press.

Cockcroft, J. D. 2010. *Mexico's Revolution: Then and Now*. New York: Monthly Review Press.

Coe, M. D., and R. Koontz. 2008. *Mexico: From the Olmecs to the Aztecs.* London: Thames & Hudson.

Corchado, A. 2013. *Midnight in Mexico: A Reporter's Journey through a Country's Descent into the Darkness.* New York: Penguin Press.

Crosby, A. W. 1972. *The Columbian Exchange: Biological and Cultural Consequences of 1492.* Westport, CT: Greenwood Press.

Dym, J., and K. Offen. 2011. *Mapping Latin America: A Cartographic Reader.* Chicago: The University of Chicago Press.

Foote, N. A., ed. 2013. *The Caribbean History Reader.* London: Routledge.

Franko, P. M. 2007. *The Puzzle of Latin American Economic Development.* Lanham, MD: Rowman & Littlefield.

Girard, P. R. 2010. *Haiti: The Tumultuous History—From Pearl of the Caribbean to Broken Nation.* Houndmills, UK: Palgrave Macmillan.

Jackiewicz, E. L., and F. J. Bosco, eds. 2012. *Placing Latin America: Contemporary Themes in Geography.* 2nd ed. Lanham, MD: Rowman and Littlefield.

Jaffe, R., and J. C. G. Aguiar. 2012. "Introduction: Neoliberalism and Urban Space in Latin America and the Caribbean." *Singapore Journal of Tropical Geography* 33:153–156.

James, C. L. R. 2001. *The Black Jacobins: Toussaint L'Ouverture and the San Domingo Revolution.* London: Penguin.

Jelly-Shapiro, J. 2016. *Island People: The Caribbean and the World.* New York: Knopf.

Joseph, S., and T. Hamilton. 2014. "Development and Dependence along the New York-Haiti Corridor." *The Professional Geographer* 66:149–159.

Katz, J. 2014. *The Big Truck That Went By: How the World Came to Save Haiti and Left behind a Disaster.* Basingstoke, UK: Palgrave Macmillan.

Kent, R. B. 2016. *Latin America: Regions and People.* 2nd ed. New York: Guilford.

Klak, T., ed. 2000. *Globalization and Neoliberalism: The Caribbean Context.* Lanham, MD: Rowman & Littlefield.

Klak, T., J. Wiley, E. G. Mullaney, S. Peteru, S. Regan, and J. Y. Merilus. 2011. "Inclusive Neoliberalism? Perspectives from Eastern Caribbean Farmers." *Progress in Development Studies* 11:33–61.

Knight, F. W. 2012. *The Caribbean: The Genesis of a Fragmented Nationalism.* 2nd ed. Oxford: Oxford University Press.

Kronik, J., and D. Verner. 2010. *Indigenous Peoples and Climate Change in Latin America and the Caribbean.* Washington, DC: World Bank.

Latimer, J. 2009. *Buccaneers of the Caribbean: How Piracy Forged an Empire.* Cambridge, MA: Harvard University Press.

Mann, C. C. 2005. *1491: New Revelations of the Americas before Columbus.* New York: Knopf.

Mann, C. C. 2011. *1493: Uncovering the New World Columbus Created.* New York: Knopf.

Martinez, O. 2016. *A History of Violence: Living and Dying in Central America.* London: Verso.

McCullough, D. G. 1977. *The Path Between the Seas: The Creation of the Panama Canal, 1870–1914.* New York: Simon and Schuster.

McMillan, B., ed. 2002. *Captive Passage: The Transatlantic Slave Trade and the Making of the Americas.* Washington, DC: Smithsonian Institution Press.

McNeill, J. R. 2010. *Mosquito Empires: Ecology and War in the Greater Caribbean, 1620–1914.* New York: Cambridge University Press.

Meyer, M. C., and W. L. Sherman. 1979. *The Course of Mexican History.* New York: Oxford University Press.

Mundey, B. N. 1996. *The Mapping of New Spain: Indigenous Cartography and the Maps of the Relaciones Geografias.* Chicago: University of Chicago Press.

Nelson, D. M. 2009. *Reckoning: The Ends of War in Guatemala.* Durham, NC: Duke University Press.

The North American Environmental Atlas: http://www.cec.org/Page.asp?PageID=924&SiteNodeID=495&AA_SiteLanguageID=1

Olwig, K. F. 2007. *Caribbean Journeys: An Ethnography of Migration and Home in Three Family Networks.* Durham, NC: Duke University Press.

O'Neil, S. K. 2013. *Two Nations Indivisible: Mexico, the United States, and the Road Ahead.* Oxford: Oxford University Press.

Palmié, S., and F. A. Scarano, eds. 2011. *The Caribbean: A History of the Region and Its Peoples.* Chicago: The University of Chicago Press.

Perera, V. 1993. *Unfinished Conquest: The Guatemalan Tragedy.* Berkeley: University of California Press.

Pollard, J., C. McEwan, and A. Hughes, eds. 2011. *Postcolonial Economies.* London: Zed Books.

Portes. A., C. Dore-Cabral, and P. Landolt, eds. 1997. *The Urban Caribbean.* Baltimore: Johns Hopkins University Press.

Potter, A. E., and A. Sluyter. 2012. "Barbuda: A Caribbean Island in Transition." *Focus on Geography* 55:140–145.

Richardson, B. C. 1992. *The Caribbean in the Wider World, 1492–1992: A Regional Geography.* Cambridge: Cambridge University Press.

Richardson-Ngwenya, P. 2013. "Situated Knowledge and the EU Sugar Reform: A Caribbean Life History." *Area* 45:188–197.

Rumney, T. A. 2012. *Caribbean Geography: A Scholarly Bibliography.* Lanham, MD: Scarecrow Press.

Scarpaci, J. L., and A. H. Portela. 2013. *Cuban Landscapes: Heritage, Memory, and Place.* New York: Guilford Press.

Scarpaci, J. L., R. Segre, and M. Coyula. 2002. *Havana: Two Faces of the Antillean Metropolis.* Chapel Hill: UNC Press Books.

Schoultz, L. 1998. *Beneath the United States: A History of U.S. Policy Toward Latin America.* Cambridge, MA: Harvard University Press.

Schwartz, S. 2014. *Sea of Storms: A History of Hurricanes in the Greater Caribbean from Columbus to Katrina.* Princeton, NJ: Princeton University Press.

Sellers-Garcia, S. 2012. "The Mail in Time: Postal Routes and Conceptions of Distance in Colonial Guatemala." *Colonial Latin American Review* 21:77–99.

Sexton, J. 2011. *The Monroe Doctrine: Empire and Nation in Nineteenth-Century America.* New York: Hill and Wang.

Smithsonian Institution. 2002. *Captive Passage: The Transatlantic Slave Trade and the Making of the Americas.* Washington, DC: Smithsonian Institution Press.

Way, J. T. 2012. *The Mayan in the Mall: Globalization, Development, and the Making of Modern Guatemala.* Durham, NC: Duke University Press.

Western, J. 1992. *A Passage to England: Barbadian Londoners Speak of Home.* Minneapolis: University of Minnesota Press.

Williams, E. 1970. *From Columbus to Castro: A History of the Caribbean.* New York: Random House.

Learning Outcomes

The region of Central America and the Caribbean consists of the isthmus between North and South America and the island nations in the Caribbean.

The region is one of intense geological activity that generates environmental hazards that have differential impact on the rich and the poor.

The wind and weather patterns of the region played an important role in the early European colonization of the region, the rise of the North Atlantic triangular trade route, and the creation of an Atlantic economy that linked Europe, Africa, Central America/Caribbean, and North America.

The Columbian Encounter of 1492 was a profound transformative moment, a pivot of major environmental, social, and political change. It had four major impacts on the region: (1) the Columbian Exchange; (2) the imposition of Spanish and European control; (3) an environmental transformation; and (4) a demographic holocaust of the indigenous people.

There is a long and enduring division, marked by political and economic inequality, between the Hispanics and the indigenous people of Mesoamerica, particularly in countries such as Guatemala, where Hispanic elites have long dominated over the indigenous Mayan.

The economy of the region was transformed after the Columbian Encounter to provide tropical resources and cheap commodities for European markets, often at the cost of brutal labor exploitation and resource despoliation. Sugar was an important factor in this colonial economy. More recently, tourism, some manufacturing, and services have diversified the economy of the region.

In the last 50 years, population has grown quickly with a marked concentration in cities.

An informal urbanism and urban primacy began to dominate the region's growth patterns.

The melding of indigenous, European, and African cultures and people groups has created a melting pot of vitality and enormous diversity in the region.

Three themes dominate the political geography of the region: the colonial legacy, the influence of the United States, and boundary disputes, including both land and maritime disputes.

The Central America and Caribbean region can be further subdivided into three distinct subregions: Mexico, Central America, and the Caribbean island nations.

Mexico's sheer size, by both land mass and population, makes it the biggest country by far in the region.

The small Central American republics of the region include the troubled nations of El Salvador, Guatemala, Honduras, and Nicaragua, along with the more stable countries of Costa Rica and Panama, as well as the former British territory of Belize.

The island nations of the Caribbean include the more populous states of the Greater Antilles, including Cuba, Haiti, and the Dominican Republic, along with the smaller nations of the Lesser Antilles. Despite their close proximity and shared colonial history, they provide stark contrast.

4

South America

The Andes mountain range forms the spine of the subcontinent. The Amazon River basin is an important lung and biological reservoir for the entire planet. European settlers and colonists marginalized the indigenous peoples. While political independence was attained for former Spanish, British, and Dutch colonies, economic independence has been more difficult to achieve. In the past decades, economic growth based on import substitution, manufacturing, and a primary commodity boom fostered greater social and political stability. But does the ending of the boom signal more difficult times ahead?

LEARNING OBJECTIVES

Survey the region's major physiographic areas and describe the influence of altitude and the ocean on human activity.

Compare and contrast the region's human transformations during the Pre-Colombian and Post-Colombian periods.

Explain the features of the region's colonial economy and neoliberal policies and relate these to its poverty reduction efforts.

Discuss the region's population distributions and cultural categories and examine change in its racial, religious, and demographic character.

List the opportunities of expanding agriculture into the region's rural rainforests and evaluate the associated costs of deforestation.

Describe the drivers of the region's urbanization and connect these to its disparate cities and urban primacy.

Summarize the region's geopolitical challenges with focus on legacies of colonialism, international disputes, and intranational conflict.

Although South America and Central America are separated into different chapters of this book, there are many connections and similarities between them. Much of South America shares a similar Hispanic legacy with the republics of Central America. And the water that surrounds the islands of the Caribbean also laps against the shores of South America. There are shared histories and geographies. The division between Central America and the Caribbean is a useful way to keep a vast amount of material manageable but one that does injustice to their many shared experiences, links, and connections. So let us begin then with a tacit admission that South America is part of a much bigger and connected America. The divisions into North, Central, and South America are conveniences that foreground difference, whereas they are only parts of a more pervasive and singular continental connection to a wider world.

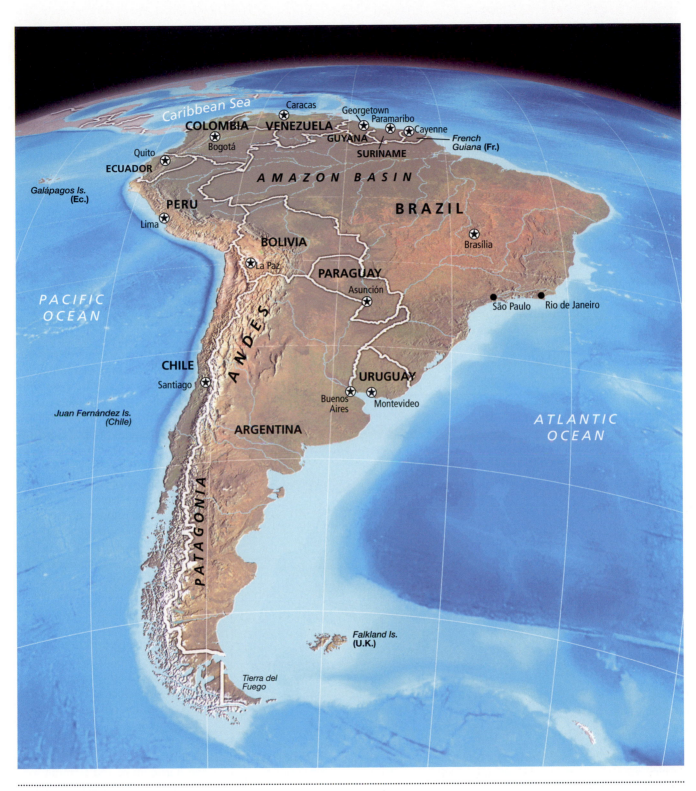

4.1 Map of region

The Environmental Context

The Backbone of the Andes

The Andes are one of the youngest mountain ranges in the world, the result of the clash of the Nazca and South American plates. It is the world's longest mountain chain stretching more than 4,500 miles all the way to the southern tip of the continent. In the north the mountain chain splits off into two separate ranges, Cordillera Oriental and Cordillera Occidental. Between them lies the fertile, intermountain land of Colombia.

The tectonic fault between the Nazca and South American plates is still very active with continuing seismic movement and the possibility of **tsunamis** off the coast. In 2010, an earthquake registering 8.8 on the Richter scale knocked out electricity to almost 90 percent of Chile's population (Figure 4.2). More than 350,000 homes were damaged or destroyed and total costs were estimated between $5 billion and $7 billion. Four years later, on April 1, 2014, an 8.2 magnitude earthquake struck just off the coast of Chile along the fault line. The seismic forces generated a 6.9-foot tsunami wave and rendered 80,000 people homeless.

The altitude changes along the Andes also create ecological diversity. Over just one 12-mile transect, the ecology changes from tropical rainforest dry oak forest (starting around 3,300 feet), dwarf forest and scrub (8,200–11,500 feet), alpine bog (11,500–14,750 feet), and frozen wilderness (above 14,500 feet). Heavy rainfall occurs on the eastern slopes, but the western coasts of Peru and northern Chile receive little rainfall.

Amazonia

One of the most distinctive features of the physical geography of South America is the giant river basin of the Amazon, which is roughly half the size of the continental United States. It contains the world's largest ecological region of tropical rainforest, typified by hot wet climate and lush forest vegetation (Figure 4.3). There are two seasons, hot and wet, followed by hot and even wetter. It is a zone of biological diversity without parallel. A 2-mile square patch of forest in Peru recorded over five thousand species of plants and animals. The subtle differences from dark forest floor to warmer and light-filled treetops create a vertical dimension for numerous ecological niches. The rainforest is one of the older ecosystems in the world. Too far south to be impacted by the last Ice Age, the ecology has developed over 120 million years, enough time for a biological complexity to unfold. The northern areas of the continental United States, in contrast, only emerged from under the ice caps less than 20,000 years ago.

We should remember, however, that this is not a pristine wilderness, but a human habitat settled for thousands of years. The earliest foragers penetrated the forest around 13,000 years ago, and by 4,000 years ago horticulture and farming were practiced. It is the rate of change and the weight of the human impact that are disturbingly recent.

The Amazon is a scene of constant conflict. Between 1835 and 1840, one-third of Amazonia's population died from slavery and enforced work. The conflicts

4.2 Map of 2010 earthquake zone in Chile

Epicenter Magnitude 8.8

2010 EARTHQUAKE ZONE IN CHILE

Estimated shaking intensity

- Severe
- Strong
- Moderate

PACIFIC OCEAN

CHILE

ARGENTINA

Valparaíso
Santiago
San Bernardo
Rancagua
Mendoza
San Rafael
Curicó
Constitución
Talca
Cauquenes
Chillán
Talcahuano
Concepción
Curanilahue
Angol
Temuco
Valdivia

continue between peasants and indigenous people, on the one hand, and powerful mineral and landed interests, on the other, in an area of the country far from media attention and effective government control.

In recent years, more ranchers, farmers, loggers, miners, and squatters have moved into the rainforest. Between 1970 and 2014, more than 15 percent of the total forest cover was cut down, an absolute loss of around 300,000 square miles. Deforestation is concentrated along highways such as the Belem-Brasilia highway opened in 1958 as the Brazilian government encouraged land-hungry settlers to move into the Amazon. More recently, the growth of soybean production for export and biodiesel production has pushed the agricultural frontier even further into the forest. Almost half of deforestation in South America is caused by forest clearance for soybean production. Global climate change is reinforced by this deforestation because living trees soak up carbon dioxide but release it when they are cut down and burned.

As the moisture and temperatures decrease southward, tropical rainforests turn into trees and grass and then grasslands (pampas). Much of the temperate grassland of the pampas was turned over to grazing and grain production in the late nineteenth century. This is an area of active fire management as farmers burn the vegetation at the frontier between forest and savannah.

Deserts and semideserts occur where there is little rain such as the Atacama Desert along Chile's coast, the Plateau of Borboreme in Brazil, and in Patagonia in Argentina. Attempts at blooming the hot deserts are hampered by problems of salinization as groundwater irrigation picks up the mineral salts in the desert soils.

El Niño and La Niña

El Niño is the systemic variation in sea temperature in the Pacific Ocean, off the coast of South America. Most years the water remains cold, but every 3–7 years, between October and March, the water warms up to 5°C higher than normal. Occurring around Christmas, the Spaniards called it El Niño with reference to the birth of Christ. There are local effects on the fishing industry as the warmer water kills the plankton that feed the fish. El Niño also affects wider weather patterns. In the United States, El Niño years are associated with more storms along the coast of California, drier conditions in Hawaii, more rain in the southeastern United States, and more snowfall in the Rockies and less in the upper Midwest. El Niño is also associated with drought in Australia and drier conditions in Fiji.

La Niña often follows an El Niño event and is characterized by a decrease in sea temperature across the Eastern Central Pacific of up to 5°C. La Niña events are associated with wetter conditions in South Africa, increased cyclonic activity off the coast of China, drought along the coast of

4.3 Forests in South America

South America, and more flooding inland. In the United States, the event is associated with more rain in the upper Midwest and drier conditions in the south. La Niña is linked to wetter conditions in Australia

Historical Geographies

The history of the population–environment relationship in America can be divided into two distinct periods: Pre-Columbian and Post-Columbian. Both periods were active periods of human transformation.

The Pre-Columbian World

The Pre-Columbian world was a complex mosaic of local cultures responsive to the constraints and opportunities of their environment. Hunters eradicated much of the mega fauna, gardened in the rainforest, and set fire to the savannahs. Local cultures embodied the environment in the words of the languages, the myths of their cosmology, and in the locale of their deities and spirits.

The Pre-Columbian world was also a world of expanding empires and competing territorial claims. The Incas are perhaps the best known. The Inca (also known as Inka) Empire ruled a vast territory stretching over 2,500 miles from northern Ecuador through Peru and Bolivia to southern Chile and Argentina. Lasting from roughly 1438 to 1553, it was the largest Pre-Columbian Empire in South America and, in 1500, at its peak, one of the largest in the world. Based in Cuzco, in present-day Peru, the Incan Empire built roads to connect its vast domain and siphon off tribute, food, and wealth from the periphery. Inca imperialism was very much about costs and benefits. The cost of direct territorial control was borne only where and when there were significant gains for the imperial core. Where profits could be made there was heavy investment in roads, temples, and agriculture. Where there was less direct benefit, Inca control was in the form of patronage of local elites. Defense was left to local allies with promise of Inca support. This strategy allowed the Inca to extend their control widely while minimizing costs. Imperial control varied from territorial strategies of direct control to indirect control in response to local conditions and changes over time. Top-level positions were reserved for Inca officials. The mid-ranking positions were often filled with members of local elites to cement political power and build up a system of patronage and loyalty. Today around 9 million Andean people still speak variants of the basic language of the Inca Empire, Quechan.

The Columbian Encounter

The **Columbian Encounter** is the name given to the interaction between the Old and New World after 1492. The most tragic was the **demographic holocaust** as indigenous peoples were killed off by the millions with the introduction of Old World diseases such as measles and smallpox. Diseases that were little more than a temporary annoyance for Europeans proved fatal for the people of the New World. It is estimated that close to 90 percent of the population of the New World died off within a century.

The Columbian Encounter also involved an exchange, often known as the **Columbian Exchange**, of plants and animals. The Old World received chocolate (cacao), maize, potatoes, sisal, tobacco, and vanilla. The potato, for example, indigenous to the Andes, created a new food source for the peasantry of Europe, allowing them to widen their source of starch and carbohydrate. It became a crop of major significance in the life and diet of Europe. The New World received almonds, apples, coffee, dates, figs, rice, tea, and horses. The year 1492 stands as a pivotal moment in the globalization of the world as two separate worlds were brought into contact, creating for the first time a truly globalized world.

The Columbian Encounter transformed the continent by incorporating a global economic system and consequent global economic cycles. Rainforest, especially closer to river frontages and along the coast, was cut down and replaced by sugar plantations, tobacco farms, and coffee farms. Pampas were turned into grazing for cattle and sheep. Mines were dug to meet global demand for copper, gold, and silver. Sugar plantations replaced rainforest, cattle grazing plains were created from woodland and savannah, and mountainsides were ripped open for their valuable minerals.

Economic Transformations

South America, like Central America and the Caribbean, became a producer and exporter of primacy produce such as beef, sugar, rubber, minerals, and other commodities. Land was cleared and is still being cleared to make way for primary products exported to the wider world (Figure 4.4).

In the colonial/neocolonial era, an export economy was developed that centered on primary products such as sugar in the tropical north, coffee in the uplands, and cattle raising in the more temperate south. The emphasis was on exploiting the environment to produce products for the global marketplace. In the early years of the colonial economy, the export activity was clustered on the coast or close to rivers. By the late nineteenth

4.4 Sugar plantation in intermontane region of Andes (photo: John Rennie Short)

century, with improvements in transportation, sugar cane and rubber plantations were more widely dispersed and cattle ranching extended into the more distant interiors.

One distinctive feature of the colonial economy was the granting of large estates to the elite. Across central and South America much of the productive and accessible land was parceled out, often through Royal charter, into large private estates. The system was referred to as **latifundia**, from the Latin for "spacious farm/estate." For centuries, these large landowners controlled the political system as well as the economy. Resistance to their rule from the urban intelligentsia and the rural poor constituted the main political dynamic of political conflict and change throughout South America. It is only in the past 50 years that the power of the landowners was successfully challenged by urban business interests and rural social movements. Small-scale farming persists and is responsible for the supply of foodstuffs for local and national consumption.

From 1914 onward, South America's export trade was much reduced by World War I and then the Great Depression. The primary produce export economy was overlain with import substitution as factories sprung up to make the manufactured goods previously imported from North America and Europe. Industrialization occurred in selected urban regions of the continent, and it was more successful in the larger countries such as Brazil and Argentina, where a large national market generated a sufficiently large enough domestic demand (Figure 4.5).

Neoliberalism

From the 1980s to recent times, many countries in South America, as in other parts of the world, followed policies of **neoliberalism** by reducing import quotas, abolishing government spending on social welfare, and adopting fiscal discipline to promote foreign trade. State enterprises were put up for sale in large privatization programs. The impacts were particularly pronounced in South America because the existing welfare support structure was so slight that even small reductions led to increased poverty for many households. Reducing subsidies had a heavy impact on the poor and low-income groups. The main criticism of neoliberalism is that it was not all that effective. Growth rates were slower in South America than in Southeast Asia where a **neo-Keynesianism** rather than a neoliberalism was pursued. From the 1980s to the 1990s, a mixture of income shrinkage and inflation created economic distress. Social resistance to neoliberalism provoked, at least in some countries, a change in political leadership and a renewed commitment to income redistribution and the promotion of social welfare.

Poverty Reduction

By the late 1990s and early 2000s, a global commodity boom, driven in large part by China's rapid and large-scale industrialization, created spectacular rates of growth throughout much of South America, replacing several decades of recession, hyperinflation, and political turmoil. China's growth sucked in raw materials. Oil, copper, beef, palm oil, and soybeans from South America now find their major market in China. Almost 90 percent of Ecuador's oil exports end up in China.

The increasing government revenues, in association with greater government control over key sectors, allowed governments to spend more on social welfare, education, and infrastructure, spreading the wealth and generating increased standards of living for the majority of the population. Across South America from 1990 and 2005, social expenditure increased from 8.8 percent of GDP to 11.3 percent while per-capita government spending increased from US$264 to US$418. As a result, in the last 10 years almost 80 million people in South America were lifted out of poverty. The leftward shift, embodied in the election of such populist leaders as Bachelet (Chile), Correa (Ecuador), Humala (Peru), Kirchner (Argentina), Morales (Bolivia), Mujica (Uruguay), and Lula and Rousseff (Brazil), was prompted and sustained by the increased revenue from the commodities boom.

All of these election results reflect specific national political situations, but they also represent a repudiation of an imposed neoliberalism and a commitment to a more direct redistribution of national income and wealth away from the very rich toward the poor. The ending of the boom raises new concerns across South America about the prospects for economic growth, income redistribution, and government finances.

4.5 Aircraft factory in São Paulo (Sue Cunningham Photographic/Alamy Stock Photo)

Social Geographies

Population Differences

The population of South America is highly concentrated along the coasts. Much of the rainforest, grasslands, and high mountains are sparsely populated. Cities along or near the coast house the vast majority of the population, reflecting the pattern of export-orientated economic development.

The population can roughly be divided into three main groups—indigenous, blacks, and whites—although, as we shall see, these are not fixed and stable categories.

INDIGENOUS PEOPLES

The indigenous peoples are made up of different tribal, language, and ethnic groups. Indigeneity is expressed through language, material culture, and the politics of identity. There are two arguments. One is that true indigeneity is disappearing as South America becomes more modern and globalized. There is some support for this position as native language speakers are declining in number, especially for the very small language groups of the rainforest. The indigenous population is often represented as the internal "other," different from the dominant, white, Hispanic elites.

4.6 Indigenous woman in Ecuador (photo: John Rennie Short)

An alternative argument points to the persistence and survival of indigenous culture and its expression as part of national identities. The election of Evo Morales in Bolivia, for example, marks the first time that an indigenous person became president of that country. There had been seventy-nine presidents before him. The indigeneity of South America is also ever present in the current population. More than 70 percent of Chileans and 50 percent of all Argentinians, two of the more European of South American countries, have at least one indigenous ancestor.

Estimates vary, but around 25 million people in South America can be classified as indigenous; that is, they belong to a group that predated European colonization. The indigenous population of South America is concentrated in an **Andean Indigenous Belt** that runs from Colombia, through Ecuador and Peru, into Bolivia and scattered parts of the Amazonian rainforest (Figure 4.6). Table 4.1 lists the indigenous population as both absolute and percentage values of the total population of South American countries. Peru and Bolivia stand out with significant indigenous populations.

The indigenous people survived by living on marginal lands that were less desired by the European colonizers. In recent years, however, their position has become more visible because of two processes. The first is the increasing expression of indigenous land rights in response to commodification and state takeovers that threaten native lands. Many indigenous groups are organizing against the loss of their lands. Second, the increasing democratization across South America allows indigenous concerns to be seen as legitimate rights. Take the case of Chile where more than 1 million people are **Mapuche** who live in a region in the central part of the country. Mapuche activism is concerned with formal recognition as a people, a demand for territory, and some form of political autonomy. Beginning in 1997, activists set fires to forests operated by private

COUNTRY	INDIGENOUS PEOPLES (MILLIONS)	AS % OF COUNTRY'S TOTAL POP.
Peru	13.8	45.3
Bolivia	6.0	55.5
Ecuador	3.4	22.3
Chile	1.7	9.7
Colombia	1.4	3.0
Brazil	0.70	0.35
Argentina	0.60	1.46
Venezuela	0.52	1.76
Paraguay	0.09	1.36

TABLE 4.1 The Indigenous Population in South America

companies on traditional Mapuche lands. The resistance continues in the form of arson attacks and renewed resistance to plans for hydroelectric dams on traditional Mapuche land.

BLACKS

Blacks are a significant element of the population of South America, numbering around 120 million people. Black South Americans are descendants of slaves brought over to work in the mines and plantations. Between 1451 and 1870, almost 12 million slaves were shipped from Africa to the New World. In the total for South America, 41 percent were sent to Brazil where sugar plantations required constant supplies of slave labor because the working conditions were so hard and the death rates so high. The slave trade was abolished in Brazil in 1850, and slavery was abolished in 1888, but blacks remain at the bottom of the socioeconomic hierarchy. Their spatial distribution today reflects the colonial plantation economy of the past. Blacks are concentrated along the tropical coasts of South America with a heavy concentration along Brazil's eastern coast. The African influence lives on in music, dance, food, and religion.

WHITES

Whites are the descendants of European colonizers. To be able to identify a European lineage is important for social status. It is not accidental that only three indigenous people have been elected to the highest offices: Benoit-Juarez (1861–1872) in Mexico, Alejandro Toledo (2001–2006) in Peru, and Evo Morales in Bolivia (2006–present). The European descendants remain at the apex of political and economic power on the subcontinent. They are concentrated in the towns and the professions, and constitute the economic and political elites.

FLUID CATEGORIES

Although we have identified three main groups—indigenous, blacks, and whites—what is distinctive in South America is the pervasive extent of racial and ethnic mixing. Europeans, black slaves, and indigenous people did not remain separate entities but intermixed (Figure 4.7). A complex classification sought to bring sense and order to the confusion. Mestizos were people of mixed European and indigenous ancestry. In Portuguese the term *mamelucos* was used. People with black and white ancestry were called *mulatto*, while ***zambos*** referred to those with black and indigenous parentage. The classification was not innocent of social implications; the more European the better, as it conferred higher status. Blacks and "Indians" were long relegated to the bottom of the social pyramid.

The racial categories, in practice, are more fluid than fixed. Becoming more urban and more Catholic turned "Indians" into mestizos. There is also the process known as the whitening of South America. This occurred most dramatically in Argentina: in the early 1880s, almost a third of the population was black; they now constitute less than 5 percent. The decline is the result of a combination of differential birth rates, loss of men in military campaigns, and higher infant mortality rates, as well as racial mixing that led to a whitening of the population.

The categories are not fixed in time but liminal, capable of change when people change dress, diet, or religious practices. The move from black or "Indian" to mestizo and mulatto is possible with changes in income and consumption patterns. Over generations, the goal of many is upward mobility and racial reclassification. According to the 2010 census, 43 percent of people in Brazil describe themselves as mixed race and only 8 percent describe themselves as black. The figures may be aspirational because a mixed-race category lays

4.7 Diversity of South America: these children playing in a river in Colombia give some sense of the diversity of the region. (Photo: John Rennie Short)

4.8 Catholic Church in Quito (photo: John Rennie Short)

claim to whiteness, whereas the category of black does not. In the same census, 48 percent described themselves as white. White are less likely to claim mixed race than blacks. In response to this "creeping whiteness," there are also counter-movements. *Indigenismo* is the reaffirmation of indigenous people and their rights, while *negritude* is a South American form of Black Power embodied formally in literary movements and more commonly in a renewed appreciation of the African influence on the music, food, and popular culture of South America. In Colombia, for example, more people are now proud of their Afro-Colombian heritage. In Brazil

the Afro-Brazilian culture is being revalued. Black slaves developed Capoeira, a martial art disguised as dance. It is now considered an important Brazilian art form.

In Venezuela, for example, there is a complex melding of indigenous tribes with Spanish colonists and African slaves and more recent immigrations of Europeans. Out of a total population of 30 million, according to official classifications, about 50 percent are mestizo, 43 percent are of European ancestry, 4 percent are black, and 3 percent are Amerindian. Analysis of DNA suggests that the genetic contributions are 60 percent European, 23 percent Amerindian, and 16 percent African.

Religion

The dominant form of religion in South America, reflecting the huge influence of the Catholic countries of Portugal and Spain, is Catholicism. Every town in South America has a Catholic church, the local priests wield considerable social as well as religious power, and the church plays an important role in the everyday life of most countries (Figure 4.8). The stranglehold of a conservative Catholicism is much weakened by liberation movements promoted by radical priests and by secular trends such as acceptance of divorce, family planning, and gay rights. The Catholic Church remains strong but not as all-powerful as it once was. There are also indigenous religions and hybrid forms. Candomble is a religion in Brazil that emerged from animist beliefs brought by Africans. Its newfound respectability, however, is under attack from the growing force of radical evangelical Christians who depict it as devil worship.

The Demographic Dividend

Population growth in South America has followed a predictable pattern through the demographic transition. A phase of limited population growth, because of relatively high death rates, was followed by rapid population growth, especially in the first half of the twentieth century, because of declining death rates and increasing birth rates. Birth rates then began to moderate with a dramatic decline in fertility from an average of 5.8 children per woman in 1950 to 2.09 in 2010. Population growth has evened out to around 1.1 percent a year, down from 1.9 percent in 1990.

Figure 4.9 summarizes the trends through a comparison of the population pyramid for South America in 1970 with 2010. Notice how the pyramid narrows

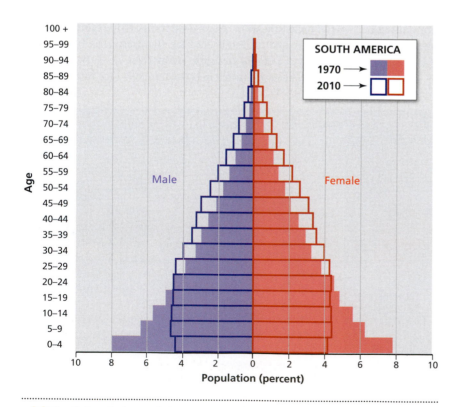

4.9 South America's population structure, 1970 and 2010

from 1970 to 2010 as birth rates decline and life expectancy increases. The very young now constitute a smaller proportion of total population. A major consequence of this shift is the demographic dividend. The population aged between 15 and 64 years now constitutes 65 percent of the total population in South America. In 1990, it was around 59 percent. As the large numbers in cohorts from previous growth spurts enter the job market, there are proportionately more people of working age compared to the very young or the very old. The net effect is a relative and absolute increase of younger, more productive workers.

The recent rapid economic growth of countries such as Brazil is in part based on this dividend. The proportion of the population in Brazil aged between 15 and 65 years, generally considered the most economically active age range, is 67 percent. In poorer countries such as Afghanistan, Bangladesh, and Cameroon, the respective figures are 55, 61, and 56.

The opportunity for this demographic dividend is limited because increasing life expectancy soon "ages" the population, reducing the ratio of workers to nonworkers. There is a short window of time when there is a rising share of people of working age. During this time, holding everything else constant, average per-capita income rises, both reflecting and prompting enhanced economic growth. Between one-third and one-half of economic growth in states such as Brazil may be due to the demographic dividend. The dividend is likely to be a significant feature of South America until at least 2040.

Rural Focus

Extending the Agricultural Frontier into the Amazon

The Amazonian rainforests act as air filters absorbing large amounts of carbon dioxide. They are the lungs of the planet. They are also an economic opportunity. Rainforests are being felled to make way for palm oil plantations and to plant soybeans and other export crops (Figure 4.10). Palm oil is a ubiquitous constituent in a variety of products from cooking to biofuels. The rich biodiversity of the forest is being replaced with plantation monoculture. The process of deforestation is increased when there are growing demands for the export crops. But even when the commodity process declines, countries try to sustain economic development by boosting production and to maintain government revenues by granting new concessions for forest clearance.

When there is a boom, there is deforestation; and when there is a slump, there is deforestation. The pressure on the rainforest is relentless whether it is a boom or a slump.

4.10 Deforestation (luoman/Getty Images)

Urban Trends

Although the large cities of South America share the similarities of large informal settlements, segregated housing markets, rapid growth, and the emergence of a middle class, there are also some differences. Brazilian cities have a Portuguese colonial architectural legacy, the cities of the Andes have a large indigenous population, and the cities of the lowland tropical region have a larger proportion of blacks. Farther south in Argentina, Uruguay, and southern Chile, where there was less slave labor, immigrants were mainly from Europe. Thus, while the black-white divide is an important part of social stratification and residential segregation in Brazilian cities such as São Paulo or Rio de Janeiro, it is less of a feature for Montevideo or Buenos Aires. Meanwhile indigenous people are a significant element in the social make-up of cities like Quito or La Paz.

An Urbanized Realm

South America is one of the most urbanized regions of the world with more than 4 out of every 5 people living in cities. The rate of growth was rapid. In 1950, fewer than 2 out of every 5 people lived in cities. This growth was based on natural increase, as the urban population increased in size, but it was also fueled by massive rural to urban migration that, by the 1980s, constituted between 40 and 70 percent of city population growth. People left the land in droves as jobs and opportunities shriveled in the agricultural sector and throughout the countryside. Jobs and economic opportunities were concentrated in the cities where people could dream of a better life for themselves and their children. Urban dominance is now so firmly established that estimates for 2050 suggest that 9 out of every 10 people in South America will live in cities.

The vast migration of often poor people into cities that were not geared to accommodate them meant an increase in informal settlements, as people were forced to build their own accommodation, often on the most environmentally hazardous sites. There was also an expansion of the informal economy as workers found few opportunities in the formal sector. Across the continent slums sprouted in city centers and in the periphery, and jobs in the informal economy grew to almost 50 percent of total urban employment.

Unequal Cities

The cities are sites of extreme disparities. At one extreme are the exclusive gated communities housing the rich and powerful, while at the other are the slum housing areas of the poor and marginalized (Figures 4.11 and 4.12). Segregation of households by race and class is marked and obvious with very different life experiences dependent on the accident of birth. Those born into poverty have more limited opportunities to be upwardly mobile, while the very rich cement their position at the apex of the social and economic hierarchy. The cities are the stage for the display of income disparities and marked differences in urban life experiences. One consequence of this visible poverty living side by side with affluence is violence and crime. Crime is an endemic problem in South American cities that leads to the clustering of the rich into fortified and defended neighborhoods. Much of downtown urban renewal and development demarcates spaces safe for investment and the more affluent people. Slums of poor displaced migrants, meanwhile, are good recruiting grounds for gang members. Often crime and criminal activity is an important

constituent of the informal economy. Urban securitization is achieved often at the expense of strong-arming the poor and the militarization of the policing of the slums. The city is the place of opportunity, the expression of vast inequalities, and the setting for various forms of resistance from community groups and labor organization.

The deep divisions are now softened by the emergence of a middle class, especially in the large and economically dynamic countries such as Brazil. With more people in the middle-income ranges, cities now exhibit more than just the extremes of residential segregation. As cities expand, elite suburbs and lower income suburbs are brought into closer proximity on the outer edges of the city. Middle-class areas now often occupy the dividing areas between rich and poor neighborhoods. The need for cheap, accessible domestic labor is also the reason for the location of lower income communities close to rich and middle-income areas.

Cities continue to expand outward, creating a more dispersed metropolitan spread. Although the pattern is the same—cities spreading outward—the precise process can differ. In Buenos Aires, for example, the outward spread is of high- and medium-income households; in Montevideo, it is mostly slums; while in Santiago, Chile, it is subsidized housing that underlies peripheral expansion.

Urban Primacy

A distinctive feature of urbanization in South America is **urban primacy**, which refers to the concentration of a nation's population in just one city. The most distinctive example is Uruguay, where more than half of all people in the country live in just the capital city of Montevideo. The equivalent for the United States would be a metropolis of 165 million people. Primacy is particularly marked in Peru, Argentina, and Chile. It is not a universal condition on the

4.11 Gated community of the wealthy, Guayaquil (photo: John Rennie Short)

4.12 Rochina favela in Rio de Janeiro (chensiyuan/Wikimedia Commons (CC BY-SA 4.0))

continent, as urban primacy does not occur in Bolivia, Brazil, Ecuador, or Venezuela.

The largest capital cities are home to the elites, the population center of gravity as well as the spatial-economic hub of the national economy. Primacy was reinforced as economies shifted from primary to secondary and then tertiary economic sectors, as rural to urban migration increased, and foreign investment connected the national to the global through the big cities. People, jobs, and investment concentrated in the major city of many countries. Economic processes from state takeovers to more recent rounds of privatization have all reinforced the centuries-long persistence of the urban concentration of economic, political, and social power.

Globalization can reinforce urban primacy. Consider the recent case of Santiago, Chile, where the metro population is 7 million in a country with a total population of 16.7 million. The city is responsible for almost half of gross domestic product. Recent forms of globalization reinforced the initial urban primacy of the colonial system. In the past decade, over 50 multination companies have located in the city, including Nestle, Packard Bell, and Unilever. The city is an export base for the wider South American market. Foreign investment from the United States and Spain flows into the city. The city also plays an important role in the cultural circuit of capital and the diffusion of managerial best practices. The city is a magnet for rural migrants in Chile but also for both unskilled and middle-income professionals from the neighboring countries of Bolivia, Ecuador, and Peru and to a lesser extent Argentina. As with other cities around the world, urban primacy embodies and enhances the creation of a global city.

Table 4.2 lists all the city regions—the wider metropolitan region rather than the formal city limits—in South America with a population of more than 3 million. São Paulo is by far the largest, and Brazil is home to ten of the fifteen biggest. Cities are spreading over municipal boundaries so that we can now identify megapolitan urban regions that are functionally one big region, even though data are collected for constituent parts that may even have separate political representations. Greater Buenos Aires includes not only the Argentinean city of Rosario but also the Uruguayan capital of Montevideo. Table 4.2 shows how the Rio-São Paulo metro region ranks as a massive center of concentrated population, and Colombia now appears as a country of four giant urban regions.

TABLE 4.2 Metropolitan Regions in South America

MEGAPOLITAN REGIONS	COUNTRY	POPULATION (MILLIONS)
Rio-São Paulo	Brazil	45.6
Greater Buenos Aires	Argentina-Uruguay	17.0
Lima-Callao	Peru	10.5
Greater Bogota	Colombia	13.0
Caracas-Valencia	Venezuela	9.0
Northeast Atlantic	Colombia	6.0
Medellin Valley	Colombia	4.5
Greater Cali	Colombia	3.8

City Focus: Rio-São Paulo, South America's Megacity

With a population over 45 million, Rio-São Paulo is the continent's largest megacity. It houses almost one-quarter of Brazil's total population and is responsible for nearly one-third of its GDP. It is the amalgam of two major cities, Rio de Janeiro and São Paulo, and over 450 smaller municipalities.

The Portuguese founded Rio in 1536. It was the capital of the Empire of Brazil for a brief period and emerged as the nation's principal port and major industrial and commercial city. The city grew along the coastline and upward into the surrounding hillsides. The city fractures along class and ethnic lines with the predominantly white upper classes separated out from the poor blacks living in **favelas**, self-built housing in high-density areas in the hillsides. The favelas are the base for criminal gangs, with names such as

Red Command, Friends of Friends, and Third Command, and especially drug cartels. Gang shootouts are a major cause of high homicide rates. Military-style police actions to pacify the worst affected favelas began in 2008 and were ramped up in the lead-up to the World Cup in 2014 and the Olympic Games in 2016. A special police riot squad, known as the "Shock Battalion" was reputed to use brutal techniques. Riot police were sent to Rocinha, the city's largest favela, after a shoot-out between rival gangs. Homicide rates have been cut in half. But the escalating violence between gangs and police as well as police brutality has reduced public support for the program.

São Paulo lacks the seaside imagery employed in the portrayal of Rio. It is an inland city and grew as a transport point for the coffee trade. Profits from the coffee trade were invested in new economic ventures and the city grew as an industrial city. It developed a reputation, in contrast to the fun-loving, sensual Rio, as a hard-working, more pragmatic place. Recent years have seen a marked deindustrialization, a suburban spread of people and retail, and the emergence of a major financial center. As with its sister city of Rio, São Paulo's rapid growth has strained the city's infrastructure and created a heavy environmental impact in the form of air pollution, inadequate sanitation, and poor living conditions in the favelas. A water crisis, also known as a hydric collapse, due in part to consecutive years of very dry, hot conditions, is leading citizens to store water at home and to drill their own wells. Even during the rainy season of 2015 in January and February, neighborhoods were without water for days at a time. There are calls for more efficient rationing of water.

The two cities have sprawled so much that they now form a vast megacity. While the urban realities are of a single metropolitan region, there is still political balkanization and uncoordinated municipal policies and practices. Rapid and unplanned urbanization has created problems of housing availability, infrastructure inadequacies, and environmental deterioration. A quarter of the population of Rio lives in the favelas. It is a megacity that has yet to imagine or organize itself as a single urban region or to provide adequate living conditions for all its citizens.

Geopolitics

The geopolitics of South America is shaped by legacies of empire, stirrings of nationalism, international tensions, and intranational conflicts. Let us look at each in turn.

Imperial Legacies

In the 1494 Treaty of Tordesillas, the Pope divided South America between Spanish and Portuguese spheres of influence. By 1600, the Spanish were entrenched in the western region of South America and the Portuguese along

4.13 Map of Rio-São Paulo

the coast of Brazil. In the coastal area between the Orinoco and the Amazon, the English, then British, French, and Dutch all laid claim to territory to establish plantation economies for sugar, coffee, and cocoa. The imperial yokes were always resisted, not only by indigenous peoples but also by local Hispanic elites in South America chafing at the power of distant European centers. The effective power of the Spanish Empire was always undermined by distance and the costs of administering a vast realm.

Emerging Nationalism and Pan-Nationalisms

From the beginning of the nineteenth century, Hispanics born in South America fought for independence from Spain. In 1810, autonomous governments were established in Argentina, Chile, and Venezuela. A significant figure in the independence struggle was Simon Bolivar (1783–1830). He was part of the movement that liberated Venezuela in 1813 and New Granada in 1818, which, in turn, became Colombia, Ecuador, and Venezuela. With others he helped Bolivia and Peru achieve independence. From 1819 to 1830, he was president of the Republic of Gran Colombia. Bolivar is a significant figure because his name lives on, not only in the name of countries—Bolivia and the Bolivarian Republic of Venezuela—but also in the political idea of **Bolivarism** that expresses an anti-imperialism rhetoric that promotes income redistribution, national sovereignty, and an antipathy toward the power of the United States in the region.

International Tensions

There are a number of territorial disputes between the different countries in South America that are a legacy of imprecise colonial boundaries. A territorial dispute between Ecuador and Peru led to outright war in 1941 and remained a simmering source of resentment, breaking out into outright conflict in 1981 and 1995. The dispute was ended by agreement in 1998.

Since 1990, territorial disputes between Argentina and Chile, Ecuador and Peru, Chile and Peru, and Brazil with all of its neighbors have been arbitrated, but unresolved issues still remain in Colombia, Guyana, and Venezuela. Colombia has a disputed maritime border with Nicaragua. Venezuela claims land west of the Essequibo River, almost two-thirds of the entire territory of Guyana. Their conflict over a maritime boundary is made all the more bitter because of large reserves of oil in ocean waters off the disputed territory. The dispute is often used by the government of Venezuela to redirect popular passions away from a focus on economic and political crisis at home.

THE FALKLAND ISLANDS/ISLAS MALVINAS

The Falkland Islands are a source of dispute between the United Kingdom and Argentina (Figure 4.14). Throughout history, numerous countries laid claim to the Falklands, including France, Spain, Argentina, and even for 2 months in 1831/1832 the United States. It came under effective British control in 1833 when the British established a colony and also laid claim to the surrounding islands of South Shetlands, South Sandwich, and South Orkney. Argentina protested the British occupation of the islands that they called Las Malvinas. When it joined the new body of the United Nations in 1945, Argentina again raised its claim to Las Malvinas.

The relationship between the two countries included attempts at negotiation and compromise as well as continuing disagreement. The islands were linked to Argentina via the only air route to the island. An Argentina company supplied oil and gas to the islands. There were also bouts of conflict. In 1976, Argentina landed an expedition in the South Sandwich Islands and fired upon a British ship visiting South Georgia and the next year cut off fuel supply to the Falkland Islands airport. The British dispatched a nuclear submarine and two frigates, to drive home the point that the islands were still British.

In April 1982, Argentinian forces invaded the islands; in response, Britain sent Royal Navy ships, including two aircraft carriers, *HMS Invincible* and *Hermes*, in a task force of 124 vessels. British Special Forces landed on the islands on May 21. The fight to retake the island proved difficult and costly, though relatively quick (Figure 4.15). At least 650 Argentinians were killed and over 11,000 were taken as prisoners. The British lost 258 men. The Argentinian forces surrendered on June 14. The defeat led to the overthrow of the Argentinian military dictatorship, while the victory catapulted the British Prime Minister Margaret Thatcher into higher approval ratings than she had before the conflict. She won the 1983 general election on the back of the victory against Argentina.

4.14 Map of the Falklands

4.15 Remains of helicopter shot down during Falklands War in 1982 (Matt Fowler Photography/Alamy Stock Photo)

After the war, the islanders were made British citizens, military garrisons on the islands were strengthened, and the Falkland Islands, along with the South Georgia and South Sandwich Islands, became a distinct British oversees territory, joining BVI, Gibraltar, and the Cayman Islands, as well as eleven other such places, under UK sovereignty.

Britain and Argentina restored diplomatic relations in 1989, but Argentina continues to make the return of the Falklands to Argentina sovereignty an important priority of foreign relations as well as a part of its constitution. The almost three thousand permanent residents continue to want to remain British, while the existence of oil deposits in the surrounding seas adds an extra dimension to competing territorial claims.

Competing maps of the Falklands are not just differences in language, with Spanish names for English in the Argentinian maps. They have different names. The main city is called Port Stanley in the British maps and Puerto Argentina in the Argentinian maps.

Intranational Conflicts

The most destructive form of conflict in South America is not at the level of state versus state but between different interests within a state.

South America has a history of military coups replacing civilian governments. Consider the case of Argentina where coups in 1930, 1943, 1955, 1962, 1966, and 1976 replaced civilian governments with military juntas. Suriname is a more recently independent country but has tried to move up in the league of military coups with two in 1989 (one in February and another one in August), and one in 1990. Bolivia experienced coups in 1937, 1943, 1946, 1951, 1952, 1964, 1970, 1971, and 1980. Ecuador's most recent

coup was in 2000 and an attempted coup by National Police, annoyed at having their benefits cut, occurred in 2010, while Paraguay's most recent was only in 1989. There were also the failed coup d'états in Venezuela in 2002 and Ecuador in 2012.

With few channels of democratic politics, elites and especially military elites obtained and held power through naked force. Between 1974 and 1983, the Argentinian military junta conducted what is known as the Dirty War against political dissidents, killing almost 13,000 people.

Then there are the insurgencies. Urban insurgency in Argentina (1969–1976), Brazil (1964–1970), and Uruguay (1968–1972) involved assassinations, kidnapping, and sabotage. In Uruguay, for example, the Tupamaros were a group determined to overthrow what they saw as an unjust and unfair system. In Argentina, the Monteneros emerged in 1970 and used kidnappings as a way to fund their activities. In Argentina the People's Revolutionary Army followed similar tactics. The insurgent movements born out of inequality and political disenfranchisement promoted and provoked violence that spiraled into more violence as the army and police responded with equal and sometimes greater force. The main victims were the innocent residents of the country caught between brutality and violence. The democratic process also suffered as brutal military regimes, and even civil regimes, bypassed democratic processes to deal with the insurgency threat. These insurgencies that so marked the recent past have now largely disappeared as a more benign cycle of democratization and economic growth has displaced the downward spiral of violence and state terror.

Social movements have replaced insurgencies. So, rather than taking to the gun and the bomb, more social organizations have emerged that seek to influence governments through lobbying, mass action, and a more formal intervention in a democratic politics. They include indigenous peoples movements in Bolivia, Ecuador, and Colombia, such as the Confederation of Indigenous Peoples of Bolivia; poor peoples' movements, such as rural worker groups in Brazil and Bolivia; and human rights groups such as those in Argentina that emerged from the mothers of families of those who "disappeared" during the Dirty War. As the societies have become more open to nonelite participation, a variety of ethnic, class, and political groups have emerged to occupy a space previously sealed from public participation.

What is heartening is that the number of coups and illegal grabs of power are becoming less rather than more frequent. Cold War rhetoric can no longer be used to justify extraordinary measures, and the United States is no longer so actively engaged in the internal politics of countries in South America. With a wave of democratization and the rise of civil society and democratic institutions, South America is becoming more settled and peaceful.

Connections

South America is connected to the rest of the world in many different ways. The subcontinent is connected to the rest of the global economy through commodity booms.

Commodity Booms and Busts

The global demand for commodities waxes and wanes, creating booms and slumps. Let us consider booms in Amazonia.

The first rubber boom, for example, from 1879 to 1912, was based on the extraction of rubber in plantations along the tributaries of the Amazon basin in Brazil, Colombia, Ecuador, and Peru. Local peoples were pressed into what constituted slave labor, and when the local labor supply was exhausted, workers were imported from Europe. Immigration and newfound wealth led to the development and growth of cities such as Belem and Manaus. Extravagant opera houses were built in both cities as rubber barons expressed their wealth in ostentatious displays (Figure 4.16). But, by the early twentieth century, the cultivation of rubber trees in other tropical regions in Africa and South East Asia undercut the monopoly of Amazonia rubber.

A second rubber boom in 1942 to 1945 was based on the huge wartime demand for rubber products and the loss of the South East Asia supply areas to the Japanese. At the end of the war, global rubber demand was again met by South East Asia, and in South America over 30,000 workers in Amazonia left the abandoned rubber plantations. The experience of the rubber boom embodies a typical pattern of primary resource development, albeit in an exaggerated form and in a telescoped time period, of quick money for a minority, often difficult conditions for the workers, and very limited long-term multiplier effects that fail to create a trajectory of long-term economic growth and development.

4.16 Opera House in Belem (Balthasar Thomass / Alamy Stock Photo)

Recent years have witnessed an agricultural boom, especially for meat and soya. The Brazilian and Chinese governments have plans to build the world's largest grain canal to link the cerrado savannah lands of Mato Grosso, which grow a third of the world's soya, to the port of Santarem on the Amazon River. There, a new container port will ship the produce to global markets. The cattle ranches and soya farms in the Mato Grosso and the port are to be linked by turning the Tapajos River and its tributaries that flow through Amazonia into a giant canal. The project involves the building of forty-nine dams that will also generate electricity. More than a million hectares of forest will be cleared, and the land of the indigenous Munduruku people will be impinged. The agricultural boom, like the previous rubber, logging, and mining booms, will have major environmental and social impacts. The city of Santarem, for example, will double its population from 300,000 to 600,000. While the project will provide jobs and profits, the dams and infrastructural investment along the Tapajos River will also lead to deforestation, loss of wildlife habitat, and further marginalization of indigenous people.

Subregions

Gran Colombia: Colombia, Ecuador, and Venezuela

Gran Colombia was the name given to a new state formed from an old Spanish viceroyalty. It initially contained territory in present-day Colombia, Venezuela, Ecuador, and Panama, as well parts of Peru and Brazil. In 1831, Colombia, Ecuador, and Venezuela became independent states. Colombia was reduced in size when Panama, with US backing, seceded in 1903. The United States gave support in return for being given land to build the Panama Canal.

The three main states have similarities, and their national flags are a reminder of their former connection; all three share the same color scheme of horizontal lines of yellow, blue, and red. They also share middle-income status in the global economy. Their differences relate to the differing trajectories of recent history and the luck of resource endowment.

Colombia is only just emerging from a crippling 50-year-old internal insurgency. The Revolutionary Armed Forces of Colombia (FARC) emerged in the 1960s as an insurgency that used violence to call attention to the vast inequalities in the country. By the early 1990s, it had control of over 50 percent of the country with 18,000 fighters in seventy different areas. The money to fund its organization came from the drug trade, extortion of companies, castle rustling, and kidnapping. The military, with enormous help from the United States, waged a successful campaign against FARC. The weakening of FARC's popular appeal and its dwindling capability allowed the government to be involved in peace

4.17 Map of Columbia, Ecuador, and Venezuela

talks that began in Havana in 2012. A peace treaty marking the end of the conflict came into full force in 2017.

The violence between FARC, the military, right-wing paramilitary groups, and criminal gangs cast a pall over the country's economic growth and civic functioning. Close to 6 million people were displaced and up to 250,000 were killed over the course of the long conflict. The drug trade was a major source of funding for gangs and insurgencies.

Colombia remains one of the more unequal societies, but now with the insurgency at an end there may be a peace dividend. Problems remain. High, though declining, crime rates remain an important issue, the appropriation of peasant lands is a continuing injustice, and the deforestation of Colombia's Amazonia through illegal mining and timber extraction is a problem. More than 120,000 hectares of forest was cleared in 2013, although this was less than the annual figure of 238,000 over the years 2005 and 2010. While the government has reduced the formal deforestation, it has difficulty controlling illegal operations deep in the jungle.

Ecuador has a similar income inequality to Colombia but a less violent recent past. Extreme poverty has declined due to a resource boom, primarily oil that generates revenue and government spending on social welfare. In the twenty-first century, Ecuador has followed a path to wider sharing of economic growth through more social welfare spending. The poverty rate declined from 37 percent in 2007 to 27 percent in 2012. Ties with the United States

remain: the country's currency is the US dollar, and US tourists are an important source of income.

Ecuador has the third largest reserves in South America after Brazil and Venezuela. Most of Ecuador's richest oil reserves are in the east of the country, in Amazonia. It is a region of incredible biodiversity as well as the site of rich mineral resources. New drilling ventures are pushing further into the forest—into the lands of the indigenous people, such as Taromenane and the Waorani, and in formerly protected areas such as the Waorani Ethnic Reserve and the Yasuni National Park. In 2013, the government redrew maps of tribal territory, reducing them in size so that they no longer covered oil reserves and closed the country's most prominent environmental advocacy group, Pachamama, because it criticized the expansion of oil drilling into the national park. The oil provides the means for Ecuador's economic growth but perhaps at the cost of a reduction in biodiversity, indigenous land rights, and ultimately of tribal survival.

Venezuela's rich oil reserve gave it a special place as one of the more affluent countries in South America. The Orinoco River basin is estimated to have deposits of at least 1.2 trillion barrels of oil. Previously, much of the wealth was unevenly distributed with a small rich elite and a large mass of impoverished. A military dictatorship in the 1950s created massive public works programs. For years it was a pro-US, relatively stable society. The Rockefellers owned much of the oil reserves as well as vast tracts of land. Things changed radically with the election of Hugo Chavez in 1999. In an example of contemporary Bolivarism, he established close ties with Cuba and espoused an anti-US rhetoric. He also embarked on an ambitious program of wealth redistribution and encouraged the more active participation of people, which included the "invasion" of occupied buildings by squatters. His immediate successor also follows a more distinctly anti-US posture.

Venezuela's fortunes are tied to the world price of oil, something out of its immediate control. Attempts to create economic alternatives such as the industrial city founded in 1961, Ciudad Guyana, on the Orinoco River have failed to diversify and reduce the reliance on oil. By 2013, oil was responsible for 96 percent of total goods exported, up from 68 percent in 1998. When the price of oil dropped precipitously in 2014, Venezuela was immediately impacted. Venezuela has the world's largest known oil reserves, but its production costs are higher than those in the Middle East. When the price of a barrel of oil drops below $100, the country faces a huge and immediate loss of revenue impacting the costs of imports, social programs, and the subsidized oil it gives to allies in the Caribbean and Central America. The drop in oil affects living standards, political stability, and Venezuela's geopolitical strength.

The state owner oil corporation the PDVSA (Petroleos de Venezuela SA) not only produces oil but also provides cheap fuel to allies, funds domestic redistributive social programs, including health and education programs, and subsidizes the price of oil in Venezuela. It is part oil company and part social welfare program, with revenues devoted more to social agenda than to reinvestment. After Chavez came to power, many of the more effective managers were replaced with political appointees with limited managerial or technical experience. The result is that as costs have risen and prices slumped, the company and the entire economy are under stress. The country is marked by violence as state power is contested not only by different political interests but also by private armies and criminal gangs.

Caracas, like many other cities in the region, is marked by obvious inequalities between, on the one hand, the wealthy districts, with their gated enclaves and membership in the Caracas Country Club where huge villas surround a golf course, and on the other hand, hillside slums. "El 23" is one of the city's largest slums. Originally built in the 1950s as a public housing project, it is now a shantytown of 100,000 living in the formerly green areas of the eighty original buildings. Crime is endemic, especially in the poorer areas, and the murder rate has tripled since 2000.

The Andean Arc: Bolivia, Chile, and Peru

All three countries face similar problems of encouraging growth and ensuring some measure of equity. They can be arranged in order of national wealth with Chile as one of the richest countries in South America, Bolivia one of the poorest, and Peru in between (Figure 4.18). Chile's wealth is in part a result of generous resource endowment, especially of copper that constitutes 60 percent of exports. A stable and progressive system of government ensures that some

4.18 Map of Bolivia, Chile, and Peru

of this wealth filters down from the elites. The country has one of the best education systems on the continent. Chile's trajectory took a less progressive turn after a military junta, with help from the United States, overthrew the democratically elected government of Salvador Allende in 1973 and, under General Pinochet, inaugurated almost 20 years of repression and neoliberalism until 1990. Economic growth was promoted but at the expense of social welfare with little regulation of business and a privatization of the health and education system. More recently, Chile has experienced the leftward shift common to much of South America toward greater emphasis on the dual goals of spending on social welfare as well as promoting economic growth.

Most of Chile's copper is exported to China, so Chile is dependent not only on the world price of copper but on the health and dynamism of the Chinese manufacturing sector.

Peru stretches from the shores of the Pacific to the Andes and into Amazonia, and it occupies an economic space between the relative wealth of Chile and the relative poverty of Bolivia. Its reliance on resource-led economic development started off with guano, essentially bird droppings that, because they were rich in nitrogen, potassium, and phosphate, were used as fertilizers. The guano export trade was a central part of Peru's role in global trade in the nineteenth century.

Its position between many different countries has exacerbated border disputes after independence. In 1932, a war was fought with Colombia over disputed territory in Amazonia. Ten years later another dispute erupted with Ecuador. Another round of conflict with Ecuador flared up in 1981 and again in 1995. Over time the disputes became brief military clashes rather than full-blown wars. Peace with Ecuador was finally achieved with a border agreement in 1998.

Rapid inflation and economic decline in the 1980s led to a sharp reduction in living standards helped in the creating support for the insurgency movement, Shining Path (Sendero Luminoso). The insurgency movement extended its reach across the country and at its peak, around 1990, had over five thousand members who waged war across Peru. Almost 70,000 Peruvians were killed in the conflict; the vast majority were innocent civilians. The Shining Path drew some of its support from the peasantry who were denied access to their cash crop of coca by Peruvian and US actions. While most of the original Shining Path leaders were killed or captured, the remnant of the movement lives on as an organized criminal enterprise with less commitment to any formal political ideology. There are now no more than three hundred to five hundred fighters. However, the war on drugs in Peru waged by the Peruvian military with massive US aid is in effect a war on the peasantry that, in turn, gives the insurgency a foothold of support in the countryside.

Inland from the coast, poor infrastructure makes it difficult to spread economic growth, and a third of the country's population lives in just one city, Lima, where informal settlements, known as *invasions*, developed on the outskirts of the city, especially to the north of the city on the fringes of the desert. Rural-to-urban migration increased from the 1980s as economic upheavals drove people from the land and into the city looking for better economic opportunities and access to public services. The housing in the city is dominated both by a Western form of modernism and more local and self-built vernacular styles. Peruvian modernity is the result of this encounter between modern architectural forms and a self-built architecture.

In Bolivia, one the continent's poorest countries, the governing regime was for many years a small, rich elite presiding over a vast majority of poor people. This class division was overlain with ethnic divisions as the elite were mainly white and Hispanic while the majority of the country's population is indigenous. It was only in 1952 that indigenous people were allowed to vote. Bolivia provides raw material for the global economy and more recently has become an important exporter of natural gas.

Evo Morales took office in 2006, the first time that an indigenous person had achieved the pinnacle of political power in Bolivia. He came to power in part due to his leadership of the coca farmers, who were upset at the eradication program fostered by the United States. It meant a reduction in their living standards. He pledged to redistribute land and restore dignity to the indigenous people of the country. In some cases the land was seized from the traditional landowners and the haciendas were divided up into multiple occupancy. In 2008, the Guarani people were given back land from wealthy farmer and ranchers.

Morales consolidated government control over much of the economy and pushed to develop Bolivia's gas fields along the foothills of the Andes Mountains. Gas royalties increased government revenues and allowed more spending on welfare. Initially there was much resistance to Morales, especially from the wealthier lowland regions where in 2008 opponents sought to build a separatist movement. It fizzled out. The revenues from the gas and oil sector allow a greatly expanded role for social welfare that buttressed Morales's support among ordinary Bolivians. Economic growth of around 7 percent allows both the business sector and the ordinary people to partake in the newfound wealth as the government embarks on programs of facilitating growth, pursuing fiscal discipline, and providing social welfare. The upward trajectory is predicated on high world prices for natural gas.

The Lands of the Pampas: Argentina, Paraguay, and Uruguay

These three countries occupy the southeastern section of the continent and have a wide climatic range from tropical in northern Argentina to polar at Argentina's southernmost point on the tip of the continent (Figure 4.19).

Argentina is the second largest country in South America. It was initially part of the Spanish empire but always peripheral due to its lack of valuable minerals. It became independent in 1816. The country experienced spectacular economic growth in the middle to late nineteenth century as primary products, especially beef and wheat, were exported to markets in North America and Europe. The development of refrigerated containers on faster ships meant fresh products could be shipped quickly and cheaply. The economic success attracted massive waves of immigrants from Europe, especially Spain and Italy. More than 6.6 million immigrants between 1880 and 1920 transformed the demographics, making the country more European than other parts of South America. By 1908, Argentina was the seventh richest country in the world as measured by per-capita income, almost double that of Spain and four times that of Brazil. In many ways it looked like Australia: a commodities-based economy successfully linked into the global economy.

A military coup in 1930 marked the beginning of political instabilities that threatened and eventually undermined Argentina's position as one of the wealthiest countries in the world. From the 1950s onward, the state was placed in the crippling vice of declining revenue because of global trade patterns, and increased expenditures, due to new claims on government spending. The gap led to bouts of massive inflation and an increased reliance on foreign debt that placed restrictions on fiscal flexibility. The resultant political instability was the backdrop for military dictatorship in 1966.

Since 1983, Argentina has returned to democracy, but the economic problems remain and the government has to rely on foreign funds. Its large internal market and skilled workforce is the basis for an industrial economy; however, there is a very unequal distribution of income and chronic debt

problems made worse by low growth rates and the lack of a demographic dividend.

Whereas Argentina has long links with the outside world—immigration and a vibrant import/export economy—Paraguay, after its independence from Spain in 1811, suffered from isolation and dismemberment. A series of dictatorships followed isolationist politics, and the land area of the country was reduced by 70 percent after it lost

4.19 Map of Brazil (with regional divisions cited in text), Argentina, Paraguay, and Uruguay

the Paraguayan War (1864–1870) against the combined forces of Brazil, Argentina, and Uruguay. It was home to the continent's longest lasting military dictatorship from 1954 to 1989, which was aided by the United States. Paraguay ranks after Bolivia as one of the poorest countries in the continent with substantial numbers living below the poverty line.

The basic character of the country is defined by the simple fact that a tiny elite owns most of the land and the majority of people are landless peasants. Most of the population is concentrated in the southeast region of the country with a primate city distribution. The economy is dominated by primary production, especially beef, soybean, and wheat. Mining products constitute 25 percent of GDP. Some growth was stimulated with a hydroelectric project on the Paran River.

Uruguay is a small country that is basically an extended city region centered on Montevideo, where 1.8 of the country's 3.3 million people live. It won independence from Spain in 1828. Similar to Argentina, it experienced massive immigration from Europe in the late nineteenth century and remains dependent on a narrow range of primary commodities including cattle, wool, and soybeans. Montevideo is a major port in South America, and Uruguay is one of the richest countries in the continent. More than one in three Uruguayans lives in the capital city. Since 1984, after decades of political turmoil and insurgence, Uruguay returned to a more normal democratic form of politics. It is one of the more liberal societies in South America that has legalized abortion, cannabis, and same-sex marriage.

The South American Giant: Brazil

The largest country, both in terms of territorial size and population, is Brazil, the giant of South America now emerging into more global significance as the fifth largest economy in the world.

Brazil has long produced primary commodities for the world market: sugar, coffee, and rubber. Each had its cycle of commodity boom and slump. Since the 1960s, the country embarked on a more ambitious plan of industrialization. The **Brazilian Miracle**, from 1960 to 1980, transformed the country from rural to urban with massive migration to the cities and the mushrooming of slums around all major cities, known as favelas. More than 12 million people live in the favelas. Within the favelas there is marked social-spatial segregation, often with the poorest areas further up the hillside in more inaccessible locations.

In the 1980s, in the wake of oil price increases and slumping world trade, Brazil experienced a debt crisis that created massive inflation and fiscal uncertainty. A policy of privatization was pursued from 1992 that concentrated on opening up Brazil to global competition and reducing government subsidies. The policies created mass inequalities that in turn generated a political backlash and ultimately the election in 2002 of Luis da Silva (Lula), a former trade union organizer and leader. Under a more liberally progressive government, Brazil replaced privatization and liberalization with a stronger commitment to social welfare, such as the hiring of doctors to work in poor remote areas. One popular program was the Family Allowance income support program that was introduced in 2003. Under this program, millions of Brazilians receive up to $55 a month. In some of the poorer parts of cities, more than half the population can receive family allowances. It is cited as part of the reason for the more than 50 percent drop in the poverty rate. Between 2002 and 2010, 40 million people escaped from absolute poverty, and almost half of the total population can now be considered as middle class.

Brazil is considered one the BRICS (see Chapter 1), defined as large developing economies with impressive rates of growth. In the case of Brazil there are a number of positive elements. Cheap power is available from hydroelectric, newly discovered oil reserves, and ethanol from plants. There is a large market for its agricultural products. Brazil also has a significant manufacturing sector making cars and airplanes as well as mass-producing many consumer goods. The country makes steel, cars, aircraft, and aluminum as well as exports coffee, bananas, and palm oil. Its large internal market and demographic dividend, of declining birth rates and increasing proportion of people of working age, provide the context for the steady economic growth and beginnings of wider redistribution of increased national wealth.

On the other hand, and there is invariably always an "on the other hand," there are signs of stress. There are barriers to increased growth, including persistent pockets of poverty, high tax burdens, inadequate infrastructure, high crime rates, bureaucratic bumbling, endemic political corruption, declining soil fertility, and limited physical infrastructure.

Brazil is experiencing the **middle-income trap** that can occur when countries, after very rapid growth from low- to middle-income status, falter due to poor infrastructure—poor roads; inadequate sanitation, education, and health facilities; and low productivity.

There are also the "problems" of democratization. Once a society is opened to democratic participation, people are more able and willing to complain about government performance. As authoritarian regimes become more democratic, they are more open to rumblings from the population. In Brazil, critics complain about corruption, poor service, and rising costs. These complaints were exacerbated with the costs for hosting the World Cup and Olympic Games.

It is difficult to speak of Brazil in the singular as such a large country exhibits marked geographic variability. Five major regions can be identified: Amazonia, Center West, Northeast, Southeast, and South.

The Amazonian basin constitutes 42 percent of the country. It is home to one of the globe's richest areas of biodiversity. The Amazonian region has long been exploited for commodity booms from rubber to, more recently, palm oil. It is also the last refuge for numerous indigenous groups. The Center West region of the country is an elevated plateau constituting 22 percent of Brazil. It consists of forest and woodland known as *cerrado*. Farming in the north gives way to grassland toward the south, where ranching and beef production are important activities. This ecosystem is the unique home to many flora and fauna which makes its ecological deterioration all that more disturbing. International public opinion focuses on Amazonia. While Amazonia is 18 percent deforested, more than 50 percent of the cerrado is deforested. Between 2003 and 2013 cropland increased by 2.3 million hectares.

The Northeast is the poorest region. Along the coast was the main area of plantations worked by slave labor. Today there is still sugar and coffee production. Away from the urban coast is the interior *sertao* interior, afflicted by periodic drought and widespread poverty. The Southeast is the fastest growing part of the country centered on the largest cities of Rio, São Paolo, and Belo. The South is a subtropical zone. It is a major farming area with 14 million cattle and 10 million sheep.

the Dutch town of Stabroek was renamed Georgetown in honor of the British King. Sugar plantations were the main economic activity, but they began to subside with the emergence of other sugar-growing areas in North America and Australia. Moreover, the abolition of slavery in 1834 reduced the labor supply. To make up the gap, planters shipped in indentured laborers from Europe and especially India, some of whom became rice farmers and later the nucleus of a professional and merchant class in the city. The territory became independent from the United Kingdom in 1966.

Guyana is one of the smallest countries in South America, at least in terms of population, with only around 750,000. Almost half are descendants of the East Indian indentured laborers brought to the country, and just under a third are descended from the black slaves. About one in ten is Amerindian, which includes at least nine separate tribes, including the Arawak, Carib, and Warrau. Most of the non-Amerindians are concentrated along the coastal zone. Georgetown contains more than one in every three people in the entire country. The economy is dominated by primary production, including gold, sugar, rice, timber, and bauxite mining in a sand belt that runs parallel to the coast. Infrastructure development away from the narrow coastal zone is limited.

Suriname, like Guyana, was a Dutch colony long dominated by slave plantations. It has a similar history of the early importance of sugar, cotton, and coffee plantations that declined due to competition and the abolition of slavery. East Indians were an important source of indentured

Fragments of Empires: Guyana, Suriname, and Guiana

All of South America was formally under the possession of either Portugal or Spain. The one exception is a region, consisting of three separate countries, along the Caribbean coast, sandwiched between the Spanish and Portuguese spheres of influence (see Figure 4.17). It was an area of intense competition between European powers eager to muscle into South America. Spain claimed the territory, but the British, French, and Dutch established trading posts and plantations along the coast. Each of the three countries shares a similar geography of coastal urban development with European influence fading into the interior where more Amerindians live in the rainforest.

Guyana was first colonized by the Dutch; by 1775 there were over three hundred slave plantations of sugar, coffee, and cotton. Like other regions in the Caribbean basin, it was the scene of intense rivalry between slaves and slave holders; in 1763–1764, five thousand slaves rose up in rebellion. It finally reverted to British control in 1803, and

4.20 Map of Guyana, Suriname, and French Guiana

laborers in the immediate aftermath of slavery's abolition and became the core of the professional class in the country. Laborers were also recruited from the Dutch East Indies and Java, in particular, giving the territory a multiracial and multireligious population. Today, more than 25 percent of the population is classified as East Indian in origin. Over 20 percent are classified as Maroon, originally slaves who escaped from the brutality of the plantations into the jungle. Almost 15 percent are Javanese, and another 15 percent are classified as Creole, the product of Dutch and African peoples.

The country achieved independence from the Netherlands in 1975. It has a small population with just over half a million, mainly concentrated along the coast. Almost half of the entire country lives in the capital city of Paramaribo.

The economy is a primary producer with heavy reliance on bauxite, which constitutes around 70 percent of export revenues. It also exports rice, bananas, and fish. Oil is likely to play a larger role in the future.

Guiana, formerly known as French Guiana, is not an independent country but a region of France and a part of the European Union.

Early French attempts to establish a colony in the region failed due to the prevalence of tropical disease. The place was so fatal for Europeans that for a hundred years, from 1852, the French transported dangerous criminals to the prison facility of Devils Island, which in fact was a prison complex strung out along the mainland and three separate islands. With poor soils and limited natural resources, the territory is kept afloat with subsidies from Paris. Rice production that achieved some measure of success was much reduced because of EU health regulations that forbid the heavy use of pesticides and fertilizers. Most of the quarter of a million population lives in the narrow coastal zone. Inland the population is predominantly Amerindian and Maroons.

Focus: The Galápagos

The Galápagos is a group of small islands located on either side of the equator off the coast of Ecuador (Figure 4.20). They are a World Heritage site, not only for their ecological diversity but also for their importance in the history of science.

The islands were created by a **geologic hotspot**, a plume of mantle breaking through the seabed. As the surface land moves, due to plate tectonics, new islands are formed. A similar process also created the Hawaiian Islands.

In 1835, the *HMS Beagle*, a Royal Navy sloop, visited the islands, tasked with hydrographic surveys, part of theRoyal Navy's plan to map the world's oceans and seas. A young naturalist, Charles Darwin, was also part of the crew. He paid his own way and kept a diary of the journey. It became a part of the official account and was published in 1839 as a book, *The Voyage of the Beagle*. Its mixture of travel memoir and field journal proved very popular. It was also the basis of the observations from which Darwin developed his theory of evolution.

He was fascinated by the fact that there were distinct species of reptiles, birds, including finches, and tortoises on the different islands. At the time most believed that species were stable, unchanged since Creation. The stability of species argument seemed to be undermined by these finches and reptiles. The data from the Galápagos Islands were a vital part of his theory of natural selection and the more general notion that evolution was marked by species differentiation over time. Darwin's finches, as they are now called, have different beaks that evolved to eat the different food sources on the different islands.

The islands became a national park in 1959 and were one of the first examples of ecological tourism (Figure 4.21). People travelled from around the world to see a beautiful group of islands, a dazzling range of flora and fauna, including giant tortoises and marine iguanas, and to visit an important site in the history of science. The number of visitors is strictly controlled. In 1986, 27,000 square miles around the islands was declared a marine reserve.

There are threats to the biological purity of the islands. Invasive species threaten the delicate ecological balance

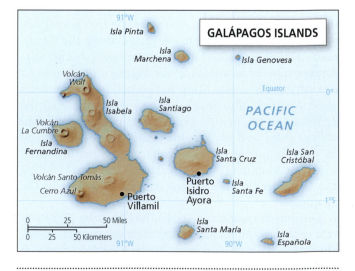

4.21 Map of Galápagos Islands

in a place with few natural predators. Illegal fishing and poaching threaten the sea cucumber and sea lions, and tourism is a mixed blessing as the more people come, the more the ecological balance is threatened. However, the demand for ecological tourism helps to protect the ecology of the islands as visitors come to see the natural wonders. And the islands are well managed by the Ecuadorean government. The local population numbers around 27,000.

4.22 Crowded fish market in Puerto Isidra Ayora, Galápagos (photo: John Rennie Short)

Select Bibliography

Anderson, M. D. 2011. *Disaster Writing: The Cultural Politics of Catastrophe in Latin America*. Charlottesville: University of Virginia Press.

Angosto-Ferrández, L. F., ed. 2013. *Democracy, Revolution and Geopolitics in Latin America*. London: Routledge.

Arana, M. 2013. *Bolívar: American Liberator*. New York: Simon & Schuster.

Brosn, M. F. 2014. *Upriver: The Turbulent Life and Times of an Amazonian People*. Cambridge, MA: Harvard University Press.

Burkholder, M., and L. Johnson. 2012. *Colonial Latin America*. 8th ed. Oxford: Oxford University Press.

Carroll, R. 2013. *Comandante: Myth and Reality in Hugo Chávez's Venezuela*. New York: Penguin Press.

Castañeda, J. G. 2006. "Latin America's Left Turn." *Foreign Affairs* 85:28–43.

Chasteen, J. C. 2011. *Born in Blood and Fire: A Concise History of Latin America*. New York: Norton.

Chatwin, B. 1977. *In Patagonia*. New York: Summit Books.

Clawson, D. L. 2012. *Latin America and the Caribbean*. 5th ed. Oxford: Oxford University Press.

Corrales, J., and M. Penfold-Becerra. 2011. *Dragon in the Tropics: Hugo Chávez and the Political Economy of Revolution in Venezuela*. Washington, DC: Brookings Institution Press.

Crosby, A. W. 1972. *The Columbian Exchange: Biological and Cultural Consequences of 1492*. Westport, CT: Greenwood Press.

Dym, J., and K. Offen. 2011. *Mapping Latin America: A Cartographic Reader*. Chicago: The University of Chicago Press.

Farthing, L. C., and B. H. Kohl. 2014. *Evo's Bolivia: Continuity and Change*. Austin: University of Texas Press.

Filomeno, F. A. 2014. *Monsanto and Intellectual Property in South America*. Basingstoke, UK: Palgrave Macmillan.

Franko, P. M. *The Puzzle of Latin American Economic Development*. Lanham, MD: Rowman & Littlefield.

Gallagher, K., and R. Porzecanski. 2010. *The Dragon in the Room: China and the Future of Latin American Industrialization*. Palo Alto, CA: Stanford University Press.

Gwynne, R. N., and K. Cristobal, eds. 2014. *Latin America Transformed: Globalization and Modernity*. London: Routledge.

Hedges, J. 2011. *Argentina: A Modern History*. New York: Palgrave Macmillan.

Hutchison, E., T. Klubock, N. Milanich, and P. Winn. 2014. *The Chile Reader: History, Culture, Politics*. Durham, NC: Duke University Press.

Jackiewicz, E. L., and F. J. Bosco, eds. 2016. *Placing Latin America: Contemporary Themes in Geography*. 3rd ed. Lanham, MD: Rowman and Littlefield.

Jaffe, R., and J. C. G. Aguiar. 2012. "Introduction—Neoliberalism and Urban Space in Latin America and the Caribbean." *Singapore Journal of Tropical Geography* 33:153–156.

Keeling, D. J. 2013. "Transport Research Challenges in Latin America." *Journal of Transport Geography* 29:103–104.

Kelly, P. 1997. *Checkerboards and Shatterbelts: The Geopolitics of South America*. Austin: University of Texas Press.

Kent, R. B. 2016. *Latin America: Regions and People*. 2nd ed. New York: Guilford Press.

Kronik, J., and D. Verner. 2010. *Indigenous Peoples and Climate Change in Latin America and the Caribbean*. Washington, DC: World Bank.

Malpass, M. A., and S. Alconini, eds. 2010. *Distant Provinces in the Inka Empire: Toward a Deeper Understanding of Inka Imperialism*. Iowa City: University of Iowa Press.

Mann, C. C. 2005. *1491: New Revelations of the Americas before Columbus*. New York: Knopf.

Mann, C. C. 2011. *1493: Uncovering the New World Columbus Created*. New York: Knopf.

Marcy, W. L. 2010. *The Politics of Cocaine: How U.S. Foreign Policy Has Created a Thriving Drug Industry in Central and South America*. Chicago: Lawrence Hill Books.

McMillan, B., ed. 2002. *Captive Passage: The Transatlantic Slave Trade and the Making of the Americas*. Washington, DC: Smithsonian Institution Press.

Muñoz, H. 2008. *The Dictator's Shadow: Life under Augusto Pinochet*. New York: Basic Books.

Palacios, M. 2006. *Between Legitimacy and Violence: A History of Colombia, 1875–2002*. Durham, NC: Duke University Press.

Pollard, J., C. McEwan, and A. Hughes, eds. 2011. *Postcolonial Economies*. London: Zed Books.

Reid, M. 2014. *Brazil: The Troubled Rise of a Global Power*. New Haven, CT: Yale University Press.

Renfrew, D. 2011. "The Curse of Wealth: Political Ecologies of Latin American Neoliberalism." *Geography Compass* 5:581–594.

Rohter, L. 2010. *Brazil on the Rise: The Story of a Country Transformed*. New York: Palgrave Macmillan.

Rumney, T. A. 2013. *The Geography of South America: A Scholarly Guide and Bibliography*. Plymouth, UK: Scarecrow Press.

Sawyer, S. 2004. *Crude Chronicles: Indigenous Politics, Multinational Oil, and Neoliberalism in Ecuador*. Durham, NC: Duke University Press.

Schoultz, L. 1998. *Beneath the United States: A History of U.S. Policy Toward Latin America*. Cambridge, MA: Harvard University Press.

Sexton, J. 2011. *The Monroe Doctrine: Empire and Nation in Nineteenth-Century America*. New York: Hill and Wang.

Smithsonian Institution. 2002. *Captive Passage: The Transatlantic Slave Trade and the Making of the Americas*. Washington, DC: Smithsonian Institution Press.

Veblen, T., K. Young, and A. Orme. 2007. *The Physical Geography of South America*. Oxford: Oxford University Press.

Weitzman, H. 2012. *Latin Lessons: How South America Stopped Listening to the United States and Started Prospering*. Hoboken, NJ: Wiley.

Learning Outcomes

South America is realistically part of a much bigger and connected "America," with many shared histories and geographies.

The environment of South America contains several distinct regions, including the Andes region, Amazonia, and smaller ecosystems such as the cerrado.

The history of the relationship between the population of the Americas and its environment can be divided into the distinct periods known as Pre-Columbian and Post-Columbian (separated by the Columbian Encounter).

The Pre-Columbian period was marked by local cultures responsive to the constraints of their local environments.

The Columbian Encounter inaugurated significant shifts in ecological relations as the ecology of the continent was transformed by the continent's incorporation into a global economic system and consequently affected by global economic cycles.

Four distinct, though merging, periods of South America's connection with the global economy can be identified that have shaped the economic geography of the continent and continue to exercise a huge influence: the colonial/neocolonial era, import substitution, neoliberalism, and the global commodities boom.

The vast majority of South America's population is housed near or along the coasts, a reminder of the importance of its export-oriented economic development.

South America's population can be divided into three main, yet fluid categories: indigenous people located mainly in the rainforest and mountain areas, black South Americans who are descendants of slaves and concentrated along the tropical coasts; and the white descendants of European colonizers who remain at the apex of political and economic power of the subcontinent.

Some of the countries in the region such as Brazil are experiencing the demographic dividend.

South America is one of the most urbanized regions of the world with more than 4 out of every 5 people living in rapidly growing cities, which have become sites of extreme disparities and inequality.

Urbanization in South America is marked by urban primacy.

The geopolitics of South America is shaped by a legacy of imperialism, emerging nationalism and pan-nationalism, and intranational conflicts.

South America's connection to the rest of the world is evident through the booms and busts of certain commodities.

South America can be further divided into the subregions of Gran Colombia, the Andean Arc, the Land of the Pampas, and Brazil.

Although there are vast similarities among the Gran Colombian states of Colombia, Venezuela, and Ecuador, including similar flag schemes and middle-income status, they have significant differences related to the differing trajectories of recent history and the luck of resource endowment.

All three of the Andean Arc countries of Bolivia, Chile, and Peru have faced similar problems related to balancing economic growth with equity. While Chile remains one of the wealthiest countries of South America, Bolivia is still one of the continent's poorest, while Peru falls in between the two.

Argentina is the second largest country in the region with trading links to the outside world, especially Europe, through immigration and economic transactions, while Paraguay was marked by isolation, dismemberment, and remains a relatively poor economy. Uruguay is dominated by the success of its primate city, Montevideo, making it one of the richer countries of South America.

Brazil is the largest country in the region by land mass and population, while also maintaining the most successful economy, propelling itself to one of the largest in the world, even being included into the acronym BRICS, used as a designation for similar world economies experiencing impressive rates of growth.

Although Brazil's economy has recently been highly successful and its demographic dividend continues to be positive, signs of stress are hampering its growth opportunities. Some of these stressors include a middle-income trap, continued corruption, and poverty.

Guyana, Suriname, and French Guiana share a similar geography of coastal urban development and European influence fading into the interior where more Amerindians live in the rainforest.

5

Europe

This is a region that has experienced major environmental transformation due to the commercialization of agriculture and the rise of industrial regions. The economy is marked by a shift toward the postindustrial, while geopolitics is dominated by the development and enlargement of the European Union. We can identify one large metro region, Eurometro. There are substantial regional variations. As a wealthy area, it attracts migrants from Africa and the Middle East.

LEARNING OBJECTIVES

Recognize the geologic origins of the region's physical landscapes and identify the impacts of the European Anthropocene.

Discuss the region's historical divides with focus on their religious, economic, and geopolitical expressions.

Explain the region's agricultural and industrial economies and survey its related economic, political, and environmental issues.

Describe the region's demographic trend of aging and connect it to conflicts over immigration.

Examine the reasons for the Common Agricultural Policy in the region and evaluate rural impacts and calls for reform.

Summarize the region's urban development and four distinct urban forms and relate these to the formation of the Eurometro.

Discuss the integration of the region into the European Union and appraise four major issues that challenge its continued success.

The Environmental Context

Seismic Activity

Much of Europe sits in the middle of the Eurasian Plate, making seismic activity a restricted phenomenon across much of the continent. However, at its edges in the eastern and central Mediterranean, the African Plate is moving against the southern edge of the Eurasian Plate, creating seismic activity. The Greek island of Santorini sits on the **South Aegean Volcanic Arc of Islands**. A huge volcanic eruption 3,600 years ago basically blew off the top of the volcano, leaving a submerged **caldera** (Figure 5.2).

Mount Vesuvius erupted in 79 CE, burying the Roman cities of Pompeii and Herculaneum in suffocating ash and poisonous gases. More than 3 million people live in the shadow of this active volcano. Mt. Etna in Sicily remains in constant activity.

Earthquakes are also part of this seismic activity. An earthquake destroyed the old city of Dubrovnik in the Adriatic in 1667, killed five thousand people, and flattened the medieval buildings. In 1755, an earthquake struck Lisbon, Portugal. Eyewitness accounts, eerily similar to the 2004 tsunami in South East Asia, describe how city residents moved out of buildings to see the harbor bottom laid bare by receding water. Minutes later a tsunami engulfed the city. Almost 20 percent of the city's 200,000 population perished. The event, occurring at a time of rapid communication, sent shock waves through Europe, denting the self-confidence of the **Enlightenment** project and its notion of progress and reminding people of the limits of human knowledge. Attempts to understand the phenomenon helped establish a more scientific physical geography.

5.1 Map of region

Earthquakes continue to cause damage. The 1980 Irpinia earthquake in Italy, registering 6.8 on the **Richter scale**, killed 3,000 people and left 300,000 people homeless. As with many earthquakes, the shock waves rippled through the social fabric as much as the underlying geology. As the results of an inquiry later revealed, only one-quarter of the $40 billion rebuilding funds went to people in need. The rest was diverted to the well connected, bribes to politicians, and payoffs to organized crime. After the 2009 L'Aquila earthquake in Italy, in which 309 people were killed, six scientists were found guilty of manslaughter for minimizing the risk of earthquakes—a reminder that predicting earthquakes is a hazardous activity at the best of times but especially when making the wrong prediction only 6 days before a 5.9 earthquake.

Physical Landscapes

The physical landscape of Europe consists of very old rocks, the result of mountain building millions of years ago, including the Ardennes, Black Forest, Massif Central in France, and the Scottish Highlands, as well as the much more recently created and hence higher mountains of the Alps, Balkans, and Carpathians.

The more mountainous and previously less accessible areas were the home of distinct regional cultures different from the people of the plains below. In Scotland, for example, the Highlands remained an area of feudal precapitalist social relations as the Lowland area witnessed rapid urbanization, agricultural commercialization, and industrialization. Transport improvements, migration, and the creation of national cultures through education and mass communications gradually succeeded in incorporating the people of the mountains more fully into the respective nation-states. Yet to this day the remnants of distinct mountain cultures are evident in traditional dress, folk tales, songs, dialect, language, and dance across the continent, whether it be in the remnants of Gaelic speaking in the Scottish Highlands, the old folk songs of the Carpathians, or the traditional dance of the Catalonian Pyrenees.

This region has also witnessed two much more recent (in geological terms at least) agents of environmental change. The first is the recent **Ice Age** that created permanent ice sheets in the northern part of Europe and active glaciers in higher altitude such as the Alps. A series of cold snaps caused successive ice ages in Earth's history, but the most recent Ice Age from 2.5 million years to 10,000 years ago, interspersed with **interglacial warming** periods, shaped much of the landscape of northern Europe. Giant ice sheets pushed south as far as 52 degrees North. Glaciers in the higher mountain areas such as the Alps, enriched with snow and ice in the colder climates,

5.2 The town of Santorini in Greece sits precariously on the slopes of the caldera of an extinct volcano (photo: John Rennie Short)

snaked down the slopes from their mountain fastness. The result was a glaciated landscape of **U-shaped valleys** and steep-sided **fjords** (Figure 5.3). The ground was striated and scraped by the massive blocks of ice, creating depressions and trenches for **kettle lakes** and **finger lakes**. At the edge of the retreating glaciers, **peat bogs** were formed along a wide band from the west coast of Ireland through central and eastern Europe. Scandinavia has an ice-sculpted landscape of fjords, lakes, and bogs that is a reminder of these colder times. The limit of glaciation runs through the United Kingdom. Northern Scotland has the glaciated landscape of steep-sided, U-shaped valleys and **lochs**, while southern England, too far south for the glaciers to reach, has the gentle rolling chalk hills of a landform untouched by destructive ice. The ending of the Ice Age and the melting of the ice caps impacted the whole of Europe as sea levels rose and valleys were flooded. The Gulf of Bothnia is still rising up as a long slow bounce back from the huge weight of the ice sheets. A landscape of **drumlins**, **eskers**, **moraines**, and peat bogs was created along the boundary line of the retreating mass of glacial ice.

Although the ice sheets receded 10,000 years ago, mountain glaciers in the high Alps remain because their high altitude keeps temperatures low. Most mountain glaciers are on

the narrow cusp between ice and water and thus provide us with a very sensitive barometer of global climate change. Since 1850, 40 percent of the glaciation in Austria has disappeared. The snow line is migrating north, and the ski season is getting shorter. Researchers using aerial photographs of the Italian Alps showed how small glaciers are experiencing retreat and shrinkage and a reduction of 50 percent from 1954 to 2003. A positive feedback loop is created as shrinkage of the glacier uncovers rock, which, when exposed to the sun, warms up, melting even more ice and snow.

The European Anthropocene

The second major environmental change is the incredible impact of humans. Europe is one of the most densely populated, though not the densest, region of the world with people thick on the ground. Not only is there a heavy weight of population but also a long history of environmental transformation wrought by centuries of urbanization and industrialization, the commercialization of agriculture, and the heavy environmental impact imposed by millions of affluent people. With an economy built on environmental transformations using sophisticated technology, Europe is at

5.3 U-Shaped fjord near Kristiansund, Norway (photo: John Rennie Short)

the forward edge of human-induced environmental transformation. The impact of humans has led some scientists to propose a new geological era known as the **Anthropocene** to reflect the huge impact that human activities have on the earth. Europe is one of the regions of the world most advanced into the Anthropocene era. The transformation is evident in the deforestation of the European forest over the past three thousand years. While much attention is concentrated on the reduction of the tropical rainforest, the temperate forests of northern and central Europe have been cut down over the centuries. Over 95 percent of the forest in Europe has disappeared. Sicily's forest cover was much reduced as early as Roman times, when the island became an agricultural region for the expanding Empire. Pasture and crop fields now cover the deforested landscape (Figure 5.4). The 100-square-mile Bialowieza Forest along the border between Poland and Belarus is one of the few fragments of the giant forest that once covered the northern plains of Europe.

If the agriculture revolution reduced the forest cover, then the **Industrial Revolution** created a new industrial landscape of coal mines, steel plants, factories, and towns. Small villages became towns, and towns mushroomed into giant conurbations. In the process the landscape was utterly transformed, forests were cleared, land and water polluted, habitats lost, and flora and fauna much reduced.

Across the entire continent, natural environments were transformed. The extent and depth of the transformation were so great as to encourage in more recent years a stronger commitment to environment improvement, habitat protection, and encouragement of biodiversity. The realization of loss of biodiversity led to the creation of protected national parks and wilderness areas. Protected areas include Camargue Nature Reserve in France, the Danube Delta, and numerous wildlife refuges such as the Donana National Park in southern Spain.

POLDERS

Nowhere is the human imprint and the environmental transformation so apparent as in the creation of the new landscape of **polders**, low-lying land that is drained of water and turned into productive land. Polders are directly reclaimed from the sea or from marshes and flood plains. Polders are found throughout Europe, including Belgium, France, Germany, Italy, Poland, Slovenia, and the United Kingdom. Polder reclamation is very important in the Netherlands, where 20 percent of the country sits below sea level and another 50 percent is less than 4 feet above sea level. The Dutch made an early start a thousand years ago and have created over three thousand polders. The 27-square-mile Beemster polder was created between 1609 and 1612 and now contains nine thousand people, in neat grid-patterned towns. The 370-square-mile Flevopolder was drained between 1955 and 1968.

The Cardinal Geography of Europe

The physical geography of Europe varies by cardinal directions. West to east: from the moist green coast of western Ireland to the hot summer and bitterly cold winters of central Poland is to take a transect from the temperate, maritime influenced climates of western Europe to the more extreme continental weather of central and eastern Europe. South to north: from the balmy Mediterranean coast of Provence in France to the bitter winter cold of Finland is to move from the warm, wet winters and hot, dry summers of the Mediterranean to the cold winters and endless summer days of the subarctic. While varied, Europe's climate does not encompass subtropical or tropical climes. The desire and demand for products of the tropics was an important impetus for European expansion into the warmer area of the world. The lure of the tropics drove European colonial and trade expansion.

5.4 Polderlands in Netherlands (photo: John Rennie Short)

Historical Geographies

There is also a human geography associated with these cardinal directions.

The North-South Divide

From south to north is to move from the predominant Catholic countries of Italy and Spain to the Protestant areas of North Germany and Scandinavia (Figures 5.5 and 5.6). This division was an extra source of tension between competing powers as when England and Spain fought for naval supremacy, a struggle often cast in terms of a religious war between Protestants and Catholics. This religious fault line was also the site of intranational divisions such as between the Protestants of the northern Netherlands and the Catholics in the south. The religious divide also was a major cause of the Thirty Year War (1618–1648) that devastated central Europe.

In much of Europe, riven by religious tensions for centuries, one long-term consequence was a secularization of public life and a commitment to removing religious references from the public discourse of civil society. The United States, in contrast, retains a more overt religiosity and linkage between religion and politics. Few European leaders, unlike presidents of the United States, will ask publicly for God's blessing.

The division of Catholic south and Protestant north also overlays differences in economic history as the Protestant northern regions were first to enter the Industrial Revolution. This led one commentator, Max Weber, to suggest that there was a connection between the different religions and economic growth. He argued that the Protestant work ethic, connected to the Protestant theology of individual redemption through work and sacrifice, was more conducive to the development of capitalism. He associated Protestantism with individual effort to accumulate wealth in contrast to the emphasis on the more collective nature of the religious devotion of Catholicism. More recent research undermines the simple connection but does draw attention to the connection between Protestant theology, which stressed an individual understanding of the word of God, embodied most clearly in translation by scholars of the Bible into vernacular languages, and the consequent encouragement of mass literacy and hence a more educated workforce. Catholicism, in contrast, placed little emphasis on mass literacy with more emphasis on an educated priestly class conversant in Latin

5.5 St Peter's in Rome is the center of the Catholic Church, a powerful influence in Europe but especially in southern Europe (photo: John Rennie Short)

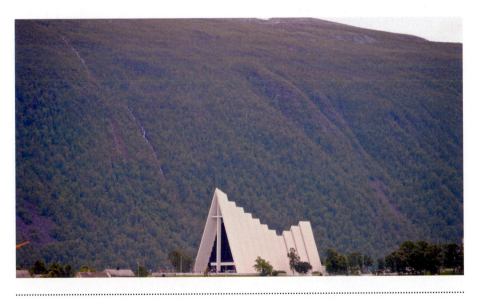

5.6 Lutheran Church in Tromso, Norway. The Protestant religion took root in northern Europe (photo: John Rennie Short)

who guided the largely illiterate faithful. The wide differences in literacy levels between Protestant and Catholic Europe from the sixteenth to the nineteenth century is one possible reason behind the economic geography of a more developed northern Europe in the early years of the Industrial Revolution.

Today, widespread secularism has eroded traditional religious belief systems, and urbanization and industrialization have spread more evenly across the continent. The north-south divide remains, however, in terms of climatic differences and is the reason behind the annual migration of holidaymakers from central and northern Europe seeking the sun. Many of the littoral regions of the Mediterranean, blessed with hot, dry summers and warm winters, are based around the seasonal tourist trade as well as more permanent relocations of those seeking to leave the colder parts of Europe.

The West-East Divide

The west-east divide also marks a transition from maritime-based to land-based societies. A distinction can be made then between the maritime-based empires of Britain, the Netherlands, Spain, and Portugal and the land-based empires of the more eastern European regions such as the Austro-Hungarian Empire.

When maritime knowledge was more limited, interaction was restricted to coastal trade or short ventures across the inland seas. The European economy centered in the south on the Mediterranean and in the north on the Baltic. Cities such as Venice and Lubeck were important hubs of long-distance trade. But when navigation advances and new geographical knowledge in the sixteenth and seventeenth centuries allowed transoceanic travel, the western ports of Britain, Portugal, Spain, France, and the Netherlands, which all had ports along the North Sea or the Atlantic Ocean, now had access to global trade routes and imperial claims. The center of economic gravity in Europe shifted in the sixteenth century from the inland sea empires of the Baltic and the Mediterranean to the western maritime regions.

Venice's wealth and power were evident in the fifteenth century, but by the nineteenth century Britain was the wealthiest country in the region.

There was also the more recent divide of East and West Europe. It was Winston Churchill who delivered a speech in 1946 that referred to an Iron Curtain: "From Stettin in the Baltic to Trieste in the Adriatic an iron curtain has descended across the Continent" (Figure 5.7). This curtain effectively demarcated the two halves of Europe from the end of World War II to 1989. East Europe was under Soviet control, though Yugoslavia was more independent from Moscow than East Germany or Poland. Trade and movement between the two blocs were limited, and missiles and armies faced each other across barbed wire borderlands. The fall of communism heralded the ending of the division that scarred Europe.

Economic Transformations

Intensification of Agriculture

In the Middle Ages, the growing of grains, horticulture, and the grazing of animals were adapted to local conditions. In the hot, dry summers of the Mediterranean, olives and grapes were produced and goats were herded. Further north, corn was grown and thick woolen sheep were raised, and close to the burgeoning towns even more intensive agriculture developed due to the proximity of a large market. Transport costs were high so long-distance trade was often by ships. Few perishables could be shipped long distances and hence the demand for Oriental and tropical spices that could mask the smell of rotting meat and help preserve produce.

Over the centuries, improvements in agriculture allowed higher yields and hence more population, which, in turn, increased demand and increased intensification (Figure 5.8). This upward spiral of agricultural extension and intensification resulted in massive environmental change.

More recently, the intensification of agriculture is aided by the application of pesticides and fertilizers. Increased yield came at the cost of negative environmental effects that include increased nutrient runoff that creates **algae blooms**, which kill off rivers and

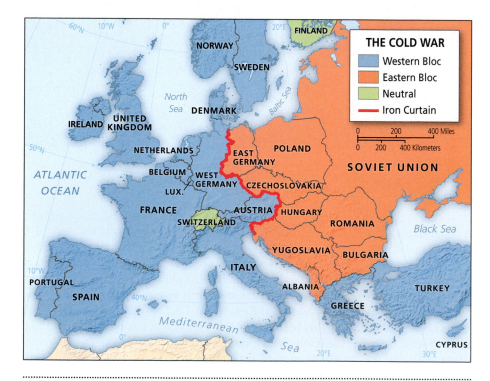

5.7 Europe divided during Cold War

5.8 A landscape of agricultural intensity in northern Italy (photo: John Rennie Short)

TABLE 5.1	Agriculture in Selected European Countries	
	AGRICULTURE AS % OF TOTAL LABOR FORCE	**AGRICULTURE AS % OF GDP**
UK	0.7	1.4
France	1.9	2.9
Italy	2.0	3.9
Portugal	11.0	2.6
Poland	12.9	4.0

Manufacturing centers in Europe, Rhine-Ruhr Valley

lakes and produce dead zones along the coast. In a counterpoint to this trend, the realization of the heavy environmental impacts of efficient modern agriculture practices puts Europe in the forefront of organic sustainable agricultural practices. The process is helped, in large part, because the many smallholdings throughout Europe, in contrast to the more agribusiness nature of US farming, often lack the large capital necessary to invest in industrial agriculture. The small farmers of Europe can shift more easily than the highly capitalized farmers of the United States, into the smaller scale, more labor-intensive practice of organic agriculture.

As economies mature, the number of people actively involved in agriculture declines, and its proportion of total economic activity is reduced. This process varies across the continent. We can make a distinction in Europe between countries such as the United Kingdom and the Netherlands, where farming is a small portion of total economic activity and constitutes a small part of the labor force. Here, farming is very efficient, highly capitalized with only a few less capital-intensive sectors such as sheep rearing and smallholding mixed farms, called crofts in the Scottish Highlands. In countries where the agricultural and industrial revolutions came much later or were not so pronounced, as is commonly found in much of southern and eastern Europe, agriculture remains a more significant part of the economy and of the total labor force. These differences are highlighted in Table 5.1.

In the United Kingdom, agriculture employs few people and is a relatively small though very efficient sector of the national economy. In Portugal and Poland, in contrast, agriculture employs proportionately more people. In countries such as France and Germany, agriculture falls in between these two extremes. One of the issues facing the United Kingdom when first considering joining the **European Union (EU)**, when it was the European Economic Commission, was the structure of subsidies that were geared up for efficient industries and inefficient agriculture. The United Kingdom, in contrast, had more efficient agricultural sectors and less efficient industrial sectors. The greater emphasis of subsidies for agriculture more than industry thus favored Germany and France over the United Kingdom.

Industrialization and Deindustrialization

Europe was the crucible of the Industrial Revolution. Beginning in the early nineteenth century in the textile towns of northern England, the mechanization of production increased the output of factories. The Industrial Revolution was also an urban revolution as factories clustered around each other; small villages became towns, and towns grew into cities.

By the early to mid-twentieth century, heavy industrial complexes were spread across the continent from shipbuilding in Glasgow and Bilbao to steel making in Sheffield and Essen. The Ruhr region was one of the largest concentrations of heavy engineering and metal production; it consists of numerous cities, including Cologne and Dusseldorf. Iron and steel production and heavy metal industries created one of the largest industrial areas in the world. By the 1970s, these older industrial areas were on the decline.

The most significant change in the economic geography of Europe is the decline of industrial employment as manufacturing becomes more efficient and some of the more basic manufacturing processes are offshored to the cheaper labor areas of Asia. **Deindustrialization** refers to the declining size of the manufacturing sector. It is most evident in those regions built around manufacturing with few alternative forms of employment. London lost significant industrial employment, but the dynamism of the service sector soaked up the employment loss. In central Scotland, in contrast, alternative forms of employment were less available to offset the loss of manufacturing and heavy engineering employment. The result is that the industrial cities and regions of the continent have faced considerable difficulties as their economic base has declined.

Deindustrialization was apparent throughout Western Europe by the 1970s, but it became particularly acute in the 1990s in the industrial areas of former Eastern Europe, such as the former East Germany, Slovakia, and Czech Republic, as older, less efficient, and polluting industries became subject to global competition and higher EU standards after the fall of communism in 1989. The city of Leipzig has a typical trajectory of a former East German city built on industry that experienced rapid deindustrialization, abandoned factories, and high unemployment. People moved to the better economic opportunities of the former West Germany.

Across the former East Germany, deindustrialization, job loss, outmigration, and subsequent population loss have led to shrinking cities and regions. Saxony-Anhalt, the center of East Germany's chemical industry, lost a fifth of its 3 million population from unification in 1990 to 2006. Former industrial cities such as Jena and Chemnitz will, on current projections, lose 20 percent of their population by 2020 as people move to the better employment opportunities of the former West Germany. Urban shrinkage is a major problem of former manufacturing cities and regions in Europe, especially postcommunist Europe, as it entails a declining tax base and the loss of younger and more ambitious people. From 1996 to 2009 all cities in Slovakia lost population.

Manufacturing is still an important part of the European economy. The deskilling of work associated with mechanization and automation freed industry and manufacturing from the traditional pools of skilled labor. Manufacturing is now widely dispersed throughout Europe. I was once walking in the limestone hills of Majorca, Spain, far from any large town when I came across a small village where most of the women in their front rooms were working on rented sewing machines making expensive ski boots.

In Germany, industry still employs a quarter of the labor force and the country's very successful export-led economy is built around the manufacturing of electronics, machinery, metals, and vehicles. The success of the German industrial sector is explained by a number of factors, including its complete renewal after 1945. Another important element is the nature of corporate organization. Most private companies in Germany have corporate boards that include workers as well as management. This is one reason behind the large amount of resources devoted to employee training and reinvestment in improving working conditions and productivity. German private companies are less subject to shareholder pressure to provide quick returns or management tendencies to reward themselves with company shares. Both of these can lead to the shifting of operations to cheaper labor areas around the world and the sacrifice of long-term sustainability for short-term financial gains. The emphasis on quick profits, high immediate

returns, and bountiful management packages that characterize the corporate world in the United States and the United Kingdom is lacking in Germany, and this facilitates longer term investment strategies and a wider definition, and more evenly distributed rewards, of company success.

Across Europe, industry employs between 15 percent and 30 percent of the workforce. Industrial employment is still concentrated in the countries of northwest Europe and parts of Spain and Italy. More sophisticated manufacturing remains at the heart of the economies of industrial Europe. Manufacturing employment of this type declines with distance from the Rhine Ruhr heartland.

BROWNFIELDS

The environmental legacy of industrialization and its aftermath is apparent across industrial urban Europe whether it is in the abandoned factories or the polluted industrial sites known as **brownfields**. In Germany there may be as many as 362,000 sites covering about 128,000 hectares, in France approximately 200,000, in the United Kingdom some 105,000 sites, and in Belgium some 50,000 sites covering at least 9,000 hectares. Environmental contamination not only has an adverse effect on the environment, it is also a significant barrier to the economic and social redevelopment of the city. In addition, there is growing evidence to suggest that many urban brownfields are situated in areas with higher concentrations of minority populations and households below the poverty level, raising important questions about environmental justice.

In countries in Central and Eastern Europe, the scale of urban pollution is often greater than in Western Europe because of the emphasis on economic growth at the expense of environmental concerns during the era of communist control from 1945 to 1989. Cities in the former East Germany have higher levels of urban contamination than cities in the former West Germany. However, the wealth and commitment of Germany to environmental improvement allow for a strong response. In Leipzig, for example, the former industrial mining site is now home to tourist-attracting marinas, beaches, meadows, forests, and lakes. To improve the air quality, since 2011, an environmental zone covering most of the city restricts cars to those with very high emission standards.

Large brownfield sites are a common feature of cities in the former Eastern bloc. Krakow in Poland has 700 hectares of brownfields within the city perimeter. Remediation started relatively late, only beginning in many cities after the fall of communism in 1989. The Czech Republic has an estimated 10,000 brownfield sites, many of them agricultural and military as well as industrial and urban. The first industrial brownfield program in the Czech Republic only began in 2002 (Figure 5.9). Renewal programs are made problematic by the uncertain nature of land ownership and the high level of "public" land where the polluting industries have long since disappeared in bankruptcy, closure, and privatization. Following the chain of responsibility to

5.9 Brownfield in Ostrava, Czech Republic (Jiří Bernard/ Wikimedia Commons (CC BY-SA 3.0))

determine who should pay for the clean-up is difficult in such murky circumstances.

Where there is a functioning land market, brownfields are redeveloped in a variety of ways: industrial reuse, commercial or residential uses, and also as green spaces such as parks, playgrounds, trails, and greenways. In Derbyshire, England, the former Markham colliery (coal mining facility) was being transformed into Markham Vale, a 200-acre business-industrial park planted with trees. The trees are a renewable energy resource, heating boilers for nearby commercial buildings.

The Postindustrial Economy

Manufacturing is still an important element in the economy of Europe and especially in the advanced sectors of electronics, machinery, and vehicles. However, across Europe the more successful regional economies have moved toward a postindustrial profile with a greater proportion of jobs connected with services. A particularly dynamic service sector is the advanced producer services of accountancy, advertising, finance, insurance, law, public relations, and management consultancy. This highly paid sector is concentrated in major cities around the world. The most important cities in Europe for advanced producer services are, in order, London, Paris, Milan, Frankfurt, Madrid, Amsterdam, Brussels, Vienna, Zurich, Warsaw, Barcelona, Dublin, Munich, Stockholm, and Prague.

A significant element of the service sector in Europe is public sector employment. The fiscal crisis of many European countries arises because of the large size of the public service even as revenues decline. The **Greek fiscal crisis** was in part a result of a bloated public sector and an unfair taxation where the rich managed to avoid paying their fair share of taxes.

Female Participation

Traditional industry in Europe, as elsewhere, was a male-dominated affair with images of work and masculinity bound closely together. In more service-based economies, women play a more significant role and so traditional gender relations are contested and shifted. In Europe, female participation rates in the formal economy are some of the highest in the rich world, because good preschool facilities allow women to work while also having children. The socialist history of Eastern Europe also encouraged women in the workforce and so the main differences are not between east and west Europe but between north and south Europe reflecting both the tenacity of traditional gender stereotypes and social welfare programs. The female participation rate in Italy, for example, is 39 percent of all women aged 15 years or more, compared to 50 percent for Czech Republic and 60 percent for Sweden.

A Welfare State

Across much of Europe, employment conditions are some of the best in the world with relatively high wage rates, good conditions, and generous benefits such as unemployment payments, maternity leave, and good health provisions. The core countries of northern and western Europe have some of the best working conditions in the world with high wages and conditions of employment declining as one moves away from the German-Nordic core. The system has provided a decent quality of life for many of the continent's citizens.

There are differences across the countries of the EU in terms of the size and effectiveness of the welfare state. Figure 5.10 highlights the difference in terms of size, as percentage of GDP, and effectiveness, in terms of coverage of the bottom 20 percent income group. Denmark has a larger welfare system than say Latvia or Poland and more effective than Spain and Greece. There are also large welfare states systems that provide less coverage to the bottom 20 percent such as Spain and Greece.

Two problems arise as Europe deindustrializes and faces competition from economies across the world. First, the security of employment means that when growth dips, new entrants to the job market are not hired, while established workers keep their jobs. The result is youth unemployment. Across Europe, youth unemployment rates are much higher than average unemployment rates. In Greece and Spain, for example, unemployment rates in 2018 for those aged 15 to 25 years were 45 and 36 percent, respectively. The respective national figures were 20 and 16 percent.

The second problem is that the generous social welfare benefits create an intergenerational transfer of wealth, as younger workers have to pay for older workers and retiree benefits that they themselves may not receive; these problems are exacerbated by the aging of the European population.

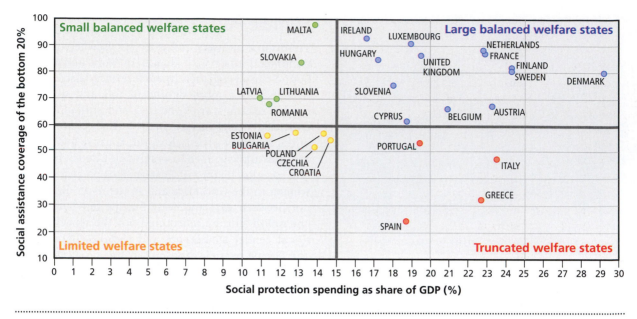

5.10 The welfare state in EU countries

Social Geographies

The Graying of Europe

Throughout much of Europe there is a distinct graying of the population. Falling birth rates and lengthening life expectancy are leading to older people becoming a larger proportion of the total population. This is part of a demographic trend marked by low fertility levels, a pronounced aging of the population, more reliance on immigration to fill job vacancies, especially at the lower wage levels, a disconnect between marriage and children as more children are born out of wedlock and more couples do not have children, and a variety of household living arrangements as the traditional mother, father, and children model is augmented by a wider variety of household structures. Traditional marriage is no longer such a dominant model of household formation. These trends are especially marked in the countries of northwest and Eastern Europe. In Norway and Sweden, 54 and 55 percent, respectively, of all births in 2017 were to unmarried women. This choice may have some negative economic consequences in some countries. In the United States, almost 41 percent of all births are to unmarried women. One in three households that experience child poverty is single parent and female headed, and only 6.4 percent are from married, two-parent families. In many northern European countries, in contrast, children born out of wedlock are generally born to stable, cohabiting couples with easier access to more generous social welfare and public health programs.

The aging of the population means that fewer people of working age have to support an increasingly higher proportion of nonworking people often subsidized through generous social welfare programs. Taxes to pay for these programs are much higher than say the United States, a decision that majority opinion in Europe supports. However, since there are limits to taxation, the welfare programs of the more generous state are under pressure from a nonworking population living much longer and the decline of the working population. In France, for example, life expectancy in 1960 was 69.8, but in 2016 it was 82.7 years. The number of dependants per 100 people of working age was 61 years in 1960 and estimated to rise to 73 years by 2050.

Immigration

Migrants from poorer countries within the EU and from overseas fill in many of the lower paid jobs. While this process meets economic needs, it raises political issues, especially during economic downturns. With fewer jobs and more claims on social welfare, popular opinion can be mobilized to blame foreigners and immigrants for taking jobs and receiving benefits. There is an anti-immigration narrative to recent political discourse in Europe. The anti-immigration rhetoric increased as more asylum seekers and economic migrants from North Africa and the Middle East tried to enter Europe. In 2015, close to 1 million people tried to enter Europe by small boats from the coasts of Africa and Turkey. Hundreds of thousands made the dangerous crossing to Greece from Turkey, then moving across Europe on their way to Germany and Sweden.

In the United Kingdom, for example, long a home of classic liberalism, immigration from the former Eastern Europe caused a backlash in the form of the

United Kingdom Independence Party that has moved from right-wing fringe status to more mainstream support. In the United Kingdom more than 13 percent of the population is now foreign born, many of them recent immigrants from East European countries of the EU. It is not just a British phenomenon. Across the countries of Europe there is evidence of a political backlash against foreign-born recent migrants.

Schengen Agreement under Pressure

Much of Europe now counts as a single territorial unit in terms of freedom of movement and capital mobility. In 1985, five members of the European Economic Community (EEC) signed an agreement in the Belgian village of Schengen that guaranteed passport-free movement across their common borders. By 2014, the agreement allowed passport-free movement across the common borders of twenty-six European countries, including twenty-two EU member states (Figure 5.11). The agreement also allows visitors to travel freely once they are admitted into the **Schengen Area**.

Two issues have arisen in the aftermath of the agreement. The first is that some countries use their privileged position to leverage mobile capital. One way to plug a fiscal gap, generate revenue, and fill the national coffers—especially for smaller, poorer countries and especially those undergoing property collapses, fiscal problems, and economic uncertainties—is to effectively sell access to Europe and EU citizenship to the wealthy. In Latvia, for example, anyone who buys property worth at least 50,000 Lats (US$96,000) in provincial cities, and 100,000 Lats (US$192,000) in Riga, receives a 5-year residency permit that allows them access to other countries in the Schengen Area. Greece, Spain, and Hungary also have programs that provide visas in exchange for money. Since 2012, Portugal has had a "golden visa" guaranteeing 2-year residence in return for a 500,000 euro investment in real estate or a 1 million euro investment that creates thirty jobs. Cash-strapped nations in the Schengen Area can use their entry opportunity as a way to attract mobile capital in return for

residency and fast-track citizenship that provides wider European mobility and the possibility of EU citizenship.

Malta proposed an Individual Investor Program that offered citizenship for a straight fee of 650,000 euros. There were neither investment nor residency requirements. After heavy criticism, both domestically and from European partners that the program was effectively selling European citizenship, the program was placed on hold and then in November 2013 a revised program offered citizenship in return for 1,150,000 euros. The Malta case highlights the problem of a unified system of territorial integrity with differing national rules for entry.

A second problem is what happens when people move into the region and then one of the countries introduces new restrictions. This is what happened in 2015/2016 when Sweden tightened the requirements for immigrants coming from the Middle East and especially Syria. Almost immediately Denmark felt the need to do the same in case migrants on their way to Sweden through Demark would be stopped at the border. The free movement of peoples implied in the Schengen Agreement is now under pressure due to the large number of refugees from the Middle East into Europe.

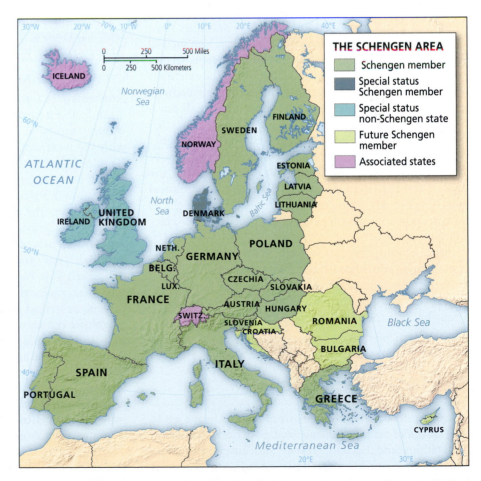

5.11 Schengen Area (photo: John Rennie Short)

Rural Focus

The Common Agricultural Policy

The **Common Agricultural Policy (CAP)** was first introduced in 1962 between the six founding members of the EEC. It was created after the searing experience of wartime food shortages and was designed to ensure food supplies at reasonable prices. Food prices were fixed while duties and quotas limited the supply of food from outside the EEC. Subsidies were paid directly to landowners. The subsidy system was initially based on the area of land grown rather than on the total crop produced. This subsidization of agriculture through price support and direct subsidies led to food surpluses, especially for butter, grain, milk, olives, and wine. The huge surpluses were referred to as butter mountains, lakes of milk, and seas of wine.

The initial system did little to protect the environment, promote sustainable agriculture, or stop the steady rural-to-urban drift of population. Critics of the system also pointed to the freezing out of developing countries from the EU market, the oversupply, the artificially high food prices, the lack of promotion for a greener agriculture, and the bias toward large farmers and landholders. There was an uneven distribution of subsidies. More subsidies went to countries with less efficient agricultural sectors such as France and Spain compared to the more efficient sectors in the United Kingdom and the Netherlands. The extension of the EU involved more agriculturally dominated economies such as that of Poland; with 2 million small holders, it will put extra pressure on CAP funding. The CAP is responsible for 40 percent of the EU budget, roughly 57 billion euros for a sector employing only 5 percent of the population and constituting only 1.6 percent of the GDP of the EU. The CAP does little for nonagricultural rural development, an increasingly important oversight, as rural areas become less dependent on agricultural production for employment opportunities.

More recent proposals for reform are directed at promoting environmental sustainability, strengthening small active farmers, and capping subsidies so that there is less public subsidization of major economic organizations simply because they produce food items.

Urban Trends

Europe is one of the most urbanized regions in the world with traces of an urban fabric that goes back to classical civilizations and beyond. Towns and cities throughout Europe have a rich historical legacy of buildings and street plans across two millennia that provide a living text of architectural changes and fashion. Every stage of urban development is writ large in the European urban scene with some cities, such as Bruges, Tallinn, Florence, and Venice as living museums of urban glories of the past.

There are a variety of urban forms in Europe: small villages that were centers of agricultural regions, ecclesiastical centers, places of learning, merchant cities, royal capitals, industrial cities, and deindustrializing cities. We will discuss four important forms and examples of each.

The Merchant City

The merchant city was a vital cog in local, regional, and global trades of flows. The earliest merchant cities were close to the coast and in some cases were the basis for commercial maritime empires in the Mediterranean and Baltic. Cities such as Tallinn were important ports for the regional trading systems of the **Hanseatic League** in the North Sea and the Baltic. The commercial enterprise of the **Venetian Empire**, for example, spread its tentacles through the Adriatic, eastern Mediterranean, and into the Black Sea and helped to create merchant cities such as Venice, Hvar, Korcula, Spilt, and Kotor. When global trade shifted to the New World, these Baltic and Mediterranean trading empires shrunk in size and importance as the center of commercial gravity shifted to the Atlantic and important trading centers were established along the long Atlantic coast and cities such as Amsterdam rose to prominence.

CITY FOCUS: AMSTERDAM

Amsterdam is a good example of a merchant city (Figure 5.12). It first developed as a pilgrimage town in the fourteenth century and then as a commercial city trading with the merchant cities of the Hanseatic League in the Baltic and the North. The **Dutch Golden Age**, with Amsterdam at its heart, began in 1578 when the Protestant Dutch broke away from Spanish control to become an independent country. The city was then free to develop as a vital node in the growing global economy. Merchants from Amsterdam helped to establish trading colonies in North America (New Amsterdam), South Africa (Cape Town), and South East Asia (Batavia now Jakarta). The city and the country were tolerant and so attracted merchants and traders, as well as English religious dissidents. It grew as a banking center, a center for diamond trade—a position it still holds—and a major trading port that connected Europe with the much wider world. The city expanded from its medieval core as successive canals ringed the city.

The Golden Age ended around the close of the seventeenth century; wars with England and then Britain and growing competition with Britain and France led to Amsterdam losing its global dominance.

Amsterdam retains a vast legacy of the Golden Age in its built form. Elegant canals sweep around in a half moon from the central railway station. Like many cities in Europe, urban renewal plans threatened this fabric but were resisted by activists and citizens in the 1960s and 1970s. Amsterdam kept its architectural fabric, and its progressive politics ensured that urban public space and accommodation in the city center were not simply the reserve of the rich. The city is not without problems; almost half the city's population has non-Dutch parents, ethnic segregation is clear, and unemployment is highest for minority groups in the city. Amsterdam, with a current city population of 810,000, combines a historical legacy with a living vitality as ordinary people continue to live and work in one of Europe's more tolerant, progressive, and beautiful cities.

5.12 Amsterdam (photo: John Rennie Short)

The Industrial City

During the Industrial Revolution beginning around 1800 in Britain, and picking up pace across western and northern Europe by 1900, new urban centers developed close to power sources such as water and coal and raw material sources such as iron and minerals. Some such as Manchester, England, grew as textile centers, others such as the Rhine Ruhr regions in Germany and Bilbao in Spain as centers of heavy engineering. The industrial city grew quickly and created problems of poor living conditions, public health scares, and environmental deterioration so marked that countermovements of public health reform, urban planning, and environmental improvement emerged that sought to manage the burgeoning cities. The planned city, the counterpart to the rapidly growing cities of the nineteenth and twentieth centuries, is an important facet of European urbanism. Industrial urbanism was concentrated in only a few regions of Europe, but their dynamic economies attracted population and so the gradual shift of the population from predominantly rural to urban that first started in the United Kingdom in the middle of the nineteenth century gradually spread throughout Europe. Many industrial cities are now declining due to deindustrialization.

Capital Cities

In a number of European countries, the largest city is also the capital city. In Austria, France, and the United Kingdom, for example, Vienna, Paris, and London dwarf the next largest cities. In some cases the growth of capital European cities was also part of an imperial enterprise as the cities like London were a capital not only of the United Kingdom but also of a vast empire that stretched around the world. The museums and banks of Copenhagen, Madrid, Paris, and London are filled with the treasure of a world incorporated into European dominance, control, and appropriation.

The Postsocialist City

Across much of what used to be termed "Eastern Europe," the legacy of Communist Party rule from 1945 to 1989 is evident in the cities. Land markets were nationalized, housing was socialized, and emphasis was placed on industrial production, often at the expense of environmental quality. Grafted onto the pre-1945 urban fabric of major cities were large ceremonial spaces given over to public demonstrations of support for Communist Party rule. Massive high-rise housing projects were built at the edge of many cities. While some of the larger cities such as Prague have weathered the transition to a market economy, other cities, especially those with a narrow industrial base, have experienced massive deindustrialization, population loss, and have legacies of extensive environmental pollution.

CITY FOCUS: WARSAW

Consider the case of Warsaw, the capital of Poland that has an urban population of 1.7 million and a metro population of 2.6 million. The city has gone through numerous transformations. The Nazis systematically destroyed the city, laying utter waste to over 85 percent of the city. Between the end of World War II and 1962, the old town was carefully recreated from the ruins in an act of architectural memorialization that embodied Polish resilience

5.13 The Palace of Culture and Science, Warsaw, Poland (photo: John Rennie Short)

and expressed Polish identity. The many marvelous examples of reassembled baroque and neoclassical architecture remind us of Warsaw's long connection with Europe.

The city emerged from almost 50 years of communist rule in 1989. Warsaw's communist legacy is still evident. The Palace of Culture and Science (formerly Joseph Stalin Palace of Culture and Science) was the tallest building in Europe until 1957 and still dominates the skyline in the city center (Figure 5.13). An exuberant example of **Soviet**

Realism, it was completed in 1955. There are also the blocks of former public housing, modernist monoliths scattered throughout Warsaw.

The command economy still lives on in an unclear and uncertain land market and the legacy of degraded and polluted urban sites. Large blocks of land lie undeveloped and undercapitalized, as investors are unsure of the property rights and legal claims. But there is also tremendous growth, especially at the edges of the city as the interstices between the modernist blocks are filled in with new housing and commerce and retail. One distinct feature is the number of gated communities. Guards and gates mark off many developments in both the central city and in the suburbs.

Like many postsocialist cities, Warsaw has an infrastructure deficit, a huge housing shortage creating an overpriced housing market, with a degraded urban environment in some places and feverish new investments in others.

The city is still very Polish, one of the few examples of a European city that over the twentieth century became more homogenous. Almost a half million Jews lived in this city prior to 1939. Now there are little more than 15,000. And yet to walk along the main pedestrian thoroughfare of Nowy Swiat is to see Starbucks and Subway, Colombian coffee houses, and Indian restaurants. Cellphone usage is endemic and universal. Poland has such a poor landline network that consumers jumped more quickly to cellphone usage. Poland joined the EU in 2004, and EU funding projects are ubiquitous throughout the city— from refurbishing seventeenth-century palaces to providing school playgrounds.

The city occupies the difficult space between command and market economies, socialist and capitalist ideologies, authoritarian and democratic systems of government, and a city still reconnecting to Europe's core.

Eurometro

Cities in Europe's core have coalesced into a major metropolitan region that links London and the southeast of England, through Paris and the Low Countries, western Germany, and Milan in northern Italy. It is the area of highest population density and highest GDP, comprising the economic and political core of the EU. It is one vast metropolitan region, and we can call it Eurometro (Figure 5.14).

Eurometro includes most cities in the EU with more than 200,000 inhabitants, and it includes over 100 million people. High-speed train connections, good roads, and cheap air travel link this metropolitan region of Europe

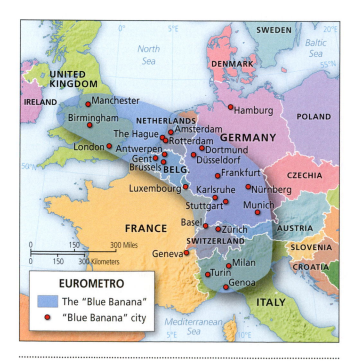

EUROMETRO
- The "Blue Banana"
- "Blue Banana" city

5.14 Eurometro

in a cohesive urban network. The bigger cities house the knowledge-intensive industries and contain the highest paid sectors, leading to growing income inequality between growing and declining sectors within the same metropolitan areas. Outside of Eurometro many regions face issues of declining population, job loss, and lack of infrastructural investment.

Geopolitics

European Union

The political geography of Europe is now dominated by the supranational EU, an economic and political alliance of most European countries. We can identify the old core of Belgium, France, Italy, Luxembourg, the Netherlands, and West Germany that formed the European Economic Community (EEC) by the Treaty of Rome signed in 1957. An important aim was to tie former combatants into an alliance that would preclude further conflicts and create a common market that could compete against the United States. By 1968, internal tariffs were removed and market integration improved. These six countries, the old core, constituted the bedrock of support for European integration. Other countries sought to join as the arrangement seemed to be working well. In 1961, Denmark, Ireland, Norway, and the United Kingdom applied to join. France vetoed the UK bid because of the United Kingdom's close ties with the United States.

European integration widened and deepened as more countries joined and as economies and political systems were meshed into a supra European system. In 1973, Denmark, Ireland, and the United Kingdom joined. In 1981, Greece was admitted, and Spain and Portugal joined in 1986. In 1993, the EU was created as a single market with the free movement of goods, people, and capital and common policies for agriculture, transport, and trade. In 1995, Austria, Finland, and Sweden joined. After the fall of the communist bloc in Eastern and Central Europe, many of the newly independent countries of the former Soviet bloc were admitted. By 2014, the EU had extended as far east as Cyprus and Bulgaria and stretched south to north from Malta to Finland. Under

consideration for entry are Albania, Iceland, Macedonia, Montenegro, Serbia, and Turkey. Bosnia and Herzegovina and Kosovo are recognized as potential candidates. However, the popular appetite for further expansion of the EU has lessened, especially into Eastern Europe, the Balkans, and Turkey as the possibility of millions of poor immigrants from these countries moving into the old core countries raises fears. The EU is now at a point of inflexion where the previous decades of continual expansion are coming up against popular resistance to EU enlargement into poorer and more peripheral countries.

Today the EU has a combined population of over 500 million. With between 20 and 25 percent of global GDP, it constitutes one of the largest single economies in the world.

The EU had proved enormously successful in integrating former enemies into trading partners and dissolved much of the mistrust that beset European affairs before the union was established. There are, however, a number of issues. Four in particular stand out.

The first is that there are marked regional differences in economic development that hamper national and EU integration (Figure 5.15). Three different types of regions within the EU were identified to guide EU investments from 2014 to 2020. More developed regions have a GDP per capita of more

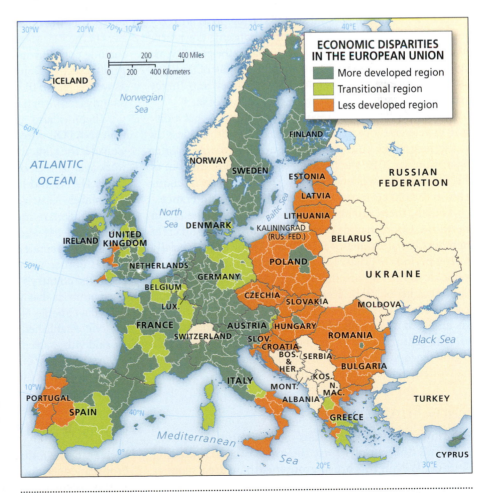

5.15 Regional disparities in the European Union

than 90 percent of the EU average. Transition regions share a GDP per capita between 75 and 90 percent, while the less developed regions have less than 75 percent. Note the concentration of more developed regions along the Eurometro corridor and in Scandinavia. Transition regions tend to be in the rural, more peripheral countries of Europe, while the less developed regions are concentrated in eastern and southern Europe. This map is used to allocate the regional policy funds that account for one-third of the total EU budget.

The second is that as the EU has widened beyond the relative homogeneity of the old core, newer entrants often have weaker economies and lower social welfare payments. The disparity generates the potential for population movement that worries some people in the old core. When the economy falters and growth slows, then the recent immigrant can become a source of disquiet and distrust. Across the richer parts of the EU, there is less support for further extension into Turkey or for bailouts of weaker economies. The Greek financial crisis from 2009 to 2016, for example, was caused by the use of the euro as a single currency across disparate economies. Greece was able to borrow at cheap rates, similar to economically frugal Germany but without the economic growth rates or the fiscal discipline to pay the loans when the economy faltered. A national economic crisis became a banking crisis for the EU, or at least the EU members that were part of the euro currency zone. Differences subsumed during periods of economic growth are exacerbated during economic downturns when the limits to integration are exposed. In the past 10 years a greater skepticism has emerged about the ability of the EU to successfully extend further into Eastern Europe and Turkey to integrate countries such as Bulgaria and Romania.

Third, there are genuine popular concerns that a distant bureaucracy in Brussels and elite political class in Strasbourg make rules for all of Europe favoring greater expansion. In 2016, the British people voted to pull out of the EU. Similar sentiments are expressed even in countries in the old core such as the Netherlands and France.

Fourth, there are also limits to the role of the EU in the world. Coordinated activity between so many different members is made difficult at the best of times but especially during periods of crisis. The EU could not intervene to halt genocide in the Balkans during the 1990s. This made it obvious that genocide in a not-too-distant place. You could reach Bosnia from Brussels in a day's car ride. So the EU has emerged as an important common economic market, burying centuries-old rivalry but still searching for its role in global geopolitics as an independent force separate from the United States, Russia, or China.

Many Europes

According to the *Oxford Atlas of the World*, Europe includes all the countries to the west of and including Ukraine. Within this broad grouping, there are a number of different Europes.

There is the Europe of the EU that includes twenty-eight sovereign states (Figure 5.16). Almost all major European countries are members of this bloc, but not all. There is a non-EU Europe. Norway and Switzerland are two rich countries that are not part of the EU. Norway with its oil bounty sees little need in tying itself to a vast multinational enterprise that includes shaky economies, and it is resistant to the idea that its fishing waters would be open to EU fishermen. Switzerland has retained its stubborn independence from European integration and involvement for over a hundred years. However, Norway still has to abide by many EU regulations, and the Swiss banking system is still subject to EU rules and global banking regulations in order for it to do business.

At the edge of the old Russian/Soviet Empire, Belarus, Moldova, and Ukraine are still under the shadow of Russia, while the Balkan states of Albania, Bosnia and Herzegovina, Kosovo, Serbia, and Macedonia are still not part of a cohesive Europe. The legacies of war, conflict, and a politics that often lacks the transparency necessary for EU membership means that these countries may have to wait some time before they are part of an enlarged EU. The appetite within the EU for the incorporation of such fragile states has diminished.

Within the EU grouping there is a further distinction between members of the Eurozone that use the euro as currency and those that retain their national currency. The euro grouping includes eighteen of the twenty-nine members of the EU. Members of the EU but not in the Eurozone include core European countries resisting monetary integration such as Denmark, Sweden, and the United Kingdom as well as recent EU members still not fully integrated such as Bulgaria, Croatia, Poland, and Romania.

There is also the **North Atlantic Treaty Organization (NATO)**, a military alliance formed in 1949 between Canada, the United States, and European countries as a bulwark against the Soviet Union. There is some similarity between NATO and EU membership, but there are also some differences. Austria, Cyprus, Finland, Ireland, and Sweden, for example, are members of the EU but not members of NATO. Austria and Finland under long-standing postwar agreements remained neutral. In the case of Finland, the shared border with the once mighty Soviet Union influenced the country's alliance during the Cold War. There are also countries that are part of NATO but not the EU such as Albania, Croatia, Norway, and Turkey. Norway and Turkey are long-established alliance members reflecting their strategic and vulnerable location during the Cold War. The post-1990 extension of NATO into former Eastern European bloc countries such as the Baltic Republics, Bulgaria, Poland, and Romania is a source of friction between Russia and the West.

Nations and States

We can make a distinction between **nations** and **states**. From a contemporary perspective, much of Western Europe seems to combine nations and states in a stable government container. There are rare examples of long-lived stability. Portugal has the most stable political boundaries and longest sense of national identity that stretches back for over nine hundred years. This is the exception. In fact, the boundaries of seemingly long-established and stable states such as the United Kingdom, France, and Spain are in fact the result of one nation incorporating other nations to form a single unitary state. France, for example, emerges from the control of the region centered on Paris to incorporate the south and west, over the years making the French of Paris the national language as Breton and Oc became marginal languages. Spain emerged from the dominance of Castile, centered on Madrid incorporating Andalucía, Cataluña, and the Basque country; this incorporation is still unfinished as Catalan independence and separatist movements remain strong. The United Kingdom grew out of an England centered on London incorporating by force and negotiation the north of England, Scotland, Ireland, and Wales into a United Kingdom. The 2014 referendum reminded outsiders that while the country may be a Kingdom, the United is sometimes more problematic. States in Europe are often made up of different nations.

The tensions between nations within states are not a settled matter. The case of Scotland, the more fractious case of Northern Ireland, Cataluña in Spain, or the regional rivalries between northern and southern Italy, and the continuing division in Belgium between the Flemish-speaking north and the French-speaking south are reminders that even the most settled of European states still have traces and living remnants of separate nations. The creation of the EU has promoted the emergence of these long-submerged nationalist sentiments because independence of a nation from a state, such as Scotland breaking away from the United Kingdom, Cataluña from Spain, or Belgium fracturing into north and south, allows these smaller units to separate from their state but still be part of unified Europe through their own membership in the EU. The EU provides the opportunity for nations to break away from their states without becoming totally irrelevant because they could remain part of a larger unified EU.

5.16 Nato and EU members

Connections

Europe is a pivotal region in world history and the contemporary global geography. It was a central hearth of the Renaissance and the Reformation, the setting for the **Scientific Revolution** and the Industrial Revolution, the crucible for modernity, the staging for overseas empires that transformed the world, and the source of world wars that remade the world. It remains one the richest areas of the world and its languages, religions, and customs are diffused across the globe. Europe's influence is global.

The Globalization of Europe

Europe has an outsized influence on world affairs. Since at least the fifteenth century, a number of different European powers became global actors. Portugal extended its reach to the Far East and South America. Spain, by 1600, was the world's first global power with an empire that stretched around the globe. In the seventeenth century, England, France, and the Netherlands extended their reach across the seas to Africa, Asia, and North and South America. By the nineteenth century, the British Empire unfurled the Union Jack across the globe, and its navy ruled the ocean waves.

There are numerous legacies of this European empire building and colonialism. There is language: Spanish is spoken throughout Central and South America. The national language of Brazil is Portuguese, the Afrikaans of South Africa is based on old Dutch, and English is the world's most dominant language in terms of its global usage and universality. There are sports: the British took rugby, soccer, and cricket with them, which explains why the best cricketing nations include Australia, India, Sri Lanka, and the West Indies, while the foremost rugby nations are New Zealand and South Africa. Then there is the legacy of cultural assets taken from their original setting and catalogued, explained, and displayed in European capitals. The British Museum and the Louvre are filled with treasures from around the world. Sometimes purchased and other times stolen, they represent a transfer of cultural assets from the rest of the world to Europe.

The world was also understood from a European perspective: the way we describe Iran and Iraq as Middle East is from the perspective of northwest Europe. There is also a deeper description of the world that measures the European as the standard—the summit—and the rest as somehow inferior, lacking, or still in the process of development. These attitudes are no longer so explicit or obvious, but they remain as a legacy of European intellectual domination and territorial appropriation.

There was also the globalization of European conflicts. When Britain and France competed for supremacy in the Seven Years War of 1754–1763, it was perhaps the first truly global war fought as it was in Europe, Central and North America, India, West Africa, and the Philippines. European conflict in the twentieth century mushroomed into world wars that engulfed the planet.

Subregions

There is no singular Europe. The geographic reality of Europe can be divided up in a number of different ways (Figure 5.17). Here I will use the EU as an important geopolitical lens with which to view the regional diversity.

The Core

The EU core consists of all those countries that were initial signatories to the 1957 Treaty of Rome. They include Belgium, France, (West) Germany, Italy, Luxembourg, and the Netherlands. They all had a traditional industrial base, more recent in the case of Italy, of heavy engineering and manufacturing. In the case of Germany, the unification of the East and West meant that the former West Germany inherited a much older manufacturing base that soon

evaporated with global competition. There is still a marked divide between East and West Germany in terms of living and environmental standards.

The core consists of a set of rich countries that are, compared to most countries in the world, democratic, with high levels of human development, low risk, low poverty rates, and a relatively high level of equality. They all share a similar democratic tradition with relatively high commitment to social welfare and limited military expenditures, although France still plays an important role in its former colonies in Africa. They share a more recent history of immigrations, from former colonies in the case of Belgium, the Netherlands, and France, as well as migration from Turkey and North Africa. This has changed the character of these countries, making them less homogenous and much more ethnically and religiously diverse. There is now a vigorous debate about identity and the limits to tolerance in some of the most tolerant and liberal countries in the world.

These countries have cohered into a monetary, political, and commercial union; however, there is also some divergence. Germany has emerged as the strongest economy built on an export industry of manufactured goods, and fiscal discipline. The other countries have had to navigate deindustrialization, an aging population, and recurring fiscal crises, with Italy perhaps the most politically and fiscally challenged of the original six, although even France is reconsidering its generous social welfare programs. The generational divide between protected workers and unemployed youth is made more stark when ethnic differences, such as in the high youth unemployment of Muslim youth in France and Netherlands, stoke social tensions.

Despite the problems, the countries in this region remain the heartland of one of the richest parts of the world and an important source of cultural leadership, economic innovation, and political tolerance.

The Inner Rings

There are three distinct parts of the inner ring.

NORDIC DEMOCRACIES

The first ring consists of the Nordic democracies of Denmark, Finland, Norway, and Sweden. Norway is not part of the EU but has a similar profile to the other three Nordic countries. All four are rich, with some of the highest per-capita incomes and the most generous social welfare programs on the planet. And despite growing inequality, they are marked by high levels of social welfare spending and social equality. Long homogenous, they have witnessed some immigration in recent years. Again some divergence: of the three EU members, only Finland is part of the Eurozone with both Sweden and Denmark remaining outside. Norway, because of its oil reserves, is now one of the richest per-capita countries in the world. Its government pension fund, known commonly as the oil fund, is

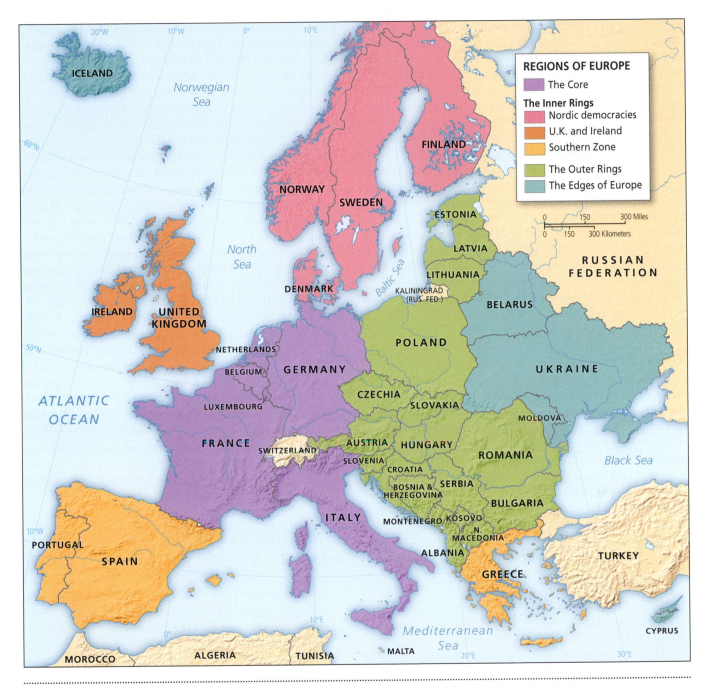

5.17 Map of Europe with core, inner rings (three types), and outer rings (different types) (see text for explanation).

conservatively valued at close to $1 trillion (in a country with just over 5 million people). It is the largest pension fund in the world that does not invest in tobacco companies, arms manufacturers, or companies with a poor environmental record. These four countries stand out as examples of social democracies at their best: affluent, fair, free, tolerant, and active members of a global community.

THE UNITED KINGDOM AND IRELAND

Another part of the inner ring is formed by the United Kingdom and Ireland. Befitting its position at the edge

of continental Europe, the United Kingdom is in a sense midway between the European core and North America. The United Kingdom has strong ties across the Atlantic, with a shared history and language and similar political cultures. Long separated from mainland European concerns, the United Kingdom has a more global and maritime character, looking across the seas rather than across the Channel. In recent years with privatization and neoliberal economics, it looks more like the United States than say Sweden or Germany. Public opinion in Britain runs ambiguous about Europe. It has long and strong ties further afield, and European integration was never that popular among the electorate. The United Kingdom is more marginal than mainstream to many EU deliberations, and its reluctance to fully commit undermines its position as an important EU member. It voted in 2016 to withdraw from the EU. It still has economic and cultural importance on the global stage. However, it is becoming a more unequal and divided nation as the rich southeast has a much higher living standard than the rest of the country. It is legitimate to think of the United Kingdom as an affluent metropolitan region of greater London and the rest of the country as a poorer periphery.

Ireland achieved rapid economic success soon after it became a member of the EU. The growth rate, based on foreign investment, was so large that it was known as the Celtic Tiger. Outmigration that had sent some of its best and brightest to the outside world was reversed. The financial crisis in 2008, caused by banks' overgenerous lending that created a property bubble, forced fiscal austerity programs. Growth has yet to achieve prerecession levels, and the talented continue to leave the island.

THE SOUTHERN ZONE

Another part of the inner ring is comprised of the southern zone of Portugal, Greece, Spain, and Malta. They have a Mediterranean climate, a reliance on tourism, and, apart from the former industrial regions of the Basque country and Barcelona region in Spain, a greater emphasis on agriculture. They are more recent members of the EU with Portugal and Spain joining in 1986, Greece in 1981, and Malta only joining in 2004. They are all members of the Eurozone, which has created problems when their domestic fiscal problems have spiraled out into wider Eurozone issues.

The Outer Rings

The outer ring is not only further from the core in simple distance but also in terms of political and economic differences.

There are different outer rings. Austria is least different from the inner-ring countries; long unaligned as part of a postwar settlement, it became part of the EU in 1995. It is similar in profile to the core EU countries.

Then, there is the new Europe of countries formerly part of the Soviet bloc. These include the Baltic republics of Estonia, Latvia, and Lithuania and the slice of eastern Europe formerly in the communist bloc: Bulgaria, Czech Republic, Hungary, Poland, Romania, Slovakia, and Slovenia. Bulgaria and Romania are less economically developed than the others. A massive deindustrialization resulted when inefficient state enterprises faced global competition after 1990. State enterprises either went out of business or privatized. In some cases the private market recoveries have been successful, but relatively high unemployment rates suggest that many more residents, especially the younger

ones, will move within the EU in search of better employment opportunities. Take the case of Romania. The country emerged from Ottoman control only in 1918 as a predominantly agrarian society with relatively late urbanization and industrialization. When communists took control in 1948, factories were nationalized and 76 percent of workers were employed in large state enterprises. A massive urbanization and industrialization in the 1960s saw cities expand and the construction of steel mills and factories making glass, garments, and trucks. After 1989 there was a massive deindustrialization. One industrial site in Bucharest employed 20,000 people in 1989 but only 400 by 2010. A truck factory in Brasov that employed 25,000 in 1990 had only 4,500 workers in 2010. Most Romanian cities peaked in population size in 1989, declining in population ever since.

Another part of the outer ring are candidate countries for EU membership: Montenegro, Serbia, FYRM (Former Yugoslav Republic of Macedonia), and Turkey. They are a mixed bag with Serbia still recovering from war and the sanctions that have limited economic development and Turkey that looks like a strong candidate for EU membership with a stable and large economy. However, growing European resistance to Turkish membership and distaste at a growing authoritarianism in the country place Turkey in a liminal zone of candidature status, unlikely to be able to join for years to come.

There is also a Balkan outer ring composed of potential EU candidates such as Albania, Bosnia, Herzegovina, and Kosovo that have yet to fully reassemble their economy and polity to EU standards.

The Edges of Europe

UKRAINE

Where does Europe end? The events of 2014 suggest that the border runs right through one country, Ukraine. With a population of 45 million, around 80 percent ethnic Ukrainian and 20 percent ethnic Russians, Ukraine sits on the borderland of the former Russian empire. It became an independent republic only in 1991 after years of being part of the USSR and centuries of Russification as Russians moved into the fertile land of eastern Ukraine.

The country is deeply divided between a Russian-speaking east and a pro-European, Ukrainian-speaking west (Figure 5.18). The division is reinforced by economic differences. Eastern Ukraine is mainly industrial with many Russian workers brought in to work the factories and industries. Western Ukraine is more rural and agricultural with a smaller proportion of ethnic Russian.

From 2004 to 2010, the government was led by the pro-western Viktor Yushchenko, who drew support from the west of the country. In 2010, the pro-Russian Viktor Yanukovych came to power. In 2014, social unrest exploded when Yanukovych decided against further integration into the EU orbit and instead signed a deal with

Russia that included $15 billion in a stimulus package and a one-third reduction in the cost of Russian natural gas. The western part of the country was the stage for mass protests. When riot police killed protestors, the unrest erupted and Yanukovych fled the country. A pro-western government was installed, but this raised the fears of Russian speakers in the east of the county. Pro-Russian military forces annexed the Crimea, and after a referendum, Crimea became a part of Russia. Russia continues to arm and support separatists in the eastern part of Ukraine.

The inability and unwillingness of the EU to do much about the deteriorating situation in Ukraine and the ability of Russia to annex Crimea and support the separatists suggests that Europe ends in the middle of the Ukraine where the furthest edge of Europe comes up against the strategic interests of a nearby Russia.

ICELAND

The most westerly part of the continent of Europe is Iceland. It sits on the very edge of two tectonic plates: the North American Plate moving westward and the Eurasian Plate moving eastward. The island also sits above a volcanic hotspot—similar to Hawaii and the Galápagos. This seismic location is the reason for Iceland's hot springs and volcanoes.

The island became an independent republic in 1944, breaking away from Danish control. It is a small country with only 327,000 people, most living in the capital city of Reykjavik. Integration with Europe increased after the country joined the European Economic Area in 1992 that gave it access to the EU. At the beginning of the twenty-first century, Iceland sought to restructure its economy to move away from boring cod fishing and to become a sexy financial center (Figure 5.19). Banks were privatized in 2003. The strategy was to move from being a country to becoming a hedge fund. Three Icelandic banks, with the largest being Kauopthing, took out $120 billion in short-term loans and then re-lent them to themselves, their friends, and their colleagues to purchase assets around the world. At their peak these three banks had paper assets worth more than ten times the country's GDP. In 2008, the bubble burst and institutions demanded their money back. The banks and the entire country were bankrupt. The crisis was so bad that drastic remedies were employed. The government coalition collapsed, banks were allowed to fail, currency was devalued, and strict capital controls were introduced. The country accepted an aid package from the International Monetary Fund. Iceland weathered the financial storm albeit with high levels of household, corporate,

5.18 The divisions of Ukraine

and government debt. Membership into the EU that was under serious discussion in 2010 was frozen in 2013. Sticking points in the negotiations were Icelandic concerns over its fisheries and lingering EU concerns with Iceland's fiscal issues. Iceland still remains, literally and metaphorically, at the edge of Europe.

CYPRUS

Cyprus sits on the opposite end of Europe to Iceland. It is not quite the farthest east as Ukraine fills that position (Figure 5.20). But it is far enough east that many data-collecting organizations and the *Oxford Atlas of the World* classify it as part of Asia. The island does straddle East and West with a complex history. An outpost of the Venetian trading empire, it came under Ottoman control from 1570 to 1878 when it became a British base in the eastern Mediterranean. More than 80 percent of the island's population are Greek

5.19 Fishing harbor in Stykkisholmur, Iceland (photo: John Rennie Short)

speakers, the remainder mostly Turkish. Most of the Greeks wanted to join Greece, while the Turks were fearful of joining with Greece. In 1960, the island became an independent state, but intercommunal violence and tension were an ever-present reality. In 1974, the Greek military junta sought to install a pro-Greek political leader. In response, the Turkish army invaded and took control of a third of the island in the north. People were relocated, 180,000 Greek Cypriots were evicted, and 50,000 Turkish Cypriots moved into Turkish held areas. The island is now divided between a Turkish Republic of Northern Cyprus recognized by no other country than Turkey and the Republic of Cyprus in the south (Figure 5.20). A UN Buffer Zone acts as a demilitarized area patrolled by UN peacekeeping forces. In recent years the wall that stood between the two areas in the main city of Nicosia was demolished, a hopeful sign of future reconciliation.

The Republic of Cyprus joined the EU in 2004. Like Iceland, Cyprus also faced a banking crisis. The two largest banks had attracted deposits from wealthy Russians and invested the money in Greek debt. The Greek debt crisis then led to the insolvency of the Cypriot banks and the inability of the government to meet its commitments. The country was bailed out by European financial organizations. In return, strong austerity measures were imposed and greater transparency was required. There was a feeling in EU circles that Cypriot banks were being used to launder Russian mob money and Russian **flight capital**. The bailout involved a levy on deposits of more than E100,000 ($129,600) to ensure that the rich Russian investors were not bailed out by European taxpayers.

In Europe but Not of Europe

There are also countries that are technically part of Europe but are not members of the EU, Eurozone, or NATO, or are likely to be in the foreseeable future. This includes Switzerland. It is a small, affluent country, but even its successful

banking sector cannot escape international regulation. Swiss banks were charged by the US government for allowing US citizens to hide their assets to avoid US taxes.

At the far outer edges are Belarus and Moldova and Ukraine. Part of the former Soviet bloc, they have yet to emerge fully into market economies and democratic societies. They occupy a zone far from the European core and very close to Russia. Belarus shares a border with Russia while Moldova has a pro-Russian breakaway state, Transnitria, within its borders.

Focus: The Regional Geography of Soccer in Spain

It goes by different names: football, futbol, and soccer. It is the most popular sport in Spain as in most European countries. The professional league, La Liga, is one of the best in the world, and the Spanish national team is regarded as one of the most talented in the world, winning the World Cup in 2010 and the European Cup in 2008 and 2012.

Soccer clubs were established in the late nineteenth century. Today the biggest teams are more than just professional sports teams; they are sites of national and regional identity, and the source of regional rivalries. Three major clubs are owned by club members rather than corporations and thus tend to reflect local sentiment more than placeless corporate interests. Athletic Bilbao is based in the Basque city (Figure 5.21). It was established by British steel workers and local students over a hundred years ago. It is an important source of Basque identity and has a formal policy of encouraging players from the Basque country. Recent squads included players such as Gorka Iraizoz, Benta Etxebarria, Xabier Etxeita, and Carlos Gurpegi, all proud Basque names. The red stripes of the club colors are suggestive of the Basque national flag. During the Franco era, from 1938 to 1977, Basque identity was crushed by the dictatorship so the team colors and Basque players became a potent form of resistance. In 2004–2005, the Basque government paid the club to have the word *Euskadi*, meaning "Basque," on their shorts. The club's contemporary away colors, of white, red, and green, reference the Basque national flag.

Barcelona (Barca to it loyal fans) is not only a football club in a major city; it is the national team of Cataluña. Founded in 1899, its motto of *Mes que un club* (More than a club) reflects its central role in modern Catalan identity. Its colors include the Catalan flag. The club has always had a difficult relationship with the Castilian central government based in Madrid. In 1925, fans booed the Spanish national anthem in protest against the governing authorities, and the football stadium was closed for 6 months. During the Spanish Civil War (1936–1939), many of its players, as well as those of Athletic Bilbao,

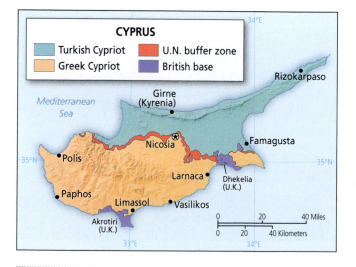

5.20 Map of Cyprus

5.21 Schoolchildren show their allegiance in Bilbao, Spain (photo: John Rennie Short)

signed up for the Republican cause and when Barcelona toured Mexico and the United States in 1937, half the team sought political asylum. In a story often told by present-day supporters, the team was visited at half time in a cup game against Real Madrid in 1943 and told by the Director of State Security to lose the game. Today Barca is one of the most successful and richest clubs in the world with a huge fan base, attracting some of the most gifted players in the world and winning trophies in Spain and Europe. The team's success is in part due to the local support of its fans down through the years that see in the club not just a professional sports team but a realization and embodiment of Catalan identity.

Barcelona's main rival is Real Madrid. Founded in 1902, its name means "Royal Madrid." It is the club most associated with the central Castilian authorities of Spain. It was given its royal appendage by King Alonso in 1920, and it was a favorite of Franco during his long dictatorship. It is the establishment team of Castille and national conservative Spain. The club, also one of the richest in the world, buys some of the best players in the world. When Real Madrid won the European Championship in 2014, the victory was all the sweeter because the success coincided with Barcelona's lack of success in that season. The league fixtures between Real Madrid and Barcelona are termed the El Clasico and are a reminder of the regional division and continuing rivalry between Castille and Cataluña.

Focus: Brexit

In 2016, a majority in the United Kingdom voted to leave the EU. A look at the geography of the vote provides insight into the political geography of the country (Figure 5.22).

There were three areas of support for remaining inside the EU.

First, a major area of support was centered on London. It is by far the most affluent part of Britain. London has emerged as a global financial center attracting expertise and investment from across the globe and around Europe. London is hard-wired into the financial circuits of the EU and global economy.

The rich and wealthy are concentrated in London and the South East, where household incomes are higher than the rest of the country. As the United Kingdom became a more unequal and divided society, the cleavage between London and the South East and the rest of the country became more marked. The rest of the country in England and Wales, and

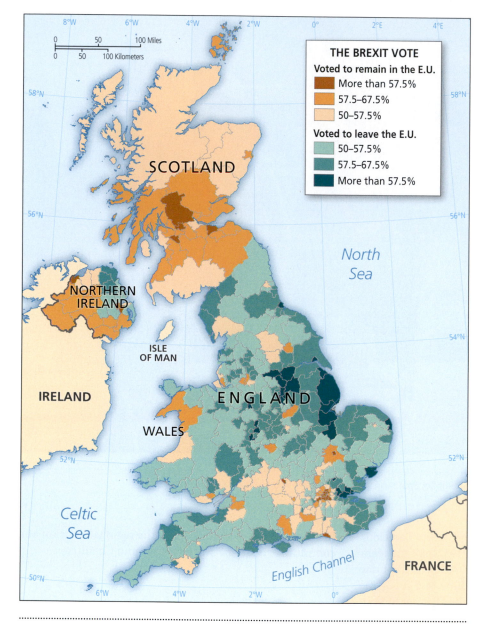

5.22 The Brexit vote

especially the lower income groups in those regions, was bypassed by the emphasis on the London money machine. The exit vote was also a vote against the dominance of London and the political establishment. The exit vote managed to capture popular resentment against the status quo just as Trump and Sanders did in the United States in 2016.

Second, all districts in Scotland voted to remain in the EU. This is a complete reversal to the 1975 referendum when Scotland voted against joining the EU, and much of the rest of the country embraced the European project. What happened? The EU provided benefits to Scotland that the London-biased UK government did not. Scotland's political culture, it turns out, is closer to the EU than that of London-based Tories. And years of punishing Thatcher rule soured many in Scotland against a reliance on the UK political system.

Third, Northern Ireland, as a whole, voted to remain in the EU. Northern Ireland is in a similar economic position to Scotland but in a very different political context. The province is split between those who want to join up with Eire and those who want to stay in the United Kingdom.

The Brexit vote revealed and embodied the deep divide in the United Kingdom between the different regions.

Select Bibliography

Albrechts, L., S. Hardy, M. Hart, and A. Katos, eds. 2013. *An Enlarged Europe: Regions in Competition.* London: Routledge.

Applebaum, A. 2012. *Iron Curtain: The Crushing of Eastern Europe, 1944–1956.* New York: Random.

Applebaum, A. 2015. *Between East and West: Across the Borderlands of Europe.* New York: Anchor.

Backman, C. R. 2008. *The Worlds of Medieval Europe.* New York: Oxford University Press.

Ballas, D., D. Dorling, and B. D. Henning. 2014. *The Social Atlas of Europe.* London: Policy Press.

Beatley, T. 2012. *Green Cities of Europe: Global Lessons on Green Urbanism.* Washington, DC: Island Press.

Benevelo, L. 1993. *The European City.* Oxford: Blackwell.

Black, J. 2013. *London: A History.* Lancaster, UK: Carnegie.

Blouet, B. W. 2012. *The EU and Neighbors: A Geography of Europe in the Modern World.* Hoboken, NJ: Wiley.

Buruma, I. 2006. *Murder in Amsterdam: Liberal Europe, Islam and the Limits of Tolerance.* New York: Penguin.

Cole, J. 2013. *A Geography of the European Union.* London: Routledge.

Crolley, L. 1997. "Real Madrid v Barcelona: The State against a Nation? The Changing Role of Football in Spain." *International Journal of Iberian Studies* 10:33–43.

Davies, N. 2014. *Europe: A History*. New York: Random House.

DeJean, J. 2014. *How Paris Became Paris*. London: Bloomsbury.

Diolaiuti, G. A., D. Maragno, C. D'Agata, C. Smiraglia, and D. Bocchiola. 2011. "Glacier Retreat and Climate Change: Documenting the Last 50 Years of Alpine Glacier History from Area and Geometry Changes of Dosde Piazzi Glaciers (Lombardy Alps, Italy)." *Progress in Physical Geography* 35:161–182.

Ford, R., and M. J. Goodwin. 2014. *Revolt on the Right: Explaining Support for the Radical Right in Britain*. London: Routledge.

Gorzelak, G., G. Maier, and G. Petrakos, eds. 2013. *Integration and Transition in Europe: The Economic Geography of Interaction*. London: Routledge.

Hill, B. 2012. *Understanding the Common Agricultural Policy*. Oxon, UK: Earthscan.

Hughes, R. 2007. *Barcelona: The Great Enchantress*. Washington, DC: National Geographic.

Jones, E. 2003. *The European Miracle: Environments, Economies, and Geopolitics in the History of Europe and Asia*. Cambridge: Cambridge University Press.

Judt, T. 2005. *Postwar: A History of Europe Since 1945*. New York: Penguin.

Ker-Lindsay, J. 2011. *The Cyprus Problem: What Everyone Needs to Know*. Oxford: Oxford University Press.

Komska, Y. 2015. *The Iron Curtain: The Cold War's Quiet Border*. Chicago: Chicago University Press.

Lees, A., and L. H. Lees. 2007. *Cities and the Making of Modern Europe, 1750–1914*. Cambridge: Cambridge University Press.

McCormick, J. 2013. *The European Union: Politics and Policies*. 5th ed. Boulder, CO: Westview Press.

Merriman, J. M. 2009. *A History of Modern Europe: From the Renaissance to the Present*. New York: W.W. Norton.

Murphy, A., T. G. Jordan-Bychkov, and B. B. Jordan. 2014. *The European Culture Area*. Lanham, MD: Rowman and Littlefield.

Pisani-Ferry, J. 2011. *The Euro Crisis and Its Aftermath*. New York: Oxford University Press.

Reid, A. 2000. *Borderland: A Journey through the History of Ukraine*. New York: Basic Books.

Shore, M. 2013. *The Taste of Ashes: The Afterlife of Totalitarianism in Eastern Europe*. New York: Broadway Books.

Szabó, G. R. 2013. "Basque Identity and Soccer." *Soccer & Society* 14:525–547.

Tsenkova, S., and Z. Nedović-Budić. 2006. *The Urban Mosaic of Post-socialist Europe*. Heidelberg: Physica-Verlag.

Vaczi, M. 2013. "'The Spanish Fury': A Political Geography of Soccer in Spain." *International Review for the Sociology of Sport*. http://footballperspectives.org/spanish-fury-political-geography-soccer-spain.

Walton, N., and J. Zielonka, eds. 2013. *The New Political Geography of Europe*. London: European Council on Foreign Relations. http://www.ecfr.eu/page/-/ECFR72_POLICY_REPORT_AW.pdf.

Wawro, G. 2014. *A Mad Catastrophe: The Outbreak of World War I and the Collapse of the Hapsburg Empire*. New York: Basic Books.

Weber, M. 1930. *The Protestant Ethic and the Spirit of Capitalism*. London: Butler and Tanner. https://archive.org/details/protestantethics00webe.

Learning Outcomes

Europe can be divided into a more Catholic southern region and a more Protestant north.

A division between Western and Eastern Europe marks a transition from the maritime-based empires of Britain, the Netherlands, Spain, and Portugal to land-based societies such as the Austro-Hungarian Empire.

The commercialization of agriculture as well as the Industrial Revolution had lasting effects on the economies and geographies of the region.

Europe has experienced periods of intense industrialization, subsequent deindustrialization, and then recent trends toward service sector development in what is called the postindustrial period.

Across much of Europe, employment conditions are some of the best in the world with relatively high wage rates, good conditions, and generous benefits such as unemployment payments, maternity leave, and good health provisions. However, there is high youth unemployment.

Europe is going through a second demographic transition in which falling birth rates and lengthening life expectancy are leading to older people becoming a larger proportion of the total population.

In recent years, relatively prosperous European countries have attracted a growing number of immigrants from less prosperous or secure surrounding countries, creating conflict and concern over the intercountry agreement of the Schengen Area.

Europe is one of the most urbanized regions in the world with a diverse range of urban forms.

Environmental contamination, especially in the form of brownfield, is spread across the older industrial areas of Europe.

The political geography of Europe is dominated by the supranational European Union (EU), an economic and political alliance of most European countries.

The EU is a source of economic growth and political stability but has run into problems associated with expansion beyond the relative homogeneity of the old core, a distant bureaucracy, and the limits of coordination among its member nations.

Europe and its members can be described in many ways, including, for example, those who are part of the EU, as well as those wanting to become members, those who comprise NATO, and those who abstain from either type of connection, such as Switzerland.

Europe has an outsized influence on world affairs resulting from its countries' historical dominance as global actors.

The geographic reality of Europe can be divided up in a number of different ways, including the rich core countries of Belgium, France, Germany, Italy, Luxembourg, and the Netherlands; the inner rings that include the Nordic Democracies, United Kingdom, and Ireland; and the Southern Zone of Portugal, Greece, Spain, and Malta.

There are a number of countries that are technically part of Europe but are not members of the EU, Eurozone, or NATO, nor are likely to be in the foreseeable future, such as Switzerland, Belarus, Moldova, and Ukraine, along with other smaller microstates.

Russia and Its Neighbors

I have named this region of the world "Russia and its neighbors." In this part of the world, Russia dominates, not only in terms of size and population but also in terms of historical importance and current geopolitical significance. Since 1989, geopolitical change has reordered this region. While there is a shift toward more market economies, the state still plays a huge economic role. There is still a heavy reliance on the export of primary products and especially oil and gas. Democratization has proved elusive.

The Environmental Context

A Sprawling Land Mass

The land mass of Russia stretches over 160 degrees of longitude, almost halfway around the world from 30 degrees West to 170 degrees East. Slow-moving rivers flow across a vast, low-lying plain, surrounded by mountains in the south and east. Movement was relatively easy across the vast grassland plain that stretches across much of the length of this land mass. The low topography allowed the Mongol horsemen to ride west and then later allowed the Russians to expand east. The northern aspect, however, and the lack of warm-water ports meant that naval power, until the Cold War, never matched the significance of land-based forces.

Much of Russia and its immediate neighbors sit on a single tectonic plate and so tectonic activity occurs at the edges of the giant plate. The Caucasus is an area of seismic activity. The Kamchatka peninsula in the Far East is part of the **Pacific Ring of Fire**.

The land mass is situated well to the north of the equator, so the climate is continental, with hot summers and bitterly cold winters. The average January temperature in Verkhoyansk in Russia is −50.4°F. It is so cold

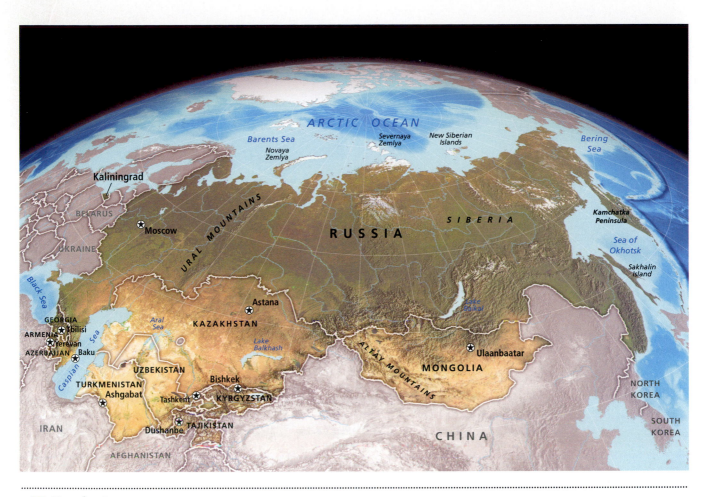

6.1 Map of region

that from the 1860s to 1900, political prisoners were sent there. Just being there was punishment enough. The city of Yakutsk, with a population of 200,000, is one of the coldest, bigger cities in the world. The average January high is −34°F; almost balmy compared to Verkhoyansk, but still very cold for most of us. There are some relative hotspots. The Crimea, for example, has a Mediterranean climate.

Four Ecological Zones

The vast plain, broken in the middle by the low-lying Ural Mountains, consists of four distinct latitudinal ecological zones, from north to south, tundra, taiga, steppe, and desert (Figure 6.2). The size of these regions is such that the first three are now known, around the world, by their Russian names. In the tundra, temperatures are too cold for tree growth, and migratory animals and birds arrive in the spring and leave in the fall before the harsh winter descends again (Figure 6.3). Then there is the taiga, a large belt of forest containing almost a third of the planet's trees; then grassland, called steppe, which allows for easy movement east and west, and in the south, desert, where rainfall

is too low to support a rich vegetation cover. The most expansive zone is the forest zone and especially the coniferous forest that covers almost a third of the land area. As you move south through this forest zone, the dense coniferous forest turns into mixed coniferous-deciduous and then deciduous forest.

The large size of these ecosystems is impressive, and that makes the consequences of climate change in this region all the more dramatic. The Russian Arctic is the warmest it has been for the past four hundred years. Thawing of the **permafrost**, soil that is frozen for at least 2 years, is leading to soil instability that threatens infrastructure such as power lines, pipelines, and building foundations. The warming permafrost also releases methane that has twenty-five times the warming capacity of carbon dioxide. Russia, with 60 percent of its land covered in permafrost, emits 32 million metric tons of methane per year or almost two-thirds of all methane emissions from permafrost regions. The draining and drying of the peat land is also creating fire hazards. Peat wildfires are increasing and in 2010, wildfires reduced the visibility in Moscow to less than 1,000 feet. The vast forest, long a

6.2 Ecological regions

6.3 Winter in Far East Russia (photo: John Rennie Short)

source of fuel, is being cut down, raising carbon dioxide levels and increasing the likelihood of even more pronounced global climate change.

The Warming Arctic

Russia's northern land boundary stretches along the Arctic. Frozen for much of the year, this region is now at the forefront of global climate change. Temperatures are rising faster in the Arctic than in any other region in the world, and the pack ice is shrinking. The rising temperatures are unlocking frozen seas, increasing the possibility of more navigable waters and more accessible resources. The North West Passage, long a chimera that attracted intrepid polar explorers, is now a reality, courtesy of global climate change.

It is 11,250 miles from Yokohama in Japan to Rotterdam using the Suez Canal, but only 7,350 miles by Russia's Northern Sea Route. In 2012, a Russian ice-breaker broke the now thinner ice to allow two German merchant ships to complete a commercial shipment across the Arctic. In 2013, seventy-one ships made the crossing. However, traveling through the Arctic Sea still requires ships with thick hulls and plenty of insurance.

Historical Geographies

The low-lying topography meant there were few physical limits to Russian expansion. And so Russia became less a bounded nation than a sprawling empire with shifting boundaries, uncertain limits, and no clear-cut sense of where it ended. The corollary was that Russia was vulnerable to invasion. This vulnerability, with the consequent need for compensatory expansion, influenced and continues to influence Russian geopolitical thought.

Russian Expansion

The most important element of the economic and political geography of this region is the history of Russian and then Soviet expansion (Figure 6.4). It began with defeat. In the thirteenth century, Mongol horsemen rode west. In 1240, they sacked the city of Kiev and ruled much of the area until the later fifteenth century. Their genetic legacy is evident today in the high cheekbones of many Russians. A proto-Russian state only slowly began to emerge in the space created by the decline of Mongol power and the collapse of the **Byzantine Empire**. From a small political unit centered on Moscow around 1400, Russians overthrew the Tatar yoke by 1480 and soon extended their reach across the vast plains all the way east to the Pacific, as far west as competing empires would allow, and south into the complex topography of the Caucasus and then up against the resistance of the Ottoman Empire and later the British Empire.

The expansion east was made relatively easily and quickly across the steppe by the lack of organized resistance. There was economic incentive to ensure better access to the lucrative fur trade. The most powerful state

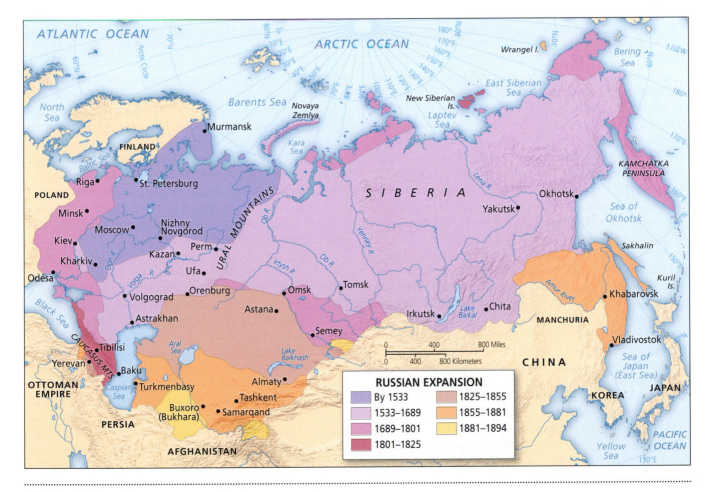

RUSSIAN EXPANSION

By 1533	1825–1855
1533–1689	1855–1881
1689–1801	1881–1894
1801–1825	

6.4 Russian Expansion

in the East, the Chinese empire had turned in on itself rather than projecting outward. By 1600, Russian control extended beyond the Urals east and south into the vast steppe. By 1650, forts were established all the way to the Pacific. Eventually Russian control stretched into what is now Alaska, which the Russians sold to the United States in 1867.

Expansion in the south and west encountered stronger resistance. Peter the Great (1672–1725) grabbed Estonia from Sweden in the Northern Wars and Azov from the Turks. St. Petersburg was established as a new capital on the Baltic, and land was taken from the Finns to make the new capital less vulnerable. In 1695, a campaign against the Tatars yielded Crimea. By 1800, Russia control extended westward into Poland and south into the Caucasus, Kazakhstan, and Turkestan.

The expansion was in part an economic mission: a search for furs and rare minerals. It was also partly geostrategic as imperial rivalries with European powers, the Ottoman Empire, and the Persian and local states fueled more annexations. The later expansion into central Asia, for example, was in part a countermove against the British moving north from their base in India. The Russians lacked ice-free ports along much of the continental land mass, and this became a strategic hindrance with the rise of naval powers such as the United Kingdom, then Japan, and the United States. The ice-free, Black Sea ports were restrictive choke points in times of conflict.

A state with long complex borders with many countries and rival empires is invariably drawn into conflict and war. Sir Richard Chancellor's remark that Russia "was a very large and spacious country, every way bounded by diverse nations," is as true today as when it was made in 1553. The more the Russians expanded, the more their borders lengthened and the more potential and actual conflict was generated. There was a tendency to **imperial overstretch** as the empire's borders continually expanded outward, each extension creating new zones of conflict and potential sites of rivalry.

There was also an ideology of empire: Mother Russia unifying all of Eurasia, bringer of civilization to the backward nations and protector of all Slavs and adherents to the Eastern Orthodoxy. Tsar Alexander III (reigned 1881–1894) vigorously pursued a policy of Russification, the active creation of one empire with one church (Russian Orthodox) and one language (Russian).

The expansion also involved the settling of native Russians throughout the Empire, creating Russian-speaking minorities in non-Russian-speaking lands. This reinforced the fact that Russia never developed as a nation-state with a stable coherent center, but emerged over the centuries as a constantly expansionist power with moveable borders and a fluid territorial identity. Russia was always more a territorial imaginary than an ethnic homeland.

But there were also setbacks: Russia was defeated by a combined French and British force in the Crimea War (1853–1856), and expansion in the Far East was halted by defeat in the 1904–1905 war with Japan and in the southern borders by British power emanating from the Indian subcontinent.

Despite the setbacks, Russian territorial expansion created a continental power. Russia, which had a landmass of only 9,000 square miles in 1462, had grown to 5.2 million square miles by 1914, covering over a sixth of the world's total land mass. This huge area touched on three major regions: Europe, Central Asia, and the Far East; it also shared borders with numerous countries.

The Soviet Empire

In 1917, everything changed. After popular protests, the Czar was forced to resign, and a provisional government was established in the then capital of St. Petersburg. Later in the year, in October, the Bolsheviks took control and established the world's first communist state. Russia became the Soviet Union. The national imaginary changed, but a global role was still imagined, this time as leader of a worldwide communist movement.

The early years of the USSR saw territorial losses as Poland broke away, taking western Ukraine and Belarus, as did Finland and the Baltic Republics. Romania annexed Bessarabia. However, the Soviets soon regrouped, invading parts of the former Tsarist Empire that tried to break away, including Armenia, Azerbaijan, Georgia, and Ukraine. In 1939, the USSR took back parts of Poland, and by 1940, in a pact with Hitler, took over the Baltic states and parts of Finland.

By the end of World War II, the Soviet Army had advanced as far west as Berlin. In the postwar settlement between the Allied Powers, the Soviets maintained control over large swathes of Eastern Europe (Figure 6.5). At its largest extent, the Soviet Empire, consisting of the USSR and client states, extended across Eurasia.

Soviet rule was soon established throughout most of the old Tsarist territories. Power was centralized in the Communist Party with very little room for a civil society. From 1936 on there was a policy of Russification, imposing the Russian language throughout the USSR and installing Russian political elites. The Soviets were never able to quench the independent sentiments in regions with a strong national identity. I was reminded of this in a poignant exhibit piece in a museum I visited in Tallinn, Estonia. It was made in a Siberian gulag around 1950. The political prisoner had fashioned a belt made of blue, black, and white electrical cable wiring, the colors of Estonia's national flag.

At the height of the Cold War, there were severe limits to the amount of autonomy enjoyed by the allies of the USSR in Eastern Europe. The Soviet Union sent troops into Hungary in 1956 and Czechoslovakia in 1968 to

6.5 The Soviet Empire

TABLE 6.1	The Relative Decline of Russia		
TERRITORY	**SHARE OF WORLD'S LAND SURFACE**	**SHARE OF WORLD'S POPULATION**	**SHARE OF WORLD GDP**
Russian Empire, 1913	17 percent	9.8 percent	9.4 percent
Russia today	13 percent	2.5 percent	1.6 percent

suppress political liberalization. After 1968, however, the grip was weakened. The political elite in Moscow realized that power could be more effectively wielded using local communist officials, encouraging ethnic identities, up to a point well short of political expression, and even the policy toward Eastern Europe thawed as Hungary embarked on market reforms in the late 1970s and early 1980s without any Soviet intervention.

In 1989, the Soviet bloc collapsed, and by 1991, new countries emerged from the jigsaw of former Soviet Republics. Most of them became independent states. However, there were several distinct ethnic regions, such as Chechnya and Dagestan, that remain part of Russia.

While Russia has shrunk in relative size from its greatest extent in 1913, it remains an important economic and military power (Table 6.1).

Economic Transformations: From Planned to Market Economies

Under the Soviet arrangement, most of the economic system was brought under state control. Private farms were collectivized and factories were nationalized. The most distinctive economic feature of this entire region is the shift from this centrally planned system to market economies. Central government remains a powerful force in the economy and especially in the large and wealthy energy sector.

This transition to market economies is not complete, and the state continues to play a large role in economic affairs; few countries have achieved open and fair market systems and corruption is rampant.

The Primary Sector

Russia and its near neighbors have long been important sites for natural resources, including oil, gas, and minerals. Under the Soviet system, all these were under state control. With the fall of the USSR, much of this resource base was privatized.

Russia is the largest exporter of refined petroleum and natural gas in the world and the second largest exporter of crude oil. Turkmenistan has the fourth largest natural gas reserves, while Kazakhstan and Uzbekistan all have large reserves of natural gas (Figure 6.6). Azerbaijan was long an important source of oil.

The privatization of these valuable resources since 1990 was the basis for the creation of the oligarchs, people who made immense fortunes. In Russia, thirty people own half of the national resources in the country, and income inequality widened so that by 2000 the richest person was 250,000 times richer than the poorest person. The privatization of large public utilities, state enterprises, and publicly owned assets provided the opportunity for quick and massive profits. The privatization of public assets was often marked by cronyism and rigging that aided the already well connected. Close ties between senior government officials, politicians, and major private companies meant that many privatizations did not result in market efficiencies but in personal enrichment.

The majority of Russian billionaires, now over a hundred, have as their original and major sources of wealth the privatization of the national resource base. Vagit Alekperov (net worth of US$13.9 billion) was a deputy minister in the Soviet oil industry. After 1989 he gained control of three ministry-controlled oil fields and established his own oil company, Lukoil. Roman Abramovich (US$13.4 billion net worth) owes his wealth to the privatization of the Soviet/Russian

public oil company that he acquired at a fraction of its real worth. The private company Yukos formed in the mid-1990s was a major oil producer until its owner Mikhail Khodorovsky began promoting democratic social movements. In 2003, he was arrested and the company was crushed with tax bills. The company was bankrupted and its assets given to two other major companies, Rosneft and Gazprom.

In the large and lucrative energy sector, the state retains a large measure of control. Gazprom, centered in Moscow, is one of the largest extractors of natural gas in the world. The Russian government owns a majority share, linking its policies closely to the geopolitical strategies of the Russian state. Similarly, the oil company Rosneft is also majority owned by the Russian government. Rosneft exploits oil in the Arctic shelf and the Black Sea in joint ventures with Exxon Mobil. Lukoil, in contrast, one of the largest oil companies formed from state enterprises, is privately owned and is listed on London's Stock Exchange. Conoco Philips owns 20 percent. Transneft, which runs the largest oil pipeline system in the world, is wholly state owned.

Agriculture

The collectivization of agriculture and large-scale modernization program was a central feature of the Soviet system. Especially after the famines of the late 1920s, there was an active policy of destroying the larger, private farms and organizing agriculture into state farms either as collectives of local villagers or as state enterprises. Both systems were wracked with inefficiencies, and the USSR had difficulty feeding most of its population. Shortfalls were made up with small-scale agriculture on very small plots in dachas (country homes and cottages) around the country. Almost a third of total food was grown and produced on these tiny, but highly efficient plots. In Russia in the post-Soviet era, the state farms were privatized but rather than a widespread allocation of land to many small-scale farmers, more than three-quarters of all land in cultivation is still run by the former state farms now reconstituted as private companies, often with former officials doing very well in the privatization process. This is particularly marked in the central Asian republics where large farms grow much of the cotton and grain. Private family farms play a small but increasing role in agricultural production. In Russia small private plots grow more than 83 percent of all potatoes, 70 percent of vegetables, and 40 percent of all meat and poultry.

Manufacturing

The Soviet system emphasized the creation of a manufacturing base. Large plants engaged in metal and machinery production employed thousands of workers. By 1989, more than 60 percent of USSR's industrial output was heavy machinery. Much of it was concentrated around Moscow, along the Volga and in what is now eastern Ukraine. There was less emphasis on light manufacturing and consumer

6.6 Oil and gas fields in Kazakhstan (Bernard Bisson/Getty Images)

goods. Since the fall of the USSR, many of the state firms were privatized often with the same result as in the energy sector, with a few well-connected people doing particularly well in the sale of public assets into private hands. These industries, often emerging from the inefficient system of central planning, are not well equipped to compete on global markets, especially against more efficient European, Japanese, and South Korean manufacturers.

Global trade in this part of the world is dominated by the export of natural resources such as oil and natural gas and the import of foodstuffs and the more sophisticated manufacturing products. As in much of the Western world, there has been a massive deindustrialization as industries have shrunk in size and many have gone out of business.

Deindustrialization hit the towns and cities of industrial regions very hard because huge plants employing thousands of workers dominated the sector. The resulting massive job loss in places with a lack of alternative jobs created major economic and social dislocation.

Pollution

Throughout much of the Soviet era, the emphasis was more on raising production levels than with longer term sustainability. Environmental standards were low and rarely enforced. The legacy of this narrowly economistic vision is an environment damaged by pollution and contamination. Nuclear contamination, and water and industrial pollution mark the environmental record of Russia and its neighbors.

Industries were geared up to maximize production and meet quotas rather than adhering to high standards of environmental protection. The result is that across this region of the world, there is a large number of contaminated sites, brownfields, and the detritus of the industrial age. Lake Baikal, for example, is one of the world's largest and deepest freshwater lakes situated in the taiga of southern Siberia. In the 1960s, the Baikalsk Paper and Pulp Mill was built on the shores of the magnificent lake and, until it closed in 2013, it pumped more than 6 million tons of toxic sludge into the lake, severely reducing water quality. Water quality was further degraded by pollution from the uncontrolled discharge of sewage from the increasing number of private homes built around the lake that did not have septic systems. There are plans to clean up the lake with federal money and to build sewage plants and water treatment plants. Government investment is vital in

order to stimulate tourism and recreation that would generate jobs and much needed revenue for the local area.

In Russia and its near neighbors, as elsewhere in the world, there is a recognition that environmental clean-up is not simply an ethical question of the need to clean up the toxic legacy of polluted and contaminated sites but also a matter of economics in providing jobs and new economic opportunities.

THE TRAGEDY OF THE ARAL SEA

There is no more poignant embodiment of environmental mismanagement than the Aral Sea. Situated between Uzbekistan and Kazakhstan, the sea originally covered 26,000 square miles and was once the fourth largest lake in the world. The water from the two main rivers that fed the sea, the Amu Darya and the Syr Darya, were diverted for irrigation projects in Uzbekistan and Turkmenistan. The water was used to grow cotton, but poor management meant that much of the water was lost before reaching the fields. The sea is now less than 10 percent of its original volume and is little more than a series of ponds (Figure 6.7). The once thriving fishing industry is destroyed, salinity levels have increased, and high winds whip up salt dust storms that shower the surrounding land in plumes of salt and dust. Fish and wildlife have disappeared. The land, exposed by the receding sea, is a toxic mixture of salt and chemicals, creating an unhealthy environment. High rates of cancer, infant mortality, and respiratory disease make this a dangerous and deadly place.

The shrinking of the Aral Sea is a particularly tragic case of environmental mismanagement and an extreme example of what happens when emphasis is placed on promoting short production without considering long-term sustainability. Attempts are being made to increase water levels. A dam was completed in 2005 and, 3 years later, water levels had risen and salinity had decreased. Perhaps the Aral Sea may return.

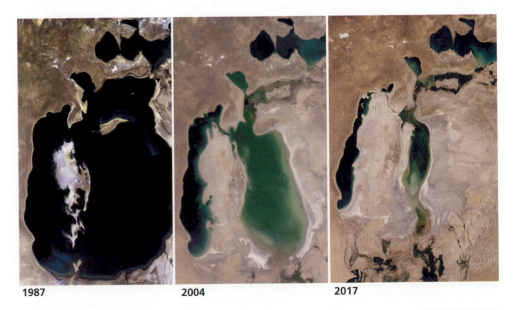

1987   2004   2017

6.7 The shrinking Aral Sea

Producer Services

Producer services such as finance, banking, law, and insurance are concentrated in the big cities and especially in Moscow and St. Petersburg. Both of these cities are important in the global flow of producer services. They are now more fully integrated into global flows with a consequent rise in the number of international firms and transnational corporations. A distinction can be made between the more dynamic economies of these two cities and the more restricted opportunities, especially for young people, in much of the rest of the region.

Eurasian Economic Union

On January 1, 2015, the Eurasian Economic Union (EEU) came into being. It is an organization for economic cooperation between Russia and former regions of the USSR's now independent states (Figure 6.8). The EEU was formulated by Russia as a way to link up former parts of the Soviet Union and as a counterweight to the EU. The combined population of the full EEU is around 183 million. It draws upon the model of the EU and includes a Eurasian Development Bank to promote the emergence of market economies and to have a common currency within 10 years.

The EEU is part of the coalition of economies into supranational organization to achieve economies of scale and economic heft in a world dominated by giant economies such as those of China, the United States, and the EU. It is also driven by Russian geopolitical strategies of reconstituting the economic might of the Soviet Union. Ukraine was envisioned as a key member until the anti-Russian protest led to the fall of the pro-Russian president in 2014. While the specific economic impacts are difficult to predict at this early stage, the current and proposed future members are likely to have considerable weight in global energy markets because of their rich natural gas and oil deposits. The hidden threat for the smaller members is Russian dominance and the possibility of a reemergence of the Russian Empire in a new economic guise.

Rural Focus

The Fergana Valley

The Fergana Valley stretches for 8,500 miles across three countries: Kyrgyzstan, Tajikistan, and Uzbekistan (Figure 6.9). It is a well-watered region traversed by two major rivers, Naryn and Kara Darya. The rivers allow cultivation in the warm dry climate. The main crops include wheat, cotton, rice, silk, vegetables, and fruit. The valley is a major source of food for Central Asia. It is one of the most populated parts of Central Asia.

While the surrounding area has an average population density of 40 persons per square mile, in the valley it is 1,600 persons per square mile.

In ancient times the valley was the site of grain cultivation and grape growing. Irrigation allowed the cultivation of a range of crops. Cotton cultivation on a massive scale was introduced under the Soviet system.

After the break-up of the Soviet Union, the Valley was divided among the three former Republics. The division has heightened ethnic conflicts between an Uzbek population stranded in Kyrgyzstan and the local Kyrgyz. The ethnic distribution is complex, undermining the neatness of national boundaries. In the valley there are Uzbek exclaves in the Kyrgyzstan, a Kyrgyz exclave in Uzbekistan, and two Tajik exclaves in Kyrgyzstan.

Ethnic violence erupted in 2010, and relations between Uzbekistan and Kyrgyzstan remain tense. Their conflicts over water and land reduce the agricultural efficiency of the region.

6.8 Eurasian Economic Union

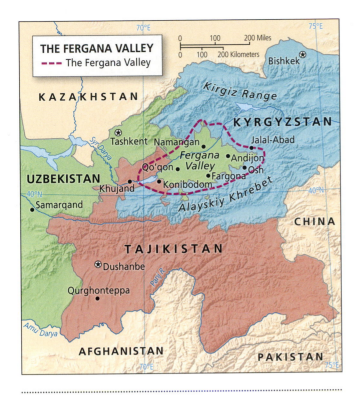

6.9 The Fergana Valley

Social Geographies

Population Movements

In this region, during both the Russian and Soviet Empires, peoples were relocated and forced to move, especially during and immediately after World War II when the Soviets doubted the loyalty of ethnic minorities. Consider the case of Crimea. From 1783 onward, Russians, Ukrainians, and Germans were settled in Crimea to weaken the local Crimean Tatar population, a Turkic-speaking Muslim group. In 1941, the entire German-speaking population of Crimea, almost 800,000 people, were deported to central Asia, and in 1944 most of the Tatar population of Crimea, 190,000 people, were deported in railway cars, half of them dying on the way. They were only allowed to return in the 1980s. In other wartime measures, the Soviets deported 93,000 Kalmyks to Siberia and the entire Chechen and Ingush population of the North Caucasus, almost 500,000 people, to what is now Uzbekistan and Siberia. When the Karelia region of Finland was ceded to the USSR in 1940, more than 400,000 Finns relocated to Finland. Relocations persisted even after the ending of the war. Three million German-speaking people of East Prussia were expelled to Germany. In 1947, around 125,000 Azerbaijanis were forced to move from the Armenia Socialist Republic to the Azerbaijan Socialist Republic. And in 1951, more than 575,000

Japanese and Koreans were removed from the Sakhalin region and the Kuril Islands.

Ethnic cleansings, migrations, forced relocation, and labor force transfers rearranged the population geography across the entire region. There are now some 25 million Russians living outside of Russia and 50 million former Soviet citizens living outside their home country. Today, there is also substantial economic migration as people from across the entire region move, sometimes permanently but often on a temporary basis, especially toward the employment opportunities in big cities such as St. Petersburg and Moscow.

A Multiethnic Russia

Russia has a multicultural, multiethnic character. There are at least 180 officially recognized ethnic groups that include the Russians, by far the largest group at 77 percent of the population, as well as numerous ethic language groups that range from the over 5 million Tatars to the 2,867 Uyghurs in Central Asia close to the border with China (Table 6.2). There are also indigenous peoples, who are listed as titular nations to autonomous areas, such as the Chechens, Karelians, Komi, and the Yakuts. Russia is considered the "title" nation, as they are the dominant ethnic and language group.

We can consider the complex social geographies in an examination of Central Asia and the Caucasus.

Ethnic and Language Groups in Central Asia

In this region there are three main ethnic/language groups: Mongols, Europeans, and Iranians (Persians). The combination of Mongols and Europeans produced the Turkic people who speak Turkish and include the Azeri, Turkmen, Turks, and Tatars. The combination of Mongols and Iranians (Persians) produced the Persian-speaking Tajiks. Then there is the combination of primary groupings with secondary ones. Turks and Mongols produced the Kazakhs, while Turks and Iranians produced the Uzbeks.

Azerbaijan, Kazakhstan, Kyrgyzstan, Turkmenistan, and Uzbekistan are mainly Turkic speaking, while Tajikistan is mainly Persian speaking. However, just to add to the linguistic complexity, Turkic and Persian speakers are found outside of their core countries, and significant numbers of Russian speakers are found in all these countries. And to add further complexity, the names of languages change as do alphabets. Consider Tajik, spoken by around 4.5 million people. The initial name of the language, Persian, was changed in the Stalinist era to break the link between Persian speakers in the Soviet Union and what was then Persia (now Iran). The Soviets wanted

TABLE 6.2 Ethnic/Language Groups in Russia

GROUP	POPULATION (MILLIONS)	LANGUAGE	MAIN AREAS
Russians	111.01	Indo-European	European Russia
Tatars	5.31	Turkic	European Russia
Ukranians	1.92	Indo-European	European Russia
Bashkirs	1.58	Turkic	European Russia
Chuvashs	1.43	Turkic	European Russia
Chechens	1.43	Caucasian	Caucasus
Armenians	1.18	Indo-European	Caucasus
Avars	0.91	Caucasian	Caucasus
Mordvins	0.74	Uralic	European Russia
Kazakhs	0.64	Turkic	Central Asian Russia
Azebaijainis	0.60	Turkic	Caucasus
Dargins	0.51	Caucasian	Caucasus

Only groups with more than 500,000 population are included.

to minimize any pro-Persian sentiment among Persian speakers inside the USSR. Before 1928 the language was written in a Persian-Arabic script. Between 1928 and 1940, it was written in the Latin alphabet (the one used by English speakers) and then, with the policy of Russification, in a modified Cyrillic alphabet. In 1989, the official policy was to write the language using an Arabic alphabet, but some still want the language written in a Latin alphabet. When even the letters used to write a language change, you can appreciate that this is a complex part of the world not only to understand but also to write and read about.

The legacy of Persian is still evident across the region in the use of the Persian word *istan* to refer to place or land.

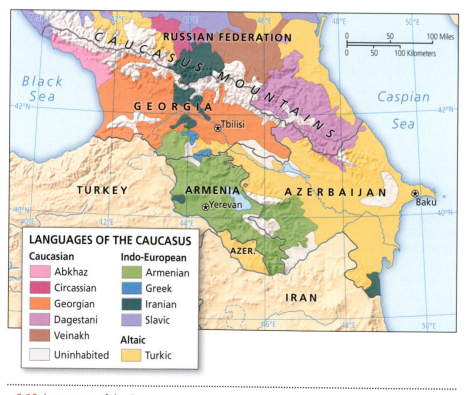

6.10 Languages of the Caucasus

Ethnic and Language Groups in the Caucasus

The Caucasus is a complex mosaic of ethnic and language groups (Figure 6.10). In terms of ethnicity, there are Altaic, Caucasian, and Indo-Europeans. There is a rich variety of language groups. The distinct Caucasian languages include Georgian, Dagestani (includes Agul and Avar), and Veininakh (Chechen and Ingush). There are also Armenian and Persian speakers (including Ossetians). States contain awkward combinations of different groups; Georgia, for example, contains Georgians as well as Ossetians and other Persian speakers. Russia contains all of the aforementioned groups.

Years of Russian and then Soviet rule have not dampened the urge for independence among some of the larger groups. The most dramatic example is the Chechen independence movement. The Chechens are a mountain people living in the Caucasus. They resisted domination in the long Russo-Caucasian war from 1816 to 1864 in alliance with the Avars in neighboring Dagestan. In 1917, Chechnya, along with Ingushetia and Dagestan, proclaimed independence from Russia. In 1921, Soviet troops attacked and their victory ensured that the region would remain under Soviet control. During World War II, the entire Chechen population was rounded up on orders from Stalin and shipped to Kazakhstan, accused of collaboration with the Germans. They were allowed to return only in 1956. In 1990, Chechen separatists declared independence and got their hands on Soviet weaponry. The Russian argument against their independence was that neither region, unlike Estonia or Georgia, was an independent entity in the USSR, independence would provoke other groups in the region to claim independence, and the region was too important to the Russian oil economy and to broader geopolitical strategy. In 1994, Russia reclaimed the territory by armed force. Russian forces regained control of the main city, Grozny, but at some costs. After Chechen rebels took over 1,600 hostages at a hospital in Russia, a truce was signed in 1996 and a brief period of a shadowy independence followed. Fearful of Islamic-inspired terrorism sweeping into other provinces in the region and in response to attacks in Russian cities, Russian troops entered the region again in 1999. The bulk of troops were only pulled out 10 years later in 2009, and there is an ongoing threat of insurgency and terrorist attacks.

Religious Revivals

There are two main religions in this region. The Russian Orthodox Church is the dominant religion in Russia, although as a whole the country is still marked by secularism, the legacy of Soviet rule. Initially part of the Eastern Roman Empire based in Constantinople, the Church emerged as an independent entity in the fifteenth century. The Church, always an embodiment of Russian national identity, had a complex relationship with the state. During reactionary times, the Church was embraced by the political elite; at other times, it was distanced, as with Peter the Great who pursued a vigorous modernization and brought the Church under government control. While the peasantry were persistent believers, the ruling elites oscillated between religious revival and criticism of what they saw as a backward religion that hobbled Russian modernization.

The Russian Orthodox Church diffused across Eurasia as the Russian Empire expanded but saw few converts in the western borderland where Catholicism remained strong or in the south rim where Islam remained important. Under the Soviet regime the Church was marginalized and rendered ineffective. However, in the past two decades there is a religious revival as it is now more explicitly associated with a Russian nationalism.

Along the southern rim of the old Russian Empire, Islam was the dominant religion. It had diffused out from the Middle East along the trade routes of the Silk Road. Later the Persian empires diffused Shia Islam into regions under its influence such as Tajikistan. Islam, like all religions under the Soviet rule, was marginalized and discouraged. In the post-Soviet world, there has been a revival of Islam in general, and in a few places more fundamentalist movements align with political movements such as in Dagestan and Chechnya.

Urban Trends

Long-Established Cities

Cities played and continue to play an important role in the economic and political life of the region. There are some very old cities. Bukhara, Samarkand, and Tashkent were key hubs on the Silk Road trade routes linking Europe and China. Later, as traders and merchants diffused Islam along the routes, they developed as Islamic religious centers. Samarkand was an established city by the fourth century BCE before it became a Russian city in 1868. Tashkent was an important trading city in the second century BCE. It was also an important Islamic center before it came under Russian control in 1865 and became the administrative capital of Russian Empire in central Asia. In the late nineteenth century, new radial streets and broad avenues were built in the style of Russian architecture.

In medieval Russian cities, the citadel or fort area was called the kremlin. In Moscow the Kremlin was built in 1156 on the banks of the Moskva River. Moscow became capital of the expanding Russian Empire as early as the fourteenth century. Long a center of Russian political life, it regained its role as capital after 1917. From 1712 to 1917, St. Petersburg was built as a brand new capital, located on the Baltic to connect more with Europe. Peter the Great thought that Moscow was too backward and wanted a new capital closer to Europe. Even to this day Moscow remains more Russian while St. Petersburg looks and sometimes feels more European.

Political power continues to play a defining role in city making. The brand new capital of Kazakhstan, Astana, was built on the windswept steppes on the orders of the country's authoritarian leader.

Soviet Urbanization

Under the Soviet regime, there was a conscious attempt at urbanization. Cities were built as administrative and industrial centers across the region. Soviet ideology promoted industry and cities as very important elements in building a modern society and an economy that could compete with the West. Money was poured into city building and industrialization.

The internal structure of the new cities drew their inspiration from the **Moscow Plan of 1935**. This blueprint for Soviet city structure stressed limited city size, the importance of ideological spaces in city centers, and state control of urban land and housing and limitation on population movements. In reality, away from the ceremonial centers drab new housing blocks were built on the edge of existing cities far from services, and there were few pollution controls.

Post-Soviet Urbanization

Since 1989, there has been marketization of urban and housing markets and greater segregation by income and class as new inequalities emerge. An urban car and mall culture similar to Europe has emerged. And many consumer services such as grocery stores and day care centers are now established in the peripheral housing estates.

Some of the industrial cities of the Soviet era, many built in the 1930s and 1950s, are now in vulnerable positions. In 2011, officials from the Russian city of Yekaterinburg submitted a nomination to the Bureau of International Exhibitions to host Expo 2020. It was part of a marketing campaign to rebrand the city. During the Soviet years, the city was called Sverdlovsk; it was a military-industrial city closed to foreigners with entry into and out of the city strictly controlled. It was a closed, supposedly secret city, hidden from international view and global scrutiny. The city's name change and Expo 2020 application embody a more general trend—the marketing of cities to achieve recognition, to change the internal and external image of the city, to connect to a global flow of positive urban images and upbeat urban imaginaries, and to find a new role in a postindustrial economy and a post-Soviet world.

In St. Petersburg, Russia, not only was the old Soviet name of Leningrad dropped, the city was rebranded as an entrepreneurial, economically competitive, globally connected city.

The larger cities now attract migrants from across the region. More than a fifth of Armenia's national income comes from Armenians living and working in Russia. Because Armenia is a member of the Eurasian Economic Union, Armenians find it relatively easy to work in Russia. They dominate construction and asphalt laying in Moscow. Armenian men are lured by the higher wages and better employment opportunities in Russia, creating a gender imbalance in villages and towns across the country as the employable men are siphoned off by the demand in Russia for labor. In many homes, extra strain is placed on women now made solely responsible for maintaining the home and children. More than a third of men have other families in Russia, draining away the amount of remittances sent back to families in Armenia. Many Armenian wives now follow their husbands to the cities of Russia.

There are also the urban consequences of the geopolitical fallout. In the larger cities across central Asian republics, there are substantial Slavic populations in central Asian cities such as Alma Ata, Dushanbe, and Tashkent that in the past were part of the USSR. Now Slavs are a minority in the new countries. Some cities ended up in other countries. Bukhara and Samarqand, long-established centers of Tajik culture and learning, are now part of Uzbekistan.

Pollution

Pollution remains a substantial problem for cities in the region. Many of the industries were developed with few pollution controls, as the emphasis was on growth and fulfillment of economic plans rather than on environmental quality. As in Europe, the result is a legacy of polluted cities and brownfields.

One of the most polluted cities in Russia is Norilsk. It was built by **gulag** labor north of the Arctic Circle as a mining center to exploit the nickel, palladium, platinum, copper, and cobalt. The smelters and mines produce a poisonous cocktail of heavy metal pollution that fouls the air and contaminates the soil.

The capital city of Mongolia, Ulaanbaatar, has the dubious record of air quality being one of the most polluted cities in the world (Figure 6.11). Between 1989 and 2006, the city's population doubled; today the city's population is just over 1 million. Currently 60 percent of the city's residents live in informal settlements in *gers*. The *ger* is a felted tent heated with inefficient coal-fueled stoves that use raw lignite coal or wood for heating and cooking. Ulaanbaatar has long and cold winters, and so heating is necessary for about 7–8 months of the year. The type of coal used is particularly soft and smoky. The city also suffers from emissions from soft coal–burning, older power plants and boilers, as well as numerous old cars with inefficient combustion. Unpaved roads also add dry dust to the air.

6.11 Air pollution in Ulaanbaatar (REUTERS/B. Rentsendorj)

The chokingly thick pollution that afflicts the city is a result of a combination of the combustion of soft coal, the congested road traffic, and windborne dust. Strong winds, particularly in spring, bring dust from the Gobi Desert to the city. Sand and dust storms, increasing as a result of desertification, add to the problem. With few trees and hardly any parkland in the city, the regularity and severity of windstorms in the city are increasing, creating dangerous levels of airborne dust. The air pollution is so bad, especially in the winter, that airplanes at the city airport are often unable to land or take off.

City Focus: St. Petersburg: What's in a Name?

Name changes can embody significant historical shifts. This is evident if we look at the history of St. Petersburg in Russia. The city went through four name changes. First, from 1709 to 1914, it was known as St. Petersburg, named by Peter the Great who extended the Russian Empire outward from its core around Moscow. Peter built a new city on the Baltic because he wanted to shift the capital from inland in Moscow to one with a window on Europe. It was part of an ambitious plan of modernization and imperial expansion. He proposed a beard tax, for example, to encourage a more modern clean-shaven look among his subjects.

St. Petersburg was built on piles driven into a marsh. It was a huge endeavor; it took over 5 years, and 25,000 workers died in the brutal conditions. From their bones arose a beautiful Enlightenment city laid out in a grid with grand avenues and beautiful buildings (Figure 6.12). A fortress, navy yards, and palaces were constructed, and in 1712 merchants and nobles were forced to move from Moscow. It became the capital of the Russian Empire. Tsarina Elizabeth and Catherine the Great expanded the city to be a center of science, culture, and art as well as the official residence of the royal family and the court. By 1800, the city had a

population of close to 220,000 and was the main intellectual center of the Russian Empire. Throughout the nineteenth century, the city also grew as an industrial city; the 1861 emancipation of the serfs increased rural migration to the city. The city was the major industrial city in the empire with 250,000 workers mostly in large factories—the ideal setting for political radicalization and the setting for some of the novels of the great Russian writers, including Dostoyevsky, Gogol, and Pushkin. In Dostoyevsky's novel *Crime and Punishment* (1866), the story of the murder of an old woman by the student Raskolnikov, the city is depicted as a place of great poverty and disorder. The vivid descriptions of the noise of the city reflect Raskolnikov's tortured mental state. The city was also a setting of social conflict and political uprisings. In 1905, a strike brought out 150,000 workers.

In 1914, the city's name was changed to Petrograd. The Russians were fighting against the Germans in World War I and wanted to remove the German taint implied in the previous name that employed the German words *Sankt* and *burg*. The name change inaugurated an intense period of political upheaval. In February 1917, a new provisional government was established. In March, the Tsar abdicated, and later that year in the October Revolution the Bolsheviks took control of the Winter Palace. A communist state was established. When Germans troops invaded Estonia, the communists felt the city was at risk, and in 1918, they moved the capital to Moscow, where it remains to this day.

In 1924, after the Soviet leader Lenin died, the city was renamed for a third time as Leningrad. By 1939, the city was a major industrial center with a population of over 3 million. During World War II, what the Russians term "The Great Patriotic War," the city was under siege from the German Army from 1941 to 1944. More than a million people starved to death and the city only regained its 1939 population in the 1960s. The city was largely rebuilt and historic renovation ensured that the rich baroque and neoclassical architectural legacy remained.

After the fall of the Soviet Union, the city was renamed in 1991 for a fourth time, returning to its original name of St. Petersburg. The name change signaled a self-conscious distancing from the Soviet past in erasing the name of the former leader. The city now has a population close to 5 million.

The beautiful architecture of imperial Russia is in the center of the city surrounded by housing for the workers (Figure 6.13). The striking contrast between the opulence of the Tsars and the life of ordinary people was a major reason behind the overthrow of the Romanoffs.

Today, almost 20 percent of the population is Muslim, mainly immigrants and their children from Central Asia. In 2014, a new mosque, inspired by Uzebeki designs, was completed. Today St. Petersburg is one of the most European and cosmopolitan of Russian cities, fulfilling Peter's great hope for a city that would link Russia to the outside world.

6.12 Imperial Palace St. Petersburg (photo: John Rennie Short)

6.13 Soviet and post-Soviet housing in St. Petersburg (photo: John Rennie Short)

Geopolitics

A New Political Geography

The break-up of the Soviet Empire into a multiethnic Russia and a complex patchwork of states resulted in a complex political geography of states with more than one nation, nations without states, intrastate rivalries, and interstate conflicts.

The break-up of the USSR after 1991 did not result in a precise demarcation of ethnic peoples as many Russians were now located outside of Russia in newly independent Republics, while many non-Russians, such as Armenians, Azerbaijanis, and Ukrainians, were now located inside Russia. The new boundaries, previously more porous and arbitrary, became solidified, locking people into distinct nation-states and creating border disputes. In 1991, half a million Uzbeks ended up in Kyrgyzstan and Tajikistan, and millions of Russians found themselves in new states. Almost 2 million Russians were now in an independent Kazakhstan.

There were also territorial disputes between the new countries. The densely populated fertile agricultural region of the Fergana Valley is a flashpoint between Kyrgyzstan, Tajikistan, and Uzbekistan. Kazakhstan, Kyrgyzstan, and Tajikistan inherited border disputes of the old USSR with China, although these were settled by 1999. By 2014, most of the disputes were settled and most countries, apart from Uzbekistan, now have demilitarized borders.

At the edges of Russia, especially in the west and southern borders, the population geography does not align with national borders. But the problem is not only with the awkward political geography of the "breakaway Republics" stranding peoples on the "wrong" side of a border that hardens over the years. Russia also faces issues of legitimacy in the Caucasus.

Russia's geopolitical role is in part a function of geography, a vast land with boundaries with many countries but also dependent on economic trends. But the price of a barrel of oil and its gas equivalent also plays a significant role in underwriting confidence and military strategy. In 1998, when a barrel of oil was only $10, the ability of the Russian government to maintain superpower military posture and tend to social programs was limited. By 2008, the price of a barrel of oil increased to $100 and enabled the Russian government, especially under Putin who came to power in 1999/2000, to embark on an ambitious program of military build-up and modernization. One result was the ability of the Russian government to effectively mobilize troops and irregulars in the standoff with Ukraine in 2004, Georgia in 2008, and with the annexation of Crimea in 2014. Government revenue from the energy sector gives Russia the confidence and the ability to adopt a more aggressive military posture.

Geopolitical Relations on the Western Edge

The three Baltic states of Estonia, Latvia, and Lithuania were all Soviet republics in the USSR, but they quickly sought and gained independence after the break-up. They all quickly followed a pro–Western European policy in order to distance themselves from Russia. In 2004, all three countries became full members of the EU and NATO, linking their economies and defense to a European future.

There are over 1 million ethnic Russians in the three countries whose combined population is 6.8 million. Lithuania has the fewest ethnic Russians or Russian speakers, although a third of the population of the port city of Klaipeda is Russian. But the situation is different in Latvia and Estonia where a quarter of the population is ethnically Russian. Almost a third of the Estonian population speaks Russian and especially in the east of the country, bordering Russia. More than 40 percent of the population of the capital city of Tallinn is Russian speaking, and more than 50% of the population of Riga, the capital of Latvia, is Russian speaking with heavy concentrations in the southeastern part of the country. The number of Russian speakers has declined since 1989 as Russians have moved back to Russia or overseas. However, the existence of a substantial number of ethnic Russians and Russian speakers means that there is still a strong Russian legacy in the Baltic republics.

Belarus and Ukraine are two countries that share a border with Russia yet have not followed the European path followed by the Balkan republics. Belarus, at the time of writing, is one of the last dictatorships in mainland Europe with stronger links to Russia than Europe. Belarus is neither a member of NATO nor the EU and is unlikely to qualify for membership. Although only 8 percent are ethnically Russian, almost three quarters of the population speak Russian at home, and Russia is one of the official languages of the state. Ukraine, as we have already noted in Chapter 5, in a sense is two countries, a pro-Russian east region and a pro-Western-speaking region in the west. The overthrow of the Russian-backed President of Ukraine in 2014 worried many Russian speakers in the east of the country who saw the shift to the West as a move away from their original homeland. Crimea, with a 60 percent Russian population, was annexed by Russia from Ukraine in 2014. The quarter million Tatars living in Crimea are now under Russian authority.

Tensions in the Caucasus

There are numerous independent states that emerged in the Caucasus in the post-Soviet world. They include Armenia, Azerbaijan, and Georgia. These were all part of the Soviet Union, though they did have some relative autonomy as Soviet Socialist Republics (SSRs). They have only small numbers of ethnic Russians, less than 2 percent of their total population.

There are also separatist movements in the Russian Caucasus. The Chechen separatist movement has morphed over the years. In the immediate break-up of the Soviet Union, the movement was mainly secular and nationalist. From 1999 on the movement was motivated by Islamic fundamentalism, and since 2007 it aligned with other groups in the area with the aim of establishing a Caucasus Emirate. The capital city of Grozny is the site of frequent terrorist attacks.

New States Emerge in Central Asia

In the aftermath of the break-up of the Soviet Union, a number of separate states were created in Central Asia. They include Kazakhstan, Kyrgyzstan, Tajikistan, Turkmenistan, and Uzbekistan. These were former SSRs that were themselves more a product of more recent Russian/Soviet cartographies than embodiments of long-held national sentiments. Take the case of Kyrgyzstan and Uzbekistan. Before 1924 these countries did not exist. The whole area was a complex multiethnic place. Annexed by the expanding Russian Empire in 1876, the Soviets created the administrative structure of separate Kyrgyz and Uzbek units in 1924. This division helped forge separate identities, although not separate economic units. In 1991, the two former regions of the USSR became separate states and an uncertain, arbitrary, and hazy distinction now became codified into two separate states sharing an international boundary. Tensions between the two states erupted in 1998 and resulted in the closing of the border.

The Far East

In the Far East, Russia has more stable borders with its neighbors. Russia shares borders with China, Korea, and the United States that have remained stable over the years (Figure 6.14). There were territorial disputes with Japan and one remains problematic.

Sakhalin is a large island off Russia's Far East coast. In direct response to Russian expeditions into the region, Japan in 1807, 1845, and again in 1865 declared sovereignty over the island. Russians settled the northern part of the island, digging coal mines and building towns. In 1905, in a peace treaty that halted a war between the two countries, the southern part reverted to Japan while Russia was given the northern part. After Japan's defeat in 1945, the entire island came under Russian control. More than 575,000 Japanese and Koreans were evacuated during the war or displaced after its end.

The Kuril Islands are a long chain of eleven islands that run from the Kamchatka peninsula of Russia to the Japanese island of Hokkaido. In 1875, Japan had control of the entire chain. During World War II, the Soviets secured the entire island chain in amphibious landings. In the San Francisco Peace Treaty of 1951, Japan was forced to renounce its claims on the islands. However, Japan did not recognize the four most southern islands as part of the chain, and Russia was not a signatory to the 1951 Treaty. In other words, there was just enough confusion to create a long-standing dispute with Japan claiming all four islands. In 2011 and 2013, Russia offered to give up the two smaller southernmost islands, but Japan refused. Both countries fight their case because there are important fishing areas and offshore gas and oil reserves. There are also strategic implications. Russia wants a Pacific presence while Japan does not want a Russia so close to its national territory.

6.14 The border between Russia and China stretches for 4,200 km. Here is the crossing at Manzhouli/Zabaykalsk. (jason_she/ Wikimedia Commons (CC BY-SA 3.0))

Connections

Russia and some immediate neighbors are very rich in natural resources. Kyrgyzstan, Turkmenistan, and Uzbekistan are rich in natural gas. Azerbaijan and Kazakhstan are rich in oil deposits.

Since the fall of the USSR, these resources are now more firmly connected to the global economy. One of the most obvious manifestations of these new economic connections is the pipelines that transport oil and gas. The plethora of pipelines led commentators to name the region "Pipelineistan."

Pipelineistan

There are pipelines for gas and oil. The natural gas pipelines from Russian and some central Asia republics to Europe go through Ukraine (three in total) and Belarus. Russia supplies EU countries with around 15 billion cubic feet a day of liquefied natural gas (Figure 6.15). Around 30 percent of Europe's total natural gas consumption is supplied by Russia. Two more pipelines are planned, one under the Baltic and another across the seabed of the Black Sea. Pipelines also carry gas from Central Asia through Russia. A proposed pipeline, the Trans Caspian, is planned to cross the Caspian Sea. Pipelines also take gas eastward. Under an agreement signed in 2014, three new pipelines will be added to the existing pipelines that link Russia, Turkmenistan, and China. Gazprom will supply China with 3.75 billion cubic feet of liquefied natural gas a day for 30 years starting in 2018. This represents almost a quarter of China's total current demand.

Oil pipelines link Russia and her near neighbors with Europe, China, and South Asia. The Druzhba pipeline is one of the longest oil pipelines in the world. It carries oil over 2,500 miles from Eastern Russia and Kazakhstan to Europe. Initially constructed by the USSR to supply oil to its allies in Eastern Europe, it is now extended to meet the huge demand from Western Europe. In Belarus the pipeline branches off with one link going into Germany through Poland while another snakes south through Ukraine and Hungary to the oil refinery at Omisalj on the coast of Croatia.

The infrastructure is expensive to build and maintain. It is estimated to cost $70 billion to build a new pipeline between Russia and China.

There is a political geography associated with these pipelines. Russia, as the main supplier of energy to Ukraine, has extra leverage in the relationship between the two countries. In 2014, Uzbekistan stopped exporting gas to southern Kyrgyzstan, preferring to export to China. It will take years and lots of money to build a pipeline from the gas reserves in northern Kyrgyzstan to the Fergana Valley in the southern part of the country. In retaliation, Kyrgyzstan may divert irrigation water destined for Uzbekistan.

Subregions

Although this part of the world has a shared history of Russian imperialism and Soviet occupation, we can also draw a distinction between four clear regional groupings: Russia, the Caucasus, the Central Asian Republics, and Mongolia.

Russia

Russia is one of the largest countries in the world and stretches from the Baltic to the Pacific Ocean and from the frozen Arctic to the almost subtropical climes of

6.15 Gas routes to Europe

the Black Sea. Its northern exposure means that little land is available for agriculture. Only around 7 percent of its land surface is arable land.

Russians use the term "title nation" to refer to the dominance of Russians and the Russian language in the Russian state. Russia is distinctive in that it is not only a state with more than one nation, though one dominates, but also a nation across numerous states as Russians are found in independent states that were previously republics in the former Soviet Union (Table 6.3).

Russia is a vast multicultural country finding a new role as it moves from a command to a market economy. It is marked by a weak civil society and depleted social capital. Its population of 142.5 million ranks in the top-ten countries by population. However, its low birth rates and comparatively high death rates, especially for men, mean it has a major demographic problem. Between 1991, the peak of Russia's population, and 2008, almost 12.5 million more Russians died than were born. Life expectancy decreased for men and did not improve much for women. Although Russia shares a situation with other countries of shrinking population, the collapse was not caused only by falling birth rates but also by declining health. Deaths from cardiovascular diseases and injuries are three times the rate in Western Europe. Immigration from the Caucasus and the Central Asian republics has offset this decline somewhat. As birth rates plummet, those aged over 65 years constitute an increasing proportion of the population, putting a strain on the welfare system. The aging has implications for long-term economic growth as Russia's global proportion of working age dropped from 2.4 percent in 2005 to 1.6 percent in 2015. There is also less manpower available for the large military establishment.

The demographic collapse that occurred between 1990 and 2008 was one of the unforeseen consequences of Russia's shock therapy of forced entry into the market economy with reduced government spending and a marked decline in the health and welfare benefits for the majority of Russians. After 2008 the decline slowed for the next 4 years, and since 2012 there has been an uptick in the birth rates as the economy has improved. Russia with roughly the same population as Japan now has twice as many births.

Russia has the unenvied position of sharing borders with fourteen other countries and with many Russians outside of Russia and many non-Russians inside of Russia. The borders can still be sources of geopolitical tensions, most severely exemplified in Ukraine and the Russian annexation of Crimea.

Russia with its vast nuclear arsenal and large military forces is still a force in world affairs, with not quite the global reach of the United States or the dynamic growth of China, but still with major world power status and the ability to project power, especially in neighboring regions, as events in Georgia and then Ukraine clearly demonstrated.

After the fall of communism, Russia did not adopt European-style socialism with generous welfare and greater equality but a free market casino system where a tiny minority benefitted most. From 2000 to 2014, average incomes increased with the commodities boom in oil and natural gas, but the country, less than 30 years from a socialist state, is now one of the most unequal societies in Europe. While it has the largest GDP per capita in the region, it is also one of the most unequal, ranking only after Turkmenistan. The oligarchs do well, while most ordinary Russians struggle to get by.

The vast revenues created by the oil boom allowed an expansion of the military payroll. But much of it also disappeared into a vortex of corruption. And if the price of oil falls, state revenues shrink. Economic problems are then often displaced by politicians into nationalist rhetoric that sees Russia as a victim of Western power and influence.

Russia has benefitted from its rich resource base. More than two-thirds of its revenue comes from the sale of oil and natural gas. With such a narrow export base, the country's economic health is dependent on the global demand for oil and gas. It has to import food and the more sophisticated manufacturing and consumer durables.

For centuries the main religion was Orthodox Christianity, but under the Soviets an explicit policy of secularization and anti-religious sentiment was pursued. There is now evidence of a religious revival, especially linked to a more overt expression of Russian nationalism (Figure 6.16). In the collapse of the Soviet Union, the Russian Orthodox Church reclaimed property from the state and also from schools and hospitals. The influx of money allowed a building program of new churches, and the Church hierarchy carefully allied themselves with the new economic and political elites. The Bible is now a mandatory subject of study in schools, and Russian Orthodoxy now melds with more strident forms of Russian nationalism.

TABLE 6.3	Ethnic Russians in the Former Soviet Republics
COUNTRY	**PERCENT ETHNIC RUSSIAN**
Russia	77.7
Latvia	26.2
Estonia	24.8
Kazakhstan	23.7
Ukraine	17.3
Kyrgyzstan	12.5
Belarus	8.3
Moldova	5.9
Lithuania	5.8
Uzbekistan	5.5
Turkmenistan	4.0

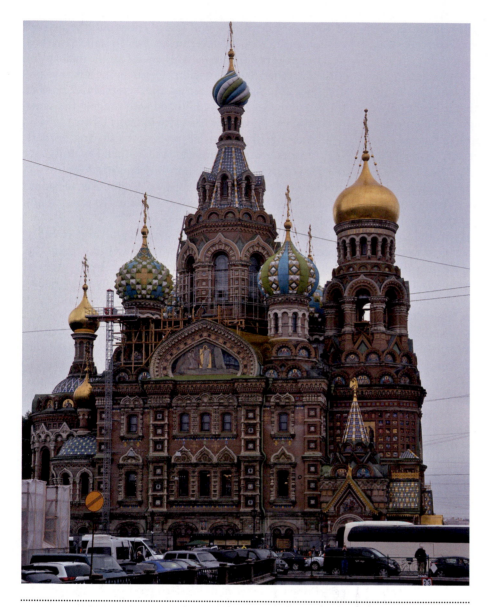

6.16 Church of the Saviour on Spilled Blood, St Petersburg, was built between 1883 and 1907. In contrast to the rest of the city, it was built in an overtly Russian nationalist style, echoed in current sentiments. (photo: John Rennie Short)

The Caucasus

The Caucasus, the name of a mountain range and the overall region, are complex both in terms of their geology and their geopolitics (Figure 6.17). The mountain chain is formed by the tectonic collision between the Arabian and Eurasian Plates. The complex topography of jagged peaks and hidden valleys has allowed a complex tapestry of ethnicities and languages—Ottoman, Persian, and Russian—to exist between the political equivalent of three tectonic plates. Complex topographies limit spatial integration and reinforce inward-looking localism. Local populations have long resisted imperial incorporation. While Russia still controls territory in the Caucasus, it struggles to retain the support of all the various ethnic groups within its borders,

especially in the regions of Chechnya and Dagestan.

Armenia, Azerbaijan, and Georgia were all SSRs in the former Soviet Union that became independent in 1991.

Armenia and Azerbaijan are close neighbors with different identities. Armenia is a predominantly Christian country and one of the first states, in 301 CE, to adopt Christianity as the official state religion. Armenians now constitute more than 98 percent of the population. Azerbaijan is predominantly a Turkic people who are Shia Muslim.

Tensions between the two erupted after independence from the old Soviet Union. In 1990, war broke out between them over the region of Nagorno-Karabakh. In the 1920s, the Soviets allocated the Armenian area of Nagorno-Karabakh to the Azerbaijan SSR. When the two countries became independent from the Soviet Union in 1991, they soon came to war over the disputed area. A war was fought between 1990 and 1994 with massive population displacement as 230,000 Armenians left Azerbaijan and 800,000 Azeris left Armenia. Although Nagorno-Karabakh, with around 140,000 living in 1,700 square miles, is de jure part of Azerbaijan, de facto it is now part of Armenia with a predominantly Armenian population as most Azerbaijanis have left or been forced out. In 1993, Turkey closed the border with the landlocked nation of Armenia in support of Azerbaijan. Armenia is now closer to Russia because of this conflict with Azerbaijan and Turkey. In 2016, conflict again erupted with shelling and border incursions.

For a number of years, Armenia straddled both East and West, being a member of the Russian-led military alliance as well as a participant in NATO. But Armenia is now closer to the Russian sphere of influence. There is a Russian military base in Armenia, Russia supplies Armenia with cheap gas, which is guaranteed until 2043, and it is the main supplier of arms and military equipment. Russia is considered an ally in the conflict with Azerbaijan.

Armenia's conflict with its neighbor Azerbaijan continues to hamper its economic development. A small, landlocked country hemmed in by less than friendly neighbors, and with few oil and gas reserves, Armenia struggles to move from the agribusiness of the Soviet era to efficient small-scale farming. Agriculture employs almost 45 percent of the entire labor force. It is a poor country with a small minority of wealthy influential people who control much of the government and the economy. Remittances from abroad are an important source of national income.

Azerbaijan is more fortunate as it has more gas and oil reserves. Pipelines link the country's oil fields to Europe and the rest of the world. Baku, the capital city, is one of the oldest oil cities in the world. Oil and gas revenue is neither evenly distributed nor invested in much-needed education or social welfare. The centralization of the old Soviet system remains, with private oligarchs replacing Soviet commissars.

Georgia was first annexed by Russia in 1801. After a brief period of independence from 1918 to 1921, it was brought under Soviet control. It gained independence in 1991. But there was a substantial Russian-speaking population especially in the region bordering Russia. In a process also seen in Ukraine, the election of a more Western-orientated leader in 2004, in a democratic movement known as the Rose Revolution, pointed the country westward rather than eastward. Two regions of the country broke away from Georgia. Russian troops invaded in 2008 and supported these breakaway regions of Abkhazia and South Ossetia. The rest of Georgia remains more pro-Western and seeks entry into the EU and NATO.

A reduced Georgia now looks to the West, hoping to join both the EU and NATO. It is hampered by the lack of oil and natural gas within its borders. It has shifted its supply of energy from Russia to Azerbaijan. Corruption and lack of tax revenues, endemic problems in this part of the world, hinder economic development. Foreign investors are reluctant to invest heavily because of the sensitivity

6.17 The Caucasus

of the geopolitical situation. The economy is still dominated by agriculture.

The Central Asian Republics

There are five "stans" in this region that are located in an area of continental climate dominated by steppe and desert (Figure 6.18). Only in the south, in the valleys, is

6.18 The Central Asian Republics

cropland agriculture economically feasible with cotton an important crop in the fertile valleys of Tajikistan and Uzbekistan. Their economies are dominated by the export of primary products, especially oil, gas, and minerals.

The region is beset with environmental problems, the legacy of Soviet weapons testing and uranium mining as well as poor agricultural practices. The salinization of much of the soil used to grow cotton is a major issue. The Aral Sea is an exemplar case of severe environmental degradation.

Under the USSR, Kazakhstan was used as a dumping ground for troublesome ethnic minorities and political prisoners. Almost a quarter of the population is ethnic Russians and more than two-thirds of the 17 million is ethnic Kazakhs whose absolute and relative numbers are increasing as Russians leave the country. Since independence there is a more active promotion of Kazakh culture and language. There are still close ties with Russia. Kazakhstan is a member of the Russian-dominated Eurasian Economic Union. Russia seeks to maintain its influence in Kazakhstan because the country is an important buffer with China and an important source of natural strategic resources, especially uranium and oil.

Kazakhstan is by far the largest of the "stans"; it is the world's ninth largest country, but it ranks sixty-third in population, an indication of the low population density across much of the steppes. Parts of the country's northern steppes were part of the **Virgin Lands Campaign** that arose from the severe food shortage facing the USSR in the 1950s. The project brought 30 million hectares under the plow. After a successful harvest in 1958, yields went down as the monoculture depleted the soil of nutrients.

The country has significant oil and gas reserves, and some of the vast revenues have filtered down to give it the second highest GDP per capita in the region, after Russia. It is an authoritarian regime ruled by the same leader since 1991. In 2015, he was elected to yet another period in office with 97.5 percent of the vote! A decline in oil prices, the major source of government revenue, may present difficulties for the regime and stoke ethnic tension between Kazakhs, mainly Muslims, and ethnic Russians.

Kyrgyzstan was made part of the Russian Empire in 1876 and became an independent state in 1991. The country has four Uzbek and two Tajikistan enclaves. It is one of the poorer countries in the region. A majority of the population is Kyrgyz, only 7 percent are Russian, although Russian is one along with Kyrgyz of the official languages. In the past 20 years the country has become more dominated by Kyrgyz as other ethnic groups have left in large numbers, responding to the political uncertainty, growing ethnic tensions, and poor economic performance.

Since the country became independent in 1991, a variety of movements have rocked the political scene. A "Tulip Revolution" in 2005 ousted the sitting president. Ethnic

unrest between Uzbeks and Kyrgyz worsened. Without large-scale reserves of oil and gas, landlocked far from sea routes, the economy of Kyrgyzstan is poorly developed. There were severe electricity shortages in 2010 that provided a constant reminder of the government's corruption and incompetence and created the basis for the Second Kyrgyz Revolution when a new government was installed. Ethnic tensions between Uzbeks and Kyrgyz increased, and 400,000 citizens were displaced. More than 40 percent of the country's GDP is in the form of remittances from people working in Russia.

In Tajikistan most people are Persian-speaking Tajiks, who came under Russian control in 1864, tried for independence in 1917, but were brought back under Soviet control. After it became an independent country in 1991, Tajikistan was wracked by a civil war between Russian- and Iranian-backed groups. More than 100,000 people were killed and 1.2 million were made refugees in a country with a population of 8 million. There was famine in 2001, and food security remains an issue in this poorest of the stans. It has a mainly Muslim population, but fundamentalist Islam has few natural holds. However, in the east of the country there have been outbreaks of Islamic militarism. There are more than 1 million migrants working abroad, many in Russia, and their remittances are a major source of national income. Bukhara and Samarqand, long-established centers of Tajik culture and learning, are now part of Uzbekistan.

The Russian Empire annexed the area that is now Turkmenistan in 1881. It gained independence in 1991, but the president for life (he died in 2006) was a former communist strongman who kept a tight rein on power. It is a despotic state with limited human rights. It is the most homogenous of the "stans" with most of the population composed of ethnic Turkmen who make up 85 percent of the population. Turkmenistan has huge reserves of natural gas. Pipelines transport the gas across the Caspian Sea to an energy-hungry Europe. It has a poor record of democratic accountability and human rights.

Uzbekistan is a Turkic-speaking country with a predominantly Sunni Muslim population that was part of the USSR from 1924 to 1991. It was a place where Soviet authorities dumped Chechens and Tatars. It is doubly landlocked, surrounded as it is by landlocked countries. It is a major cotton-growing region, a role assigned to it by Soviet central planners as well as an exporter of natural gas, minerals, and tobacco. The shift from a command to a market economy is slight, with the state retaining important controls. It is a form of police state with extensive state powers of arrest and imprisonment, little formal democracy, and an authoritarian regime that squats over its citizenry. Foreign investment is small. Uzbeks constitute 4 out of every 5 in the population, and a policy of Uzbekistanization further marginalizes the minority of Tajik speakers. Uzbeks also constitute the second largest ethnic group in

Tajikistan, Kyrgyzstan, and Turkmenistan. Tashkent is the largest city and capital, though it is more cosmopolitan than the rest of the country with significant numbers of Russian speakers and Tajiks. Also the city was an important hub of Islamic culture. The city sits in an active fault zone. In 1966, an earthquake destroyed almost 80,000 homes and made 300,000 homeless.

Mongolia

Mongolia counts as one of Russia's neighbors not only because of geographic proximity but also because of political connections for most of the twentieth century (Figure 6.19). It was a close ally, but unlike many of the other countries in this region, it has no history of Russian formal imperialism or Soviet occupation.

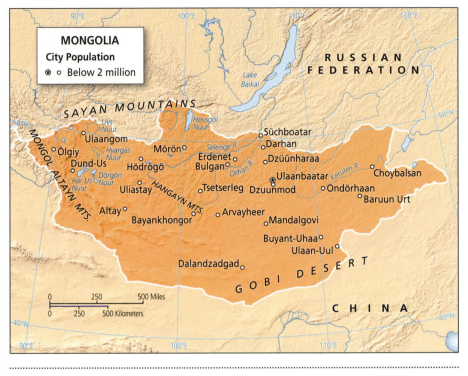

6.19 Mongolia

It is the world's largest landlocked country and is fully enclosed by China and Russia. It is semidesert and grassland with few permanent croplands and a continental climate of hot summers and severely cold winters. It was the center of a **Mongol Empire** that under Genghis Khan and Kublai Khan extended its reach east into China and far west into eastern Europe and Mesopotamia. With the collapse of this empire, it came under Chinese control. Inner Mongolia is still part of China, while what was once known as Outer Mongolia became independent in 1921, with Soviet help, and a communist state in 1924. It followed a pro-Society policy, even during the Sino Soviet split. The ex–Communist Party still holds considerable political power. The economy was traditionally based on pastoral farming, but the exploitation of rich mineral reserves has attracted foreign, especially Chinese investment. China is Mongolia's largest trading partner, although Russia supplies most of its energy needs. Landlocked and surrounded by Russia and China, it follows a pro-Western foreign policy. Ethnic Mongols constitute more than 95 percent of the country's population.

Desertification is a major issue in Mongolia. From 1970 to 2007, 887 rivers, 2,096 springs, and 1,166 lakes have dried up, turning more than a quarter of the country into desert. One culprit is climate change, which has increased the mean annual temperature and decreased the annual rainfall. Since 1960 the number of dust and sandstorms has increased between three and four times. Desertification accelerated by overgrazing, especially by goats, now affects 78 percent of Mongolia to varying degrees. The livestock population has increased just as the available land is being reduced. Desertification is most pronounced in the south and west, while pasture degradation is evident in the northern tier of the country.

Focus: Exclaves in the Post-Soviet World

Exclaves are political fragments of a state not physically connected to that state and surrounded by the territory of other states. They are found all over the world but have a particular potency in the aftermath of the break-up of the USSR into distinct nation-states.

Azerbaijan has four exclaves in Armenia, the largest being the Nakhchivan Autonomous Republic that covers 580 square miles and holds a population of 410,000. Tajikistan has two exclaves in Kyrgyzstan and one in Uzbekistan. Kyrgyzstan has an exclave in Uzbekistan, which in turn has four exclaves in Kyrgyzstan. One of them, Sokh, has a population of 70,000, and to add to the complexity, most of them are Tajiks.

One of the biggest is Kaliningrad, an exclave of Russia, situated far to the west of the main Russian territory and surrounded by the Baltic, Lithuania, and Poland (Figure 6.20). It is a part of Russia completely surrounded by EU and NATO countries. The exclave has a population of

6.20 Exclave of Kaliningrad

almost 450,000. For centuries it was part of East Prussia with a strong German influence. After World War I, it was an exclave of Germany. At the end of World War II, more than 2 million Germans left the region before the advance of the Red Army. When the region came under Soviet control, more Germans were evicted and the main city of Konigsberg was renamed Kaliningrad. After the war, the territory was divvied up among Lithuania, Poland, and the USSR. The Soviet Baltic fleet was headquartered in Kaliningrad. The strategic importance of the region—it contains three air bases as well as a major port—meant that after 1989 the Russians were not going to give up the territory and so it remains a part of Russia, an important naval port, surrounded by NATO and EU countries. In order to encourage economic activity, the Russian government reduced taxes and custom duties for manufactured goods in the region. It is now an important manufacturing hub, making televisions and cars. The region also trades with EU members.

The post-Soviet break-up into separate countries was not a clean tear along distinct ethnic and national lines, but a messy affair with minorities stranded in new states. Exclaves represent the realities of population geography compared to the sometimes arbitrary nature of national boundaries in the post-Soviet political fragmentation.

Select Bibliography

Abazov, R. 2008. *Palgrave Concise Historical Atlas of Central Asia*. London: Palgrave.

Alexievich, S. 2016. *Secondhand Time: The Last of the Soviets*. New York: Random House.

Applebaum, A. 2003. *Gulag: A History*. New York: Random House.

Aslund, A. 2012. *How Capitalism Was Built: The Transformation of Central and Eastern Europe, Russia, the Caucasus and Central Asia*. Cambridge: Cambridge University Press.

Balmaceda, M. 2013. *Politics of Energy Dependency: Ukraine, Belarus, and Lithuania between Domestic Oligarchs and Russian Pressure, 1922–2010*. Toronto: University of Toronto Press.

Beckwith, C. I. 2011. *Empires of the Silk Road*. Princeton, NJ: Princeton University Press.

Black, L. 2004. *Russians in Alaska: 1732–1867*. Fairbanks: University of Alaska.

Blinnikov, M. S. 2011. *A Geography of Russia and Its Neighbors*. New York: Guilford Press.

Bobelian, M. 2009. *Children of Armenia: A Forgotten Genocide and the Century-Long Struggle for Justice*. New York: Simon & Schuster.

Buckler, J. A. 2005. *Mapping St. Petersburg: Imperial Text and City Shape*. Princeton, NJ: Princeton University Press.

Clover, C. 2016. *Black Wind, White Snow: The Rise of Russia's New Nationalism*. New Haven, CT: Yale University Press.

Crumley, M. L. 2013 *Sowing Market Reforms: The Internationalization of Russian Agriculture*. New York: Palgrave Macmillan.

Cummings, S. N. 2013. *Understanding Central Asia: Politics and Contested Transformations*. London: Routledge.

de Waal, T. 2010. *The Caucuses: An Introduction*. Oxford: Oxford University Press.

de Waal, T. 2013. *Black Garden: Armenia and Azerbaijan through Peace and War*. New York: NYU Press.

Eberstadt, N. 2011. "The Dying Bear: Russia's Demographic Disaster." *Foreign Affairs* 90:95–108.

Ellman, M. 2006. *Russia's Oil and Natural Gas: Bonanza or Curse?* London: Anthem Press.

Feifer, G. 2014. *Russia: The People behind the Power*. New York: Hachette.

Frankopan, P. 2016. *The Silk Roads: A New History of the World*. New York: Knopf.

Galeotti, M. 2014. *Russia's Wars in Chechnya 1994–2009*. Oxford: Osprey.

Golden, P. B. 2011. *Central Asia in World History*. Oxford: Oxford University Press.

Golubchikov, O. 2010. "World-City-Entrepreneurialism: Globalist Imaginaries, Neoliberal Geographies, and the Production of New St Petersburg." *Environment and Planning A* 42:626–643.

Green, D. 2015. *Midnight in Siberia: A Train Journey through the Heart of Russia*. New York: W. W. Norton.

Gustafson, T. 2012. *Wheel of Fortune: The Battle for Oil and Power in Russia*. Cambridge, MA: Belknap Press.

Hiro, D. 2011. *Inside Central Asia*. New York: Overlook Duckworth.

Hoffman, D. E. 2003. *The Oligarchs: Wealth and Power in the New Russia*. New York: Public Affairs.

Hosking, G. 2012. *Russian History: A Very Short Introduction*. Oxford: Oxford University Press.

Josephson, P. R. 2014. *The Conquest of the Russian Arctic*. Cambridge, MA: Harvard University Press.

Kasekamp, A. 2010. *A History of the Baltic States*. Houndmills, UK: Palgrave Macmillan.

Kelly, C. 2014. *St Petersburg: Shadows of the Past*. New Haven, CT: Yale University Press.

Kotkin, S. 2001. *Armageddon Averted: The Soviet Collapse*. Oxford: Oxford University Press.

Lincoln, W. B. 1993. *The Conquest of a Continent: Siberia and the Russians*. New York: Random House.

Lucas, E. 2014. *The New Cold War: Putin's Russia and the Threat to the West*. New York: Palgrave Macmillan.

Myers, S. L. 2016. *The New Tsar: The Rise and Reign of Vladimir Putin*. New York: Vintage.

Pomfret, R. 2014. *The Economies of Central Asia*. Princeton, NJ: Princeton University Press.

Purs, A. 2012. *Baltic Facades: Estonia, Latvia and Lithuania Since 1945*. London: Reaction.

Rayfield, D. 2012. *Edge of Empires: A History of Georgia*. London: Reaction.

Riga, L. 2012. *The Bolsheviks and the Russian Empire*. Cambridge: Cambridge University Press.

Seegal, S. 2012. *Mapping Europe's Borderlands: Russian Cartography in the Age of Empire*. Chicago: University of Chicago Press.

Sunderland, W. 2006. *Taming the Wild Field: Colonization and Empire on the Russian Steppe*. Ithaca, NY: Cornell University Press.

Vatansever, A. 2017. "Is Russia Building Too Many Pipelines; Explaining Russia's Oil and Gas Export Strategy." *Energy Policy* 108:1–11.

Volkov, S. 2010. *St. Petersburg: A Cultural History*. New York: Free Press.

Learning Outcomes

Russia dominates this region.

There are four distinct climate zones of tundra, taiga, steppe, and desert.

Russia's border with the Arctic puts it at the forefront of global climate change but also in touch with a huge, mostly untapped supply of rich resources and huge commercial opportunities.

Russian expansion was partially abetted by its low-lying topography that made it easy to traverse but also left it susceptible to invasion.

Russia's historical expansion was in part an economic mission: a search for furs and rare minerals. But it was also partly geostrategic as rivalries with neighboring countries and empires fueled more annexations.

Under the Soviet system, most of the economic system was brought under state control. Now there is a shift to market economies.

Russia and its neighbors' economies have long been based on primary resources, as evidenced by Russia's status as the largest exporter of refined petroleum and natural gas and second largest exporter of crude oil, while its close neighbors also have vast reserves. The privatization of these resources led to the creation of the powerful oligarchs.

Manufacturing and industry were especially encouraged during the Soviet era. Producer services are dominant in the major cities.

The Eurasian Economic Union was created to counter the EU's power.

There are more than 180 ethnic groups, and numerous language groups, in this region.

Cities played and continue to play important roles in the economic and political life of the region. The Soviet Union emphasized urbanization; subsequent marketization has led to cities marked by segregation, new inequalities, and problems of pollution.

The break-up of the Soviet Empire into a multiethnic Russia and a complex patchwork of states resulted in a complex political geography of states with more than one nation, nations without states, intrastate rivalries, and interstate conflicts.

Along Russia's western border with Western Europe, there are three main types of states with millions of ethnic Russians: Finland; countries that emerged after 1991 to follow a pro-European path such as Estonia, Latvia, and Lithuania; and those that are in a more liminal position, such as Belarus and Ukraine.

Russia is a major power; however, its dependence on a narrow export base leaves it highly susceptible to world economic fluctuations.

The Caucuses, named after the prominent mountain range in the area, is made up of the former SSRs Armenia, Azerbaijan, and Georgia.

The Central Asian Republics of the region are made up of the five "stans" that are dominated by steppe and desert climates and the export of primary products, and are beset with environmental problems.

Mongolia is distinct from the other areas of the region due to its geographic differences. On the border with China it retains close ties to Russia.

7

East Asia

This is one of the fastest growing regions in the world. Beginning in the 1970s, first Japan and then South Korea and Taiwan emerged as major manufacturing centers in the global economy. Since the 1990s, China experienced spectacular rates of growth. Rapid industrialization and urbanization have transformed the environment, improved living standards, and changed social conditions. Population growth is leveling off as birth rates plummet. Selected cities in the region have achieved megacity status. China is now a powerhouse of the global economy, and it has emerged as a major world power.

LEARNING OBJECTIVES

Identify the region's four physical geography zones and relate each to its human settlements.

Discuss the influence of China's civilization on the region's culture and geopolitics and interpret the rise of the Japanese Empire in that context.

Summarize the region's rapid industrialization and role in the global shift and evaluate impacts to its environmental quality and agricultural activities.

Distinguish between the region's ethnic homogeneity and the presence of ethnic minorities and examine its diversity of demographic and religious patterns.

Outline the initial use of China's hukou system and connect it to its current use as a control for rural-to-urban migration.

Survey the rapid urbanization across the region and recognize resulting land supply problems.

Explain the intraregion conflicts that shape the geopolitics of land and seas.

The Environmental Context

There are four different zones of physical geography in this region. In the far west of China is the largest plateau in the world, the Plateau of Tibet, which extends over 1 million square miles at an average height of 13,000 feet. The Himalayas, the highest mountains in the world, rise up to over 26,000 feet on the plateau's southern edge, a visible reminder of tectonic forces as the Indo-Australian Plate crashes into the Eurasian Plate. This zone is the source of some of China's largest rivers.

In central China, there is a complex mosaic of basins and of hills between 5,000 to 10,000 feet high.

The peninsula of Korea and the island of Japan form another zone. Although both are mountainous with limited arable land, they differ in geologic age. While the Korean peninsula is a long established landscape, Japan is the result of the much more recent tectonic activity along the active boundary edge of the Pacific and Eurasian Plates. Japan has active volcanoes and sits on the edge of one of the most unstable plate boundaries in the world, part of the **Pacific Ring of Fire**.

The fourth zone is the riverine systems that traverse China from west to east.

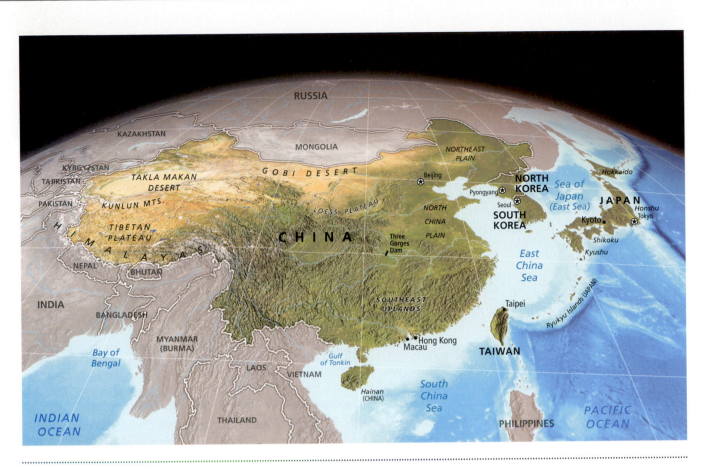

7.1 Map of region

Three River Basins: Hearths of Civilization

The riverine system consists of three giant river systems that create the rich farming lands of eastern China (Figure 7.2). The rich soils of these river drainage basins, managed and engineered for centuries with canals, banks, and dikes to maximize agricultural production, were the sites of the world's earliest civilizations and remain centers of concentrated population density.

Most of the population in China is concentrated in these fertile plains where farming has been practiced for over thirty centuries, evident in the careful terracing of the loess deposits in the north and the terraced rice paddies in the south. An intricate pattern of canals and ditches crisscrosses the land. A division can be made between the cooler wheat-producing areas of the north that also grow maize, millet, and potatoes, and the warmer and wetter rice-growing areas of the south, where a sophisticated aquaculture involves a complex mosaic of rice paddies, fishponds, and mulberry trees.

The longest river, the Chang Jiang (Yangtze), flows through the middle of China over 4,000 miles, all the way from the Himalayas to the sea, close to Shanghai. Its drainage basin covers almost one-fifth of China's land surface, and it is a major source of irrigation water and a vital east-to-west transport route connecting the interior to cities on the coast. The construction of the giant Three Gorges Dam was designed to lessen the possibility of flooding. There are plans to transfer water from this river system to the thirsty grain-growing regions to the north.

In the north of China, the second longest river, the Huang He (Yellow), deposits around 1,750 million tons of sediment each year. In its middle passage the river slices through the soft **loess**, soil from windblown deposits, creating such a heavy sediment load as to give it the English name of Yellow River.

In the south is the third largest river, Zhu Jiang (Pearl River). Because of the summer monsoon rains, it has a larger discharge than the Huang He. At its mouth, it joins with other rivers to form the **Pearl River Delta** region, one of China's industrial growth regions. The Pearl River Delta is now a **polycentric city** of eleven compact cities, including Hong Kong, Macau, Huizhou, and Guangzhou, linked by 2,500 miles of transport within a 1-hour travel time of each other. This metro region is projected to increase its population to 65 million by 2025.

7.2 Three river basins of China

The rivers systems that bring the bounty of soil fertility also deliver the threat of flooding. China's rivers are sources of tragedy as well as agricultural wealth. Despite the construction of complex irrigation systems and flood defenses, the mighty rivers regularly burst their banks after winter snowmelt and heavy spring rains and, in the south, torrential

monsoons. The Huang He River was nicknamed "China's Sorrow" for the amount of damage it created over the centuries. In 1887, more than 1 million people lost their lives when the water flooded over the dikes around the city of Zhengzhou. In 1991, flooding of a tributary river killed 2,000 people and made millions homeless. The disaster was one of the main reasons behind the construction of the Three Gorges dam, which opened in 2012, as a way to avoid future flooding and generate power for China's power-hungry industries. The project was controversial, with critics claiming that it was a very expensive project that destroyed an area of natural beauty and flooded arable land. More than 1.3 million people lost their homes in the project. Supporters of the scheme argue that the dam harnesses the river flow into electricity generation while leveling the seasonal flows of the river and lessening the risk of downstream flooding.

Climate and Weather

The arid steppes and deserts of the interior have a continental climate of hot summers and very cold winters. The climate is wetter closer to the coast. From north to south the climate shifts from temperate to subtropical and even tropical in the very far south on the island of Hainan. In the winter, cold air moves south from the north and west, while in the summer and especially in the subtropical south, the winds from the warm sea bring monsoonal rain and the risk of typhoon, the local name for what is referred to elsewhere as a hurricane.

Typhoons are driven by rising sea temperatures and are a late summer phenomena, from June to October across the region. They impact the east coast of China as well as Japan and Korea. A typical typhoon will begin near the Marian Islands, pick up energy as it moves north, driven by prevailing winds across warm seas. If large enough,

7.3 Track of Typhoon Usagi

it will punish the coastlines of China, Japan, and Korea with high winds, heavy flooding, and tidal surges. In 2013, the typhoon season was very active with Typhoon Usagi making landfall in Guangdong (Figure 7.3), Typhoon Man-yi in Japan, Typhoon Soulik in Taiwan, and Typhoon Dansas hitting both Japan and South Korea. Typhoon Usagi alone wreaked $4.6 billion worth of damage. Typhoons are regular, almost predictable occurrences, and now early good warning systems allow people to evacuate before the storm hits. Buildings and infrastructure, in contrast, remain fixed in place. The damage figures are only going to increase with more coastal development in the delta regions of Shanghai and Guangdong. The warming seas are also likely to increase the frequency of major storms and lengthen the typhoon season.

Seismic Activity

A major environmental hazard in this region is seismic activity along the border of the Pacific and Eurasian Plates. In 2008, an earthquake recording 7.9 on the Richter scale struck in Sichuan province only 50 miles from the 8 million population city of Chengdu. It was the strongest earthquake to hit China since 1950 and killed 87,000 people, injured 370,000, and made 5 million people homeless. One of the most heartbreaking features of this disaster was the collapse of a large number of poorly built schools and the resultant deaths of schoolchildren. The major cities of Shanghai and Tokyo with a combined population of 50 million are especially vulnerable to earthquakes. An earthquake hit Taiwan in 2016.

Underwater plate slippage creates **tsunamis**. On March 11, 2011, an earthquake on the ocean floor 230 miles northeast of Tokyo, Japan, registered 9.0 on the Richter scale. The resulting tsunami waves lashed over the coastline, eradicating towns and villages. More than 20,000 people were killed, and the total damage is estimated at $300 billion. The destructive tsunami waves crashed into the Fukushima Daiichi nuclear power plant. Contaminated steam was released in order to avoid a giant explosion. Subsequent explosions also resulted in radiation leakage that spread across 700 square miles and forced the evacuation of 100,000 people. The radioactivity released from the Daiichi nuclear power plant will last for many years, and Japan's optimistic reliance on nuclear power is now permanently shattered. All nuclear plants were closed as Japan rethought its energy policy.

Soil Erosion

Soil erosion is most pronounced in China. Overfarming, overgrazing, and deforestation expose the topsoil to erosion and, in northern China, to inundation of windblown desert sand. The exposed land is gullied by rain.

In the sparsely settled western region of China, desertification is a concerning issue while in the more populated east and center, the very rapid urbanization and industrialization

have meant increasing vulnerability to soil erosion and land degradation. Rapid industrial growth has increased the demand for timber and led to the clearance of large swathes of forest, contributing to soil erosion and desertification.

Historical Geographies

The Long Fall and Recent Rise of China

China is one of the world's centers of early civilization. The country exerted a huge influence on the surrounding area, extending its language and culture throughout the region. Even today Japan's script is based on Chinese characters. For centuries China also exacted tribute in the form of allegiance to the Emperor from political entities in Burma, Korea, parts of Japan, Thailand, and Vietnam.

China became China, as we know it today, under the **Qing Dynasty** (1644–1912 CE) when the approximate boundaries of contemporary China were created. In the first half of the Dynasty, the boundaries of the state were enlarged with annexations in Mongolia and Tibet. One of its most effective leaders was the Emperor Kangxi, who ruled for 61 years from 1661 to 1772. During most of his reign, China controlled one-third of the world's wealth. He quelled internal opposition, blocked Russian advances in the north, and expanded the empire to the northwest into Manchuria and Mongolia and controlled Taiwan and Tibet. By the turn of the nineteenth century, however, China was declining due to internal forces and external threats. There was a shift in the global balance of power and wealth away from China. The entry of the Western powers in association with China's decline led to what the Chinese refer to as a century of national humiliation that started with the war with Britain over opium. The Chinese wanted to restrict the import of the drug, because of its debilitating effects on the population, but the British wanted the profits from importing it from South Asia. The Chinese forces were no match in what became known as the **Opium War** (1839–1842). China was forced to "open" up its ports to European traders. China lost control over its national space as treaty ports were placed under effective European rather than Chinese sovereign control. Hong Kong was ceded to the British. The Taiping Rebellion, a nationalist backlash to foreign incursions and influence, cost the lives of 20 million Chinese. A second Opium War (1856–1860) had the same results. More Chinese ports were opened up to foreigners. China had to pay vast indemnities, and the European powers nibbled at China's territory. British surveyors in India extended Britain's imperial boundaries into China, and Russia annexed part of outer Manchuria to allow them access to the Pacific Ocean and to build Vladivostok.

After defeat in the Sino-Japanese War of 1894–1895, China had to cede control over Taiwan, the Pescadores Islands, and the Liadong peninsula in Manchuria. The treaty also forced China to open up four new treaty ports and to pay substantial reparations to Japan.

In 1910, the Qing Empire collapsed. Postimperial China was characterized as a "loose sheet of sand" as warlords fought for supremacy, a bitter civil war developed between nationalists and communists, and Japan annexed Manchuria and then invaded China. It was only in 1949 that some form of territorial integrity was established. After a shaky start, and a century of economic dislocation and political turmoil, a resurgent China only emerged after the economic reforms of the 1980s. Today, China is a global economic power and a regional military power that seeks to reclaim former territories such as Taiwan, still independent but not internationally recognized, and to assert its claims to South China Seas.

For outsiders it is crucial to remember that after a century of turmoil and weakness, China's political leaders see national stability and international standing as important issues. These are important not only for making up for a century when it was on the wrong end of asymmetrical relations with other powers and subject to dismemberment and annexations but also to reclaim its position as one of the world's preeminent centers of power and influence. Its newfound economic strength provides a strong base to make this claim.

Although China's economic growth is phenomenal, we should also recall its centuries-old political-cultural stability. Despite the vagaries of a century of disruption by outside powers and internal dissension, China is remarkable for the long-term stability of its national boundaries and for the persistence of the idea of China. It is perhaps best considered as a **civilization-state** with a history dating back two millennia. As a civilization-state, China has shared values centering on **Confucianism**, family values, shared language and food, and a long tradition of centralized hierarchical control. The Chinese term for nation-state can be translated as nation-family, a reminder that the state is seen less as an outside agent but more a defender and guardian of Chinese civilization.

The Legacy of Imperial Japan

In 1853, two ships under the command of US naval officer Matthew Perry sailed into Tokyo Bay. The Japanese had never before seen steam-powered steel ships. It marked the opening up of Japan to foreign influence and the end of a regime that promoted two centuries of seclusion. A new regime, known as the Meiji Restoration, embarked on a path of rapid modernization, industrialization, and militarization. Building a manufacturing base as quickly as possible was a central goal. Shipyards and factories were constructed, an iron and steel industry was quickly established, and a national railroad system was built. Japan relied on imports of raw materials. An industrial military economy was established that was the basis for Japan's military success against China, Korea, and Russia.

The reliance of its industrial growth on imported raw material was a major reason behind Japan's expansionist policies and its role in World War II (Figure 7.4). Japan moved quickly to assert its role in the region. It fought against China and Russia, eventually annexing Korea and Taiwan. Korea was a colony of Japan from 1910 to 1945. In 1931, Japan annexed Manchuria, setting up a puppet regime and in 1937 invaded China.

7.4 Japanese Empire, 1942

In 1941, it launched an attack on the United States and undertook an invasion of South East Asia. Overseas expansion was seen as a way to avoid economic stagnation, compete with the other major powers, and assert Japan's role in the wider world. The main victims were other countries in the region. In its war with China, the Japanese military killed 10 million Chinese. Approximately 300,000 were executed in one city over a 6-week period in what is known as the Rape of Nanking.

Japan's colonial legacy lives on in territorial disputes, such as the Diaoyu/Senkaku Islands between China and Japan and naming controversies. It is impossible to understand the East Sea/Sea of Japan controversy between South Korea and Japan, in which South Korea claims that the Sea of Japan should also be known as the East Sea, without an understanding of Korean resentment against Japan's colonial record in Korea or to explain China's reaction over the islands without reference to the collective memories of Japan's brutal war and atrocities against China. These memories are often stoked by nationalism. In 2014, a new holiday was celebrated in China, "Victory Day of the Chinese People's War of Resistance against Japanese Aggression."

Economic Transformations

In the past 50 years, East Asia was the scene of one of the quickest and most impressive shifts in economic fortunes. Countries that suffered from endemic famine have today become major hubs of a global economy.

Rapid Industrialization

At the heart of this transformation was a rapid industrialization and the global shift of manufacturing away from Western Europe and North America.

First Japan, then Taiwan, South Korea, and more latterly China underwent a rapid economic transformation that grew their economies, transformed their societies, and impacted the world.

Japan's second round of industrialization began after its defeat in World War II, when a new industrial order was established often with US advisors eager to try out new methods of industrial organization and production. More expensive, technologically advanced products soon replaced the cheap goods of the 1950s and 1960s. The country is now an important global producer of manufacturing, especially motor vehicles and machinery. Japanese car companies, such as Acura, Honda, Mazda, and Toyota, are now global brands that produce almost 30 percent of all motor vehicle production in the world and 70 percent of all manufacturing robots. The manufacturing sector is concentrated in the capital area around Tokyo and Yokohama. This region is responsible for 40 percent of Japan's GDP. The Kansai region in the west is responsible for 16 percent of national GDP, an economy the size of South Korea, Taiwan, and Singapore combined. It is marked by new sectors such as video games. The area around Nagoya in the Chubu region is responsible for 20 percent of industrial production. With the rising costs of production in Tokyo because of higher land and labor assembly costs, some firms are moving their operations to Chubu.

During Japanese occupation, South Korea effectively became a primary producing colony. After political independence in 1945, it embarked on a rapid path of modernization thorough industrialization. The steel and shipbuilding of the 1960s and 1970s were soon augmented by more technologically sophisticated industries such as automobiles, electronics, and chemicals. As in Japan, industrial production was and still is concentrated in just a few cities. Seoul and surrounding cities are responsible for almost half of the nation's GDP. The result is a marked regional imbalance between the smaller cities and rural areas and the Seoul metro region, where there are rising property prices and congestion costs.

China is a more recent entry on the industrial global stage. The Communist Party, which came to power in 1949, quickly established manufacturing as a key element of national strategy to build up China's economy in the face of international threats from the more dominant United States. Although there was a major push from 1958 to 1960 to industrialize, even as late as the 1960s more people worked on the land than in the city. In the early 1980s a major policy shift encouraged private markets to flourish. In 1984, fourteen special economic development zones were established in cities along the coast. They were meant as test beds to encourage private markets and encourage contact between Chinese and foreign companies to generate technology transfers. They are the basis for the urban coastal industrialization of modern China.

The problem for all three economies is that they are losing manufacturing jobs as the cost of labor rises and there is global competition from lower cost areas. Because of the high costs of domestic labor, and to counter **protectionism** in export markets, Japanese companies now build manufacturing plants outside of Japan. Overseas production of Japanese companies has now increased from around 8 percent in 1994 to 20 percent in 2010. Manufacturing jobs in Japan have declined from around 150,000 in 1994 to 100,000 in 2010.

Although China currently remains the major site for clothing manufacturing in the world, the much lower wages in countries such as Cambodia, Bangladesh, and Vietnam are attracting factories that produce goods for such global companies as Nike.

7.5 Air quality in Beijing (Feng Li/Getty Images)

Environmental Pollution

Environmental pollution is caused by very rapid urban and industrial development. As the countries embarked on very rapid industrialization, they all experienced problems of severe air pollution caused by the burning of fossil fuels and emissions from manufacturing industries. There is a distinct temporal-spatial sequencing with Japan, then South Korea, and finally China going through the trajectory from breakneck growth, limited concern with environmental issues and lack of regulation, to a greater environmental sensitivity and more pronounced regulation. In Japan, the worst environmental effects were experienced from the 1950s to the mid-1970s when there was rapid industrial growth and few environmental restraints. A factory making vinyl since the 1940s pumped mercury compounds into Minimata Bay on the west coast of Kyushu. By the 1960s the mercury had moved up the food chain, causing devastating illness and death to local people. Minimita highlighted the negative effects on human health of

TABLE 7.1	Annual Mean Suspended Particles in Selected Cities	
CITY	**PARTS PER MILLION OF M10/M3**	**DATA DATE**
Los Angeles	33	2012
London	22	2011
Beijing	121	2010
Tokyo	23	2009
Seoul	49	2010

See text for discussion of technical terms

unrestrained economic growth and limited environmental protection. An environmental agency was established in 1971, and Japan shifted toward nuclear power plants to provide electricity. South Korea has a similar history, though it is delayed by one to two decades. In 1991, a factory in Taegu dumped 300 tons of the carcinogen phenol into the Nakdong River, threatening the water supply to more than 10 million people.

As the economies have matured, people have become wealthier and placed greater value on urban environmental quality. As a result, the air and water quality has improved in cities across Japan and South Korea.

China is still at the stage of rapid industrialization with a greater emphasis on job creation and economic growth than environmental protection. However, the severity of air, water, and land pollution is now a source of some debate, and the middle-class urban residents, in particular, now want the government to do something about cleaning up the environment. Between 1981 and 2004, the amount of carbon dioxide emission increased by 40 percent, the amount of solid wastes produced by industry increased by 104 percent, and wastewater discharge increased by 44 percent. The costs of environmental pollution on the health of China's population are now estimated between 4 and 10 percent of GDP. These spiraling health care costs constitute a major drag on national revenue, family incomes, and economic growth potential over the long term. More recently, China's cities have embarked on programs to improve environmental quality by a large margin.

Air quality in Beijing is notoriously bad, a result of pollution from factories, exhaust from cars, and sand storms (Figure 7.5). A good measure of air quality is the annual mean of suspended particles in the air, expressed as parts per million. The World Health Organization guideline level for safe air quality is 20. Table 7.1 lists the figures for the fine particulate matter of less than 10 microns (M10) in a cubic meter (m3) in the three megacities in the region, as well as London and Los Angeles for comparison. Note how Tokyo's air quality is better than that of Los Angeles and close to the figure for London. Beijing has the poorest air quality by a large margin.

Agriculture in China

Agriculture plays an important economic and cultural role in East Asia. In China it has a very long history, with the first characters used to represent rice dating back over 3,000 years to the time of the Zhou Dynasty when large-scale irrigation was first implemented. Rice cultivation was the material basis for China's emergence as a world center of early civilization, and it is still a significant source of employment and economic activity. More than 40 percent of China's labor force is employed in agriculture, and it is responsible for around 10 percent of GDP.

After 1949, private land was effectively abolished, and all farmers were grouped in communes of collectively managed farms. By 1973 there were almost 50,000 communes covering most of China's arable land. The aim was to build up a rural socialist self-reliance. In practice, the system led to low incomes, limited investment, and poor distribution of food. In 1978, the system was replaced by a market system, a rare example of a bottom-up change, in which farmers were allowed to farm their own allocation of land in exchange for turning some of the produce over to the state. In 1984, 15-year leases were introduced and then extended to 30 years in 1999. Farmers can now grow what they want and may respond to the vagaries of changing domestic and international demand.

Problems remain even with the almost total marketization of the agricultural sector. Land ownership is a major issue. The maximum of 30-year leases still limits very large-scale investment. The large number of small farms makes it more difficult to achieve economies of scale. Pollution, due to lack of regulation of pesticide and overuse of fertilizers, also creates problems of tainted and contaminated produce. One-fifth of all farmland is contaminated, and 10 percent of China's rice is now compromised with **cadmium** from fertilizers.

China is the world's largest producer of rice, wheat, pork, eggs, cotton, fruit, and vegetables. The nation is moving quickly from family plots to agribusiness and from villages to company towns as individual landholdings overseen by village government are consolidated into company management. The companies then pay families to lease their land and often provide accommodation in nearby towns. This process lowers costs, makes it easier to manage pollution, and shifts people off the land.

Agriculture still plays a large role, especially in areas close to the large and burgeoning centers in the eastern coastal zone where farmers report higher income than those in the more interior regions as they have easier and cheaper access to domestic and foreign markets. The agricultural sector has moved from a commune system to mini entrepreneurs and larger scale agribusiness.

Agriculture in Japan and South Korea

Only 4 percent of Japan's labor force is involved in agriculture—a steep decline from 50 percent in 1920. While large companies that are now household names dominate Japan's industry, Japan's agricultural sector is a small family farm affair responsible for only 1.3 percent of the nation's GDP. The average farm is only 4 acres in size, compared to the US average of 500 acres. Much of Japan's agriculture is concerned with market gardening, to feed the large and often proximate cities. Small family farms of market gardens are important elements of the periurban landscape as fields and farms are now located between the sprawling tentacles of the metropolis. The farms are intensively farmed using mechanization and fertilizers. Their market is the shifting and fickle urban dwellers of nearby cities. For some of the luxury items, such as **Kobe beef**, the market is global.

Rice production plays a huge role in the cultural life of Japan. The seasonal cycle of rice planting, cultivation, and harvesting is embodied in painting, music, myths, and national legends. Rice is a typical offering at a Shinto temple. The shared rice bowl is a symbol and metaphor of Japan's communal culture. While an important part of Japan's past and cultural identity, it is nevertheless of declining economic significance. The number of farmers declined from over 12 million in 1960 to 2 million in 2015. Farmers are an aging population; more than 70 percent are aged over 60 years. However, rice is so central to Japan's view

7.6 Small vegetable plots at edge of Seoul, South Korea (photo: John Rennie Short)

of itself that rice farmers are given strong protection from overseas competition. Rice produced in Japan, because of labor and land costs, is six times more expensive than rice grown in Thailand or California. Japan's rice farmers are perceived as carriers of Japanese traditional values and are protected from these cheaper imports.

Agriculture in South Korea also employs few people, less than 1 percent of the labor force, and it is responsible for only 3 percent of GDP. Most farms are relatively small, on average only 1.4 hectares. Farmers closest to the main towns and cities have the ability to apply more intensive techniques to meet the market garden demand of urban consumers, especially fruit and vegetables (Figure 7.6). Rice is still the dominant crop and covers almost 50 percent of cultivated land and overall the farming population is aging.

Social Geographies

Ethnic Homogeneity

One important feature of population in this region is the ethnic homogeneity. Both Japan and Korea (South and especially North) are marked by an ethnic homogeneity that they maintain with strict immigration policies. More than 98.5 percent of Japan's 127 million population are ethnic Japanese with only very small numbers of Korean and Chinese. An indigenous people, the Ainu who inhabit the northern and central islands of Japan number between only 25,000 and 200,000 and fewer than 100 can speak the language. In South and North Korea the official statistics suggest only a tiny proportion of non-Koreans, although there are probably a larger number of undocumented workers in South Korea.

TABLE 7.2 Ethnic Groups in China	
ETHNIC GROUP	**NUMBER (IN MILLIONS)**
Han	1,220
Zhuang	16.9
Hui	10.5
Manchu	10.3
Uighur	10.0
Miao	9.4
Yi	8.7
Tuja	8.3
Tibetan	6.2
Mongol	5.9

Unlike China, which grew by annexation of other lands and peoples, the boundaries of Japan and Korea remained relatively static. In both Japan and Korea, there is a more one-to-one relationship between the ethnic nation and the state. Of course, the division of Korea into North and South has added a new wrinkle.

Ethnic Groups in China

In China, the position is different. Han Chinese dominate, as they constitute more than 91 percent of China's vast population of 1.3 billion. However, we should be careful of the designation of Han. It is not a centuries-old ethnic reality but a nineteenth-century invention; in fact, it was used to describe an amalgam of races to contrast with the Qing rulers, who were Manchus, and the foreign Westerners.

Han Chinese have spread across into the furthest reaches of China. In the past 25 years, millions of Han Chinese have settled throughout greater China, diluting the local influence and affecting the language and culture of the non-Han. In Tibet, Inner Mongolia and Xinjiang ethnic minorities persist despite the central authorities' attempts to Sinicize them.

There are at least fifty-six different ethnic groups officially recognized by the Chinese government. The largest ethnic groups are identified in Table 7.2.

The largest ethnic minority, with a distinctive language, is the Zhunag, who are concentrated in southern China. The Hui are mainly Muslims concentrated in the northwest and central plains. They are similar to Han Chinese, differing only in their religious affiliation. The Manchu give their name to Manchuria. They are found throughout the country in at least thirty provincial regions, with half of them, almost 6 million in total, in Liaoning province in the northeast of the country. Most Manchu no longer speak nor write their mother tongue, a Mongolian-based language. And millions of Han Chinese now live in Manchuria, many of them for generations; 3 million settled in the 1920s alone. Manchu make up less than 10 percent of the 110 million people in the region. And while their national identity card contains a designation of Manchu, few can speak Manchu.

Inner Mongolia is home to the Mongol people. Some of them are traditional nomadic people of the grassland. This was the hearth area for the **Mongol Empire**. It became part of China under the Qing. Today, the Han Chinese dominate in the towns, and only 17 percent of the 24 million people who live in Inner Mongolia are Mongols.

The Uighurs are a Turkic-speaking people who live in China's outermost region of Xinjiang, where they constitute around 43 percent of the total population of 22 million. Most of them are Muslim. The region was only brought under Chinese control in 1949. As China links the region with the rest of the country, with a new $23 billion bullet train, it also opens the way for more Han

Chinese to settle. The region has rich resources of oil, natural gas, and minerals. The main city of Urumqi is now segregated by ethnicity, with Han Chinese in the north and Uighurs in the south of the city. With a distinctive ethnicity and religion, located far from central government, the Uighurs present a possible threat to China's territorial integrity and a region of potential resistance to Han domination. Many Uighur fear that their culture will be destroyed by the invasion of Han Chinese. In 1997, mass arrest and execution in the city of Ghulja marked a decisive defeat for the separatists. In 2009, clashes between Uighurs and Han Chinese resulted in two hundred deaths. On October 28, 2013, a car driven by Uighurs plowed into crowds in Tiananmen Square, killing tourists. It was a symbolic act that brought Uighur resistance right into the political heart of China. There is now a vicious cycle as Chinese authorities clamp down with more surveillance and more overt policing that tends to provoke even more Uighur resentment. In May 2014, a bomb blast killed over thirty people in a market in Urumqi. In the same month, more than ninety people were killed in the town of Elishku when police fired into a Uighur crowd. The independence struggle is sometimes clothed in the rhetoric of a more conservative Islam. China banned the wearing of beards and veils. In July 2014, crowds protested the arrest of young girls in the town of Alaqagha for refusing to remove headscarves. Uighur resistance to Chinese rule and the onset of modernity that marginalizes Uighurs is now expressed in a conservative Islam as a symbolic form of resistance. The Chinese government feels the need to squash separatist sentiment, but if it is too vigorous, it generates a backlash. If there is not enough exertion of central power, the more space is created for separatists to make their claims. The rumblings on China's western border will therefore continue.

Long isolated from the rest of the world by inaccessible mountain ranges and a hermitic society, Tibet developed as a feudal Buddhist **theocracy**. It came under Chinese control under the Qing Dynasty. After the fall of this dynasty, Tibet drifted away from Chinese control into relative obscurity and became a source of wonder for Western travellers looking for the premodern in a world of change. Tibet was the basis for the imagery of Shangri-la. In 1950, Chinese troops annexed Tibet, and communist China reclaimed the territory. The Dalai Lama, both the spiritual and political leader of the country, was forced to flee the country after a failed uprising against Chinese control in 1959. The Chinese embarked on a policy of secularization (more than six thousand monasteries were destroyed), modernization (with new roads and railways), and incorporation into China through the in-migration of Han Chinese. In the capital city of Lhasa, there are now more Han than ethnic Tibetans. Tibet has become a celebrity issue, although China resists vigorously any international calls for Tibetan autonomy.

Demographic Trends

Just as we can locate the three countries along a continuum of economic development with Japan developing as a modern industrial economy first, followed by South Korea and then China, so we can also locate them along a continuum of demographic change with Japan in the lead followed by South Korea and then China. Although the timing of the trajectory varies, the entire region is now characterized by low fertility rates, rapidly aging population, and a shrinking labor force.

CHINA

China with its vast population and recent history of population growth is now merging into population stability. Up until the mid-1960s, China had a high birth rate with an annual population growth rate of around 2.3 percent. Since then both birth and death rates have declined, and population growth is now less then 0.7 percent. China has a huge population, close to 1.4 billion, but is quickly approaching a steady state of zero growth. This raises the issue of labor supply as labor becomes less plentiful, thus more expensive. China is moving away from its role in the global economy as a source of very cheap labor. The working population has fallen each year since 2012, although the total is still a staggering 916 million. By early 2030, the number of people aged over 60 years will constitute a quarter of the total population (Figure 7.7).

The mortality rate was rapidly reduced by the introduction of comprehensive public health measures by the communists in 1949. Prior to this there were no formal public health care systems. It only started in 1949 with the introduction of vaccinations, improved sanitation, and the official campaign to resist four pests: rats, flies, mosquitos, and sparrows. The net effect was to eliminate deadly diseases, reduce infant mortality rates, and increase life expectancy. After 10 years of the "barefoot" doctor program that encouraged health professionals to visit the underserved rural areas, average life expectancy increased from 51 years in 1965 to 65 years in 1975.

The one important footnote to this general improvement in life expectancy was the great famine of 1959–1962 when an estimated 35 million died due to the forced collectivization of agriculture, the decline of food production, and the inability of people, because they were forced to remain where they were, to relocate to where food was more plentiful.

Birth rates were reduced by the active intervention of the government. Family planning measures were first promoted as official government policy in 1956 and reaffirmed in 1971 with grassroots programs of education at the village and neighborhood level. In 1979, the one-child policy was formally introduced. It was especially implemented in urban areas. Families with only one child received preferential treatment in access to schools and

7.7 China's population is getting older, but still keeping fit. A woman practices tai chi in a park in Shanghai, a common sight in urban China. (Photo: John Rennie Short)

in other parts of the world, favor smaller rather than larger families. The policy was so successful in reducing the birth rate that China is now in the final stages of the demographic transition with a profile of declining birth rates, currently around 1.4, well below the United States, and an aging population similar to the more advanced economies. In 2013, China relaxed its one-child policy, but for many in the urban middle class, maintaining small family size is now so ingrained and connected to improved economic opportunities that the rule change is unlikely to generate a baby boom. In 2015, the policy was officially ended when the government announced that all married couples would be allowed to have two children. It was in large part a response to declining economic growth rates and increasing wages. Despite its huge population, China still has labor shortages, and the price of labor is increasing. The one-child policy created over 130 million people who were the only child in their families. The subject of parental devotion, the boys were commonly referred to as Little Emperors because they were so spoiled. Despite the change it is unlikely to increase the average family size for China's burgeoning middle class.

The public health system was spectacularly successful in the early years of communist control, but, as elsewhere, it is now under strain as the population ages and health care costs increase. When the old collectivist system crumbled with the privatization of state-run firms, health care became a private matter. More is being spent on health care as the population ages. This ultimately creates a drag on domestic demand as people are saving for health care rather than

nurseries, housing, and government benefits. The policy had high levels of compliance in part backed up by coercion. China's economic growth has created a more prosperous middle class who, similar to their counterparts purchasing goods and services. In response, health care insurance for rural residents was introduced in 2003 and for urban residents in 2007. It is an official goal to insure the entire population by 2020.

JAPAN AND SOUTH KOREA

Japan, and to a slightly lesser extent South Korea, is at the furthest extent of the demographic transition with very low birth rates, long life expectancy, and a graying of the population.

Japan is in fact losing population because 1.3 million people die every year, but only 1 million babies are born each year.

Japan and South Korea have some of the lowest birth rates in the world with an average of only 1.2 births per woman. Women of reproductive age are not getting married and not having children. In both South Korea and Japan, there are obstacles for women with children entering or reentering the job market, so women are deciding not to get married and not to have children. While out-of-wedlock births are increasingly common in Western Europe and North America, severing the ties between marriage and childbirth is culturally less acceptable in Japan and South Korea.

The population in Japan is aging because of sharply falling birth rates and increasing life expectancy. With a total population of 127.3 million, Japan now has the largest **dependency ratio**, which is the number of people aged 65 years and over, per 100 persons aged 15–64 years. The figure for Japan is 62 years, compared to 37 years for China. By way of comparison for countries in the middle stages of the demographic transition, such as India, the figure is 16 years. At 85.5 years, Japan has one of the longest life expectancies in the world.

Japan's population declined in 2005 for the first time in the modern era, and estimates have the total heading further downward. This demographic collapse is very marked in rural areas as young people move into the cities, leaving behind an increasingly aged and aging population, sometimes below the threshold for the efficient provision of basic service such as health, education, and transport and placing fiscal strain on the ability of the workforce to maintain an increasingly large and long-living, nonworking population. Japanese cultural attitudes preclude an encouragement of immigration as a way to fill job vacancies. South Korea is closer to Japan, with low birth rates and increasing life expectancy, than to China. All three countries, however, are either at or close to the final stage of the demographic transition, with Japan leading the world into the experience of a graying, declining population.

Belief Systems

The dominant religion in this region is **Confucianism**. It is as much a moral code as a religion. It is based on the teachings of Confucius (551–479 BCE), who lauded family loyalty, social harmony, and respect for the social hierarchy and political order. It promotes trust in the government. It has widely pervaded social codes in East Asia.

In China, Japan, and Korea (North and South), albeit in different political contexts, there is a shared commitment to hard work, family cohesion, sacrifice, and a basic belief in government. It has provided the ideological basis for the **development state** and the consequent rapid economic growth organized by the government.

Among the more formal religions, Buddhism diffused from its hearth in India across trade routes. There are approximately 10,000 Buddhist temples in China with many different strands. Tibetan Buddhism is particularly distinctive.

Buddhism even made its way to Korea but was replaced by the Joseon Dynasty with Confucianism. In recent years, however, Confucianism has made a revival in South Korea (Figure 7.8). There are also many Christian, mainly Protestant churches, in South Korea initially brought by US missionaries.

Shintoism is the dominant religion in Japan. It is a traditional religion constituted in thousands of shrines and temples dedicated to the worship of numerous gods.

7.8 Seoul: lanterns in Buddhist temple celebrate the birth of Buddha (photo: John Rennie Short)

As the religion of the traditional agricultural society, it has many festivals devoted to ensuring fertility in spring and fall. The cities of Japan are filled with small neighborhood shrines used for weddings and occasional visits.

North Korea had no formal religion, but the regime has a cult of personality that draws on Confucian notions of social order to prop up a dynastic dictatorship.

There are also local religions practiced by some of the ethnic minorities in China. Islam diffused along the routes of the Silk Road from its hearth in Arabia all the way across Eurasia. Islam is more common in the western region of China along the old trade routes. The Uighurs tend to follow Islam, which makes them even more distinctive in contrast to the dominant Han Chinese.

Rural Focus

The Hukou System

Hukou is a household registration system with deep roots in China's past. Registering households was long employed so that the imperial center had knowledge and control over people's movements. It was reintroduced in 1958 as a form of internal passport to monitor and to limit rural-to-urban migration. People who moved without permission did not receive food rations, health, or housing. It created a sharp division between households registered in rural areas and those registered in urban areas. The city dwellers had better access to public services such as health, housing, and employment. The system limited rural-to-urban migration and restricted the size of Chinese cities. Many of the cities in much of the developing world, in contrast, expanded quickly with rural migrants.

Economic liberalization after 1980 meant expanding job opportunities in coastal cities. People moved to these cities to find work and income, but they were often denied access to health, housing, and education. In the burgeoning cities, many rural migrants were marginalized second-class citizens. Between 120 and 250 million people were characterized as a "floating" population living in temporary housing, often in suburban villages, separated from family members with difficult and limited access to health and educational facilities. They were a huge reservoir for cheap labor as China's economy boomed. Often migrants from the same province or district would live together as initial migrants provided information and help that encouraged subsequent migrants from the same region. Migrants from the province of Sichuan tend to move to the city of Shenzen. In Beijing, migrants from Zhejiang and Henan group together in enclaves that act as mutual aid organizations.

The authorities have loosened the rules of the *hukou* system. Recent provisions allow more rural residents to buy temporary residence permits and have reduced the cost of these permits. In 2014, it was announced that the system was to be abolished for smaller cities and towns, reduced for medium-sized cities, but still held in place for the large metropolitan areas. Whether the system remains in force or is abolished, massive rural-to-urban migration will still occur, formally registered or not. **Circular migration** with people working in cities but still retaining official rural residence will continue as an important household strategy to maximize income and minimize risk. Rather than moving permanently to the city, many rural migrants maintain their connection with the village as they still have assets and resources in the countryside, including family and friends, connections, maybe even land and housing. Even with the easing of registration, many migrants will maintain dual places as a form of household coping strategy.

If there was no registration system and rural-to-urban migration was made easier, most of the demographic models suggest increased growth in the four largest metros of Beijing, Chongqing, Guangzhou, and Shanghai, as well as a second group of thirty-one smaller cities in the heartland, including Fujian, Hunan, and Kwantung. A peripheral group of cities, including Harbin in the north and Kunming in the west, shielded from competition by distance, would also receive more migrants.

Urban Trends

Rapid Urbanization

In the past 50 years, East Asia witnessed an urban explosion. All three of the major countries in the region experienced a major redistribution of population as people moved to the cities and cities grew into metropolitan areas. The vast rural-to-urban migration was and still is prompted by the lack of economic opportunities in the countryside and the growth of employment opportunities in fast-growing cities.

Japan was a predominantly rural country until 1950. Today it is predominantly an urban society with more than 80 million living in cities. Table 7.3 lists the ten largest metro areas in the country that contain 66 percent of the country's entire population. The largest by far is the metro region centered on Tokyo that has a population of almost 38 million. The metro economy of producing goods and service valued at $1.9 trillion, more than a third of the national total, is the largest in the world.

South Korea also experienced rapid urbanization associated with its transformation from predominantly rural and agricultural to urban and industrial. In 1950, the urban population constituted only 14.5 percent of the

TABLE 7.3	Ten Largest Metro Areas in Japan
METRO AREA	**POPULATION (MILLIONS)**
Greater Tokyo	37.8
Kyoto-Osaka-Kobe	18.7
Nagoya (Chukyo)	8.9
Fukuoka	5.5
Sapporo	2.6
Sendai	2.2
Hiroshima	2.0
Okayama	1.6
Kumamoto	1.4
Niigata	1.3

7.9 China has experienced rapid urban growth in the last two decades. A new subway in the inland city of Nanchang being built. (photo: John Rennie Short)

total population. By 2015, it was close to 90 percent. Most of the growth was concentrated in the large and rapidly industrializing cities.

In 1950, China was a predominantly rural society with more than 90 percent of the vast population living in the countryside. Initially, the cities grew rapidly after communist rule was established as industrial development was encouraged and there were few controls on people moving from the countryside. Within a decade, however, controls were placed on the ability of rural residents to move to the cities. From 1961 to 1963, there was a reverse movement from urban to rural as the government shipped almost 50 million people from the cities to the countryside. China's urbanization picked up pace in the 1970s when the percentage of people living in urban areas doubled. Today 51.2 percent of the population is considered urban, although the real figures are probably much higher. The urban growth is particularly marked along the east coast in large cities. In recent years, however, some of the inland cities such as Nanchang have also witnessed spectacular growth rates (Figure 7.9). There are at least fifty-three metro regions with a population greater than 1 million. They contain 30 percent of the nation's population and are responsible for 53 percent of GDP and 62 percent of non-farming GDP. The big urban areas are responsible for most economic growth.

There are many issues associated with this rapid urbanization. Here we will focus on just two: land supply and environmental degradation.

The Problem of Land Supply

Land supply was always a concern for Japan and Korea with limited flat land, but it is an increasing problem even in the vast land of China because of the concentration of

urban growth. Between 1990 and 2005, the urban land surface increased by 153 percent. This rapid extension reduces the amount of land available for agriculture and places pressure on local ecologies. Land supply is exacerbated by differences in landholdings between urban and rural areas. While urban land is effectively public land owned by the government, rural dwellers have effective title to their holdings. Urban conversions at the city fringe, with feverish rate of growth, spark off political conflict between agents of growth keen to obtain cheap land and rural residents trying to keep control of their landholdings. Many of the local political disputes in China center on land use conflicts between agents of urban growth, including developers, builders, and local politicians, on the one hand, and rural residents, on the other. Two-thirds of all "social disturbances" recorded by the government relate to land disputes between farmers and developers/local governments.

City Focus: Tokyo

Tokyo is the largest urban agglomeration in the world with a population of 37.8 million (Figure 7.10). It began its life as a humble fishing village, called Edo, ruled by warlords. By 1700, it had grown to over 1 million people. The city was regulated with the discipline of a rigidly hierarchical society. Around the castle on the high ground, in an area known as Yamanote, were the villas of the elite and the feudal lords who were forced by the rulers to live some of the year in the city. People came from all over the country to work in the city. They lived in the marginal low-lying lands and built sprawling villages of high density. The area was known as Shitamachi, the lower town, a place of single-story tenements, narrow alleys, and unplanned streets.

After the Meiji Restoration in 1868, the city grew to 2 million and the emperor's official residence was moved from Kyoto to Tokyo, marking the complete dominance of the growing metropolis in the life of the country.

Tokyo grew as a series of cellular villages, essentially replicating the pattern of Shitamachi. Local neighborhoods were the center of life. This pattern was recreated after two episodes of destruction. In 1923, much of the city was destroyed in an earthquake. The devastation reached apocalyptic extremes when resultant firestorms whipped up by high winds raced through the city. Over 40,000 people were burned to death on one 40-acre site. Tokyo was destroyed again in 1945 when the US Air Force firebombed the city. More than 100,000 were killed, 1 million were injured, and 1 million rendered homeless. Because of these two catastrophes, Tokyo has few old buildings.

The city has extended outward as accretions of distinct neighborhoods centered on local shopping centers. The city has multiple centers. Kasumigaseki is the center for the national government; Aoyama is the place for up-market shopping; and Ginza is the place for entertainment, while youth shoppers congregate in Karajuku and Shibuya. There is clustering of retail and economic activity around major metro stations, such as Shinjuku, where three subways and seven railways transport 3 million people a day.

7.10 Map of Tokyo

Geopolitics

The geopolitics of this region is shaped by the dialectic of conflict-cooperation between the adjacent powers and also the United States.

Bordering China

China is a vast country that borders many states, including Russia, North Korea, Mongolia, Kazakhstan, Kyrgyzstan, Tajikistan, Pakistan, India, Nepal, Bhutan, Myanmar, Laos, and Vietnam. The complex history of these border regions embodies the shifting fortunes of China. Many of China's territorial disputes arise from the century of humiliation when China was forced to sign unequal treaties with outside powers and lost effective sovereignty over territory. Let us consider two case studies: Manchuria and India.

The North East region of China consists of the three provinces of Heilongjiang, Jilin, and Liaoning, forming part of what is known as Manchuria. They are located beyond the north edge of the Great Wall and only became part of China under the Qing Dynasty. In the nineteenth century, as the empire was weakened by internal uprisings, other powers probed at the empire's weak points. The Russians annexed an area north of the Amur River and then the Ussuri River, eventually annexing 350,000 square miles. In 1931, the Japanese Army invaded, and the region then came under Japanese control until 1945. More than a quarter of a million Japanese settlers moved into the region. The region is now part of China, but the legacy of war persists. Since 1945, two thousand people have died, killed by unearthed Japanese chemical weapons. Border conflicts with the USSR/Russia continued over disputed territory, only settled in 2004, with the transfer of Yinlong Island.

Some disputes remain unsettled, such as the border areas with India. There are, in fact, two disputed areas along the India-China border: Arunachael Pradesh, which is part of India but claimed by China; and Aksai Chin, which is part of China but claimed by India. British imperial officials created the boundary lines between India and China. The Macartney-McDonald Line of 1893 allocated Aksai Chin to China. The McMahon Line, drawn in 1914 between Tibet and Britain, gave Arunachael Pradesh to India and annexed 25,000 square miles of territory that previously belonged to China. Today, China claims the 32,000 square miles of Arunachael Pradesh, while India claims that China is illegally occupying the 14,380 square miles of Aksai Chin. China and India fought an intense short-lived war in 1962 over these contested border territories.

South and North Korea: A Frozen Conflict

Korea was a unified country for centuries. The **Joseon Dynasty** ruled from 1392 to 1910 in one of the longest continuous political systems. At the end of World War II, Korea was caught up in a wider geopolitical struggle between East and West. The Korean peninsula was divided by the USSR and United States into separate zones of occupation—the Soviet zone north of the 38th parallel, while the US zone lay to the south. The initial plan was to place the country under the trusteeship of the United Nations. However, competing national interests, eager to grab their chance, resulted in the partition of the country. In 1948, the Republic of Korea (South Korea) was established in the south under the leadership of Syngman Rhee, while Kim Il Sung established the Democratic People's Republic of Korea (North Korea). Neither regime was democratic.

In 1950, the incursions of North Korean troops escalated into the Korean War that enmeshed the United States as well as other powers since the UN was involved in fighting for the South. The North Koreans pushed all the way to the south of the peninsula but were pushed back by South Korean, US, and UN forces. By 1953, the competing forces were back where they started—facing each other across the 38th parallel. More than 2 million troops were killed or wounded, and 2.5 million civilians were either killed or wounded. Armistice negotiations began in 1951, and an agreement was reached in July 1953. No formal peace treaty was signed. The 1953 Armistice was an agreement between military commanders of the UN, North Korea, and China to cease hostilities, exchange prisoners of war, and establish a demarcation line with a demilitarized zone stretching 2 kilometers in either direction from this line. The military ceasefire froze relations on a permanent war footing. A formal peace treaty that could provide some form of political resolution to the conflict remains in principle a tantalizing possibility, but in practice the continuing uncertain and unresolved arrangement constitutes a dangerous situation.

The two Koreas have yet to sign any peace treaty and so remain technically at war; national sovereignty remains a cloudy, contested issue. The "temporary" armistice line has become a fixed element in the Korean landscape, a narrow but deep scar that divides North from South.

Conflict in the East China Sea

There are three distinct conflicts in the East China Sea.

First, China and Japan have competing claims for the Diaoyu/Senkaku Islands and the consequently differing demarcations of each country's **Exclusive Economic Zone** **(EEZ)** (Figure 7.11). EEZs, first introduced in the 1982 UN Convention on the Law of the Sea, are areas in which states can claim exclusionary rights over marine resources. They have a limit of 200 nautical miles (370 km) from the coastal baseline, the seaward edge of its territorial sea. It may extend beyond the 200-mile limit into the continental shelf. The islands were long part of imperial China. In the fifteenth century, they were named *Diaoyutai* (meaning "fishing platform"). They were annexed by Japan in 1895 by the Treaty of Shimonoseki in the wake of its victory in the Sino-Japanese War. From 1945 to 1971, the islands were administered by the United States and then placed under Japanese control. Both the Republic of China (ROC), commonly referred to as Taiwan, and China agree that they are part of Taiwan. Japan maintains it has sovereignty and thus an extended EEZ.

Second, the dispute between South Korea and China centers on a disputed EEZ demarcation in the Yellow Sea. The dispute is heightened by fishing activities and claims over a submerged rock called Socotra Rock, Iedo, and Suyan. The Socotra Rock remains submerged even during low tide. South Korea claims it as part of its continental shelf and in 2003 built a marine research station on the reef.

Third, the conflict between Japan and South Korea centers on a small group of islands and a contested name. Dokdo, which the Japanese refer to as Takeshima, consists of two small, rocky islands surrounded by approximately thirty-three smaller rocks. In total it amounts to just fewer than 2 square kilometers. After the formal takeover of

7.11 Disputed islands in East China Sea

Korea by Japan in 1910, the island, as with all of Korea, came under Japanese control. In 1905, a local government in Japan, Shimane Prefecture, unilaterally incorporated Dokdo/Takeshima into its territory. With Japan's defeat by Allied forces in 1945, its title through colonial domination was effectively renounced. Today Dokdo is under effective Korean control. Even as Japan–Korea relations improve, Dokdo remains unresolved. The issue is not just one of political posturing. The rich fishing stocks and the existence of the valuable resource of gas hydrate make the competing claims all the more economically relevant. Even the smallest of islands can become opportunities to extend 200-mile EEZs.

There is also the issue of what to call the sea between Korea and Japan. Koreans for centuries have called the sea the East Sea; it was only in the late nineteenth century that the term "Sea of Japan" was used. It was not indigenous to the area but a European name later adopted by the Japanese. As a colonial power, Japan was able to rename it as the Sea of Japan. This was standardized by international agreements drawn up while Japan was the colonial power in Korea. In 2007, the two Korean states made representation to the ninth UN Conference on the Standardization of Geographical Names. The South Korean request and official government position is for a dual naming system of East Sea/Sea of Japan.

Conflict in the South China Sea

The South China Sea has vast oil and natural gas reserves and vast fisheries. The world's busiest sea lanes pass through the South China Seas. China claims most of the area, including the fishing areas, the oil and gas reserves, and the vital sea lanes. China's claim incurs on the EEZs, as presently defined, of Vietnam, the Philippines, Malaysia, and Brunei (Figure 7.12).

There are disputes over the various islands in the sea. The Paracel Islands are claimed by both China and Vietnam. When China moved a giant oil rig to the region in 2013, it was seen as a direct threat by the Vietnamese and undermined China–Vietnam relations. The Scarborough Shoal is claimed by both China and the Philippines. Six countries—China, Vietnam, Taiwan, Malaysia, Brunei, and the Philippines—make claim to the Spratly Islands, more than 740 reefs, islets, atolls, and islands. And all but Brunei have built or claimed outposts in the island. However, China's creeping invasion began in 2013 when it started to build on existing rock outcrops and reefs to create 1,200 acres of new land. Vietnam and Philippines have undertaken similar ventures, but China has outpaced every other country combined. Sand is pumped up from the sea floor and spread across the island and even submerged reefs are then concreted over to become bases capable of providing docks for ships and runways for planes. The Chinese have militarized the disputed islands.

Connections

The Silk Road: Old and New

The Silk Road emerged over 2,000 years ago as a major trade route that linked Beijing with Cairo, Constantinople, and Rome. There were in fact a number of routes, including one across the Taklimakan Desert, a southern one that fringed the foothills of the Himalayas, and a northern route that fringed the Tien Shan Mountains. There were numerous offshoots, so the term "Silk Roads" is more accurate even though more than silk was carried along this artery of global trade. Goods, including spices, porcelain, jade, gold and silver, ideas, technologies, and even viruses, were traded along this Eurasian network. Parts of the route were closed off when the Ottomans finally gained control of Constantinople in 1453 and supplanted the Byzantine Empire. The closure of the overland route between China and Europe led to the decline

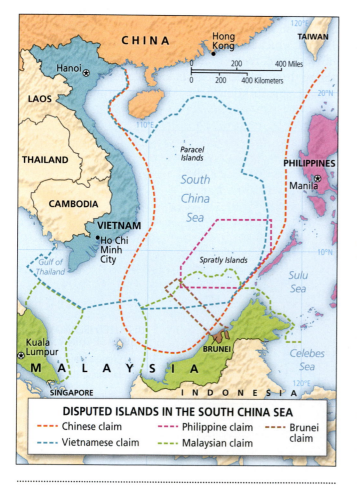

7.12 Disputes in South China Sea

of merchant empires such as Venice that relied on the overland trade with China and the Middle East, and the rise of powers such as the Portuguese and Spanish that had access to the Atlantic Ocean and could more easily mount seagoing trips to China. Christopher Columbus persuaded the King and Queen of Spain that he could sail ships there, too.

There was also cultural exchange as religious beliefs diffused along the route leading to the spread of Buddhism and then Islam from their original sources in India and Arabia, respectively. Buddhism made its way from South Asia into China and even Korea and Japan. The giant Buddhas of Bamiyan in Afghanistan built in the sixth century were a reminder of the region's pre-Islamic past. Sadly, the Taliban destroyed the magnificent monuments in 2001 for that very reason. Islam also diffused along the route. To this day the ethnic Uighurs in China are Muslim. Hybrid artistic styles emerged, including a Grecian-Buddhist style as Greek aesthetics were combined with Buddhist styles to create beautiful artifacts such as the Bamiyan Buddhas.

The term "Silk Road" is so redolent in meaning that is it often employed for contemporary infrastructure projects. China's plans to build a bullet train to Xinjiang and eventually to connect with Turkey and Bulgaria are described by officials as building a "New Silk Road." In 2014, China pledged $40 billion to a Silk Road Infrastructure Fund to improve connectivity across Asia. The Fund will be used to invest in roads, railways, and airports to link Asia and Europe. The project now known officially as the Belt and Road Initiative will employ China's vast foreign exchange reserves and provide a platform for the export of Chinese goods and technology, while also enhancing economic links and promoting trade with Asian and European countries. A Maritime Silk Road involves investments in ports throughout the Indo-Pacific. In total, China's new silk roads will link seventy countries comprising 30 percent of world GDP.

Subregions

China: A Spectacular Rise

When the communists took control over the vast territory in 1949, the economy was nationalized, landlords were executed, private companies were abolished, and all land was nationalized. The only political party was the Chinese Communist Party (CCP). China struggled to industrialize under the early years of communist rule. Under the Mao regime, from 1949 to 1976, the **Great Leap Forward** and then the **Cultural Revolution** undermined the economic development of China. China embarked on the Great Leap Forward from 1958 to 1961 to rapidly industrialize and modernize. It was a disaster. The economy in fact shrank and almost 20 million died of starvation. During the Cultural Revolution from 1966 to 1971, emphasis was placed on ideological purity; many were persecuted for being traitors to the class struggle, and the economy faltered. After Mao died in 1976, a new reformist era was inaugurated. The private market was encouraged, private wealth accumulation became acceptable, and increasing income inequality became more pronounced.

ECONOMIC REFORMS

Since the economic reforms of the 1980s, the economy has expanded enormously and more than 250 million people were lifted from poverty into the middle class. China has emerged as one of the world's success stories. Table 7.4 provides the basic picture of spectacular growth and improved living conditions for millions of people. China followed in the footsteps of Japan and then South Korea with strategies of large state involvement and the protection of local industries.

In 1978, China's per-capita income was one-third that of most countries in sub-Saharan Africa, and the average Chinese household income was $200; by 2018, it was $10,200. The percentage below the poverty line of $1.90 a day declined from 88 percent in 1981 to just less than 1 percent in 2018. Not only is the scale of change impressive, so is the pace. It took 50 years for the United States to double its per-capita income: it took China 10 years. China is one of the economic success stories of the contemporary era.

Over the past 25 years, life has improved for the vast majority of China's population. There are significant spatial impacts of these changes. The most obvious is the rapid urbanization and industrialization of cities along the coastal belt and especially in the three largest metro areas of Beijing, Guangzhou, and Shanghai. China's population is concentrated in these cities along the coast. People move from the countryside to work in these manufacturing centers. Much of the rural-to-urban migration is a circular migration, with rural migrants still maintaining connections to their rural regions and villages. China's urbanization is quite distinct with a vast "floating" population of rural migrants living in the margins of urban society.

The regional disparities are marked. In the cities of Beijing, Shanghai, and Tianjin, GDP per capita is double the national average and four times that for Guizhou and Gansu.

TABLE 7.4 China's Impressive Growth

	1989	2018
GDP per capita	$403	$10,200
Private car ownership	100,000	172 million
Life expectancy, male	68	75
Life expectancy, female	71	80
Urban population	291 million	832 million

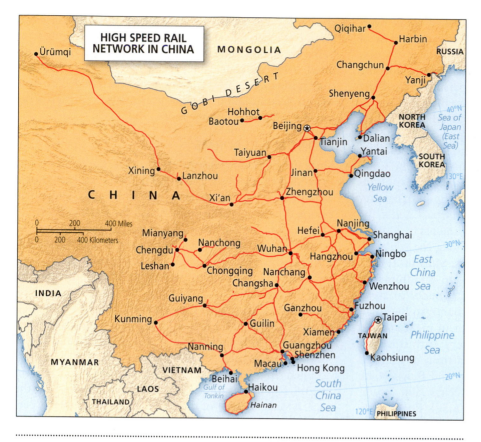

7.13 High-speed rail network in China

that run from north to south. On the Beijing-Shanghai line, opened in 2010, trains cover the 1,000 miles in under 5 hours. Beijing is also linked to Harbin in the north, with branch lines to Shenyang and Dalian. Two more lines, Hangzhou to Shenzhen and Beijing to Hong Kong, are still under construction. Four east-west, high-speed links are planned. On the line from Shanghai to Chengdu, trains at their fastest will travel at approximately 150 miles per hour (Figure 7.14). Despite the highly publicized example of corruption in their construction, the high-speed railways represent China's investment in the future and a way to cohere China's geo-economic and political space. The vast infrastructure expenditures are also a form of **Keynesianism**, as they also stimulate the economy and provide employment. The high-speed rail network is one of the largest public works programs in the world.

A SHRINKING CHINA

The "shrinking" of China is brought about by transport improvements that reduce the amount of time necessary to cover distance. The most dramatic change is the introduction of high-speed rail networks. The plan is to increase the total length of railways from around 5,000 miles in 2008 to 7,500 miles by 2020 and to construct 7,500 miles of high-speed railway (Figure 7.13). There are four high-speed lines

7.14 High-speed train (photo: John Rennie Short)

PROBLEMS OF RAPID GROWTH

So far the CCP has ensured political loyalty by delivering competent governance and economic success. If either should falter, however, the balancing act may become more difficult. Corruption is a threat to the legitimacy of the system. It is estimated that corruption costs 3 percent GDP. Since 1990, 18,000 corrupt officials have fled with over $120 billion and, in a 2-year crackdown period from 2012 to 2014, more than 100,000 officials were charged with corruption. The anti-corruption strategy was also a way for different factions within the ruling elite to settle scores. Crony capitalism and corruption at a time of rising inequality and reduced economic growth is an explosive mixture. As labor shortages push up wages and China no longer is a site for cheap mass production of consumer goods, then the transition to a less export-orientated economy may generate economic turmoil.

As the three decades of spectacular growth turn into the merely impressive growth rates of 6–7 percent per year, the tensions mount. China will try to narrow the income and life chances gap between urban and rural residents and those living on the coast and those in the interior; pursue industrial restructuring from an emphasis on export-led growth to domestic consumption; clean up the toxic legacy of pollution; and adapt to the challenge of an aging population and declining labor supply.

GLOBAL TRADER

China's imports from the rest of the world when adjusted for inflation increased from $95 billion in 1990 to $1.8 trillion in 2012. Many of these imports are in the form of primary products and raw material to feed the vast appetite of China's factories. China is the biggest importer of iron ore and coal from Australia. That country's long commodity boom since 1990 is based on Chinese economic growth. China is also Africa's biggest trading partner and imports $199 billion worth of goods, including minerals, oil, and agricultural goods. China has supported a number of infrastructural projects, including highway construction and port upgrades. In Zimbabwe, to take just one example, China is the biggest importer of Zimbabwean tobacco and the single biggest source of foreign investment. China is eager to exploit the country's resource base. In the Marange Diamond fields, Zimbabwe has a valuable source of diamonds, the second largest reserve of platinum in the world, and substantial deposits of copper, gold, and silver. The Chinese state-owned company, China International Water and Electric, farms 250,000 acres of maize.

To secure the continued supply of raw material, China is an investor as well as a trading partner. Political and military ties often come in the wake of increased trading connections. China added Zimbabwe to its list of approved tourist destinations, provides military hardware, and supported the Mugabe regime when most Western countries sought to distance themselves. China's economic power is now creating a global geopolitical presence as the country is embedded in the economic and increasingly political fabric of the global community.

A recent case of China's growing role in the global economy, beyond simply importing vast amounts of raw materials and exporting manufactured goods, is the emergence of the **Asian Infrastructure Investment Bank (AIIB)**. When the US Congress refused to support an increased role for China in the International Monetary Fund, China established the AIIB that, despite US protestations, has now attracted a large number of partners, including traditional US allies such as Australia, Brazil, Saudi Arabia, and the United Kingdom. The AIIB has pledged to provide $40 billion worth of funds for infrastructure improvement, much of it targeted at a new Silk Road of improved communication and transport across Central Asia linking China with Europe. For China it

is a way of utilizing its vast foreign exchange, currently valued at a staggering $4 trillion, and a way to signpost its growing international prestige. The AIIB signals China's emergence as an active stakeholder in the global economic order.

HONG KONG

China has two special administrative regions, Hong Kong and Macao. Both were annexed by European powers and only recently returned to the PRC.

Hong Kong became a British colony after China's defeat in the Opium War (1839–1842). Britain laid claim to the island and then annexed the neighboring island and part of the mainland. The city grew as a trading port and people flooded in from the mainland. From the 1950s to the 1970s, the city was an important manufacturing center and since the 1970s has emerged as an important banking center with banks eager to display their wealth status and cultural capital by commissioning some of the world's most famous architects, known as **starchitects**, to design their corporate head offices. The city is a wonderful display of contemporary architecture. Over 7 million people live at a density of approximately 17,000 per square mile (Figure 7.15).

Hong Kong was returned to China in 1997 when it was established as a special administrative region, under the slogan of "one country, two systems." The slogan was meant to indicate that the PRC would respect the distinct character of Hong Kong as a private market-based economy. The Chinese authorities promised to grant universal suffrage by 2017.

7.15 High-density living in Hong Kong (photo: John Rennie Short)

Since the handover, the role of Hong Kong has changed in two ways. First, it is no longer so different from other cities in China. In 1984, Hong Kong constituted the equivalent of 12 percent of China's total GDP. By 2013, it had dropped to 3 percent. In 1984, it had the equivalent of 230 percent of Shanghai's GDP but declined to 65 percent in 2013. The larger cities on mainland China now look more like Hong Kong.

In 2013, more than 50 million people from the mainland visited HK. Property prices doubled from 2009 to 2014 as mainlanders paid cash for Hong Kong property. Many in Hong Kong resented the newly rich mainlanders and their differing customs, and they were widely mocked for their lack of sophistication and the designation "locusts" was used to describe their frenzy of consumption.

The two-systems notion was tested in the summer of 2014 when demonstrations erupted after the Chinese government announced that the next chief executive of the territory would come from a slate of candidates approved by the central authorities. Protestors, many of them young students, took to the streets to demand free and open elections. The Chinese authorities cracked down heavily in the protest. Chinese authorities see the two-system designation as allowing Hong Kong to be a separate economic entity, but not a separate political entity.

MACAU

Macau is another special administrative region of China (Figure 7.16). Originally part of China, from 1557 onward it was used as trading port by the Portuguese, who paid an annual rent until 1887 when they achieved formal control. It formally became part of China in 1999. Neither as big as Hong Kong—its population is still only 635,000—nor as economically successful, its recent growth is based on manufacturing and on tourists from the nearby mainland. More than 20 million tourists visit the island, mainly from mainland China and Hong Kong. Casinos are a huge attraction, and Macau now rivals Las Vegas as a global gambling center.

Taiwan: Economic Power, Uncertain Politics

Taiwan was annexed in the early expansionary era of the Qing Dynasty and soon populated by Han Chinese. When the CCP took control over mainland China in 1949, their defeated enemies, the nationalists, set up government in Taiwan. Two million mainland Chinese moved to Taiwan, where 6 million people already lived. The mainland Chinese soon established themselves as the political and economic elite.

The Republic of China (ROC), as it was known, in contrast to the People's Republic of China (PRC), was initially seen in the West as a legitimate state, and it was given a seat at the United Nations. It received and continues to receive US economic and military support. In 2010, the United States sold $6.4 billion in arms to the island.

7.16 Map showing Hong Kong, Macau, and Taiwan

After 1971, when the PRC was accepted into the United Nations, Taiwan lost international legitimacy. The ROC continues to claim sovereignty over mainland China, including Outer Mongolia, but few countries recognize this claim. The PRC is now seen as the legitimate sovereign state. China claims Taiwan as one of its twenty-three provinces, and all official Chinese maps display Taiwan as a province of PRC.

Both states exist in a state of guarded and slightly uncertain political relations. The official CCP position is that Taiwan is a part of the PRC; this position is nonnegotiable, but the CCP seems content to play the long game and wait for reunification without forcing the issue in the short to medium term. The truce exists as long as there is no international recognition of Taiwan as independent from the PRC. Since 1992, the ROC has dropped the pretention of retaking the mainland. The policies are known as "mutual nondenial."

Despite this political uncertainty, Taiwan emerged in the 1970s as one of the Asian tigers—an economy of dynamic growth based on the export of manufactured goods. The island's growth was spectacular. In 1960, GDP per capita was around $160. By 2011, it had increased to $37,000. Taiwan became a modern industrialized society and citizens enjoyed long life expectancy and a high quality of life.

Despite the political differences, Taiwan and China are economically very close. Taiwanese companies invest billions in China and use mainland China as a source of labor-intensive industrial production. These powerful and strengthening economies make political friction less likely to spill over into outright aggression. However, the situation remains delicate and vulnerable to deterioration as well as improvement. For example, when Tsai Ing-wen was elected as President of Taiwan in 2016, her refusal to endorse a one-China policy was interpreted as provocative by the PRC. When China–US relations worsen, Taiwan also becomes more of a flashpoint.

Korea: Divergent Paths

Since the ending of the war, the two Koreas have followed two very different paths (Figure 7.17).

SOUTH KOREA

South Korea has undergone massive changes since 1945, including independence from Japan, rapid industrialization, and a greater democratization. While political development was relatively slow—with South Korea only emerging as something resembling a democratic system in 1987 after decades of military dictatorships and authoritarian regimes—economic development has been more spectacular. From 1960 to 1990, it was the second fastest growing economy in the world powered by export-led growth, producing goods, from cars and appliances to electronics and computers, sold around the world. From a weak and impoverished agricultural society at the beginning of the twentieth century, South Korea is now a prosperous urban industrial nation with an expanding and prosperous middle class. It is now the thirteenth largest economy and the tenth largest trading nation in the world. Its decisive insertion into the global economy does, however, make it very susceptible to global cycles of growth and decline. Economic crises in 1997–1998 and 2008 exposed the country's reliance on widespread borrowing and its heavy dependency on overseas export markets.

The economy is dominated by large family-controlled firms known as **chaebol** that have very close links with government. These multinational conglomerates, where families own vast enterprises by controlling the parent company, are powerful economic and political agents in South Korea with ownership across multiple sectors, though not in banking. Because they have strong political connections and are considered too big to fail, they exert enormous influence on the life of the country. Critics blame them for income inequality, political cronyism, and for blocking the expansion of more innovative small and medium-sized businesses. Supporters argue that they allow quick decision making and the pursuit of long-term growth plans rather than short-term shareholder gains.

NORTH KOREA

North Korea, in contrast, is a story of continued repression and economic hardship for the majority of its citizens. The Korean War was particularly severe on North Korea because of the US bombing campaign. Before the war, much of the country's industrial base was located in North Korea. During the war, an indiscriminate 3-year bombing campaign by the US Air Force against the North killed almost 20 percent of the total population, reduced much of the industrial infrastructure to rubble, and destroyed towns, villages, and dams. Flooding of farmland was widespread. This punitive campaign was condemned around the world, but the United States retains little historical memory of the campaign. However, it is the central feature of North Korean political propaganda used to promote the sense of being threatened and the fear of attack from South Korea and the United States.

While South Korea shifted from an authoritarian rule to military dictatorship to practicing democracy, North Korea has remained struck in an oppressive authoritarian system where power is concentrated in a family

KOREAN PENINSULA
City Population
- ⊛ ● 5–10 million
- ⊛ • 2–5 million
- ⊛ ○ Below 2 million

CHINA

Tumen R.

Ch'ongjin

Hyeson

Yalu R.

Changjin R.

Kanggye

Sinuiju

Hamhung
Hungnam

Taedong R.

Sunch'on

NORTH
KOREA

Wonsan

Pyongyang

Tongjoson-man

Sojoson-man

Korea
Bay

Namp'o

Imjin R.

DMZ

Haeju

Kaesong

Chuncheon

Goyang
Puch'on
Incheon
Ansan

Seoul
Songnam
Wonju

Suwon

Gyeonggiman

Han R.

SOUTH
KOREA

Cheongju

Kum R.

Daejeon

Naktong R.

Pohang

Gunsan
Jeonju

Iksan

Daegu

Ulsan

Gwangju

Changwon

Busan

Mokpo

Yeosu

Sea of
Japan
(East Sea)

Yellow
Sea

Western Channel

Korea Strait

Eastern Channel

Jeju Strait

Jeju-do

JAPAN

7.17 Map of Korean peninsula

dictatorship of Kim Il–Sung (ruled 1948–1994), Kim Jong-il (1994–2011), and Kim Jong-un (2011–). The regime has a cult of personality organized around the conflict with South Korea and the United States. The country is a giant military state that can draw upon 10 million men with over 5 million active frontline and reserve personnel, the largest in the world. It also has nuclear capability and regularly fires off missiles in periodic shows of strength and resolve.

The small political elite live in the capital city of Pyongyang. Party loyalists are rewarded with better living conditions than the majority of people in the country. In the 1990s, economic mismanagement and devastating floods caused major famine; the number of fatalities range from 400,000 to 3 million. Hunger is rampant between May and August before rice and corn are harvested; it is known as the barley hump. Malnutrition is rife, and so is nutrient deficiency. The average North Korean eats only half the number of calories that a South Korean eats. The dire food situation prompted some reforms. Since 2012, farmers "only" have to give 40 percent to the state and get to keep the rest. However, land is still not privately owned, so farmers are unwilling to devote too much to land that could be confiscated at any time. The country relies on exports of minerals to keep the economy afloat and fund the vast military. More than 90 percent of all exports are to China. Anthracite coal and iron ore are exported to China, although coal will not meet China's new sulfur standards. Other mineral resources include tungsten, zinc, and rare minerals such as molybdenum. Deposits of all minerals are valued at $6 trillion and help to bankroll the regime, one of the most secretive and brutal in the world.

Japan: Rising Economic Power, Declining Demographics

A snapshot picture of Japan from 1850 to the present would show how a country previously secluded from the rest of the world vigorously embraced industrialization and modernization, became an important regional power with global pretentions, invaded and dominated surrounding countries, and then rebuilt itself after devastating defeat to become a global economic power. More recently, economic growth has slowed; the population has aged while the geopolitical neighborhood has become a lot less friendly.

Japan's post war economic growth was based on rapid industrialization and the export of manufactured goods. At first the country produced relatively cheap low-tech products, but over the years it has specialized in more technologically sophisticated and expensive products. Japanese electronic products, widely criticized in the 1960s for being cheap and shoddy, by the 1990s were considered top of the line. Japan had

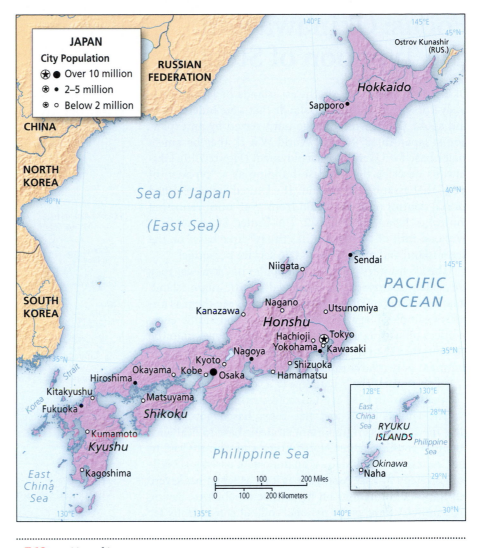

7.18   Map of Japan

successfully moved up the **value-added chain** to produce innovative products for the global markets. Cameras and cars were emblematic exports. By the early 1990s, however, economic growth stalled. New competitors were emerging, in the case of South Korea and China, or remerging in the case of the US car industry. In a property and share market collapse in 1990, the stock market lost 50 percent of its value, and two decades of subsequent anemic growth allowed China in 2010 to overtake Japan as the second largest economy. Japan is still an affluent country. The population has one of the longest life expectancies in the world, a result of affluence, diet, and excellent medical care.

Japan has a distinct **core-periphery** structure (Figure 7.18). At its center is the Tokyo-Yokohama metropolitan area on the island of Honshu. Also on this main island is the grid-patterned Kyoto with its rich architectural and garden design history; it was the capital of Japan from 794 to 1868, although Tokyo was the

effective power center from around 1600. Other cities on Honshu include the major ports and manufacturing centers of Osaka and Nagoya, the center of an 8 million population metro area. The islands of Shikoku and Kyushu constitute an inner ring. Shikoku is a small island filled with temples, although some cities have a Honshu-like modernity. Kyushu is Japan's westernmost island and was the first point of contact with Portuguese traders in the sixteenth century who brought Christianity to the islands.

The outer ring, of more peripheral areas of the country, includes Hokkaido and the Ryukyu Islands. In the northernmost part of Japan, the island of Hokkaido has one of the smallest populations. In the southernmost part of Japan are the semitropical Ryukyu Islands. These islands were ruled by Japan after an invasion in 1609 and then annexed in 1879 when Japanese was made the official language. The island chain includes the major US bases in Okinawa.

Focus: Okinawa and the Typhoon of Steel

Okinawa is one of the southwest islands of Japan previously known as the Ryukyu Islands that were independent and spoke their own language until seized by an expansionist Japan in 1879. After World War II, they became a major base for US military deployment (Figure 7.19). Okinawans describe the US base building as a "typhoon of steel" that assaulted the island. The United States retained formal control over the island from 1945 until returning control to Japan only in 1972. The United States controls over one-fifth of the land; it has thirty-three bases on the main island and more than 28,000 service personnel.

The US military bases are unpopular in Okinawa. They are widely scattered and dispersed on the main island, creating a large intrusive footprint. Locals see them as another example of Japan's colonial attitude and indifference, in effect offloading unpopular US military bases onto the island rather than in the core region of Honshu. Local resentment is regularly inflamed by criminal transgression by US personnel. The brutal rape of a young girl by US service personnel in 1993 crystalized resentment and anger.

The United States and the Tokyo-based government view the island as an important strategic base to deal with potential conflict against North Korea and China. Close to the Chinese coast, Okinawa is an important base for US military strategy in East Asia. And as long as Japan has severe self-imposed limitations on its defense capabilities, having a US military base, on its national soil but conveniently far from Tokyo, is a form of security insurance for Japan.

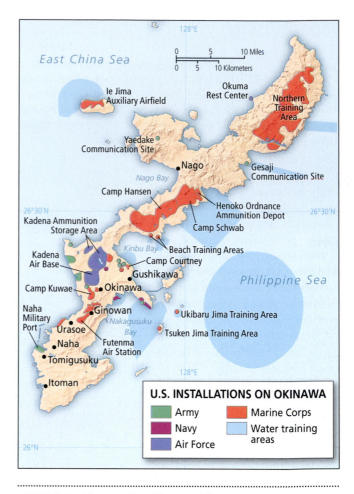

7.19 United States military bases in Okinawa

Select Bibliography

Anagnost, A., A. Arai, and H. Ren. 2013. *Global Futures in East Asia: Youth, Nation, and the New Economy in Uncertain Times.* Palo Alto, CA: Stanford University Press.

Beeson, M. 2014. *Regionalism and Globalization in East Asia: Politics, Security, and Economic Development.* Houndmills, UK: Palgrave Macmillan.

Birmingham, L., and D. McNeill. 2012. *Strong in the Rain: Surviving Japan's Earthquake, Tsunami, and Fukushima Nuclear Disaster.* New York: Palgrave Macmillan.

Bosker, M., S. Brakman, H. Garretsen, and M. Schramm. 2012. "Relaxing Hukou: Increased Labor Mobility and China's Economic Geography." *Journal of Urban Economics* 72:252–266.

Cai, K. G. 2012. *The Political Economy of East Asia: Regional and National Dimensions.* New York: Palgrave Macmillan.

Chang, I. 1997. *The Rape of Nanking: The Forgotten Holocaust of World War II.* New York: Basic.

Chellaney, B. 2010. *Asian Juggernaut: The Rise of China, India and Japan.* New York: Harper.

Dillon, M. 2012. *China: A Modern History.* New York: Palgrave Macmillan.

Du, Y., and J. Kyong-MacClain, eds. 2013. *Chinese History in Geographical Perspective.* Lanham, MD: Lexington.

Fan, C. C., and M. Sun. 2008. "Regional Inequality in China, 1978–2006." *Eurasian Geography and Economics* 49:1–18.

Fenby, J. 2013. *The Penguin History of Modern China: The Fall and Rise of a Great Power, 1850 to the Present.* London: Penguin.

French, P. 2014. *North Korea: State of Paranoia.* London: Zed Books.

Gifford, R. 2007. *China Road: A Journey into the Future of a Rising Power.* New York: Random House.

Hayton, B. 2014. *The South China Sea: The Struggle for Power in Asia.* New Haven, CT: Yale University Press.

Henshall, K. G. 2012. *A History of Japan: From Stone Age to Superpower.* New York: Palgrave Macmillan.

Hessler, P. 2001. *River Town: Two Years on the Yangtze.* New York: HarperCollins.

Holcombe, C. 2011. *A History of East Asia: From the Origins of Civilization to the Twenty-first Century*. New York: Cambridge University Press.

Huang, Y., and R. Tao. 2015. "Housing Migrants in Chinese Cities: Current Status and Policy Design." *Environment and Planning C: Government and Policy* 33:640–660.

Hui, W. 2014. *China: From Empire to Nation State*. Cambridge, MA: Harvard University Press.

Ingham, M. 2007. *Hong Kong: A Cultural History*. Oxford: Oxford University Press.

Jacques, M. 2009. *When China Rules the World: The End of the Western World and the Birth of a New Global Order*. New York: Penguin.

Lankov, A. N. 2013. *The Real North Korea: Life and Politics in the Failed Stalinist Utopia*. Oxford: Oxford University Press.

Lipman, J. N., B. Molony, and M. E. Robinson. 2012. *Modern East Asia: An Integrated History*. Boston: Pearson.

Mathews, G. 2011. *Ghetto at the Center of the World: Chungking Mansions, Hong Kong*. Chicago: University of Chicago Press.

Melvin, C. 2016. *Digital Atlas of the DPRK*. http://38northdigitalatlas.org.

Meyer, M. 2015. *In Manchuria: A Village Called Wasteland and the Transformation of Rural China*. New York: Bloomsbury.

Miller, I. J., J. A. Thomas, and B. L. Walker, eds. 2013. *Japan at Nature's Edge: The Environmental Context of a Global Power*. Honolulu: University of Hawai'i Press.

Ng, J. Y. 2014. *No City for Slow Men: Hong Kong's Quirks and Quandries Laid Bare*. Hong Kong: Blacksmith.

Osnos, E. 2014. *Age of Ambition: Chasing Fortune, Truth and Faith in the New China*. New York: Farrar, Straus and Giroux.

Ren, X. 2013. *Urban China*. Cambridge: Polity Press.

Rossabi, M., ed. 2004. *Governing China's Multiethnic Borders*. Seattle: University of Washington Press.

Samuels, R. J. 2011. *Securing Japan: Tokyo's Grand Strategy and the Future of East Asia*. Ithaca, NY: Cornell University Press.

Schell, O., and J. Delury. 2014. *Wealth and Power: China's Long March to the Twenty-First Century*. New York: Random House.

Shin, H. B. 2010. "Urban Conservation and Revalorization of Dilapidated Historic Quarter: The Case of Nanluoguxiang in Beijing." *Cities* 27:543–554.

Short, J. R. 2012. *Korea: A Cartographic History*. Chicago: University of Chicago Press.

Spence, J. D. 1974. *Emperor of China: Self Portrait of Kang-hsi (Kangxi)*. New York: A.A. Knopf.

Steinfeld, J. 2015. *Little Emperors and Material Girls: Sex and Youth in Modern China*. New York: I. B. Tauris.

Sun, M., and C. C. Fan. 2011. "China's Permanent and Temporary Migrants: Differentials and Changes, 1990–2000." *The Professional Geographer* 63:92–112.

Szechenyi, N., ed. 2018. *China's Maritime Silk Road*. Washington, DC: Center for Strategic an International Studies.

Totman, C. 2014. *Japan: An Environmental History*. New York: I. B. Tauris.

Veeck, G., C. W. Pannell, C. J. Smith, and Y. Huang. 2016. *China's Geography: Globalization and The Dynamics of Political, Economic and Social Change*. 3rd ed. Lanham, MD: Rowman and Littlefield.

Wei, Y. D., and I. Liefner. 2012. "Globalization, Industrial Restructuring, and Regional Development in China." *Applied Geography* 32:102–105.

Weightman, B. A. 2011. *Dragons and Tigers: A Geography of South, East, and Southeast Asia*. Hoboken, NJ: Wiley.

Winstanley-Chester, R. 2015. *Environment, Politics and Ideology in North Korea*. Lanham, MD: Lexington Books.

Woolley, P. J. 2005. *Geography and Japan's Strategic Choices: From Seclusion to Internationalization*. Washington, DC: Potomac Books.

Wortzel, L. M. 2013. *The Dragon Extends Its Reach*. Washington, DC: Potomac Books.

Wu, F. 2015. *Planning for Growth: Urban and Regional Planning in China*. London: Routledge.

Xie, Y., and X. Zhou. 2014. "Income Inequality in Today's China." *Proceedings of the National Academy of Sciences* 111:6928–6933.

Yu, W., J. Hong, Y. Wu, and D. Zhao. 2013. "Emerging Geography of Creativity and Labor Productivity Effects in China." *China & World Economy* 21:78–99.

Zhu, Y. 2007. "China's Floating Population and Their Settlement Intention in the Cities: Beyond the Hukou Reform." *Habitat International* 31:65–76.

Learning Outcomes

East Asia is one the fastest growing economic regions of the world. The region covers the four vastly different areas of the Plateau of Tibet, the Himalayas, the peninsula and islands of Korea and Japan, and the river basins of China.

Many of the world's earliest civilizations developed in the three river basins of the Yangtze, Yellow, and Pearl Rivers.

The region continues to grow despite environmental hazards and losses from soil erosion, flooding, environmental degradation, and typhoons.

At the heart of the region's recent, rapid economic success is the rapid industrialization and global shift of manufacturing. Agriculture still plays an important, yet diminished role across the region.

Japan and Korea are ethnically homogenous while China has a more diverse ethnic population, dominated by the Han.

China's population growth is now stabilizing, while Japan and South Korea are now characterized by low fertility rates, rapidly aging populations, and a shrinking labor force.

All three of the major countries in the region experienced a major redistribution of population over the last 50 years as people moved to the cities and cities grew into metropolitan areas.

The geopolitics of the region is shaped by intraregion conflict and cooperation as well as the influence of the United States.

China's long border is a site for cooperation as well as conflict. There are three distinct conflicts in the East and South China Seas.

China is one of the economic success stories of the contemporary era, nearly doubling its per-capita income recently over a period of only 10 years, a feat that took the United States nearly 50 years.

After the Korean War, rapid industrialization and democratization led to the success of today's South Korea, while North Korea's repressed, authoritarian society has led to a poor economy and widespread malnutrition.

Japan, although beset by population decline, has a high standard of living and a very strong economy based on the export of manufacturing products.

8

South East Asia

Since it gained independence from colonial powers, this region has emerged as one of the new powerhouses of the global economy. It is a region of spectacular economic growth and also one of marked economic variation. Alongside sites of spectacular economic growth, pockets of persistent rural poverty still remain. At one extreme are the boomtowns of Kuala Lumpur and Singapore, and at the other are the poor rural areas of Laos and Cambodia. Population growth has leveled off, but cities continue to grow. Religious and ethnic differences create a complex social and political geography and there is growing conflict in the South China Sea. The region has one of the world's most affluent states, Singapore, as well one of the largest Muslim countries in the world, Indonesia.

LEARNING OBJECTIVES

Survey the region's environmental challenges of tectonics, weather, and climate change, and connect these to impacts on human settlement.

Discuss the region's precolonial empires and distinguish its colonial experience.

Describe the importance of agriculture in the colonial period and outline its later waves of industrialization.

Recognize the social dimensions that shape the region and examine contemporary demographic trends.

Identify the island of Bali, describe its distinctive religious and agricultural landscapes, and relate this to the history of the wider region.

Define urban primacy and then connect it to the region's urbanization, poor urban environments, and large informal sectors.

Summarize intrastate, interstate, and international conflicts that shape postcolonial geopolitics in the region.

The Environmental Context

South East Asia is impacted by at least four environmental challenges: seismic activity, monsoonal storms, sea level rise, and deforestation.

Seismic Activity

This region sits on a precarious location on an arc of instability. Much of South East Asia lies along the fault line of the Australian Plate pushing against the Eurasian Plate. The Philippines sits astride the boundary of the Eurasian and Philippines Plates where volcanoes such as Mayon and Pinatubo continue to erupt on a regular basis.

There is a long history of volcanic activity in the region. Around 74,000 years ago, a giant volcanic eruption occurred at Mount Toba in Sumatra, Indonesia. Almost a billion tons of ash and dust were thrown up into the atmosphere, darkening the skies and reducing global temperatures by around 9°F for up to 5 years. The long volcanic winter severely reduced early human populations to around 10,000. In 1257, a massive volcanic eruption on the island of Lombok, Indonesia, created a 29-mile-high plume of ash, rock, and gases that spread widely, cooled the entire world, and reduced agricultural output. Mass deaths from starvation at the time were experienced as far away as London.

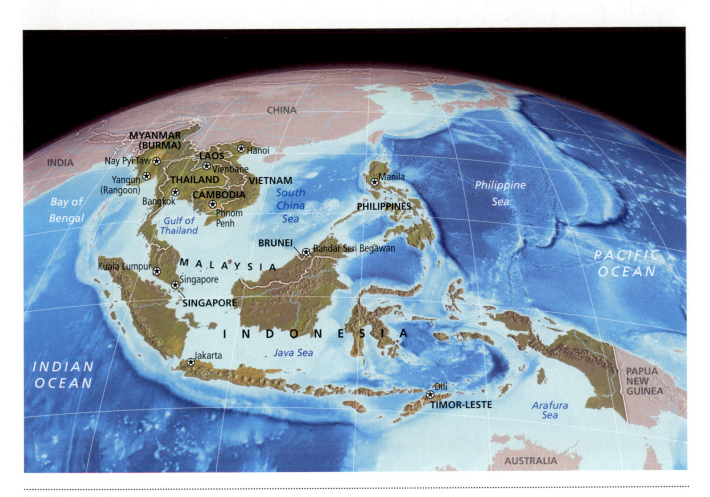

8.1 Map of region

The region saw another massive volcanic explosion when the island of Krakatoa exploded in August 27, 1883. The resultant tsunami killed approximately 36,500 people as tidal waves drowned towns and villages along the coast. The air pollution caused by all the dust and ash blotted out the sun's heat and light and reduced the world's temperature by about 1°F.

In 1991, Mount Pinatubo on the island of Luzon in the Philippines erupted: 37,000 acres of farmland were destroyed, 60,000 people were evacuated, and the US airbase Clark was abandoned.

In December 26, 2004, movement along the plate boundaries caused another major disaster. The Sumatra-Andaman earthquake occurred 19 miles below sea level off the coast of Sumatra. Along a 1,000-mile zone, plate movement raised the surface of the seabed 50 feet, displacing millions of gallons of water that caused the tsunami of postearthquake tidal waves that overwhelmed coastal communities in countries fringing the Indian Ocean. More than 230,000 people died, 125,000 were injured, and almost 1.7 million were displaced from their homes. Many of the major metro areas in South East Asia, such as Manila and Jakarta, are located on the coast, making them especially vulnerable to tsunamis.

There are major disasters of global proportions, but there are also less newsworthy but still disruptive effects. In 2015, Mount Sinabing in northern Sumatra came back to life and spewed ash over much of the region, forcing more than 10,000 people to evacuate. The main airport at Denpasar on the island of Bali, Indonesia, is regularly closed due to volcanic ash making air travel dangerous. In November 2017, 100,000 people were ordered to evacuate from their homes on the island because of the eruption of Mount Agung (Figure 8.2). More than 60,000 travelers were stranded on the island during the busy tourist season because 445 flights had to be cancelled.

Monsoons and Typhoons

Much of the region is monsoonal with a wet monsoon from May to October and dry conditions from November to April. The major storms are **typhoons**, the name given to hurricanes in this part of the world, which tend to occur during the wet **monsoon**. Typhoons are related to upper ocean temperatures in the lower latitudes of the western North Pacific. There has been an increase in the storm season and storm intensity due to an increase in the SST of more than 1.6°C compared to 2000. Typhoons are increasing in size and the typhoon season is lengthening, making the coastal communities increasingly vulnerable.

Typhoons cause widespread damage, especially in the poorer countries where there are limits to large-scale evacuation. The 2008 typhoon that hit Myanmar and the inadequate government response to widespread destruction undermined the declining public trust in the military junta. Typhoon Haiyan, with a storm surge of 10 feet and winds of 195 mph, ripped across the island nation of the Philippines in November 2013, killing 3,600 people and displacing 2 million as the storm tracked through vulnerable, high-density neighborhoods. The total damage amounted to $13 billion.

Climate Change and Flooding

The coastal cities of this region are vulnerable to flooding from sea level rise due to global climate change. The risk of coastal flooding is highest for Vietnam, Thailand, Myanmar, and Indonesia. In 2011, a major flood affected Bangkok, Thailand. Millions of acres were inundated and thousands of factories, which made electronic components in global supply chains, were flooded. The city is sinking about 0.8 inches a year, making the increased monsoonal storm that much more dangerous (Figure 8.3). Strategies include relocating much of the city to higher ground and/or building a massive sea wall with an estimated cost of $3 billion, approximately 1 percent of the country's GDP.

In total, 64 million people in cities in South Asia, a third of the total urban population, are at severe risk from coastal flooding.

8.2 This region experiences violent seismic activity. Mt. Agung in Bali is an active volcano. (Michael W. Ishak/Wikimedia Commons (CC BY-SA 3.0))

River flooding is also a problem. Over 76 percent of the urban population in Cambodia lives in areas subject to high risk of river flooding. There are social causes to increased urban flooding. Many rivers, especially in the major urban areas, are now engineered into straight flowing rivers that, with heavy rainfall, are more vulnerable to flooding. If the channels are not regularly maintained, then sediment can build up, making flooding more likely after heavy rainfall. Informal settlements also play a part as unplanned settlements along riverbanks in cities lead to the blocking and narrowing of channels. The greater risk of urban flooding is due to increased intensity and duration of major storms in association with the rapid and unplanned urban growth.

8.3 Bangkok is low-lying and vulnerable to flooding. (Photo: John Rennie Short)

Deforestation

There is also the deforestation of the tropical rainforest that is prevalent across much of the region. Deforestation occurs through the burning of forest cover to establish plantations, especially palm oil plantations, and through logging. The burning is a major cause of global air pollution as plumes of smoke, full of particulate matter, rise up into the air and are spread far and wide by the wind. In October 2015, NASA satellites detected at least 1,729 fires across Indonesia. Most were in the carbon-rich peat lands in Sumatra and Kalimantan. The fires were aided by the dry conditions of an **El Niño** event, which in this part of the world leads to drier conditions. The result was a spike in air pollution to levels that forced school closures. More than 40 million Indonesians had to breathe polluted air. I visited the island of Bali at the time and could taste and smell the acrid air. The air quality was so bad that the country's president returned from a visit to the United States in order to assess the damage and to be seen as doing something in the wake of the widespread pollution.

In Indonesia more than 6 million hectares of forest were cleared between 2000 and 2012 for agricultural commodities. There are plans to clear another 14 million hectares by 2020. In Burma, logging, both legal and illegal, produces few benefits for the wider population, as it is not used as the basis for a woodworking sector that could provide added value and generate employment and increased incomes for workers. Instead, the forest is cut down, and logs are shipped straight to East Asian countries and especially Japan. The main beneficiaries are landowners or those with political connections who can control access to the forests.

Deforestation results in a loss of **biodiversity** because the rich variety of ecological niches of multiple forest layers, which provide a range of habitat for many different species, is replaced by the monoculture of vast palm oil plantations. Three-quarters of bird and butterfly species are lost when a **tropical rainforest** is turned into oil palm trees.

Historical Geographies

Early Empires

The precolonial world was marked by interimperial rivalry, as kingdoms rose and fell, as well as interimperial contact and trade. Some of the dynasties were relatively brief while others lasted for centuries, such as the Ayutthaya Kingdom, which lasted from 1351 to 1767 in present-day Thailand. In northern Thailand, the Kingdom of Lanna, meaning "land of many farms," lasted from 1296 to 1768.

The empires developed in the low-lying river valleys where it was easier to exert central control. As one writer noted, early civilizations can't climb hills. The uplands were less easily controlled, and a distinction emerged between valley empires and hill tribes; a difference that exists today is the distribution of different ethnic groups.

The rise of the empires was linked to the switch from a dry, **swidden agriculture**, involving slash and burn, to wet rice cultivation and the vast mobilization of labor necessary to build and maintain irrigated farming systems. Sophisticated irrigation systems increased agricultural productivity. The empires had a complex social hierarchy. At the base were the farmers who provided the food for the whole community. At the top of the hierarchy was a kingly figure whose legitimation was based on divine or semidivine status, a personal embodiment of the link between the profane and the sacred world. Impressive architectural structures and planned city layouts were the built form of this social hierarchy and religious cosmology. There are three huge temple complexes in this region: Asia Angkor in Cambodia, Bagam in Myanmar, and Borobudur in Indonesia. Let us look at just one of them.

ANGKOR

The most famous precolonial city in the region is Angkor. It was the capital of the Khmer Empire from the ninth to the fourteenth century that stretched across 400,000 square miles covering parts of modern-day Myanmar, Vietnam, Laos, Malaysia, and Cambodia. The city displays the monumental size and elaborate decoration of the precolonial city in South East Asia with elements of both Hindu and Buddhist designs.

The complex was built under the rule of Khmer kings who ordered the building of a vast complex of more than three hundred temples to honor themselves and their gods. Most of the kings were Hindus and so most of the temples reference the Hindu religion with statues of gods such as Shiva. At its height, in the late thirteenth century, more than 100,000 people lived in the city.

Angkor sits on a vast complex of 1,150 square miles, similar to the metro areas of contemporary Los Angeles. The largest temple is Angkor Wat, built in the early twelfth century to honor the Hindu god Vishnu (Figure 8.4). The site, which covers almost three-quarters of a square mile, is one of the world's largest religious monuments. It consists of a central site surrounded by a moat and a 2-mile-long wall. In 1431, the King of Thailand sacked the city, and the empire moved its capital to the site where now sits the present-day city of Phnom Penh, 200 miles away. The city and the temples lay abandoned and the rainforest returned, covering much of the city. It was declared a World Heritage site in 1992 and now attracts up to 4 million visitors a year, bringing much-needed **foreign exchange** into a poor country.

8.4 Angkor Wat (Bjørn Christian Tørrissen/Wikimedia Commons (CC BY-SA 4.0))

8.5 Colonial empires in South East Asia

The Colonial Experience

The lure of spices and other exotic goods attracted Arab and later European traders. The Portuguese were the first Europeans to sail around the Cape of Good Hope in search of trading opportunities. In 1511, they sailed into Malacca, and a year later, they landed in Indonesia, eager to establish contact with the fabled Spice Islands. The spice trade was so valuable that the region was the scene of intense rivalry between European powers.

By the beginning of the twentieth century, the region, with the exception of Thailand, was divided up between four major powers (Figure 8.5). The Dutch controlled Indonesia. The French held Indochina consisting of Cambodia, Laos, and Vietnam. The British held sway over Malaysia, Myanmar, and Singapore. The Philippines was a Spanish colony until the Spanish American War in 1898 when it became a colony of the United States.

A significant turning point was World War II and especially the Japanese victories over the colonial powers. Independence movements, already in evidence before World War II, were heartened and emboldened by the colonial defeats. All the European powers sought to reestablish colonial control after the war, but the independent movements proved too strong and all achieved independence. In some cases, it was relatively easy and quick, as in the case of Myanmar; in other cases, as with Malaysia and Indonesia, independence occurred only after insurgency movements. The most tragic case was in French Indochina. The French tendency to see their colonies as part of France, and hence less able to see their eventual shift to independence, reinforced their commitment to hanging on to their colonies. France sought to reestablish itself as a legitimate world power by clinging onto its empire in Indochina—a tendency reinforced by the poor showing of official and military France in World War II. Vietnamese resistance fighters at Dien Bien Phu defeated the French in 1954. It should have marked the beginning of Vietnamese independence from colonial rule, in line with what was happening in Indonesia and Myanmar. But at this crucial juncture, the United States saw the world in **Cold War** terms and interpreted a Vietnamese resistance movement as a communist plot, part of a global strategy to achieve world domination. While other countries in the region achieved independence, the path for Vietnam, Cambodia, and Laos was littered with millions of dead and wounded.

8.6 Rice field in lowland Vietnam (photo: John Rennie Short)

Economic Transformations

This is a region of economic divergence. Some selected areas have grown dramatically and shifted to postindustrial economies of advanced producer services, as in the case of Singapore and Kuala Lumpur, while others, including much of Laos and Cambodia, remain largely rural and poor. A country such as Vietnam is located between these two extremes as the economy grew at a rate of over 6 percent over the past 15 years. There is a marked difference between landlocked Laos and the booming city of Singapore or even Ho Chi Minh City. Overall, this is one of the fastest growing areas in the world with sites of spectacular economic growth as well as pockets of persistent rural poverty.

Agriculture

The region is an important agricultural region, especially for the wet rice cultivation in the river basins of the Mekong and Irrawaddy and the fertile volcanic soils of Java and Bali (Figure 8.6). There is dry rice cultivation in the upland areas.

The region was transformed by colonial annexation and the subsequent reorientation of the economies for export markets. Traditional agriculture was replaced in selected sites with tropical plantations providing raw material for processing in Europe. Plantations of bananas, coconuts, coffee, pepper, rubber, tobacco, and sugar were established close to accessible locations along rivers and coasts. Local farmers also responded to the new economic opportunities of colonial annexation by increasing rice production through the extension of cultivated land and increasing the yield of existing areas.

After the Dutch gained control over much of Indonesia, they forced local farmers to cultivate export crops and pay taxes with crops at fixed prices. Later, a more private system of coffee, tobacco, and rubber plantations was underwritten with Dutch military force. Java became a major coffee producer, hence the name java for coffee still used in the United States. The island is the fourth largest coffee producer after Brazil, Vietnam, and Colombia.

In Indochina, the French used enforced labor and heavy taxation to make people work on plantations and produce exports goods. Sugar plantations were established in Laos, Cambodia, and Vietnam while Spain, and later the United States, encouraged sugar plantations in the Philippines. Currently, a new plantation system is being introduced as palm oil plantations are being created across the entire region.

Agriculture is still important. It provides much-needed foreign revenue. Thailand is one of the world's largest exporters of rice. And in the poorer countries, it is the main source of income. More than two-thirds of the labor force in Cambodia, Laos, and Myanmar is employed in agriculture, which is responsible for more than one-third of total GDP. Agriculture in these predominantly agrarian societies has low productivity and there is a 30 percent postharvest food loss due to poor infrastructure. The poor living conditions and restricted economic opportunities for small-scale farmers in these regions is one reason behind the rural-to-urban migration as people leave the countryside for the booming cities.

GREEN REVOLUTION

The **Green Revolution** played a significant role in boosting agricultural yields in this region. Rice yields in Malaysia quadrupled from 1960 to 1990. It allowed Thailand to become the world's largest exporter of rice. Annual rice production in the Philippines increased from 3.7 million tons to 7.7 million tons in two decades with the introduction of new strains of rice. In Vietnam, the revolution was delayed by the war with the United States, but new rice strains increased yields by 5 percent a year from 1980 to 2000. By 1989, Vietnam was able to export rice.

The Green Revolution was predicated upon greater use of fertilizers and pesticides that produced some long-term environmental damage and public health risks that are only now becoming more apparent. There is also more recent evidence of a falloff in rice yields, an indication that the heavy reliance on pesticides and fertilizers is neither a safe nor sustainable solution. Farmers are coming up against the limits to the long-term effectiveness of pesticide and fertilizers, and so the current revolution is tackling how to increase yields without damaging the environment or threatening public health. Around the region there are more examples of **organic farming**. Development issues now focus more on smallholders and sustainable forms of agriculture.

8.7 Petronas Towers in Kuala Lumpur. The city is an important hub in the global economy. (Photo: John Rennie Short)

Manufacturing and Services

This region has experienced three distinct waves of industrialization. The first began in the late 1960s and early 1970s in Singapore and in selected cities of Malaysia, Thailand, and the Philippines. It initially involved low skill levels of manufacturing such as textiles and cheap apparel. Companies were attracted by the cheap labor. A global shift in manufacturing was made possible because declining transport costs allowed companies to shift production to the cheaper labor areas of the world. This global shift impacted South East Asia as well as East Asia.

The second wave involved more advanced forms of manufacturing such as televisions, videos, and highly complex components in personal computers. In Kuala Lumpur, Malaysia, and especially around Bangkok, Thailand, this second wave provided higher paying jobs and helped to enlarge the middle class.

A third, more recent wave involves countries previously closed to the global economy. Cambodia, Laos, and especially Vietnam have opened up to foreign investment and global trade. They are attracting the routine manufacturing that was once located in China. As labor costs increase in China, manufacturing is shifting to these cheaper labor countries.

These three waves have transformed local economies, especially in selected towns and cities, and shifted national economies from predominantly rural to more mixed economies with manufacturing employment playing a bigger role.

As the economies mature, services constitute an increasingly larger proportion of total employment. In Thailand, services account for 53 percent of GDP, while agriculture only accounts for 10 percent. In Singapore, the respective figures are 84 percent and 1 percent, while for Cambodia they are 43 percent and 29 percent.

Producer Services and Global Cities

There are selected areas of advanced producer service concentrated in the larger towns and cities. Manila, Kuala Lumpur, Singapore, and Bangkok have all emerged as important centers of advanced services (Figure 8.7).

Singapore and Kuala Lumpur, in particular, are hubs in the global flows of money, ideas, and skilled personnel. Singapore is one of the world's busiest ports and also a major financial center. Kuala Lumpur is another. These cities house impressive signature buildings to mark their global status. Kuala Lumpur, for example, has the Petronas twin towers that at 1,453 feet high are symbols of growth and confidence. They were the tallest towers in the world from 1998 to 2003. Some of the cities have an interesting architectural feel as a postcolonial, postmodern city emerging from colonial and modernist legacies. The skyline of Singapore is filled with impressive works of signature architects. Both cities want to become hubs serving neighboring countries with the rest of the world.

Social Geographies

There are at least three important dimensions to the social geography of this region: religion, ethnicity and nationality, and gender.

Religion

The geography of religion roughly corresponds to the major division of mainland and insular South East Asia. The mainland areas include Cambodia, Myanmar, Laos, Thailand, and Vietnam, where Buddhism has replaced Hinduism as the dominant religion. In predominantly Buddhist South East Asia, religion is entrenched in architecture, public life, festivals, rituals, and religious iconography that give meaning and organization to social space.

In peninsular South East Asia, consisting of Indonesia, Malaysia, the Philippines, and Singapore, maritime trade routes brought ideas and practices as well as goods and merchandise from other parts of the world. Islam spread through the region along the trade routes of Arab sailors (Figure 8.8). In much of this part of South East Asia, Islam replaced Hinduism and later Buddhism to become the dominant religion except in the Philippines, where Christianity is strong, a result of centuries of Spanish influence.

Traditionally, Islam in this region was of a more tolerant form. More fundamentalist strains have only appeared relatively recently, fueled by Gulf state financing of **madrassas** and the Saudi promotion of Wahhabism interpretations. In Indonesia, only one province, Aceh, has **sharia law**.

Overlaying this basic division of mainland Buddhism and insular Islam are other religions. Christianity is found throughout the region. The Philippines is a predominantly Catholic country, and Christian minorities can be found scattered throughout the region. There is a strong legacy of colonial ecclesiastical architecture. In the less fertile uplands throughout mainland South East Asia, hill tribes with a distinctive ethnicity, language, and dress have a greater adherence to animist beliefs.

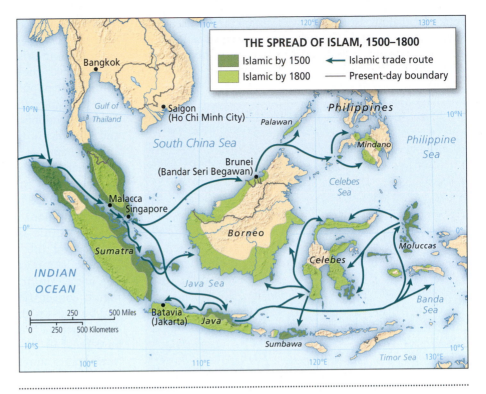

8.8 The spread of Islam

Ethnicity and Nationality

Religious affiliation is often linked to ethnicity and nationality. The majority of Thais, for example, are nominally Buddhist. Given the importance of religion and ethnicity in shaping national identity, tensions can arise in countries with ethnic or religious minorities.

Consider the case of Myanmar and the Muslim minority population of around 800,000 to 1.3 million, known as the Rohingya. They are Muslims in an overwhelmingly Buddhist country of 55 million people. For centuries they lived peacefully, with only occasional outbreaks of communal violence, but more recently they have been persecuted by the central authorities and subject to more local forms of persecution. The Rohingya are denied citizenship in Burma and have limited access to education, health care, and land rights. The military government considered them illegal immigrants from Bengal and since 1982 rendered them stateless. In recent years the persecution has increased. Intercommunal violence erupted in 2012, leaving 200 dead with 140,000 displaced, as a nationalist Buddhist monk-led movement whipped up enmity. Authorities have cordoned off the Rohingya in camps. More than 135,000 were kept in these camps in the western state of Rahkine. Living conditions became so bad that many decided to flee the country, at tremendous risk. Between 2012 and 2015, more than 120,000 fled persecution from the majority Buddhist society, leaving the country on boats for Thailand, Malaysia, and Indonesia. In 2017, another 600,000 Rohingya fled across the border into Bangladesh to escape persecution and violent attacks. Most neutral commentators described a form of **ethnic cleansing** as Rohingya villages were targeted for destruction.

8.9 Trader in Hoi An, Vietnam (photo: John Rennie Short)

Myanmar is an extreme example, the better to highlight the cleavages. But these cleavages exist in less dramatic form in other countries in the region. In Thailand, for example, the three southernmost provinces of the country are populated by Muslim Malays, some of whom seek separation from a Thailand dominated by Buddhist Thais. In Indonesia, there is tension between Christians and Muslims in the Moluccas and in Irian Jaya between animist Papuans and Muslim settlers who moved in from the other island of Indonesia. In the Philippines, conflict between Christians and Muslims in the island of Mindanao since 1960 has killed 160,000 people and displaced 2 million.

One ethnic minority group figures largely in the commercial life of the region, Chinese immigrants. Chinese communities, some of centuries-old standing, are a significant minority throughout much of the region concentrated in urban areas with employment and business niches in trade and industry. China ruled over much of Vietnam for centuries, and so Chinese have lived in Vietnamese cities for centuries. Chinese immigrants moved to Malaysia and Singapore under the British Empire. The Chinese presence is most marked in the major cities. In Singapore, 72 percent of the population is Chinese. In Malaysia as a whole, the Chinese constitute 24 percent of the national population, but in the important trading city of Penang, they constitute 45 percent. Occasionally the economic success of the Chinese provokes resentment, similar to the Jews in medieval Europe. When economic conditions worsen, anti-Chinese sentiment can be inflamed. There are around 2 million Chinese in Indonesia, and they are concentrated in the bigger cities. In 1998, as food shortages worsened, anti-Chinese riots occurred in the cities of Medan and Jakarta. More than a thousand people died in the rioting.

Gender

Another significant dimension of social geography is gender. Work is highly gendered, especially in the more traditional and rural societies where women play a significant role in agriculture and small-scale trading (Figure 8.9). In cities such as Penang, Singapore, or Kuala Lumpur, women play a wider economic role as more women are opting to have few or no children in order to maintain their career paths (Figure 8.10).

8.10 Young women in Penang, Malaysia (photo: John Rennie Short)

Demographic Trends

Population is concentrated in megacities and fertile agricultural areas and is much sparser in the upland areas further from the coast. In Indonesia, the differences are clear when comparing Java and Irian Jaya, where the densities are 126 people per square kilometer and 7 people per square kilometer, respectively.

An important demographic trend is the slowing down in the rate of population growth from the high levels of the 1960s and 1970s. The region is moving into the later stages of the demographic transition. In Indonesia, growth has declined from over 2 percent in the 1970s to around 1 percent today. In Thailand, in 1970 the average household size was 5.7 people, but by 2010 this had declined to 3.2; this is the result of increased affluence and one of the most successful family planning programs in the world. These programs were successful in reducing the birth rate because they were bottom-up implementations with local communities involved in transmitting the message for families to have fewer children. Public health measures have increased life expectancy. In Thailand, life expectancy is now 74 years of age. Indonesia is witnessing a demographic dividend as more of the population is now in the prime years of economic participation.

Singapore is perhaps most advanced along the demographic transition. The fertility rate is now only 0.79 as more women are electing to have fewer children or not having children. There are exceptions to this general rule: the poorer countries of Laos and Cambodia still have high birth rates, as does the predominantly Catholic country of the Philippines.

Rural Focus

Bali

Bali is a small island in a very large country. It is also a mainly Hindu island, more than 84 percent, in the world's largest Muslim country. It was a holdout in the replacement of Hinduism by Islam. The anomaly reveals the complex historical geography of the region.

In 1343, the Hindu Majapahit Empire, based in Java, founded a colony in Bali. The court cultivated the arts and crafts, music, and dance. The island retained its Hindu religion with the help of local leaders, while Islam became the dominant religion in the surrounding islands. Bali also held out against the Dutch, only becoming a Dutch colony after a military defeat in 1906. And so the island has this distinct cultural blend. Bali retains its centuries-old tradition of Hindu religion, dancing, craft, and music. The religion is referred to as **Bali Hindu**, a distinct form that combines Hinduism and Buddhism as well as ancestor worship and animism. There is a marvelously rich and varied musical and dance tradition on the island.

8.11 Terraced rice fields in Bali (Elena Ermakova / Alamy Stock Photo)

Bali has a spectacular landscape of terraced rice paddies (Figure 8.11). They are organized as **subaks**, a self-organized collective of farmers who share irrigation water from a common source. There are over 1,200 subaks that have survived for centuries and have evolved into sophisticated complexes of water sharing and labor cooperation. Yields are impressive. Fertilizers are not required as the volcanic soil is so naturally rich in potassium and phosphorous. There are 81,000 hectares of subaks, but they are being reduced by around 1,000 hectares a year to development. The subak landscapes are designated as a World Heritage site because of their beauty and their social significance as a 1,000-year-old form of sustainable, self-managed, agricultural cooperation. But the designation provides little protection against the march of development as hotels and resorts are built across the rice paddies to meet the demands of the 3.3 million tourists a year. A figure that, despite some fluctuation, is likely to rise in the future, putting more pressure on the subaks.

Urban Trends

This region is becoming more urban as people move to the cities. In 1975, the urban population of Indonesia was only 20 percent. Today it is closer to 50 percent. In Myanmar, in the last 25 years, the urban population has grown from 13 percent of total population to 25 percent. Across the region people have relocated from the countryside to the city. Cities have grown even as population growth has leveled off to around 1 percent.

Urbanization is characterized by the growth of large cities. Three of the twenty largest urban areas in the world are located in South East Asia: Jakarta (third at 31 million), Manila (fourth at 24 million), and Bangkok (nineteenth at 15 million). South Asia is emerging as a region of megacities.

Urban Primacy

The growth of very large cities is accelerated by **primacy**, the tendency for one or two cities to dominate the economic and political life of a country. In some cases, there is a close relationship between city and state, as in the case of Singapore, which is a city-state. In other cases, a legacy of colonial power centered in cities influenced subsequent urban growth. Yangon, Jakarta, and Manila were centers of colonial power and subsequently played an outsized role in the life of the country.

The most primate city distribution in the world is in Thailand. In 2015, it is estimated that out of a population of almost 68 million, just over 15 million lived in the urban region of Bangkok. The population of the next largest city region of Chiang Mai was under 1 million. By 2015, the city was responsible for almost half of all GDP. Thailand is one extended urban region centered in Bangkok.

In some countries, more than one city dominates the national economy. In Vietnam, for example, it is the two largest cities of Hanoi and Ho Chi Minh that, combined, have around 15 percent of the nation's total population of 95 million.

Urban Environments

Many of the cities in the region suffer from poor environmental conditions. Rapid growth in association with weak or poorly implemented pollution standards has resulted in poor environmental quality. The air pollution in Manila is so bad that it kills around 4,000 people a year with thousands debilitated by respiratory diseases. Water quality is often compromised, especially for the poorest. However, there are examples of increasing environmental quality. In Bangkok, the number of vehicles increased from 600,000 in 1980 to 6.8 million by 2015. Yet Bangkok's air quality is improving despite the greater number of vehicles because

a more affluent citizenry can now afford to buy newer vehicles with better emission controls.

Water pollution, traffic congestion, and air pollution remain significant problems, especially in the poorer countries experiencing rapid urban growth. In some cases such as Singapore and Kuala Lumpur, environmental standards are very high: a result of affluence and commitment to maintaining their global city status by ensuring the quality of the urban environment.

The cities of the region are also vulnerable to a range of environmental hazards, including the risk of tsunamis and earthquakes, typhoons, and flooding.

Informal Cities

Many of the cities of the region, especially in the poorer and rapidly growing countries, have a large informal sector both in employment and housing. In the Philippines, around 50 percent of the people work in the **informal economy**, ranging from waste pickers and street vendors to domestic and construction workers. Between a quarter and a third of the total population of Manila lives in informal settlements (Figure 8.12). Although the government regularly razes these slums and relocates the people, the pull of the capital is strong because the relocation sites are remote and distant; so the people return.

The term "informal" does not mean temporary or transient. In fact, more than three-quarters of slum dwellers have resided in the same place for more than 5 years. Many of the settlements are over 50 years old with the majority being built between 20 to 40 years ago. Over 100,000 families are living in danger zones such as alongside railroad tracks, garbage dumps, and canals and rivers subject to flooding. There are also climate-change refugees, who have been forced to move from typhoon-ravaged rural areas.

Urban eviction and demolitions continue throughout the region. But more progressive city and national governmental authorities see the slums less as problem areas and more as innovative sites for change through harnessing the entrepreneurial spirit of the slum dweller by upgrading the physical environment and regularizing legal claims to residency. These policies help to reduce urban poverty.

City Focus: Bangkok

Bangkok dominates the country of Thailand. It is the economic, political, and cultural center of the country and the world's foremost example of a primate city.

The city grew up on the banks of the Chao Phraya River that snakes its way around the heart of the city, after a new regime founded its capital in the eighteenth century. The city contains significant traditional buildings, including Wat Pho, home to the world's largest reclining Buddha and the ornate Grand Palace. These historic sites attract tourists from across the region and around the world. Throughout the city, Buddhist temples, large and small,

8.12 Informal settlements along the Pasig River, in Manila (SeaTops / Alamy Stock Photo)

and statues of Buddha, grand and modest, are interspersed with modern buildings and contemporary architecture.

Since 1960, the city's growth rate has been twice that of the rest of the country, fueled by rural-to-urban migration within Thailand and foreign immigration. Foreign migrants, initially from Burma, Cambodia, and Laos and latterly from Korea and China, are drawn to the city. The country has almost 2.3 million foreign migrants, the majority located in the capital city region. There is a long-established Chinese community, and the city's Chinatown is an important shopping area.

The city faces environmental issues of poor air quality and the risk of flooding. Most of the city is little more than 15 feet above sea level (see Figure 8.13). The city is being restructured at the periphery as Western-style suburbs are built and in the center as high-density apartments are built for the more affluent middle- and upper-income groups.

The rapid growth outpaced infrastructure by the 1980s, but more recent construction of motorways, bridges, and mass transit has solved some of the endemic traffic congestion. In Bangkok, neighborhoods contain the commercial and the residential, the old with the very new. The city has a distinctive skyline. The high-rise buildings are not clustered but widely spread along the main route ways, so the city has a vista of high-rise buildings stretching across the entire urban landscape.

The business of the red light areas of the city was significantly boosted during the Vietnam War when the city was a rest and recreation center for US troops stationed in Vietnam. Today, it attracts male sex tourists from across Asia, Europe, and North America.

In Bangkok, as in many cities in the region, there is the juxtaposition of very modern dwellings alongside slum settlements. Klong Toey in Bangkok is one of the city's largest slums, housing 100,000 people on a 1-square-kilometer site built on a swamp. A railway line runs right through the community with heavy-goods trains passing directly outside people's homes. Migrant workers drafted to build the port built the slum in the 1950s. They built their homes directly outside the construction zone. When construction was completed, the government claimed the land back and so the settlement became illegal. It remains a point of entry for poor rural migrants. Twenty thousand people have moved into the neighborhood in the past 10 years. Criminal gangs run the neighborhood. Overhead, the Skytrain carries affluent citizens and tourists above the squalor. And yet the city is a place of relative affluence. Only 12 percent of Thailand's poor live in the cities. Most poor people are in the rural areas.

The city is an important global city not only for the country but also for the wider region of South East Asia. Foreign companies operating in the region often seek out a Bangkok location in order to be effective.

Concentrating economic and population growth in one relatively small area has created a backlash from a national body politic divided between a prosperous city and

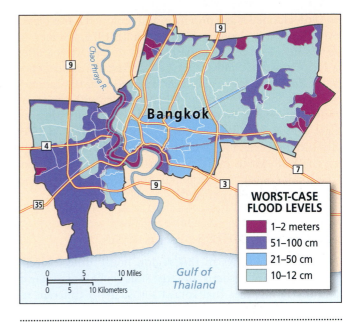

8.13 Map of Bangkok

an impoverished rural majority. A major political divide is between an urban middle class based in Bangkok and millions of poor rural people.

Geopolitics

There are two forms of geopolitics: tensions within the state and tensions between states.

Postcolonial Reshufflings

Even after decolonization, there was a reshuffling of national boundaries. Consider the example of Malaysia. The country achieved independence from the United Kingdom in 1957. It originally contained Singapore. In 1964, Sarawak and North Borneo became part of Malaysia, eager, as they were, to distance themselves from Indonesia. In 1965, Singapore broke away to become an independent state. The distrust was mutual: Singaporean political leaders, mainly Chinese, were worried about being overwhelmed by an Islamic Malaysia, while Malaysia was concerned with the economic power of a Chinese-dominated city in their national midst. In 1984, Brunei, the tiny oil-rich sultanate in Borneo, achieved independence to become one of the richest per-capita states in the world.

The postcolonial assembly and reassembly of national boundaries also occurred in Indonesia. The country became independent in 1949 after a bitter liberation struggle against the Dutch. There were colonial outliers, including West Papua, retained by the Dutch, and East Timor, a Portuguese colony. Indonesian forces invaded West Papua, in 1961, claimed it as a province, and renamed it Irian Jaya. Since 2007, it is also known as Papua. In 1975, Indonesian forces invaded the Portuguese colony of East Timor, after a local resistance movement had declared independence from Portugal. The region was annexed into the Indonesian state, but there was a violent conflict with the resistance movement. While Indonesia was majority Muslim, East Timor was overwhelmingly Catholic. Only in 1999 did Indonesia relinquish control, after a bloody campaign by the Indonesian military against the local population. The small country with a population around 1.25 million became independent in 2002.

Divergent States

National boundaries in this region house different religious and ethnic groups. In large states such as Indonesia and the Philippines that sprawl across many different islands, different peoples with complex histories and multiple allegiances and identities can pose a problem for national unity.

The newly independent Indonesia, for example, inherited a patchwork of very different ethnic and religious groups across the largest archipelagic nation in the world. The nation stretches over 3,000 miles across 18,000 islands. There are ethnic and religious divides. The native people of Papua are Melanesian, mainly animist, very different from the Muslim Indonesian. The recent influx of Indonesians is the basis for conflict, especially as the Papuans are much poorer. A similar situation occurs in Kalimantan, where the animist Kayaks are marginalized by the recent immigration of Indonesian immigrants. In Sulawesi and Maluku, tension between Christians and Muslims can erupt into violent conflict. Shared religion is no guarantee of stability. In Aceh, and especially Banda Aceh, more conservative Muslims seek religious purity by breaking away from what they see as a too secular central government in the Javanese power center.

In the Philippines, there is a powerful Muslim separatist movement in the south of the country, centered in Mindanao. Its origins lie in the US colonial period when Muslim identity was reinforced by administrative procedures such as lumping together all Muslims as one category and taxing their land. After independence in 1935, Christian settlers were encouraged to settle in Mindanao. Disparities in government service provision, with Christians doing better than Muslim communities, heightened the difference between Christian settlers and Muslim farmers. In 1972, an armed insurgence fought the Filipino state. A Muslim autonomist region was created in 1990. A 1996 peace agreement failed to take hold, and the conflict continues between Filipino troops often backed by US forces and the Moro Islamic Liberation Front.

Across the region most nation-states are multicultural. In some cases there is an outright recognition of multiculturalism. The affluent city-state of Singapore has four official languages—English, Chinese, Hindu, and Malay—that give recognition to the different elements of the society. The Chinese dominate in numbers and economic power, but there is strict enforcement of quotas in employment and public housing to maintain the ethnic balance. If a Hindu family moves out of a public housing project, then a Hindu family has to move in. In Singapore, multiculturalism is embraced and is the basis for city branding, as the area of Little India is now an important tourist attraction. In other places, multiculturalism is resisted. In Malaysia, there is a direct preference for Malays over Chinese in public service employment.

Even in Thailand, which has a longer history of territorial integrity and ethnic homogeneity than its neighbors, there are differences. In the south there are three provinces conquered by Thailand in 1785 that are populated by Malay Muslims. Since the 1960s, there have been demands for more autonomy, and a separatist movement that has morphed into demands for an Islamic caliphate. Between 2004 and 2014, 6,000 have died and 10,000 have been injured in the conflict. Despite improvements, this border region, compared to the rest of the country, has lower incomes, lower educational levels, and more limited employment opportunities.

In Myanmar, there is a major conflict between the majority Burmese and the long-running separatist movements of the hill tribes, including the Kachin, Karen, and Shan (Figure 8.14). These ethnic minorities make up between 30 and 40 percent of Burma's population of 52 million. At least 350,000 people have been displaced by the insurgency war between the Burmese Army and ethnic minorities.

International Conflicts

International disputes in the region have taken two main forms. The first is conflict between nation-states in the region. These are relatively rare. Vietnamese troops invaded Cambodia in 1979, and China invaded Vietnam in 1979. Much more damaging is the region's role as an arena for US intervention.

When French colonial power was undermined by Vietnamese nationalism in the postwar world, politicians in the United States saw it as part of a worldwide communist threat. In fact, it was a nationalist movement similar to what was taking place in Burma, Indonesia, and Malaya. But by seeing it as a scene for the global struggle, the United States got involved in the region. From 1955 to 1975, the United States engaged in a war to prop up unpopular South Vietnamese regimes dominated by rich Catholics in a country of poor Buddhists. At the height of the engagement, more than one-half million US troops

8.14 Shan women in Myanmar (photo: John Rennie Short)

the border, and the area remains in dispute. The International Court of Justice has confirmed Cambodian sovereignty, but Thailand still controls the area on the ground.

Many territorial disputes have been resolved. In 1985, Vietnam and Cambodia settled their dispute over an island in the Gulf of Thailand, and in 1990 the Laos-Vietnam border dispute was settled. Some remain outstanding such as between Vietnam and Cambodia. While capable of minor conflict, they do not provide the basis for sustained conflict.

The major international territorial dispute that has emerged in recent years is maritime. The conflict in the South China Seas involving China, the Philippines, Vietnam, and others was discussed in the previous chapter (see Figure 7.12 in Chapter 7).

were based in South Vietnam, and bombing campaigns were conducted against North Vietnam as well as Cambodia and Laos. From 1 million to 3.1 million Vietnamese, around 250,000 Cambodians, and 100,000 Laotians were killed. Almost 60,000 US troops were also killed. The war devastated the region. In Laos from 1963 to 1973, the US air force in covert operations dropped 2 million tons of bombs, more than half a ton for every single person in the country. Unexploded US ordinance is still a major hazard in the rural areas of Laos. In Cambodia, the US air force carpet-bombed the country, devastating much of the country. The US bombing set the stage for the **Khmer Rouge** in 1975 that, in turn, killed 2 million Cambodians.

Since the defeat of the United States in 1975, the region has become much more peaceable, slowly reverting to a normal state of affairs as war-damaged societies are rebuilt and defoliated areas brought back to life. The United States and Vietnam now have normal diplomatic and trading relations. Normality has returned to a region devastated by war.

Boundary Disputes

Most of the boundaries of the states in South East Asia are compromised arrangements formed in the wake of decolonization. In some cases, tensions remain. The French drew up the boundary between Thailand and Cambodia in 1907. An area around Preah Vihear was allocated to Cambodia. In 2011, there were military skirmishes across

Connections

The Bamboo Network

Bamboo is the Chinese symbol for durability. The **Bamboo Network** is the name given to the extensive connections between Chinese family businesses in South East Asia. As early as the sixteenth century, Chinese migrants from southern China moved into the trading posts throughout South East Asia, including Penang, Singapore, Bangkok, Jakarta, Manila, and Ho Chi Minh City. The immigrants grouped together in clans of extended families in self-help organizations. These organizations were the basis for many of today's firms in the cities in the region. The typical company was mid-sized and family owned. After the economic reforms in China in the 1980s, the networks reestablished connections with mainland China and investment flooded into China. In the past 10 years, the flow of capital has reversed as Chinese capital flows along the Bamboo Network into South East Asia. Today's Bamboo Network extends into London, Los Angeles, and New York as well as Bangkok and Jakarta.

We tend to think of global flows in terms of multinational companies, but family connections across the globe are often important conduits of global trade. An old Chinese maxim says that bamboo is strong because it bends but does not break.

Subregions

There are a number of different ways to make regional subdivisions of this world region. A broad division is often made between mainland and insular South East Asia, with the latter consisting of Malaysia, Singapore, Indonesia, and the Philippines. We could also, for example, make a distinction between the predominantly Buddhist societies of Cambodia, Laos, Myanmar, Thailand, and Vietnam; the more Muslim states of Brunei, Malaysia, and Indonesia; and the Catholic Philippines. We could also make a division based on the different experiences of French, British, and US colonies with that of uncolonized Thailand. Here we will combine elements of all of the above and identify five different surregions in South East Asia: the former French colonies of Cambodia, Laos, and Vietnam; Myanmar; the former British colonies of Malaysia, Singapore, and Brunei; Thailand; and the island states of Indonesia and the Philippines.

Cambodia, Laos, and Vietnam

These three countries form a coherent physical geography of a mountain spine, coastal plains, and fertile river floodplains (Figure 8.15). The Mekong, with headwaters in the Tibetan plateau, flows through all three countries. Laos lies almost entirely within the lower Mekong Basin. By the time the river reaches Cambodia, it is a large, slow-moving river. It is a major source of fisheries and the natural irrigation source of the fertile rice-growing areas in both Cambodia and lower Vietnam. More than 40 million people in the lower Mekong earn a livelihood by fishing. It is a major source of biodiversity but is under threat from damming upstream and deforestation that increases the risk of extreme runoff and flooding. The Red River flows from China into Vietnam; like the lower Mekong, it provides rich alluvial soils suitable for rice-intensive cultivation.

A distinction can be made across the region between dry cultivation of the upland areas and wet rice cultivation in the fertile river basins. The fertility of the basins is the basis for the long imperial history of the region.

The three countries share a **monsoonal climate** that consists of a dry, cool season from November to February/March and a wet, warm, and very humid season from April/May to November.

The physical geography also influences the social geography. In Laos, the Lao people live mainly in the lowlands, while ethnic minorities such as the Hmong, Mon, and Mien live in the highland areas. The majority of the population, close to 86 percent, in Vietnam is Viet, also known as Kinh. They are concentrated in the fertile deltas of the Mekong and Red Rivers and along the coast. In the upland areas of dry rice cultivation in the north and central highlands live distinct ethnic groups such as the Hmong.

There are other substantial minorities in each of the three countries. Chinese are found in the towns and cities, a reminder of China's long and close ties to this region. Under the French, the Vietnamese played an important role helping the French run their empire in Laos and Cambodia, and so Vietnamese are found in these two counties. In Cambodia, there are close to half a million Chinese and 1 million Vietnamese who dominate the professions and business.

8.15 Map of Cambodia, Laos, Myanmar, Thailand, and Vietnam

All three countries were under French control from the late nineteenth century until 1954. The French introduced a brutal plantation system of agriculture. The colonies were stripped of their assets with the wealth transferred back to France. The French reluctance to give these countries independence set the scene for the Vietnam War and subsequent political conflict.

A landlocked Laos and war-damaged Cambodia are rural and poor, though there are signs of increasing industrialization, especially around the main cities of Vientiane in Laos and Phnomh Penh in Cambodia. A major problem in landlocked Laos and Cambodia is the lack of infrastructure.

Since the early 1990s, Laos has pursued a goal to become the "battery" or major source of power in South East Asia for hydroelectric projects. Fifty major projects by 2015 involved the resettlement and displacement of villagers on affected land. While some of the commercial famers received some form of compensation, the ethnic minorities who practice swidden agriculture were not and were targeted for removal.

Since the mid-1980s, all three countries have moved away from a **command economy** toward a more **market economy**. The move is most pronounced in Vietnam. In 1986, the Vietnam government began a series of reforms, known as *doi moi*, that encouraged foreign investment and a greater role for the private market. The changes have improved living standards, especially since the mid-1990s when the reforms began to take hold and have real economic effects. From 1990 to 2010, annual per-capita income change averaged close to 6 percent compared to 4.9 for India and 3.4 for Bangladesh. National income per capita is now close to that of India and more than one and a half times that of Bangladesh. Vietnam is one of the fastest growing economies in Asia, experiencing growth rates similar to those in Japan in the 1970s and China in the 1990s (Figure 8.16). The cheap skilled labor attracts investment from multinationals and encourages local entrepreneurs to make things for the global market. Vietnam is now the world's third largest shoe exporter after China and Italy. However, there are complaints about the poor conditions for workers and the suppression of trade unions.

Of the three countries, Vietnam is leading the way in opening to the outside world, with tourism increasing and foreign investment growing.

Myanmar

Myanmar became independent in 1948. It could have been included in South Asia as it was essentially part of Britain's vast Indian empire. It sits on the edge: it is part of South Asia in terms of its colonial history, but part of South East Asia in terms of its strong Buddhist culture.

The British conquered Myanmar after three separate wars in the nineteenth century that lasted from 1824 to

8.16 Rapid economic growth in Vietnam is embodied in a construction boom in Ho Chi Minh City. (Photo: John Rennie Short)

1885. As in subsequent political struggles, Buddhist monks played a major part. The colonial capital was established in Rangoon, now Yangon.

Myanmar was a major rice producer; its teak forests were exploited, and its gems and minerals were mined. It was a classic example of primary production exploitation of a colony. The British left very little lasting cultural legacy, unlike Pakistan and India, for example, where the English language is still an important means of communication for elites and cricket is still played.

The country was only briefly democratic before a repressive military dictatorship took over and ruled from 1962 to 2011. In 1989, they renamed the country Myanmar and changed Rangoon to Yangon. Emphasis was on extractive industries such as mining and timber with military generals controlling large tracts of land. The military regime built a brand new capital, Naypyidaw, in the middle of the jungle to minimize the political strength of the population in Yangon, which at close to 6 million people is by far the largest city in the country and long a site of resistance to the junta.

The first free elections were held in 1990, but when the popular vote gave Aung San Suu Kyi the victory, the military junta imprisoned her. There were periodic bouts of resistance. In 2007, the **Saffron Revolution** was led by young monks.

By 2010, under international pressure and in order to attract foreign investment from the West, and to avoid further sanction from the United States, the junta started to democratize and embarked on some **economic liberalization**. The first meaningful elections were held in 2010, although the military reserved 25 percent of seats and kept powerful cabinet posts as well as the leading political offices.

The military's lack of response to a devastating storm in 2012, when 85,000 people were displaced, further

8.17 Teak from the forests of Burma being loaded onto ships in Yangon (photo: John Rennie Short)

undermined popular support. In November 2015, Suu Kyi's party won an historic electoral victory with the vast majority of the popular vote and over 80 percent of the seats. The election was, as one Burmese said to me, "our chance of freedom."

Myanmar is one of the poorer countries in the region, and its main exports are still primary products (Figure 8.17). Economic development was hindered by sanctions against the military government since lifted, but the long-running ethnic conflicts between the Burmese Army and the separatist movements of the hill tribes still cast a pall. The Karen, an ethnic group of 6 million in Myanmar and 1 million in neighboring Thailand, have been fighting the Burmese central government since very soon after independence in 1948. The roots of the conflict lie in the colonial period when the British excluded ethnic Burmese from the military and so the minorities were associated with colonial rule and seen by the Burmese as privileged servants of the British. The newly independent state pursued a policy of Burmanization, especially in education. Burmese identity defines the national rhetoric. At least 35,000 people have been displaced by the insurgency war. After decades of fighting, eight different ethnic groups signed an agreement with the government in October 2015. However, two of the largest groups, the Kachin and Wa, with ties to drug trafficking syndicates and with tens of thousands of troops, did not sign; fighting continues in Kachin-held areas. The Kachin areas in the far north of the country bordering China are rich in natural resources such as jade. And the control over these lucrative resources lies at the heart of the disputes between the Buddhist central government and the largely Christian Kachin.

The 1.3 million Muslim Rohingya, living in Rahkine province, are a marginalized and oppressed minority. More than 600,000 Rohingya fled across the border to Bangladesh in 2017.

8.18 Map of Brunei, Malaysia, and Singapore

Brunei, Malaysia, and Singapore

These three countries share a similar colonial experience but have diverged in the postcolonial era (Figure 8.18). Brunei is a tiny but immensely rich sultanate, closer to the emirates of the Middle East in economic profile and religious attitudes. Malaysia is a fast-developing economy but with a politics that oscillates between a secular openness and a religious fundamentalism. Singapore is now one of the richest countries, not only in the region but in the world.

Brunei is an example of the legacy of British informal rule that kept local rulers in power as long as they showed allegiance to the Crown. Brunei was ruled by a local sultan. In 1959, the sultan was made the supreme leader with the United Kingdom taking care of foreign and defense affairs. In 1984, the country became totally independent. In 2013, the Sultan announced the use of sharia law to take effect in 2016.

It is a small country. Most of the 410,000 population lives in the western part, and only 10 percent lives in the more mountainous east. Two-thirds of the population is Malay Muslims, a further 10 percent are Chinese, and there is a significant expatriate population, similar to the Gulf States.

More than 90 percent of GDP comes from oil and natural gas. The huge revenues and small population allow the government to maintain a generous welfare system. Living standards are among the highest in the world, but the almost total reliance on primary products makes the country vulnerable to the **resource curse** and to fluctuating energy prices.

Malaysia is an upper-middle-income country that experienced economic growth through exports of primary produce, import substitution, and the creation of a manufacturing base. Between 1957 and 2005, the average annual growth rate was 6.5 percent. It consists of two separated regions: peninsular Malay that borders Thailand and Singapore, and Malaysian Borneo that encompasses Brunei and is surrounded on the landward side by Indonesia. Malaysia has important maritime claims in the South China Sea.

The majority of the population is Malay and Muslim; they comprise 67 percent of the roughly 31 million people. The British encouraged the migration of Chinese and Indian workers—Chinese laborers in tin mining and Indians on the rubber estates. Even today Chinese make up a quarter of the total population, and Indians less than 8 percent. The Chinese and Indians are now concentrated in the towns and cities, the Malays in the rural areas. Banking and commerce are dominated by the Chinese and Indians, while the Malays are concentrated in fishing and forestry. In Penang, once a trading port for pepper and spices and now an industrial area known for high tech as well as banking, 45 percent of the population is Chinese, 10 percent is Indian, and 43 percent is Malays. There are close to 3 million migrant workers in Malaysia.

The capital city is Kuala Lumpur, situated inland. It has grown substantially as the economic and political center of the country, and it is now a metro region of close to 6 million people. On top of the rich colonial legacy of buildings, new flashy projects have been constructed and none are more ambitious than Petronas Towers.

There is also medical tourism and an active program of attracting the retired wealthy. In 1996, it introduced what was termed the Silver Hair Program that changed its name to Malaysia My Second Home in 2002. The program guarantees a residency for 10 years that is renewable. Malaysia advertises its high standard of living and relatively low living costs.

One major tension in Malaysian political discourse is the tension between conservative Islam and modernity. Islamic-based parties regularly call for the introduction of **sharia law**, while others argue for a more secular society and a greater insertion into the contemporary world. Malaysia struggles to combine Islam with modernization and representative democracy.

Singapore is a small city-state of little more than 700 square miles. It developed as a small port, albeit a vital one in the expanding British Empire east of Suez. In 1918, Raffles, the man who gave his name to the famous hotel in the city, signed, on behalf of the British East India Company, a treaty with the local Sultan. A map of 1825 shows a tiny settlement clustered around a river on the southernmost side of the island. It soon became Britain's main naval and military base in the region. The British surrender to the Japanese imperial Army in 1942 was described by Winston Churchill as the "worst disaster." The British reclaimed the region, but their legitimacy was undermined by the ignominious defeat to an Asian power. In the first election held in 1955, the People's Action Party, a pro-independence movement, won in a landslide. For 2 years it was part of Malaysia, but in 1965 it became an independent state.

Singapore has experienced huge economic growth. Part of its economic success lay in its political stability. Ostensibly democratic, it was effectively ruled from 1965 to 1990 by the authoritarian Lee Kuan Yew, who guided the city-state to become a business-friendly tax haven where ethnic tensions were defused. Singapore was a classic development state with an emphasis on economic growth more than democracy and an unwavering commitment to the creation of a prosperous middle class and an efficient government. In this regard Singapore was spectacularly successful. Its rule of law, lack of corruption, and political stability attracted billions of dollars of foreign investment and made it an important financial center not only in South East Asia but also across the global economy.

Multiculturalism was actively pursued, especially after race riots in the 1960s. Emphasis was placed on social stability. For example, to ensure that Muslims would be less influenced by more militant conservative clerics funded by Gulf States, all mosques in the city have to use English as the language of instruction.

In the 1970s, emphasis was placed on cheap manufacturing, but the economy soon moved up the value-added ladder to become a player in high-end manufacturing and finance. There are now over seven thousand multinational enterprises in the country, and it is now the world's fourth largest global financial center with a massive concentration of banks and financial services.

The city was transformed into one of the most modern cities in the world. The government bulldozed slums and rebuilt housing. The city now boasts a dazzling array of

8.19 Spectacular architecture in Singapore (photo: John Rennie Short)

modern architecture (Figure 8.19). More recently there is a self-conscious creation of a global city, a green city, and a cosmopolitan city. In many ways Singapore is a success story. With one of the highest average incomes, there is also social stability. There is no gun crime and few problems with drugs or corruption. The city has safe streets, good schools, and low tax rates.

The population of close to 5.5 million has one of the lowest fertility rates in the world at 0.79 children per female. The population is mainly Chinese (72 percent), while Malays and Indians make up 13 and 10 percent, respectively. The four official languages of the nation are English, Malay, Mandarin, and Tamil, an indication of the ethnic variety as well as an indication of the origins of the Chinese and Indian migrants.

Singapore makes a conscious attempt to lure the wealthy and the super-rich. It is turning itself into the Switzerland of South East Asia, attracting wealthy Europeans and Americans, billionaire Australians, and Chinese magnates. The attractions are obvious: it has banking secrecy laws; social and political stability; easy access to Asia, Australasia, and the Middle East; low personal taxation; no inheritance tax; and no capital gains. Singapore's success in attracting the wealthy creates a virtuous cycle as the presence of the wealthy increases and deepens the range of services such as fine dining, spas, and financial management services that, in turn, attract even more wealthy households.

Insular South East Asia

Both Indonesia and the Philippines are large island nations that have a shared geological and climatic context, colonial history, and a shared problem of holding together large and diverse archipelagic states (Figure 8.20).

INDONESIA

Indonesia is the largest archipelago nation in the world, consisting of more than 18,000 islands (12,000 uninhabited) that stretch 3,000 miles from the Pacific to the Indian Ocean. The country became independent in 1949 and inherited the sprawling Dutch colonial territory. A new national language was created from the mixture of Dutch, Javanese, Arabic, and Malay fused in cities such as Medan and Jakarta.

The first internal conflict was based on politics rather than on ethnicity and religion. In 1965–1966, there were mass killings of between 200,000 and 500,000 people during anticommunist **pogroms**. The victims were mainly peasants who were supporters of Communist Party supporters or voters. The authoritarian General Sukarno who ruled from 1949 to 1968, and described the slaughter as "an absolutely essential cleaning out," was replaced in a coup by another general, General Suharto, who then ruled from 1968 to 1998 in a similarly authoritarian manner. Wealth was heavily concentrated in the ruling political elite, and popular resentment led to his overthrow. It was only in 2004 that the country elected its

8.20 Map of Indonesia and Philippines

president in fair and free elections. And it is now steadily embarking toward greater democratization.

Indonesia is the fourth largest country by population. With the dual benefits of demographic dividend and abundant resources, the country has yet to fulfill its full potential because it is hampered by corruption, poor infrastructure, and the fluctuating nature of primary commodity products. While there has been some industrialization, Indonesia has not reached the level of growth achieved by Singapore or Malaysia.

Indonesia is the third largest emitter of carbon dioxide in the world, mainly from forest fires used to create plantations, especially palm oil plantations. Across the country and beyond, air quality is compromised by such persistent fires.

The bulk of the large population, almost 60 percent, lives on the small island of Java. A moderate Islam is followed (Figure 8.21). In the westernmost and poorer region of Aceh, a more conservative Islam is the basis for a strong separatist movement.

THE PHILIPPINES

The Philippines, like Indonesia, is a nation-state made up of many islands, over seven thousand, that stretch north and south over 16 degrees of latitude, from around 4 degrees North to almost 20 degrees North.

Its location in physical space as well as the social context makes this a country subject to disasters. The island chain sits in a warm part of the western Pacific, where

8.21 Visiting a mosque in Medan, Indonesia (photo: John Rennie Short)

sea surface temperatures are high enough to provide the raw material for typhoons that regularly move from east to west, pummeling the Philippines (see Figure 7.3 in Chapter 7). A combination of steep topography, frequent severe rainfall, and poor housing makes the islands subject to devastating landslides. Filipinos, especially the poorest, have moved to risky low-lying or steep areas because that is where accommodation is available and is cheap. The poor-quality housing, lack of infrastructure,

and poor local government disaster preparedness turn events such as typhoons, landslides, and storms into disasters. When Tropical Storm Washi struck Mindanao in 2011, it killed more than 1,500 in landslides not so much because of the severity of the storm but because the rain undermined the slopes where people had settled in unsafe areas.

The Philippines became a Spanish colony in the early sixteenth century and an important link in Spain's global trading network. The majority of Filipinos, more than 82 percent, is Catholic, a legacy of the centuries of Spanish rule and church influence on public life.

In 1899, a Republic was declared. It was short lived as the United States invaded the islands and imposed colonial control after the brutal 1899–1902 War, which tied up two-thirds of the US Army in a conflict that resulted in the death of 1 million Filipinos out of a total of 6 million. English became the language of government, education, and business.

The island achieved independence in 1945 and was then ruled by a government dominated by rich landowners and ambitious American business interests. Close to four hundred families, often forming provincial ruling dynasties, controlled 90 percent of the country's wealth. Democratic government was only realized in 1983 when the US-backed dictator Ferdinand Marcos and his shoe-loving wife, Imelda, were overthrown by a popular movement, known as People Power, when ordinary people took to the streets to protest corruption and economic mismanagement.

With just over 100 million people, it is one of the more populated countries in the world. There is a large **diaspora** with more than 10 million to 11 million Filipinos working overseas, as domestic servants and skilled professionals in Hong Kong, the Gulf States, North America, and Europe. Remittances from overseas workers are an important source of income for many families and communities in the Philippines. It is estimated that remittances constitute around 15 percent of total GDP. The top five destinations for Filipino workers are the United States (3.53 million), Saudi Arabia (1.02 million), Malaysia (0.99 million), the UAE (0.88 million), and Canada (0.72 million).

It still has an important plantation economy and exports sugar cane, coconut, rice, and other tropical produce. More than 30 percent work in agriculture. In the past 30 years, electronics assembly and garments, footwear and pharmaceuticals have made the Philippines a newly industrializing country.

THAILAND

Thailand stands out as the only country neither colonized by a European power nor occupied by the Japanese during World War II (Figure 8.22).

The region has a long history as the setting for empires and kingdoms. The Ayutthaya Kingdom that lasted from 1351 to 1767 and was centered in Bangkok is considered a golden period for architecture, arts, and crafts, leaving a rich legacy of architectural complexes and artworks.

Theravada Buddhism entered the country through contact with India and Sri Lanka. Today over 95 percent of the population adheres to this form of Buddhism, and there are at least 200,000 monks at any one time. Most young men are supposed to spend some time as a monk following a strictly regulated life of poverty and mindfulness (Figure 8.23).

Over 80 percent of the 64 million population is Thai and Buddhist. There are four regional dialects: Central, Northeast, Northern, and Southern, with Central considered standard Thai. Ethnic Chinese represent around 11 percent with smaller

8.22 Map of Thailand

tourist industry is an important economic sector, providing employment and foreign revenue.

Democratic government is undermined by military coups. The country was effectively ruled by military dictatorships from 1932 to 1973, and the Thai military regularly intervenes in national politics. The military staged a coup in 2014 and at the time of this writing remains in power.

Focus: Vietnam's Two Big Cities

Hanoi was the dynastic capital of Vietnam from the eleventh to the very early nineteenth century. When it was made the capital of the French empire in Indo China, modern buildings and city layouts reminiscent of urban France were laid across the old precolonial city. Alongside the old dense quarters, where guilds of workers gave their names to streets such as Tinsmith Street and Cloth Street as well as Sugar Street and Conical Hats Street, the French laid out straight-line boulevards and constructed iconic buildings to house government functions. The city was occupied by the Japanese Army from 1940 to 1946. From 1954, it was the capital of North Vietnam and heavily bombed by the US air force during the Vietnam War

8.23 Monks in Chiang Mai, Thailand (photo: John Rennie Short)

(called the American War by the Vietnamese). It became the capital of unified Vietnam in 1975. It is a modernizing and industrializing city. An engineering base was established during the war to provide arms and material in the fight against the Americans. Economic growth was stimulated by reforms, termed "doi moi," in 1986 that allowed more room for entrepreneurial activity and opened up the city to foreign investment. It is now a major center for textile manufacturing and metal manufacturing. In 1994, 73 percent of all trips in the city were made by bicycle (Figure 8.24). By 2015, almost 90 percent are made by

numbers of Malays in the south and hill tribes in the north, including the Lisu, Mien, Lahu, Karen, Hmong, and Akha. Only in the three southern provinces with a significant population of Malay Muslims is there a significant threat to the stability of the Thai state.

Thailand was a major center for the introduction of high-yielding rice plants during the Green Revolution. Since the 1970s it also became a manufacturing center initially for low-value products but now is a major manufacturer of high-value computer parts. It is also a major destination for tourists from around the world. Thailand's

8.24 Busy traffic in Ho Chi Minh City (photo: John Rennie Short)

motorcycles. As the economy has grown, people are more able to afford motorized transport. Among young men in Vietnam the commonly used phrase, "no motorbike, no girlfriend," gives an added incentive.

Ho Chi Minh, formerly Saigon and still called that by the locals, was originally the site of a small fishing village inhabited by Khmer people. From the seventeenth century, Vietnamese started moving in, and in 1858 France annexed the city. Like Hanoi, it has a rich legacy of French colonial buildings, including Notre Dame cathedral, central market, City Hall, the main railway station, and major government buildings. During French rule, it was referred to as the Paris of the Orient because of its French feel. Broad, wide boulevards mark the historic French colonial center. It was a cosmopolitan city with many foreign traders, including Chinese, Japanese, and Westerners. In 1954, it was made the capital of South Vietnam, and a new layer of government bureaucracy was added. When a North Vietnamese tank crashed into the Presidential Palace in 1975, it signaled the end of the regime and the unification of the country. The next year, the city was renamed Ho Chi Minh City. It is now the largest city in the country.

Select Bibliography

Acharya, A. 2013. *The Making of Southeast Asia: International Relations of a Region.* Ithaca, NY: Cornell University Press.

Aung-Thwin, M., and M. Aung-Thwin. 2012. *A History of Myanmar Since Ancient Times.* London: Reaktion.

Baker, J. 2012. *Crossroads: A Popular History of Malaysia and Singapore.* Singapore: Marshall Cavendish Editions.

Cotterell, A. 2014. *A History of South East Asia.* London: Marshall Cavendish.

Dayley, R., and C. D. Neher. 2013. *Southeast Asia in the New International Era.* Boulder, CO: Westview Press.

Goldoftas, B. 2006. *The Green Tiger: The Costs of Ecological Decline in the Philippines.* New York: Oxford University Press.

Green, W. N., and I. G. Baird. 2016. "Capitalizing on Compensation: Hydropower Resettlement and Commodification and Decommodification of Nature-Society Relations in Southern Laos." *Annals of Association of American Geographers* 106:853–873.

Gupta, A. 2005. *The Physical Geography of Southeast Asia.* Oxford: Oxford University Press.

Hayton, B. 2014. *The South China Sea: The Struggle for Power in Asia.* New Haven, CT: Yale University Press.

Hellwig, T., and E. Tagliacozzo, eds. 2009. *The Indonesia Reader: History, Culture, Politics.* Durham, NC: Duke University Press.

Hillmer, P. 2010. *A People's History of the Hmong.* St. Paul: Minnesota Historical Society Press.

Inglis, K. 2013. *Singapore: World City.* Hong Kong: Tuttle.

Kaplan, R. D. 2014. *Asia's Cauldron: The South China Sea and the End of a Stable Pacific.* New York: Random House.

Kim, A. M. 2008. *Learning To Be Capitalists: Entrepreneurs in Vietnam's Transition Economy.* New York: Oxford University Press.

King, V. T. 2013. *Environmental Challenges in South-East Asia.* London: Routledge.

Kratoska, P. H., R. Raben, and N. H. Schulte. 2005. *Locating Southeast Asia: Geographies of Knowledge and Politics of Space.* Singapore: Singapore University Press.

Lim, D. 2014. *Economic Growth and Employment in Vietnam.* Hoboken, NJ: Taylor & Francis.

Lockard, C. A. 2009. *Southeast Asia in World History.* Oxford: Oxford University Press.

Ng, W. H. 2012. *Singapore, the Energy Economy: From the First Refinery to the End of Cheap Oil, 1960–2010.* Hoboken, NJ: Taylor & Francis.

Popkin, S. L. 1979. *The Rational Peasant: The Political Economy of Rural Society in Vietnam.* Berkeley: University of California Press.

Rigg, J. 2013. *Southeast Asia: A Region in Transition.* London: Routledge.

Short, J. R., and L. M. Pinet-Peralta. 2009. "Urban Primacy: Reopening the Debate." *Geography Compass* 3:1245–1266.

Simone, A. 2014. *Jakarta: Drawing the City Near*. Minneapolis: University of Minneapolis Press.

Singh, S. 2012. *Natural Potency and Political Power: Forests and State Authority in Contemporary Laos*. Honolulu: University of Hawaii Press.

Suárez, T. 1999. *Early Mapping of Southeast Asia*. Singapore: Periplus.

Tarling, N., ed. 1999. *The Cambridge History of Southeast Asia: Volume 1*. Cambridge: Cambridge University Press.

Tarling, N., ed. 2000. *The Cambridge History of Southeast Asia: Volume 2*. Cambridge: Cambridge University Press.

Tran, L., and S. Marginson. 2014. *Higher Education in Vietnam: Flexibility, Mobility and Practicality in the Global Knowledge Economy*. New York: Palgrave Macmillan.

Walton, M. J. 2014. "The 'Wages of Burman-ness': Ethnicity and Burman Privilege in Contemporary Myanmar." *Journal of Contemporary Asia* 43:1–27.

Weightman, B. A. 2011. *Dragons and Tigers: A Geography of South, East, and Southeast Asia*. Hoboken, NJ: Wiley.

Winzeler, R. L. 2011. *The Peoples of Southeast Asia Today: Ethnography, Ethnology, and Change in a Complex Region*. Lanham, MD: AltaMira Press.

Wood, G. D. 2014. *Tambora: The Eruption That Changed the World*. Princeton, NJ: Princeton University Press.

Learning Outcomes

South East Asia boasts a cultural and economic diversity.

The region is impacted by environmental challenges: seismic activity caused by the clashing of three tectonic plates, frequent typhoons that impact heavily populated coastlines, river flooding as well as increases in coastal flooding due to global climate change–induced sea level rise, and deforestation for logging or by burning forest cover to establish plantations for farming.

The precolonial world of the region was marked by rivalries and contacts between various powerful kingdoms.

The rise of the empires was linked to the switch from a dry, swidden agriculture, involving slash and burn, to wet rice cultivation and the vast mobilization of labor necessary to build and maintain irrigated farming systems.

Hinduism, Buddhism, and Islam diffused along the trade routes of the region.

The most famous precolonial city in the region is Angkor, the capital of the Khmer Empire from the ninth to the fourteenth century that stretched across 400,000 square miles covering parts of modern-day Burma, Vietnam, Laos, Malaysia, and Cambodia. The city was marked by its monumental size, decoration, and Hindu and Buddhist designs.

Highly prized commodities such as spice and other exotic goods brought European traders, and the intense competition led to colonial occupations.

Although independence eventually came to all of the region's countries, some separations were more hard-fought than others, leaving some countries better positioned than others for future prosperity.

The region is one of economic divergence as some selected areas have grown dramatically and shifted to postindustrial economies of advanced producer services as in the case of Singapore and Kuala Lumpur, while others, including much of Laos and Cambodia, remain largely rural and poor. Vietnam falls somewhere in between.

Agricultural production is still dominant in this region. The poor conditions and economic returns from farming have increased rural-to-urban migration.

The Green Revolution has improved agricultural productivity in the region, but it has had negative side effects and may not be a sustainable solution for food production.

This region has experienced three distinct waves of industrialization. The first began in the late 1960s as investors built manufacturing facilities based on cheap labor; the second involved more advanced manufacturing and higher paying jobs; and the recent third wave involves Cambodia, Laos, and Vietnam, countries that were previously closed to the global economy.

Chinese immigrants play an important role in the commercial activity of the region.

The region's population growth rate has slowed from the high levels of the 1960s and 1970s, and it is now entering the later stages of the demographic transition.

Urbanization characterizes the recent history of the countries of South East Asia, as cities become megacities often propelled by urban primacy and reinforced by the centralizing forces of globalization.

Water pollution, traffic congestion, and air pollution, along with the risk of tsunamis, earthquakes, typhoons, and flooding, remain significant problems, especially in the poorer countries experiencing rapid urban growth.

The United States played a large role in the geopolitics of this region with a history of political interventions and military incursions.

Boundary disputes have been a source of consternation in the region since the end of colonial rule, but the main source of disputes today is maritime, especially the dispute between China and Vietnam, the Philippines, Malaysia, and Brunei in the South China Sea.

9

South Asia

With more than one in five of the world's population, this region contains one of the largest concentrations of people in the world. The region shares a history of British colonialism. The predominantly rural societies are quickly urbanizing, cities are growing especially as sites of manufacturing and producer services. There is a large informal sector. Religious tensions continue, and the conflict between India and Pakistan dominates the geopolitics in the region.

LEARNING OBJECTIVES

Examine the active geological, hydrological, and meteorological processes that shape the region.

Outline the region's early empires and describe the European colonial period through to its independence and partition.

Explain the importance of agriculture across the region and discuss the clustered growth of manufacturing.

Evaluate the challenges of agriculture in rural India and relate these to land access and productivity.

Summarize the religious divides in the region and recognize the impacts of continued population growth.

Identify factors of the region's rapid urbanization and connect these to its major urban problems.

Survey the main geopolitical rivalries and sources of dispute within the region.

The Environmental Context

Making Mountains

This is an active geological region. The Indian Plate is moving northward beneath the large Eurasian Plate, creating the massive, 1 million square mile Tibetan plateau and the dramatic mountain ranges of the Himalayas that reach to almost 30,000 feet, the highest terrestrial mountain range in the world. More than 80 percent of the plateau is covered in permafrost; it is referred to as the third pole, because of the massive amounts of ice and snow. The region is warming at a rate of 0.5°F per decade, and more than one-fifth of the area of permafrost has melted in the last three decades. The plateau and high mountains house the headwaters of four giant river systems, the Ganges, Indus, Irrawaddy, and Mekong, which play such a significant role in the life of South and South East Asia.

The fault line between the two plates is also the scene of earthquake activity. Nepal was hit with a massive 8.2 magnitude quake in 1934. In 2015, it was struck again with a 7.8 magnitude quake that caused extensive damage in Kathmandu, killing over 7,500 people, injuring 15,000, and impacting a quarter of the entire national population. Hundreds of thousands were made homeless, and 1.5 million needed urgent food assistance. Sacred buildings

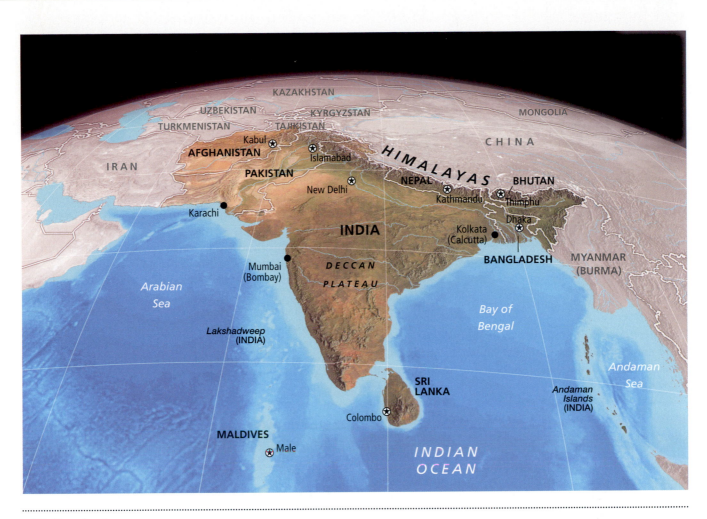

9.1 Map of region

that had stood for centuries were demolished. Across an area 75 miles long and 30 miles wide, the ground rose by 3 feet. The damage was spread over 5,600 square miles. Avalanches swept away villages and base camps in the high mountains. The quake was felt in India and Tibet. High-rise buildings collapsed in Dhaka, Bangladesh.

River Systems

Two giant rivers traverse South Asia, the Ganges and the Indus. Since they pass through different countries, river management issues are an important source of both cooperation and also conflict as dams that generate power for one country reduce the river flow for countries downstream and logging in one area causes flooding across all of the countries in the basin.

The Ganges flows 1,500 miles from the Himalayas through India into Bangladesh, where it enters the Bay of Bengal as a complex dendritic pattern of myriad streams and channels. More than 400 million people live alongside

its banks. The flow is highly seasonal as it is affected by ice melt in its upper reaches and monsoon rain in the middle and lower reaches. It is a sacred site for Hindus who bathe in acts of purification at pilgrim sites such as the ancient city of Varanasi (Figure 9.2). It also has major economic significance as it provides irrigation for crop production. The river is heavily polluted. Untreated sewage and industrial waste make it one of the most contaminated major rivers in the world.

The Indus River originates in China and flows through India and the length of Pakistan. The Indus Valley was the site of early civilizations. Today, it plays an important role supporting agriculture and food production in Pakistan. It is that country's principal source of potable water. India's construction of dams has reduced the water flow in downstream Pakistan. Industrial development along downstream banks has increased pollution levels in the river. Heavy monsoon rains increase the risk of flooding. In 2010 and 2011, floods caused extensive damage. In the 2010 floods, 1.4 million acres of cropland were

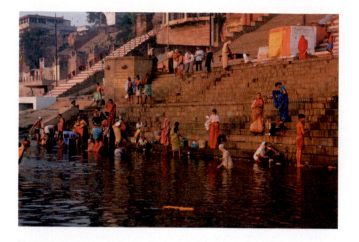

9.2 The Ganges at Varanasi (Danita Delimont / Alamy Stock Photo)

destroyed, two thousand people were killed, and a million were made homeless. In the 2011 floods, 1.7 million acres of cropland were flooded, and over four hundred people were killed.

The river systems are the lifeblood of their regions, providing commercial opportunities and cultural meaning. They all face similar challenges. Global climate change is reducing the ice cover in the Himalayas, threatening the rivers at their very source. Dam construction, deforestation, pollution, and industrial development all threaten the long-term health of the rivers. Because of shared interests, they can be sites of international cooperation; but because of competing interests, they can also be scenes of international conflict.

Monsoons

Most of this region is impacted by the annual **monsoon** weather system. Each year the system has a major impact on Bangladesh, Bhutan, India, Pakistan, and Sri Lanka. The monsoon is caused by temperature differences between the land and ocean. In the wet monsoon, from June to October, the land is warmer than the ocean and so wind blows from the Indian Ocean onto the Indian subcontinent, bringing rain and moisture. In the dry monsoon, from October to May, the sea is warmer than the land and so drier winds blow from the land to the sea.

The monsoon rains bring problems, such as flooding and inconvenience, but also life and lushness. The coming of the monsoon was traditionally welcomed as a sign of the ending of summer's oppressively high heat as well as the arrival of water, vital to crops such as rice, jute, and sugarcane. The first rains of the monsoon season are celebrated with recitations, dances, and songs. The rains revive plants and crops parched by summer heat and provide moisture to soils sunbaked by the high summer

temperatures. In Bangladesh, fragrant flowers such as jasmine begin to blossom and fill the air with the scent of a green renewal, and guava and pineapple fill out to juicy succulence.

Historical Geographies

The Indus and Ganges basins were the location of some of the world's oldest civilizations. The ancient cities of Harappa and Mohenjo-Daro in present-day Pakistan, were flourishing cities as early as 3,000 BCE. And in the Ganges basin, sophisticated city cultures with influences from the Indus cultural region were apparent by 1000 BCE.

The history of the region saw the rise and fall of empires. The most extensive was the Mughal Empire that lasted from 1526 to 1857, and at its height in 1707 controlled much of the subcontinent. It diffused Islam throughout much of the region. Its most famous building is the Taj Mahal, which was built by Shah Jahan who ruled from 1628 to 1658 (Figure 9.3).

9.3 Taj Mahal (Konstantin Kalishko / Alamy Stock Photo)

Coming under British Control

It is impossible to understand the political geography of this region without some reference to the imperial legacy.

The Indian subcontinent was a powerful draw for European merchants. The Portuguese established trading connections along the coast in Goa, Cochin, and what is now Colombo in Sri Lanka. The British East India Company received a royal charter in 1600. Initially its interest was mercantile rather than political, looking only to trade and make money. It established ports along the coast in Madras (Chennai) in 1639 and Bombay (Mumbai) in 1688. In order to secure its control over goods and trade, the company began to annex areas inland. By the end of the eighteenth century, control passed from the company to the British government. In 1858, the country came under formal British government control, and in 1877 Queen Victoria was declared the Empress of India. India, with its wealth and riches, was the jewel in the crown of the British Empire. The British did not control all of India. There was a patchwork territory of formal control intermixed with princely states, where local rulers were kept in power, such as Assam, Baluchistan, Bhutan, Hyderabad, Kashmir, Mysore, Sikkim, and Nepal.

Under the British Empire, the region was incorporated into the global economy as a source of raw materials and as a captive market for manufactured goods.

British control was weakest at the furthest edges and especially in the North West frontier zone where the expanding Russian Empire was casting a powerful shadow. The competition between the two empires was called the **Great Game**. The British fought numerous wars to secure Afghanistan. The First Afghan War (1839–1842), the Second Afghan War (1878–1880), and numerous military expeditions from 1847 to 1908 all failed to discipline the Afghans. The country became known as the graveyard of empires. It was a depiction that should have given the Soviets in the 1980s and then the United States in the 2000s pause for thought. The region was at the limit of British power, and so to solidify their position they drew a line in 1903, known as the Durand Line, that pushed British control up into the mountains in order to provide a buffer against any possible Russian threat and far enough away to provide some measure of security for the main settlements in British India in the river valleys (Figure 9.4). The line split the Pashtun cultural region between British India (later Pakistan) and Afghanistan. An area known as the North-West Frontier Provinces was established on the British side that recognized local autonomy. Under an independent Pakistan, it is now known as Federally Administered Tribal Areas and it is still fiercely independent.

Partition

The break-up of the British Empire in 1947 established the new states of Bhutan, Burma, Ceylon (Sri Lanka), India, Pakistan, and Nepal. There were problems of accommodating the different religions. The two largest religious communities were Islamic and Hindu. Some Muslims did not want to be part of greater India, as they feared being swamped by a vast Hindi majority. The Radcliffe Boundary Commission hastily drew a partition line between what is now India and Pakistan using colonial maps to make a quick fix. These maps were colonial administrative boundaries that failed to provide detailed information or noted religious shrines. There was very little "ground truthing" or reference to the complex social geography. The result was problematic. A two-territory Pakistan was created, one in the east, the predominantly Muslim Bengal, and one in the west, West Pakistan, centered on Karachi. The line in the west was drawn through the Sikh heartland. In the wake of **Partition**, there was conflict and displacement. At least 12 million in Punjab and 2 million in Bengal fled to their respective religious states. Cities such as Delhi and Calcutta experienced a huge influx of refugees. Within a year of Partition, Delhi received 2.2 million new residents, mainly Sikhs and Hindus from Punjab, and lost 0.7 million, mainly

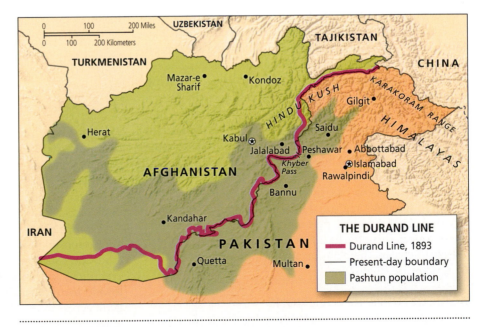

9.4 The Durand Line

Muslims, forced to flee to West Pakistan. It was the beginning of the huge informal settlements in the city.

The princely states were allowed to choose between India, Pakistan, or independence. Too small for independence, most opted for the closest state and chose according to their majority population. The majority Muslim states, such as Baluchistan, opted for Pakistan while the majority Hindu states, such as Mysore, for India. Sikkim was a princely state that became a protectorate of India from 1950 until 1975 when it was annexed. Things did not go smoothly in Kashmir, where a Hindu Maharaja ruled over a Muslim population. A bid for independence was halted by a Muslim invasion of Pashtun tribesmen, and in response, the state asked for help from India. The conflict escalated into the first Indian-Pakistan War; there were two more in 1965 and 1999. A resolution for troops to withdraw and a popular vote has never been implemented. The region is now split between a Pakistan-controlled area in the northwest and an Indian-controlled central and southeastern section.

Economic Transformations

Agriculture

Agriculture plays a significant role in the economy and culture of this region. Despite rapid and massive urbanization, the majority of the population still resides in rural areas and villages. In Bangladesh, 65 percent of workers are employed in agriculture, 81 percent in Nepal and 85 percent in Bhutan. In India, two out of every three of the country's vast population of over 1 billion are rural dwellers.

Various forms of agriculture are practiced. Much of South Asia is densely populated and so shifting cultivation is restricted to areas such as the uplands of Bangladesh or the high mountains of the Himalayan region. **Shifting cultivation** involves utilizing fields and gardens over a wide area through rotation in time. This occurs in forest and upland areas of low-density population. Not all of shifting cultivation is subsistence. Cardamom, a high-value spice, is grown in Sikkim for the market.

Population growth across South Asia has meant that shifting cultivation has given way to fixed cultivation except in the more marginal lands of low population density. Across the region, permanent agriculture predominates, especially in the fertile, watered, lowland areas. Over the past 50 years there has been steady commercialization of agriculture (Figure 9.5).

AGRICULTURAL REVOLUTIONS

The most dramatic change of recent years was the **Green Revolution**, the name given to a series of innovations especially marked from the late 1940s to the 1970s that increased agriculture production through the introduction of high-yielding varieties of crops, irrigation, and liberal helpings of pesticides and fertilizers. It began in Mexico and was then implemented in South Asia, especially India. The Green Revolution did improve yields but at the cost of environmental damage and increased negative health impacts on local communities. It is now seen as a deeply flawed experiment with some positive consequences such as increased crop production and cheaper food for consumers.

The use of genetically modified crops—referred to as GMOs (genetically modified organisms)—has increased. In 1996, worldwide GMOs were planted on only 17 million hectares; by 2018, the figure had increased to 190 million hectares. In India more than 7 million farmers plant disease-resistant cotton on 26 million acres. It requires less pesticide than traditional cotton. The debate about GM crops is especially virulent in India as it smacks of neocolonialism by the large foreign multinationals such as Monsanto. Yet Golden Rice, a form of rice genetically enriched with vitamin A, could help the 190 million children under 5 years old who suffer from a deficiency of vitamin A. The debate is less about the benefits of GMOs than about the monopoly power of the multinationals who own the seeds. Monsanto, for example, sells seeds that are genetically modified to work only with Monsanto pesticides.

Food Insecurities

Hunger still stalks South Asia. On any world map of food insecurity, South Asia is a region of high risk; not quite as bad as the extreme risk noted for sub-Saharan Africa but still a problem area where hundreds of millions are at risk from food insecurity and hunger. The Global Hunger Index identifies India as a country where the situation is "alarming" while for Bangladesh, Bhutan, Nepal, and Pakistan it is rated as "serious." Because of the huge population size of India, this region is home to one-third of the world's poor and hungry.

A Gendered Practice

There is a very marked gender division of labor in much of South Asia. Women play an outsized role in food production and food preparation. In India, the percentage of employed females in agriculture is 75 percent compared to 53 percent of men. Similar figures are reported for Bangladesh, Nepal, and Pakistan.

Women play a major role in the rural societies. They are responsible for child rearing, growing and preparing food, and collecting water and firewood (Figure 9.6). However, many of the rural development programs are geared toward men rather than women, who often have more difficult access to credit, supplies, and services. Women work longer and often much harder than men, which can trap them in a cycle of back-breaking labor with limited access

9.5 Weighing grain in Kerala, India (photo: John Rennie Short)

to education and health services. Rural development programs are only just beginning to accept the important role that women play in rural economies as the gendered division of society often devalues and ignores their contributions.

Economic Modernization

South Asia, in comparison to East Asia, still has a higher proportion of people engaged in agriculture and significantly fewer employed in manufacturing. This is important because it is manufacturing that allowed the rapid growth of countries such as China, Japan, and South Korea.

There are significant clusters of manufacturing employment, especially in the big cities. In Pakistan, the auto industry is concentrated in Port Qasim. In Bangladesh, Dhaka is the center for textile manufacturing. In India, there are four major industrial regions, centered in Delhi, Kolkata, Chennai, and Mumbai. They meet

local demand and for years were protected from overseas competition with high tariff barriers on foreign competitors.

There are certain sectors where the region is globally competitive. In Bangladesh, India, Pakistan, and Sri Lanka, the textile sector produces garments for global markets. For Bangladesh, India, and Sri Lanka, textile exports are the principal form of export earnings. India exported over $2 billion of cotton clothing to the United States in 2013. Textile factories produce goods for overseas

9.6 Women at work in Kerala (photo: John Rennie Short)

markets in **just-in-time production** in which new trends and fashions are quickly met. Working conditions in the factories are often poor but pay is more than in the agricultural sector, especially for women. In the Indian state of Tamil, Nadu textile factories employ more than 120,000 young women attracted by the possibility of earning a marriage dowry. Recruiters persuade young village girls to work in factories. Many of them are lower caste and Dalits. The average wage is $2.50 a day and exploitation of workers is rife. The poor conditions are revealed by the continual fires in factories and by a spectacular building collapse of a textile factory complex in Dhaka, Bangladesh, in 2013 that killed over 1,100 people and injured another 2,000.

In India there is a vibrant information technology service sector. Bangalore is now a center for domestic and foreign companies developing software. An educated English-speaking workforce is now as globally competitive and cheaper than similar workers in Silicon Valley, California. It began with call centers where foreign companies would employ English-speaking Indians to handle customer inquiries and complaints. Time differences allowed companies that added India call centers to provide more round-the-clock operations. Now there is a large business in which companies in North America and Europe outsource business services such as accountancy, customer interaction, documentation, or technical drawings to sites in India where educated, English-speaking labor is cheap and the time difference allows a 24-hour business coverage. Call centers soon developed into technical sites where foreign companies could get much of their work done overnight. In a globalized world, India provides English-speaking technical assistance to Europe and North America.

Compared to East Asia, most countries in South Asia have more people employed in services and agriculture than in manufacturing, which traditionally was the platform for the creation of a middle class. There is a missing manufacturing middle in the economies of South Asia.

The economies of all the countries in the region are hampered by a number of deficits. There is a physical infrastructure deficit of poor transport and irregular power supplies. There is also a cultural deficit of poor educational levels and constraints on women entering the formal economy. There is also the vicious cycle whereby a large pool of people too poor to afford purchase of goods and services stunts domestic demand and thus limits a major source of economic growth. More than 250 million Indians, for example, are considered destitute, and a further 280 million are classified as poor. They lack enough purchasing power to stimulate growth. This still leaves close to 480 million in the Indian version of the middle class, a burgeoning group whose consumption patterns provide a major stimulant to economic growth. How to increase the size and purchasing power of the middle class to break out of this vice of poverty reinforcing poverty is a major challenge for all the countries in the region.

Rural Focus

Farming in India

There are 260 million farmers in India. And they constitute a significant political and cultural as well as economic force. Plans by the Indian government to expand infrastructure developments to aid industrialization often flounder on the inability to purchase land from farmers. In the high-density farming regions, land is not only an immediate economic resource for farmers but also a long-term hedge against uncertainty and risk. Land grabs of farmland and tribal forests by government and corporate interests have become more common after economic liberalization in 1991. Forced acquisition is one reason behind the rise of rural resistance in such states as Andhra Pradesh, Bihar, Jharkhand, and West Bengal.

Most arable land in India is cultivated, and agriculture drinks up 70 percent of all fresh water. India has 20 percent of the world's population but only 5 percent of the world's potable water. Exporting one kilogram of rice, for example, means exporting 2,000 liters of water.

The Green Revolution was spectacularly successful in boosting yields. It turned India from a food importer to a food exporter. In 1966, India imported 11 million tons of grain; now it exports 200 million tons. The Green Revolution generated a more commercialized and capitalized form of farming that relied on chemical fertilizers, pesticides, and lots of water. The capitalization of agriculture comes with some financial risks as well as benefits for farmers. Unlike in the United States, there is a lack of price support or insurance for farmers who often have to turn to moneylenders who charge high rates of interest.

The geographer Sarah Jewitt has been studying rural India since 1993. She contrasts villages in the Ranchi district in Jharkhand with villages in Bulandshahr in Uttar Pradesh (Table 9.1). In Ranchi there is little irrigation, so there is limited ability to benefit from the Green Revolution. Yields are low and food insecurity is a recurring issue. The lack of transport infrastructure makes it difficult to take agricultural products to market. Poor health and food insecurity lead to low life expectancy of only 62 years. Households survive by casual labor and seasonal migration to work in brickfields in nearby towns and cities. Compare this with villages in Bulandshahr in Uttar Pradesh that were able to adopt the high-yielding strains of wheat and rice. The traditional yields of wheat were 1,000 kg/ha, but the new varieties and fertilizers increased yields to 4,000 kg/ha, with even the smaller farmers benefiting. Those owning more than 2 hectares now grow cash crops such as sugar cane. The result is a more affluent rural population who can afford to build more expensive brick houses. The extra yields of the Green Revolution allowed enough food to sustain population growth,

TABLE 9.1	Key Statistics for Ranchi and Bulandshahr Districts	
	RANCHI	**BULANDSHAHR**
% Mud-built houses	71	32
% Brick-built houses	19	65
% Villages unconnected to paved road	75	1

and even land fragmentation, which would have made low-yielding small plots uneconomical. Food insecurity has lessened and the villages now have better road access. Agriculture has become capital intensive with greater use of agro-chemicals. In the summer monsoon rice, sugar cane and maize are grown, and during the drier winter months irrigated watered wheat is grown

The full development of agriculture in India is hampered by inadequate infrastructure. It is neither easy nor cheap to move produce to markets. Poor transport makes it expensive and difficult to move produce from areas of supply to the cities of demand. There is not enough investment in ensuring the supply chain so that produce is delivered fresh. A third of India's fruit and vegetables rot because of inadequate storage.

Social Geographies

The Geography of Religions

This is a hearth region to at least three major religions. Hinduism is one of the oldest religions in the world. It is a belief system that stretches back to the early civilizations of the Indus Valley over 6,000 years ago. It diffused south across India into South East Asia. It is a monotheistic region as it pictures one life force, but it is also polytheistic with numerous gods and deities (Figure 9.7). A significant social aspect of Hinduism, which was reinforced under the British, is the **caste system**. These are social categories that determine not only position in the social hierarchy but also appropriate occupation, behaviors, and mores such as eligible marriage partners.

Buddhism also developed in north India, around the sixth century. It spread across East and South East Asia. A fundamental belief is that it is desire, in whatever form, whether the desire for wealth or fame or sensory pleasure, that causes misery. Bliss is achieved through not having desire. Retreating from the material world through permanent or temporary periods as a monk or nun is a common practice. Buddhism slowly lost influence in India but gained influence in South East Asia. There are two main forms: Mahayana Buddhism found in China, Japan, and Korea and Theravada Buddhism, which ties more closely

9.7 Hindu temple in Colombo, Sri Lanka (photo: John Rennie Short)

to the Buddha's original teaching. It is particularly strong in Sri Lanka, Myanmar, Thailand, and Laos.

Islam is a much more recent entrant into the region. In 1527, the Mughal Empire was established. The Mughals extended their influence and their Islamic faith throughout much of the subcontinent. Many Hindus converted to Islam, eager to escape the caste rigidity or to adopt the religion of the ruling powers. The legacy is that India is a country with one of the largest Muslim populations in the world. There are officially 177 million Muslims in India, a figure only exceeded by Indonesia (204 million) and Pakistan (178 million). The Muslim population in South Asia is concentrated in distinct areas, including Pakistan, which was an entry point for Muslim armies and cultural exchange from the Middle East; Bangladesh, where the rural peasantry converted en masse from Hinduism to Islam; and India, in several locations: a northern corridor from Kashmir to Kolkata that encompasses the Ganges Valley, the center of the Mughal Empire; and along the coast of southwest India, the result of Arab traders down through the centuries. At the local level throughout much of north and central India, Hindus and Muslims inhabit the same villages, living side by side. This made the communal violence during Partition all that more savage and personal as neighbors turned on neighbors.

Sikhism combines Hindu mysticism and Muslim monotheism. It is indigenous to this region and was founded in the fifteenth century in the Punjab. Practicing Sikhs have a distinctive appearance with turbans and beards for men. By 1800, a vast Sikh Empire stretched over Pakistan and India with a capital in Lahore. The entry of the British undermined this empire, and by 1900 the Sikh heartland had become the Punjab region of British India. Today, there are around 25 million Sikhs, the vast majority in the Indian state of Punjab and smaller numbers in Haryana in India as well as a diaspora in Asia, North America, and Europe.

These religious identities overlap and sometimes conflict with national identities. We can consider some of the conflicts that involve different religions and different sects of the same religion.

RELIGIOUS RIVALRIES

After the partition of British India in 1947, the national/religious difference between Pakistan and India was reinforced when Muslims fled to Pakistan and Hindus fled to India. The division, into a predominantly Muslim Pakistan and predominantly Hindu India, is not precise because, as we noted previously, there are close to 177 million Muslims in India.

From Pakistan Muslim fundamentalists have waged war against India in Kashmir and sent terrorists into the heart of India. In India there is the rise of a Hindu fundamentalism with certain parties eager to make the country a Hindu state. One political party, the BJP, is particularly associated with Hindu nationalism. Intercommunal violence continues to occur. In 2002, Hindu pilgrims on a train clashed with Muslims at a train stop in the Gujarat. Fifty-nine were killed and later Muslim communities were attacked by Hindu mobs. In the city of Varanasi, more than one thousand were killed in riots after Hindus tore down a mosque in 1992. In 2006, Muslim extremists set off bombs that killed over twenty people. The threat of violence continues, fed by discrimination—the living standards of Muslims in India are far below the rest of the country—and the enduring military stand-off with Pakistan.

Another source of conflict that overlaps religion and ethnicity is the conflict in Sri Lanka between Hindu Tamils and Buddhist Sinhalese. The Tamils live in southeast India and northeast Sri Lanka (Figure 9.8). After Sri Lanka gained independence in 1948, the Tamils

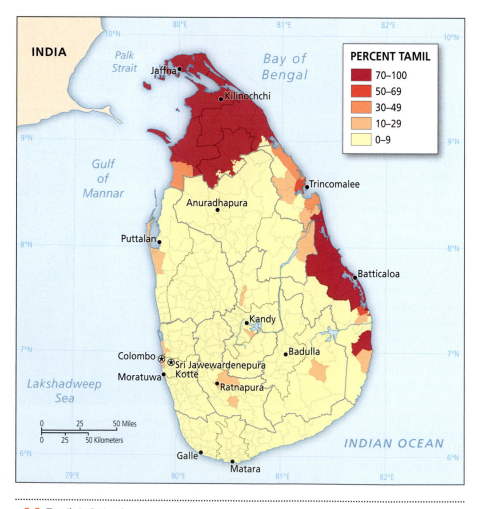

9.8 Tamils in Sri Lanka

faced discrimination and in 1983 a militant Tamil group attacked a military base. The resultant civil war was a brutal affair with war crimes on both sides and a total of 100,000 casualties and hundreds of thousands displaced as the war zone moved across the country. Worst hit was the Tamil area in the north. The civil war that began in 1983 was only officially ended in 2009, with the final defeat of the Tamil Tigers by the Sri Lankan military.

There are also intrareligious differences. In Pakistan, terrorists from the predominantly Sunni Muslim sect, between 80 and 90 percent of the total population, have waged a campaign against the Shia minority (between 10 and 20 percent). Between 1987 and 2007, as many as four thousand people were killed in violence between Sunni and Shia.

Even a shared religion does not trump other differences. Bangladesh was originally part of a Pakistan composed of East and West Pakistan that was created with Partition. The idea was that the two parts, separated by over a thousand miles, would be unified by their shared Islamic religion. That proved not to be true. Economic and ethnic differences resulted in Bangladesh breaking away after a vicious civil war from West Pakistan in 1970–1971.

The Population Explosion

This region is one of the main engines of world population growth. The highest rates of population growth were in the 1960s and 1970s when mortality rates declined but birth rates were still high. The fertility rates in Bangladesh increased from 6.30 in 1950 to 6.91 in 1975. The result was a huge population increase across the region. The population of India increased, too, from 361 million in 1950 to 1.3 billion by 2018.

One state in India alone, Uttar Pradesh, has 225 million people, 33 million more that it had 10 years ago. More than 26 million children are born in India each year. India's population is larger than that of the entire continent of Africa. The rapid rates of growth are not predicted to stabilize until 2050, when India will overtake China as the most populous country in the world with an estimated one-sixth of the total world population.

The three countries of Bangladesh, India, and Pakistan constituted only 17 percent of the world population in 1950, but they now form 22 percent of a greatly increased world population.

IMPACTS

The rapid and large population growth has a number of impacts. The first is increased pressure on the environment as the population expands and densities increase. The pressure is exerted in different ways. In farming communities, the land is often divided up between male heirs. With more males surviving, the land can be fragmented into unsustainable small plots.

There are also more specific pressures such as declining tree cover in Pakistan caused by more people cutting down timber to use as fuel. In 1947, on the eve of Partition, the population was only 37 million, and there were tree stands in the north, extensive coniferous forests in the west, and mangrove forests along the southern border with India. However, as the population increased and, because timber was the main source of fuel, more trees were cut down for fuel. Deforestation was also due to the legal and largely illegal economy of timber harvesting. The net impact is a reduction in the nation's tree canopy. Pakistan now has only a 2 percent tree canopy. The loss of trees is one cause behind the devastating floods and landslides that occurred in 2011 and 2012. The country is losing 65,000 acres of forest each year.

In western India around Mumbai, the city's construction boom is based on the mining of sand from nearby rivers. The increase in legal and illegal sand mining has an environmental impact as it reduces the recharge of rivers and depletes the groundwater. Sand dredgers cut through mangrove swamps. Between 400 million tons (the legal figure) and 1 billion tons (legal plus illegal) of sand are mined from rivers across India, resulting in flooding of rice fields. The largely unregulated industry, vital for the construction industry, has long-term deleterious effects.

Across South Asia, it is the same story told in slightly different ways. Rapid population growth in association with unregulated economic growth and lax regulatory regimes is creating an environmental crisis and increased vulnerability to flooding and global climate change. The environmental impact per capita is less in South Asia than it is in North America or western Europe, since these richer residents exert a heavier impact due to their consumption patterns, but because of the large population of this regions, the overall environment impact is large and pervasive. Very rapid population increases, along with unregulated economic growth, wreak a heavy toll on environments.

DEMOGRAPHIC DIVIDEND

The rapid population growth is beginning to decelerate because of declining birth rates as women are having fewer children. In India birth rates have almost halved from 40 births per 1,000 people in 1950 to 19 by 2018. In 1950, the average woman in India had 5.9 children, but by 2018 this had declined to 2.3. In Pakistan the respective figures were 6.6 and 2.6, and for Bangladesh the decline was from 6.3 to 2.1.

All of the countries are experiencing a decline in both infant mortality and death rates. Due to medical improvements and better public health, the number of babies who died in infancy has declined. The infant mortality rate in Bangladesh declined from 165 per 1,000 births in 1950 to 39 in 2018. The crude death rate also plummeted from 48 per 1,000 in 1950 to 5 by 2010.

All of the countries are now experiencing what is known as the demographic dividend, an increase in proportion of the economically productive population (Figure 9.9). In Bangladesh, the population aged between 15 and 64 years is now 64 percent of the total population. In 1954, it

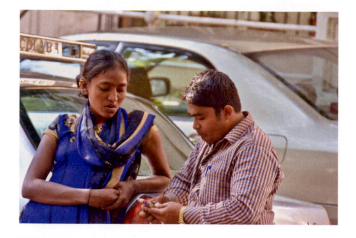

9.9 The demographic dividend: young people in Mumbai (photo: John Rennie Short)

TABLE 9.2	Ten Biggest Metro Regions in South Asia
CITY	**POPULATION (MILLIONS)**
Delhi	27.2
Mumbai	23.6
Karachi	22.1
Dhaka	17.1
Kolkata	16.2
Bangalore	10.8
Lahore	10.5
Chennai	10.3
Hyderabad	9.2
Ahmedabad	7.6

was only 54 percent. The babies in the cohorts of rapid growth during the 1960s to 1980s are now maturing into their productive years. And they constitute an increasing portion of the total population with the potential to be a powerful demographic engine of economic growth as they enter the labor force. However, it is also the case that the demographic dividend has only limited payoff in weak economies. In some cases, Pakistan in particular, the large number of young unemployed males is a main recruiting ground for fundamentalism ideologies. The demographic dividend is only a dividend if there are economic opportunities. Without opportunities the increasing number of young people without jobs or much of an economic future can be a source of political instability.

Urban Trends

The Growth of Cities

While the majority of the South Asian population remains rural and involved in agriculture, one of the dominant trends of the past 50 years is the massive rural-to-urban migration. City populations grew by attracting millions of migrants from the limited economic opportunities in the countryside. The surge of population in South Asian cities is impressive. Karachi's population increased from 1 million in 1950 to 22 million by 2018. Almost a third of the total population in the region now lives in cities, close to half a billion people. The annual urban growth rate is over 3 percent, and there are over sixty cities with a population of 1 million or more (Table 9.2).

The rate of urban population growth is so fast that it has overwhelmed the ability of the private sector or government to respond adequately, and it has placed strains on infrastructure and led to the growth of informal settlements and informal economies. The result is the makeshift and informal nature of many cities in South Asia. This infrastructure problem limits the economic development of cities. Chinese cities, for example, have fared much better in attracting foreign direct investment because they have a better infrastructure.

Without markets or government, people quite literally built their own city on the marginal sites and in more inaccessible locations but sometimes right up against the boundaries of the neighborhoods of the rich and powerful. South Asian urbanization has at first encounter a chaotic quality in the mix of land use, the press of population, and the constant traffic that jars the senses and assaults the nose. But beneath the dirt and rubbish are vibrant self-built neighborhoods and an informal economy that employs almost two-thirds of all workers in the city. The self-built city neighborhoods and informal economies are both a marker of poverty and also a symbol of resilience and energy. With markets unwilling and governments unable to provide, the people provided for themselves. Behind the informal urbanism is an incredible energy and dynamism.

Urban Problems

There are major problems in the cities, including, but not just limited to, air pollution, garbage, traffic jams, water and power supply, and adequate sewage management. More than 2.5 million people died in India in 2015 due to air pollution. In Pakistan, 23,000 adult deaths each year are directly caused by urban air pollution, and there are 5 million cases of respiratory cases in children aged under 5 years. In Delhi, 85 percent of the population lacks curbside pickup of trash. The city produces 10,000 tons of trash a day, shifted by rag pickers.

There are environmental laws on the books but weak implementation and lack of enforcement mechanisms. Delhi produces 690 tons of plastic bags a day despite an official ban on bags since 2012. The city also has some of the worst air pollution in the world. On some days it registers pollution levels ten times that of Beijing. In November 2017, pollution levels spiked to dangerous levels, and city authorities closed six thousand schools, trucks were banned from entering the city for a week, and a halt was placed on major construction sites. Later, in December 2017, the visiting Sri Lankan cricket team halted an international match with India in the city because pollution levels were so high.

The cities of South Asia are places of paradox. Levels of pollution are high and access to basic services such as fresh water or sewage control is limited. Yet the cities are a site of economic opportunity as well as a source of infection, disease, and endemic poor health that limits and constrains people's ability to fully participate. They house the very rich, protected by high walls; security guards live lives of opulence and privilege, freed from the rules and constraints of the vast majority. The same city is also home to those living in very poor conditions on the margins of the formal economy. It is not uncommon in the slums of the same city of Delhi for more than three people to live in a 60 square foot hut beside lakes of untreated sewage.

Cities are engines of growth and places of opportunity as well as environmental hazards and sites of oppressive poverty. Places of hope and despair, sites of liberation and punishment, they contain the paradoxes of private and public life in South Asia in the most exaggerated and extreme forms.

The largest cities have become so large they deserve their own section in this book.

City Focus: Dhaka

The city of Dhaka in Bangladesh is situated in the heart of the Ganges-Brahmaputra deltaic region, a low-lying plain crisscrossed by many streams and rivers in the middle of a vast river system that meanders and splits into different channels as it approaches the ocean (Figure 9.10). It is a watery city, only around 20 feet above sea level. It sits in the track of annual monsoons that bring moisture-laden winds from the Arabian Sea and the Bay of Bengal. From June until September, monsoonal downpours drench the city. The average total monsoon rainfall is almost 79 inches. In 2010, over 13 inches of rain fell in just one 24-hour period.

The city was an important node in global trading as English, French, and Dutch merchants set up trading stations. It emerged as an important administrative city in the British colonial control of the Indian subcontinent. It is the capital and largest city of Bangladesh.

9.10 Location of Dhaka

Dhaka is one the faster growing megacities in the world. In 1950, its population was only 336,000. By 1980, the figure had risen to 3.2 million and to 17.1 million by 2015. The city doubled in size from 1990 to 2005. The 2025 estimate is for a city population of 20 million. More than half live in slums, self-made settlements widely distributed in the marginal spaces in the city, on empty sites and flood plains. The very poor often have nowhere else to go but to sites that are particularly vulnerable to flooding.

The city is also an important manufacturing center, producing goods such as textiles for the global economy. The export of textiles and clothing is the biggest single source of foreign exchange with more than 4 million people employed in the industry.

The city is greatly impacted by global climate change. Increased snowmelt in the Himalayas, the source of the rivers, increases storm runoff and the possibility of flooding. The people in this land of braided streams and muddy lowlands are particularly vulnerable. People from rural Bangladesh relocate to Dhaka each year. Many of them are environmental refugees driven off the land by rising sea levels, flooding, and storm damage that have taken away their land and livelihood. As the country's primate city, Dhaka attracts all the people from the countryside displaced by rising sea levels, severe storms, flooding, and economic marginalization.

The rapid growth, especially of informal slum areas, has made the city more vulnerable to flooding because of the reduction in green space, the increase in impermeable surface, water logging, and settlement on flood-prone areas. From 1960 to 2008, half of the wetlands and a third of the water bodies in and around the city were lost to urbanization. These had acted as sponges to soak up storm water. Now, the monsoon rains bring not only relief and moisture; they also bring urban flooding. The urban hydrological system is overwhelmed by the rate and scale of urbanization. As a result, the flooding of Dhaka is now a regular occurrence. Disastrous flooding has occurred in the city in 1998, 2004, and 2011. In 2004, flooding marooned 2.3 million people as all the roads were under water and businesses were shut down for almost a month. Waterborne disease affected more than a half million people. Dhaka ranks, according to the *Economist*, as the world's second least livable city, ranking only after Harare.

City Focus: Karachi

Karachi is the largest city in Pakistan, one of the largest in South Asia, and one of the largest Muslim cities in the entire world (Figure 9.11). Precise estimates vary, but most agree that the population of the entire metro region ranges from between 18 million to around 24 million. The city contains 10 percent of the national population of Pakistan. It is a flood-prone city that spreads over 60 square miles from the Indus delta into the drier interior.

It is a port city and Pakistan's main point of global trade with the wider region and rest of the world. It was captured by the British East India Company in 1839 and became the main exporting site for the entire Indus valley region. By 1900 the city population was approximately 100,000. The city grew

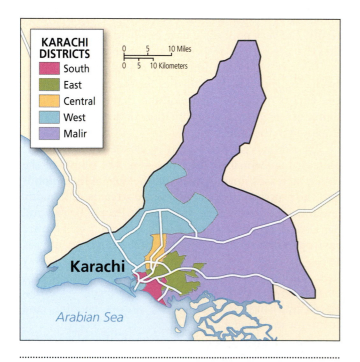

9.11 Map of Karachi

as an important hub of the British Empire. After the country achieved independence in 1947, the population was 435,000. The city was made capital of the new state, and Muslim refugees, especially Gujaratis from the nearby state in India, flooded in after Partition. Despite the rapid and large-scale departure of Hindus, the population of the city quickly reached 1.43 million by 1951. Although the political capital of the country was moved to the new city of Islamabad, the city retains economic preeminence. It has a large industrial sector, including textiles, steel, and auto manufacturing as well as a burgeoning financial and service sector. It is Pakistan's main urban connection to the flows of the global economy. It is also the media capital of the country. It is responsible for a quarter of national GDP, 25 percent of federal government revenues, and 62 percent of income tax.

The city was a destination for refugees from Muslims fleeing from India during Partition and more recently Pashtun refugees fleeing the Russian invasion in Afghanistan and subsequent civil wars. The city has a rich mix, including 7 million Pashtuns, as well as people from Sind, Punjab, Kashmir, and Baluchistan. Many languages can be heard in the city, including Urdu, Gujarati, Punjabi, Pashto, Sindhi, and Balochi. English is the second language of the wealthy and connected. Ethnic and linguistic differences are often the basis for discontent and urban unrest. There is an increasing amount of weaponry; many guns were shipped in during and immediately after the wars in Afghanistan. There are armed gangs and criminal enterprises as well as religious extremists. One Karachi friend said to me, "It may be hell at times, but it is never boring."

Like all major South Asian cities, it has a substantial portion of people working in the informal sector and living in informal settlements. Almost three-quarters work in the informal sector, often tied by subcontracting to the formal sector. More than 60 percent live in **katchiabadis**, temporary unauthorized settlements on government land. Residents and developers simply seize vacant and public lands and build homes. Residents of these settlements have proved resilient. In Orangi, a district in the northwestern part of the city, more than 2.5 million people live in Asia's largest informal settlement. It originated in the 1960s and has grown with migrants and refugees from other regions of Pakistan as well as Afghanistan and Burma. In the 1980s, the residents designed and built their own sewage system that was eventually connected to the municipal system. In response to an indifferent state and weak private market, the residents of Orangi have made their own city by building clinics, roads, and schools.

Karachi is located on the coast and sits on the Indus floodplain. It faces threats of massive flooding because of erosion in the Indus delta and along the coast. Official estimates suggest the entire city will be flooded by 2060 if nothing is done. Sea level will rise 2 inches in 50 years and inundate the lower delta region. The city's average temperature is estimated to rise by 3 degrees by 2050.

City Focus: Mumbai

Mumbai, formerly Bombay, is the largest city in India with a metro-wide population of 23 million. It sits on a narrow spot of land roughly 7 miles wide and 25 miles long. It grew as a busy seaport with Arab traders plying their goods in ships driven by the monsoonal winds. Its trading links expanded from a regional to a global role when first the Portuguese and then the British made it a colonial control point in their commercial empires. The opening of the Suez Canal in 1869 linked it even more closely and quickly to London and the British Empire. The Bombay Stock exchange, established in 1875, is the oldest in Asia.

It retained its economic importance after Independence in 1947. Migrants from rural India flooded into the city seeking better employment opportunities. By the 1980s, more than 300,000 people were employed in textile factories in the heart of the city. More than fifty giant textile factories occupied 6,000 acres in the downtown. Many of the factories have closed and the city's economy has expanded to finance, health service, and high-tech industries. It is home to many of India's millionaires, and the demand for land and property occasionally pushed Mumbai land prices to the highest in the world (Figure 9.12). In 1996, commercial land prices were higher than those of Manhattan.

It is one of the globally connected Indian cities networked into flows of capital, ideas, and people. It houses the national stock exchange of India, the Reserve Bank of India, and is headquarters of major Indian and foreign multinational companies. Mumbai is the commercial and financial heart of the country. The railway system is one of the largest commuter systems in the world with over 7 million passengers every working day.

Mumbai is a city of contrast. It comes in at number six on a worldwide list of cities with the highest number of billionaires, and yet almost two-thirds of the city population lives in slums, eking out a living from makeshift accommodation along streets and railway lines. And many live in cramped self-built dwellings without access to fresh water or proper sewage disposal. Katherine Boo gives an evocative description of the life in one slum in her book, *Behind the Beautiful Forevers*. She describes life in Annawadi, a neighborhood of 335 huts, home to more than three thousand people surrounded by luxury hotels next to the international airport. The place was built by laborers from Tamil Nadu brought in to repair the runway. Only three people in the entire slum have a permanent job. The slum has its own subdivisions, the better-off older area with access to public toilets, an area built by and inhabited by **Dalits**, and a place where people sleep in the open air on top of the garbage bags they have collected. The area regularly floods and is surrounded by piles of illegally dumped waste.

Slums such as Annawadi are being razed to make way for new development and new investments as the city elites reimagine a globally connected city more on the line of Shanghai and Seoul. There are plans to convert the old textile factories, and in 2014 more than 300,000 were made homeless as the bulldozers swept away much of the informal city to make way for the new gleaming city. The goals of a global city, however, are undercut by problems of infrastructure such as fresh water, adequate sewage removal, and regular power supplies.

The name was changed in 1995 from Bombay to Mumbai, not only as an act of Indian reclamation from an English colonial name but also as an expression of Hindu nationalism. Almost 20 percent are Muslim. The majority is Hindu. Hindu-Muslim violence has erupted at various times. The city was attacked in 2008 by an Islamic terrorist group from Pakistan that killed 173 people and wounded over 300. It is widely believed that the Pakistan intelligence agencies were behind the attack. Indo-Pakistan relations have yet to fully recover.

Geopolitics

The Conflict Between India and Pakistan

The main source of political conflict in the region is the military stand-off between India and Pakistan. The two countries have fought three wars and still spar over control in Kashmir. The Pakistan Inter-Services Intelligence (ISI) was believed to be behind the terrorist activities such as the 2008 attack on Mumbai. In conversation, an Indian colleague referred to Pakistan not by its formal name but by the description "nursery of terrorism." The

9.12 New building in Mumbai (photo: John Rennie Short)

conflict between the two countries makes it difficult for the United States to navigate between the two. The US government would like more economic connections with India, a huge market with an expanding middle class, but requires military and strategic assistance from Pakistan, a geopolitical ally in its war against terrorist organizations in the region.

Both countries possess around one hundred nuclear warheads, and in 2014 Pakistan test-fired a new ballistic missile, Shaheen–III, that can carry a nuclear warhead up to 1,900 miles. A retired Pakistan air force commander was reported as saying, "Now India doesn't have its safe havens anymore."

Productive dialogue between India and Pakistan has stalled, and in the disputed territory of Kashmir there are regular exchanges of fire across the Line of Control (Figure 9.13). The enduring conflict with India is one reason for the huge role of the military in the life of Pakistan, both as a major economic force—it controls between a third and a half of the overall economy and owns banks, companies, and land—and as a political agent. The military intelligence organization, the ISI, is often described in Pakistan newspapers as the "invisible government" because of its outsized and secretive role in Pakistan foreign policy.

This is a geopolitical hotspot of the world.

9.13 Line of control in Kashmir

Border Disputes

There are a variety of border disputes in this region.

The 1,510-mile long border between Pakistan and Afghanistan has long been a contentious site. The Afghans do not accept the border, the result of British imperial cartographers that bisects the ethnic Pashtun region. Afghanistan claims the Pashtun areas in Pakistan including the Federally Administered Tribal Areas and part of the North West Frontier Province. The border region is beyond the effective reach of either national government control and hence a space vulnerable to control by terrorist groups such as Al Qaeda.

There are tensions and military build-up along the border between Bangladesh and Myanmar. A major issue is the over 1 million Rohingya, a Muslim Burmese minority persecuted in Myanmar. Most Burmese see the Rohingya as Bengali immigrants who came during British rule. Rohingya are fleeing Burma but are denied citizenship in Bangladesh and live in refugee camps along the border.

It was only in 2015 that Bangladesh finally settled its 41-year border dispute with India when both countries agreed on the disposition of 110 border enclaves involving 37,000 people in territory passing from Indian to Bangladeshi and 14,200 people "moving" from Bangladesh to Indian territory. The exchange involves a tiny piece of territory, a 1.7-acre jute field, which is the world's only third-order enclave; it was a part of India inside Bangladesh territory (second-order enclave) that was inside Indian territory (first-order enclave). Residents in all the agreed-upon enclave swaps were given the opportunity to stay where they were or to move. The two countries agreed on their maritime boundaries in 2014.

India has numerous border issues. One flashpoint is the disputed border with Pakistan in Kashmir. As both countries have nuclear weapons, their enduring conflict provides a source of worry and disquiet in the international community. There are disputes with China over Aksai Chin in the west and Arunachal Pradesh, a region that China describes as "South Tibet." Arunachal Pradesh is remote and connected to India only by a narrow land corridor. While 7 million people visit India, only 10,000 make their way to this remote Buddhist state.

Bhutan also has a border issue with China over the disputed region of Doklam. When the Chinese recently started to build roads in the region, India, a close ally of Bhutan, sent troops to protect Bhutan's claim. This border dispute pits China against India rather than the tiny state of Bhutan.

Connections

South Asian Diasporas

The peoples of South Asia have spread far and wide across the globe. Under the British Empire, between 1834 and 1917, hundreds of thousands of Indian indentured workers were transported from India to work in plantations and commercial enterprises. Almost half a million were shipped off to work in the sugar plantations of Mauritius, a quarter of a million sent to British Guiana, and significant numbers to Natal, Trinidad, Fiji, Jamaica, Suriname, and Reunion Island. South Asian communities were established in the Caribbean, Fiji, South East Asia, and East Africa. The communities survived and prospered to play an important part of the region's economy. South Asians now dominate the small and medium-sized businesses, and their business success has sometimes prompted a backlash, as in Fiji and Uganda.

Today three South Asian counties—India, Bangladesh, and Pakistan—are in the top ten of the world's largest **diasporas** (Figure 9.14). The ranking is Mexico, India, Russia, China, Ukraine, Bangladesh, the United Kingdom, and Pakistan. We should be careful with the numbers since the definitions vary by country. When the children of Indian citizens in the United States become US citizens, are they still part of the diaspora? The data difficulties highlight the shifting and sometimes unstable nature of national identity and cultural belonging.

The Indian diaspora, consisting of people and the descendants who migrated from what is now India to other parts of the world, now numbers 20 million with more than 1 million in at least eleven different countries. In the official classifications, the diaspora consists of nonresident Indian (NRIs) citizens living aboard and persons of Indian origin (PIOs) who are citizens of other countries. The diaspora maintains its links through religion, language, and popular culture such as Bollywood movies that circulate widely across the world. Indian immigrants in the United States have proved very successful. More than 45 percent of Asian Indian immigrants in the Washington, DC, metro area, for example, have a graduate degree and their median household income in 2015 was $128,500. They are a "model minority," educated, affluent, and hard-working.

The Pakistan diaspora consists of around 7 million people, including 5 million born in Pakistan and 2 million born to Pakistani parents oversees. The main destinations

are the United Kingdom (2.12 million), Saudi Arabia (1.2 million), the United States (700,000), and Canada (300,000). Remittances from the diaspora back to Pakistan now constitute 5 percent of GDP, with the highest levels flowing to Khyber, Pakhtunkhwa, and Punjab.

The Bangladeshi diaspora is now concentered in the Middle East. Bangladesh is a major supplier of labor for the booming petro economies of the Gulf. More than 2 million reside in Saudi Arabia and 1 million reside in the UAE. More than half a million live in the United Kingdom and in Malaysia.

While the numbers of Nepalese working and living overseas are small in absolute numbers compared to the big three, they form a hugely important part of the Nepalese economy. Around 3.5 million of the total 30 million population work overseas in Saudi Arabia, South Korea, UAE, Qatar, Kuwait, Malaysia, Oman, and Japan. Their remittances back home constitute a quarter of the country's entire GDP.

Skilled and unskilled South Asians now form a significant part of the global labor supply.

Bollywood

Bollywood is the name given to the movie-making complex in Mumbai, formerly Bombay. There are other production centers in India, but Bollywood is by far the largest and best known. Bollywood produces 15 percent of India's film output but accounts for 40 percent of national film revenues. Other film centers include Tollywood, which mainly produces films for Bengali speakers, based in Hyderabad.

Movies are a popular medium in India. They can tell stories for the illiterate as well as the literate and thus have immense appeal. Bollywood has a long history. By the 1930s more than 200 movies per year were being shot there. Today Bollywood produces around 250 movies a year and employs a total of 175,000 people. Its annual income is in the order of $500 billion.

A range of movies are made, including historical romance, epics, and action thrillers. The romantic musical is a common form with chaste lovers set against exotic locations, mass dancing routines, and big musical numbers.

Bollywood was and still is influenced by movie making around the world. The Hollywood musicals of the 1920s to 1950s with their large ensemble of dancers in big production numbers played an important role in the early development of the Bollywood musical. Bollywood styles in turn now help shape world cinema and now inform modern Hollywood musicals.

Because banks and financial institutions were banned from investing in films, the bulk of money came from the Mumbai underworld. It was only in 1998 that the Indian government granted Bollywood "industry" status. Before that time, the film industry did not receive government support and was burdened with punitive taxation. The move to grant of this new status is an attempt to cut

9.14 South Asian construction workers in Dubai (photo: John Rennie Short)

off the supply of what in India is described as "black" money. A third of investment now comes from NRIs.

While Bollywood sells more tickets each year than Hollywood, roughly 3.6 million compared to 2.6 billion for Hollywood movies, the annual revenue is much less; only $1.3 billion compared to $51 billion.

Bollywood now constitutes 9 percent of national GDP and serves as an important cluster of creative industries. Thousands of skilled technicians are employed directly and in a host of ancillary industries, such as television, tourism, pop music, and advertising, all centered on the movie industry.

Subregions

Afghanistan

Afghanistan is a very poor landlocked country. It occupies a complex geopolitical space (Figure 9.15). It has borders with China, Iran, Pakistan, Tajikistan, Turkmenistan, and

Uzbekistan. The CIA lists it as part of South Asia. It could easily be considered part of the Middle East. It could also be located in Central Asia and as a country so close to Russia it could be part of the Russia and Its Neighbors chapter (Chapter 6) in this book. For many years, after all, it shared an active border with Russia and then the USSR. The complex geography is embodied in the heterogeneous population of Hazara, Pashtuns, Tajiks, Uzbeks, and many other smaller groups that make up Afghanistan's population. The country's distinctiveness requires a special focus on its geopolitics rather than being rolled up into a discussion of broader South Asian trends.

Afghanistan has a complex geopolitical location at the center of competing powers. For years it was an important crossroads between East and West. Buddhism and Islam moved through its roads. In order to assert control in the region, the British fought two wars. The first Afghan War from 1839 to 1842 resulted in ignominious defeat as a force of 16,000 was reduced to just one! A British journalist at the time referred to it as a "monstrous piece of Folly." In the second Afghan War from 1878 to 1880, the British were more successful and defeated the local ruler. After the Third Afghan War in 1917, the country achieved independence.

Afghanistan earned a reputation as the graveyard of empires. The potency of the nickname would again be revealed when the Soviets, long Afghanistan's main trading partner and political ally, invaded the country in 1979. An insurgency, funded by Gulf States and the United States and supported by Pakistan, waged a relentless guerrilla campaign against the Soviets, who like the British before them, had difficulty in the terrain and with the obdurate opposition. The Soviet Empire soon collapsed after the final withdrawal in February 1989.

After the Soviet withdrawal, the nation descended into chaos and fighting between warlords and the different ethnic groups, including Pashtuns in the south, Uzbeks and Tajiks in the north, and Hazara in the west. The country was devastated by war, civil strife, and incessant destruction that reduced many neighborhoods to rubble. Out of this chaos a Pashtun religious group, the Taliban, emerged in 1994 to take over the entire country. They were funded by the Gulf States,

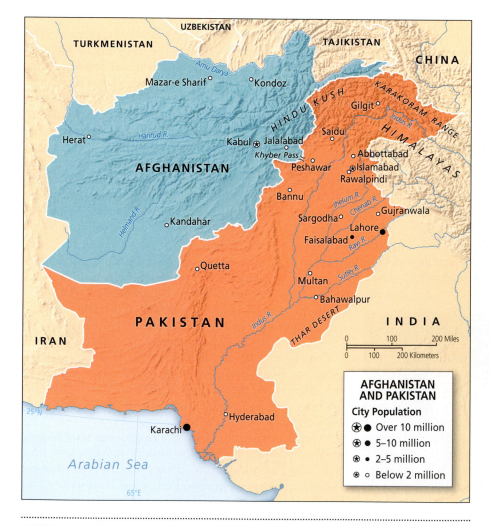

9.15 Map of Afghanistan and Pakistan

were eager to see an Islamic theocracy along strict Sunni lines, and were supported by Pakistan, which was eager to have a strategic redoubt in their ongoing conflict with India. By 2000 the Taliban had control over most of the territory. They gave comfort and support to Osama bin Laden. In late September 2011, the United States began a campaign to overthrow the Taliban. They found lots of local allies disenchanted with the Taliban's brutal rule. Within 2 months, local allies and US forces overthrew the Taliban. It was not a simple victory because the Taliban retained strongholds in villages, especially in the east and south of the country, and Al Qaeda operatives still moved between Pashtun areas in Afghanistan and Pakistan. While the United States was successful in toppling the Taliban, they were not able to eradicate support for the Taliban.

Afghanistan remains one of the poorest countries in South Asia with low life expectancy, around 50 years for both men and women. More than a third of the population lives below the Afghan poverty line.

Corruption is rife. There is little to show for the billions of dollars of foreign aid. But as numerous observers have noted, the correct way to see the country, and others like it, is not as a failed democracy, a system that has not worked. It does work, but as a successful authoritarian kleptocracy where the elite overwhelmingly benefit from the foreign aid and development funds. The one sector that does work well in Afghanistan is the narcotics trade of poppy growing and opium production; it is now worth close to $2 billion making Afghanistan the world's largest producer of opium. It could be described as a narco state. Most of the heroin consumption in Europe derives from Afghanistan.

Pakistan

Pakistan emerged from the Partition of India in 1947, first as two separate territories, East and West Pakistan, and then after East Pakistan separated, simply as Pakistan. More recently, as Islamic fundamentalism has taken a deeper hold, it calls itself the Islamic Republic of Pakistan.

Pakistan was traditionally a primary producer. Despite an increase in manufacturing, it remains an economy dominated by agriculture and services. Cotton is the basis for its large textile sector. Its main global economic connection is the textile sector concentrated in Karachi but also found in Faisalabad, Hyderabad, and Khaipur.

The official national language is Urdu, spoken in urban areas and understood in rural areas. English is used in business and government and is the main language taught in colleges. Arabic is widely used, because reading the Koran in Arabic is a religious obligation for observant Muslims. There are also a variety of regional languages, including Baluchi (southwest), Sind (southeast), Punjabi (along the eastern border), and Pashtun (in the northwest).

Although mainly a rural society, the cities have grown rapidly. Karachi is the biggest city, the main industrial center, and the key port for Afghanistan as well as Pakistan. Lahore is the sociopolitical center. There are regional capitals: Peshawar is the capital of the northwest tribal areas, and Quetta is the main urban center of Baluchistan. The capital of the country is Islamabad, a brand new city completed in 1966. Situated in the north of the country, it was chosen to counter the huge influence of Karachi, to connect with the Army headquarters in Rawalpindi and to locate it close to the disputed border of Kashmir to press home Pakistan's claim. The chosen site reveals a number of themes in Pakistan: the central role that the conflict with India plays in the life of the country, the huge role of the military that constitutes a major political and economic force in the country, and the need to provide a focal point for a nation undercut by marked regional differences.

Despite the active creation of a national consciousness, especially through the employment of Sunni Islam as a unifying force, the country is still marked by regional tensions. In the northwest tribal areas, the Pashtuns have more cultural affinity with fellow Pashtuns in neighboring Afghanistan than people in the south of Pakistan. Baluchistan was a separate kingdom until it was incorporated into Pakistan in 1947. A separatist movement gained momentum in the 1960s, was quashed in the 1970s, and has reemerged in recent years around demands for increased royalties for natural resources. The province is rich in resources but remains one of the poorest provinces in the country. A tiny elite is very wealthy, there is a small middle class, and the vast majority is poor people. The poor and the middle class are more supportive of greater autonomy. Baluchistan, like the northwest tribal areas, highlights a feature of Pakistan geography: the country emerged from separate and distinct regions, some of which, such as Baluchistan, Punjab, and the northwest, had greater links with similar regions in neighboring countries, respectively, Iran, India, and Afghanistan.

Pakistan supported the rise of the Taliban in Afghanistan and was one of the only three countries in the world—the other two were Saudi Arabia and the United Arab Emirates—to officially recognize the Taliban as the legitimate government of that country. A Taliban-controlled Afghanistan provided strategic depth for Pakistan in its struggle with India, a deep redoubt in case of a war with India, shoring up its flank. All this was to change, however, after 9/11 when the United States overthrew the Taliban and waged a war against Islamic militants in the region. Pakistan is now in the awkward position of accommodating the United States, the major provider of aid and military assistance, while also maintaining links with client groups, including the Taliban.

Pakistan's support for terrorist organizations such as Lashkar-e-Taiba, responsible for the attack on Mumbai in 2008, undercuts any attempts at rapprochement with India while support for militant groups makes the United States and Pakistan untrusting allies.

The Pakistan Taliban gained control of the northwest in 2007 and introduced sharia law, banned dancing and music shops, and destroyed four hundred schools. In 2009, the Pakistan Army launched a campaign against Taliban control on Swat Valley, only a 4-hour drive from Islamabad and hence vulnerable to Taliban takeover. Pakistan's military offensive against the Taliban pushed farmers out of the Tirah Valley, where more than 100,000 people make a living by growing marijuana. In 2014, the Pakistan Army launched a major offensive against the Pakistan Taliban in North Waziristan. More than half a million people were evacuated or left the region. Over the past decade Pakistan has lost 50,000 soldiers and civilians by taking on these militant groups, once supported and now challenged. The Pakistan Taliban slaughtered 150 students at Peshawar in December 2014.

The Himalayan States: Nepal and Bhutan

There are two states that straddle the Himalayas. Nepal is the largest of the two, situated in the high peaks between India and China (Figure 9.16). It was an absolute monarchy

9.16 Map of Bangladesh, Bhutan, India, and Nepal

that became an independent country only after Partition. More than four out of the ten are Hindu, and one in ten is Buddhist. Severe poverty prompted a Maoist insurgency that began in 1996 and cost the lives of 12,000 people. A peace deal forged in 2006 created a secular federal republic. In 2008, the monarchy was abolished, and in 2012 the former insurgents joined the Nepal Army. Since democratization, politics is often marked by short-lived governments and political turbulence.

Nepal is one of the poorest countries in the world with low life expectancy. More than a quarter of the population lives below the poverty line, and the unemployment rate is 46 percent. Of those working, more than 70 percent work in agriculture. The limited economic opportunities force many Nepalese to seek jobs abroad. Around 1,500 people left the country each day in 2014 in search of work overseas. The country is now very dependent on remittances that constitute more than a quarter of total GDP. The remittances come at a price, though. More than 725 Nepalese died in 2013 working abroad. They do the hard, difficult, and dangerous jobs in construction such as the building of soccer stadia in Qatar. Flights back to Kathmandu regularly contain the coffins of workers killed in the construction sector, where they have few protections and work with limited safety regulations.

The country is centered on the capital of Kathmandu, a primate city where rural migrants are attracted to the better health, education, and employment opportunities. People move to the city to escape the war-torn and economically distressed countryside. More than 1.7 million people now live in and around the city, a 60 percent increase since 2001. The centrality of the city to the life of the country made the 2015 earthquake that flattened the city all the more devastating, rendering the state even more dysfunctional and hampering rescue and response.

Bhutan is a small, poor, landlocked country between India and China. It was under the informal control of the British and then India until it achieved greater independence in 2007. The majority of the country's 733,000 population is Buddhist, with around 25 percent Hindu. Ethnic Nepali, who constitute around a third of the population, face discrimination and many were forced out; there are 23,000 refugees in camps in Nepal. The main export is hydroelectric power to India, Bhutan's main trading partner. Tourism is tightly controlled in order to preserve the indigenous traditions. The boundary with China has never been finally fixed and so negotiations continue.

Bangladesh

In the partition of India in 1947, Bangladesh was part of a greater Pakistan, first as East Bengal from 1947 to 1955 and then as East Pakistan. East and West Pakistan were separated by over 1,300 miles of Indian territory. Although linked by the common religion of Islam, the two wings were very different with most of East Pakistan peopled by Bengalis, in contrast to the predominantly Urdu-speaking West Pakistan, where political, military, and economic power was concentrated. The mounting grievances led to violence in 1970 and to all-out war in 1971 as the Pakistani Army killed between 300,000 and 1 million people. The conflict created mass displacement and long streams of refugees seeking safety—one in ten left their homes. Indian forces joined with Bangladeshi irregulars to defeat the Pakistani Army, and Bangladesh became an independent country in 1971 (see Figure 9.16).

More than 70 percent of exports are textiles. Garment factories in Bangladesh, concentrated in Dhaka, are part of global supply chains that supply clothing stores in North America and Europe. It is an $18 billion a year industry and an important part of economic growth in the country.

Most of Bangladesh is located on the flood plain of the Ganges (Figure 9.17). It makes the country subject to flooding from rivers as well as from the monsoonal rains and sea level rise. Dhaka is one of the most vulnerable cities to climate change as the ice melts in the Himalayas; more violent storms produce higher rainfall, and global climate change leads to higher sea levels.

Most but not all of the country is close to sea level. The Chittagong Hill Tracts is a hilly area bordering India and Burma and home to hill tribes, known as the Jumma. When the Bangladeshi government tried to settle the region with Bengalis, the tribal peoples formed a resistance movement that was successful in gaining more autonomy for the state. However, forced acquisition and a land grab of indigenous land for commercial plantations continue to provoke resistance and more than a third of the Bangladesh Army is deployed in this small region. Tobacco cultivation is the main formal economic activity. Deforestation from plantations and logging is a major environmental hazard.

The overwhelming majority of the population is Muslim. It is one of the few Muslim majority democracies, with rising living standards and increased life expectancy, higher literacy, and more gender equality. The last four governments were headed by women.

The religious homogeneity does not preclude religious tensions. There is a tension caused by differing interpretations of Islam. Traditionally a moderate and secular society, more people, especially in rural areas, now support a more conservative version of Islam and want to see sharia law. The battle between more secular and more fundamentalist interpretations frames rising political violence. Immigrant workers returning from the more conservative Gulf region and Gulf charities in the country promoting religious education strengthen the more fundamentalist Islamic interpretation. Political differences are shaped by competing interpretations of Islam.

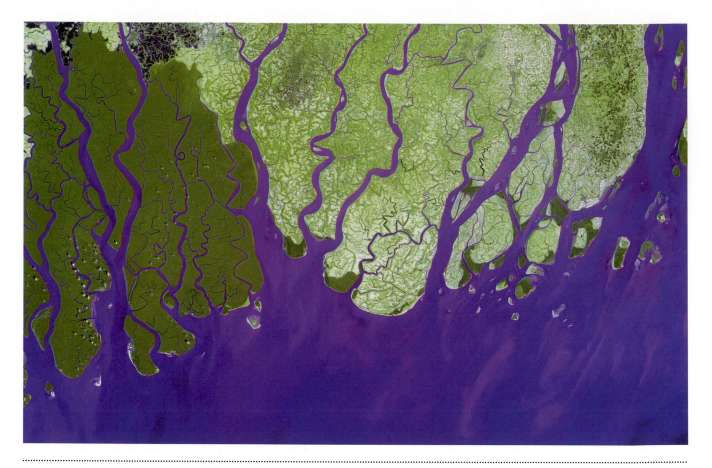

9.17 Ganges flood plain (Image courtesy of NASA)

India

India is a huge country with four major ecological regions: the mountains of the north; the Great Plains that stretch, parallel with the mountains, providing the main bread basket of India; the Deccan Plateau that is around 11 feet high in the south rising to 300 feet in the north where the tropical turns into the semiarid; and the coastal plains and islands (See Figure 9.16). Most of the population is concentrated in the Great Plains and coastal areas.

India is one of the most populous nations in the world. By 2050, it is estimated that the country will have 1.5 billion people and will overtake China as the most populous country in the world. More than 27 million children are born each year. The weight of population pressure is summarized by these startling facts: while India has only about 2 percent of the world's land area and 4 percent of fresh water, it has to feed 17 percent of the world's population. There is a tremendous strain on the resources of India and the ability to feed the entire population.

The vast majority of people are Hindu, but there is a substantial Muslim population in a belt running from Kashmir through the Great Plains to the border area surrounding Bangladesh. There is also a significant Muslim population in the region around Mumbai. Sikhs are concentrated in the Punjab.

The territory of present-day India is in large part a legacy of British colonial holdings. Regions such as Arunachak Pradesh or the Andaman Island chains were British-controlled territories that passed over to India with independence. Colonial borders and holdings became part of Indian territorial disposition.

The British left their mark in several ways. We can consider just two: railways and language. The railway system was built by the British to enhance colonial control and the exploitation of the territory. Railways are now an integral part of the country. The railway system employs a million people and is the largest employer in the country. Between 6 and 7 million use the railway system in Mumbai each working day.

In a country with diverse languages and dialects, the language of the colonial power, English, is now an important form of communication across this vast country. It is understood by most middle-class Indians and is used extensively in education and government. The widespread usage of English has allowed an easy connection to global business and commerce and is the linguistic platform for the diaspora of educated Indians working in high tech and medicine in North America, the Middle East, and Western Europe.

GEOGRAPHIC DIFFERENCES

India is not a homogenous country. There are twenty-one official languages, and the 100-rupee banknote lists fifteen different language scripts. There are substantial political differences, for example, from the more communist-leaning political culture of Kerala to the Hindu nationalist parties in Mumbai.

Along any dimension of social and political geography, there are huge differences. Consider just three states. Bihar is one of the poorest states, with the 104 million population mainly Hindu. Kerala with a population of 35 million is one of the more affluent states where the caste system is weaker and the Hindu nationalist parties have a hard time competing against the Communist Party for political control (Figure 9.18). Goa is one of the smallest states with only 1.5 million people. As an important hub of the global spice trade, it was bitterly contested between various powers. The Portuguese ruled the region from 1510 to 1961. It only became part of India in 1961 and retains a considerable Portuguese legacy in the form of ecclesiastical architecture and in religious affiliation. One in four people is Christian. Its unique quality and wonderful beaches have made it an important tourist destination—part of India, but different. International cruise lines are an important source of employment for Goans, with one out of every three families having at least one member working on an international cruise line.

If we take literacy rates, for example, there is wide chasm between Kerala, where literacy rates according to the last Census of India was 93.1 percent, a figure higher than many developed countries, and Bihar with 63.8 percent.

9.18 Election posters in Kerala (photo: John Rennie Short)

Kerala spends more on social welfare and education than most states in India. The position of women is also better than in most states, a product of successive communist state governments and the system of matrilineal inheritance. The improved life chances for women are part cause and part effect of the below-average childbirth rate in Kerala. It is 1.8 child per woman compared to 3.4 for Bihar. Kerala's birth rate is close to Finland, while Bihar's is closer to Swaziland. There are substantial differences across this vast land and huge population.

ECONOMIC REFORMS

Since 1991, India has embarked on an economic policy of deregulation, privatization, and an active encouragement of foreign trade and investment. These policies, in association with the demographic dividend, are responsible for the impressive annual growth rates of around 7 percent. A new capitalist economy and way of life had to be built quickly from the shell of a centrally planned oligopolistic economy. The speed of change is dizzying and has provided new opportunities, especially for the young and educated, who are less tied down by cultural norms and preferences and quick to imbibe the new capitalist ethos of consumption and display. The opening up of the economy to foreign investment and global trade has also widened income disparities with most benefits going to a minority, the creation of an absolutely large if relatively small middle class, and the lives of the poorest either unaffected or worsened.

One impact of economic liberalization is the transformation in selected cities that became connected to global flows of goods, people, and ideas. India's urban population is set to grow from 377 to 600 million in the next 15 years. Bangalore, Delhi, and Mumbai were just some of the cities more integrated into global flows. Malls appeared and coffee store chains proliferated across the constantly restructuring landscape of major cities. A globally financed urban development process reshaped the new Indian city as millions in slum settlements lost their homes. In 1 year in Mumbai, redevelopment displaced 90,000 homes and made 300,000 homeless (see Figure 9.12). Urban renewal involved the construction of apartment blocks and shopping malls, and infrastructure improvements, as well as the displacement of a million poor residents. Cities are engines of growth and home to the burgeoning middle class as well as the poor. In India, 1 in 8 people under 6 years of age lives in a city slum. The major cities are the eye of the storm that is swirling across India as it connects with the global economy.

World's Largest Democracy

India is the world's largest democracy. Caste based voting is a prominent feature of politics, and there are rising expectations as people want more from their governments, more opportunities, and easier access to the economic miracle of spectacular growth. Previously, parties remained in power at the national and state level by not so much promising growth as through distributing benefits. To take just one example, the highly subsidized electrical power distribution system is used by politicians to ensure political support. They condoned the theft by businesses and consumers of state-distributed power. The result was a massive theft of electricity, as politicians turned a blind eye to consumer and business stealing power. Almost a quarter of all electricity is either lost or stolen. Power is very cheap for most consumers. The state electricity companies, required by politicians to keep costs low, thus have few funds to invest in equipment and plants. In 2012, there was a massive electricity outage across India. The frequency of power outages places economic modernization in a precarious position. How to balance consumer preference and political patronage with sound policies for sustained economic growth is a dilemma for many countries, including the United States, but it is particularly acute in a rapidly modernizing democracy of over a billion people. Populism and patronage often trump sound management.

The caste system, an integral part of Hinduism, continues to affect life chances, economic opportunities, and political attitudes. The strength of the caste system varies across the country being much weaker in Kerala, after decades of communist governments that actively worked to promote the welfare of the poorest, than in states such as Bihar. Even in states such as Bihar where there is a more entrenched caste hierarchy, there have been changes. In 1978, quotas were introduced for what was referred to as "backward castes" in government employment. One group, classified as EBCs (extremely backward classes)—it includes weavers, potters, earth diggers, boatmen, fishermen, tailors, dancers, carpet makers, barbers, and sweets makers—is 20 percent of the state's 83 million population and constitutes 35 percent of the electorate and thus holds considerable political power. In an important state election in 2015, this group was an important political bloc handing victory to an opposition party and humiliating the Hindu nationalist party, the BJP, which controlled the federal government. The caste system, while it may be eroding because of economic changes and urbanization, is alive and well in the political life of the country. Mobilizing and appealing to castes is an important way to achieve political power at the local, state, and federal level.

Sri Lanka

A series of kingdoms shaped the precolonial history of Sri Lanka. The longest lasting was the Anuradhapura Dynasty, which ruled from 380 BCE to 993 CE. The most recent was a Sinhalese kingdom based in Kandy, which only succumbed to British forces in 1803. Buddhism arrived in the island around 250 BCE, and today the island is filled with Buddhist shrines and temples. The majority of the Sinhalese population practice Buddhism.

Sri Lanka attracted attention from foreign traders because of its spices and its important strategic location on the maritime trade routes that linked South and East Asia with the rest of the world (Figure 9.19). The Chinese Admiral Zheng sailed to the region in the early fifteenth century. The first Portuguese ships landed in 1501, and later Dutch, French, and English traders visited the island. With its incorporation into the British

9.19 Map of Sri Lanka

Empire, it became a source of primary commodities as coffee, tea, and rubber plantations were created in the lush countryside. Working conditions were brutal and so labor was imported from India.

It was, and still is, an important primary producer with plantations of tea and rubber in the higher altitude wet zone in the southwest of the country. Sri Lanka is one of the world's largest exporters of tea. The thirsty tea culture of Britain was slaked by tea from the Sri Lankan foothills. Rice is also grown in the inland wetlands and along the coast. Population densities mirror agricultural productivity with the highest densities in this wet zone and around Colombo, along the west coast, and around Jaffna in the far north.

The main environmental hazards in the country are disasters: landslides, especially in central highland areas; floods in all the river basins; cyclones; and monsoonal storms. The island's coast is also vulnerable to tsunamis. There was extensive damage and loss of life along a 500-mile coastal zone in the 2004 tsunami that impacted the entire Indian Ocean basin.

The main geopolitical issue facing Sri Lanka is the ethnic tension between two main groups that are divided by religion and language. The Buddhist Sinhalese constitute about 85 percent of the population. The remaining 15 percent are Hindu Tamils who are concentrated around Jaffna and along the northeast coast. There are also Tamils in the central plantation areas of Nuwara Eliya, descendants of indentured plantation laborers brought over from India in the nineteenth century.

A Tamil separatist group, the Liberation Tigers of Tamil Eelam, known popularly as the Tamil Tigers, fought an insurgency campaign from 1983 to 2009. The north and east of the island came under Tamil control. The civil war was deadly. More than 100,000 people died and the country was considered too dangerous for large-scale investment. The civil war halted economic development. The civil war only came to an end in 2009 with the final defeat of separatist forces. Close to 100,000 Sri Lankan Tamils remain in refugee camps in the southern Indian state of Tamil Nadu.

The main town of Colombo mirrors the past and indicates a possible future. The city contains the legacy of the past with the neoclassicism and Victorian confidence of British colonial architecture. The city also houses the various groups in Sri Lanka: Buddhist shrines sit alongside Hindu temples and Islamic mosques. There are also fine examples of tropical modern as in the houses of Geoffrey Bawa (1919–2003). When I visited the city in 2015, signs of new foreign investment were evidenced, from Indian property companies to US consultancy firms all establishing a presence as the country returned to political normalcy.

The harbor of Colombo is one of the deepest in the world, and Korean and Chinese investment has built state-of-the-art container port facilities (Figure 9.20). Colombo is a major trans-shipment port as huge carriers from East Asia

9.20 The deep water harbor of Colombo is now a major container port (photo: John Rennie Short)

unload their cargoes, which are then loaded to smaller ships that connect with smaller ports around the Indian Ocean. Colombo is once again at the hub of global trade routes.

The Maldives

The Maldives is a chain of islands off the southwest coast of India where the Indian Ocean meets the Arabian Sea. Around 350,000 people live across 192 islands. Like India and Pakistan, they were part of the British Empire and before that a stepping stone for Arabic traders, whose legacy lives on in the dominance of Sunni Islam and the use of Arabic in government administration. In 1887, the islands became a British protectorate that allowed some form of internal autonomy but left British control over foreign policy. The island chain was strategically important for British geopolitical strategy in the Middle East and South. The islands received independence in 1965, and 3 years later a Republic replaced the Sultanate.

The country's geographical location is both a blessing and a curse. It is situated beside rich fishing grounds, and the fishing industry employs a third of the workforce. The island chain is spread out in a series of atolls in translucent clear and largely pristine water. Since 1972, tourism has emerged as the single largest industry. More than 600,000 tourists visit the islands, almost double the island's population, providing 90 percent of all government revenues.

The islands were devastated by the 2004 tsunami when 14-foot waves crashed over a country less than 8 feet above sea level. Six islands were totally destroyed, and fourteen had to be evacuated. Climate change, and especially sea level rise, is a major issue for this low-lying island nation. In the worst-case scenarios of sea level rise, many of the islands could be underwater by 2025. Many of the islands will have to be evacuated, and the people of the Maldives will become climate change refugees.

Focus: Cricket in South Asia

One of the enduring legacies of the British Raj is the sport of cricket. The British brought the game with them when they annexed and colonized much of the subcontinent. It proved popular with the locals. So popular that cricket is now less of a sport and more of a secular national religion. Across the region, in manicured green ovals of private clubs or on dusty patches of ground, cricket playing is the most important participant and spectator sport (Figure 9.21).

Cricketers who play well for their national teams are some of the best-paid professional athletes; they are cultural icons and some of the more instantly recognizable personalities in the region. The national teams compete with the other cricket-playing countries—including Australia, England, New Zealand, and South Africa—in five-day test matches as well as the very popular one-day international competitions. Cricket is also played by women in South Asia. Despite an increasingly stricter interpretation of Islam, there is a women's cricket team in Pakistan.

Competitions between India and Pakistan are particularly charged events, attracting thousands and generating great passion. Sri Lanka, despite its relative population size

9.21 Playing cricket in Mumbai (Juergen Hasenkopf / Alamy Stock Photo)

compared to the population heavyweights of India and Pakistan, has fielded an extraordinarily successful international cricket team since the 1990s, winners of the Asia Cup in 2014 and runners-up in many international competitions. The country hosted the 2011 Cricket World Cup.

Bangladesh only became a full member of the International Cricket Council in 2000, which made it eligible to play other countries in five-day test matches. The women's team won the silver medal in the cricket tournament in the 2010 Asian Games.

Select Bibliography

Anjaria, J. S., and C. McFarlane. 2011. *Urban Navigations: Politics, Space, and the City in South Asia*. New Delhi: Routledge.

Bass, D. 2013. *Everyday Ethnicity in Sri Lanka*. New York: Routledge.

Boo, K. 2012. *Behind the Beautiful Forevers: Life, Death, and Hope in a Mumbai Undercity*. New York: Random House.

Bose, S. 2013. *Transforming India: Challenges to the World's Largest Democracy*. Cambridge, MA: Harvard University Press.

Bose, S., and A. Jalal. 2011. *Modern South Asia: History, Culture, Political Economy*. New York: Routledge.

Bradnock, R. W., and G. Williams, eds. 2014. *South Asia in a Globalising World*. London: Routledge.

Brosius, C. 2014. *India's Middle Class: New Forms of Urban Leisure, Consumption and Prosperity*. London: Routledge.

Chapman, G. P. 2009. *The Geopolitics of South Asia*. 3rd ed. Burlington, VT: Ashgate.

Chester, L. P. 2009. *Borders and Conflict in South Asia*. Manchester, UK: Manchester University Press.

Dasgupta, R. 2014. *Capital: The Eruption of Delhi*. New York: Penguin.

Datta, A. 2012. *The Illegal City: Space, Law and Gender in a Delhi Squatter Settlement*. Farnham, UK: Ashgate.

Ganti, T. 2012. *Producing Bollywood: Inside the Contemporary Hindi Film Industry*. Durham, NC: Duke University Press.

Gayer, L. 2014. *Karachi: Ordered Disorder and the Struggle for the City*. Oxford: Oxford University Press.

Guneratne, A., ed. 2012. *Culture and the Environment in the Himalaya*. London: Routledge.

Guneratne, A., and A. M. Weiss. 2014. *Pathways to Power: The Domestic Politics of South Asia*. Lanham, MD: Rowman & Littlefield.

Hannant, M. 2019. *Midnight's Grandchildren: How Young Indians Are Disrupting the World's Largest Democracy*. New York: Routledge.

Hasbullah, S., and B. Korf. 2013. "Muslim Geographies, Violence and the Antinomies of Community in Eastern Sri Lanka." *The Geographical Journal* 179:32–43.

Hyndman, J., and A. Amarasingam. 2014. "Touring 'Terrorism': Landscapes of Memory in Post-War Sri Lanka." *Geography Compass* 8:560–575.

Ives, J. D. 2004. *Himalayan Perceptions: Environmental Change and the Well-being of Mountain Peoples.* London: Routledge.

Jewitt, S., and K. Baker. 2006. "The Green Revolution Re-assessed: Insider Perspectives on Agrarian Change in Bulandshahr District, Western Uttar Pradesh, India." *Geoforum* 38:73–89.

Jewitt, S., and K. Baker. 2012. "Risk, Wealth and Agrarian Change in India: Household-Level Hazards vs. Late-Modern Global Risks at Different Points along the Risk Transition." *Global Environmental Change* 22:547–557.

Jha, P. 2014. *Battles of the New Republic: A Contemporary History of Nepal.* London: Hurst.

Jones, S. G. 2009. *In the Graveyard of Empires: America's War in Afghanistan.* New York: W.W. Norton.

Klem, B. 2014. "The Political Geography of War's End: Territorialisation, Circulation, and Moral Anxiety in Trincomalee, Sri Lanka." *Political Geography* 38:33–45.

Laruelle, M., and S. Peyrouse, eds. 2011. *Mapping Central Asia: Indian Perceptions and Strategies.* Burlington, VT: Ashgate.

Lieven, A. 2011. *Pakistan: A Hard Country.* New York: Public Affairs.

Lorenzen, M., and R. Mudambi. 2013. "Clusters, Connectivity and Catch-up: Bollywood and Bangalore in the Global Economy." *Journal of Economic Geography* 13:501–534.

Lyn, D. 2009. *In Afghanistan: Two Hundred Years of British, Russian and American Occupation.* New York: Palgrave.

Mallick, B., and J. Vogt. 2012. "Cyclone, Coastal Society and Migration: Empirical Evidence from Bangladesh." *International Development Planning Review* 34:217–240.

Metcalf, B. D., and T. R. Metcalf, 2011. *A Concise History of Modern India.* New York: Cambridge University Press.

Niang, C. T., M. Andrianaivo, and K. S. Diaz. 2013. *Connecting the Disconnected: Coping Strategies of the Financially Excluded in Bhutan.* Washington, DC: World Bank Publications.

Punathambekar, A. 2013. *From Bombay to Bollywood: The Making of a Global Media Industry.* New York: NYU Press.

Rabbani, M. R. 2017. *Invisible People.* Islamabad: Sang-e-Meel.

Rahman, M., T. Tan, and A. Ullah. 2014. *Migrant Remittances in South Asia: Social, Economic and Political Implications.* Basingstoke, UK: Palgrave Macmillan.

Rahman, S., I. A. Begum, and M. J. Alam. 2012. "Re-examining Green Revolution Diffusion and Factor/Input Markets in Bangladesh after Market Liberalization." *Singapore Journal of Tropical Geography* 33:241–254.

Rashid, H., and B. Paul. 2014. *Climate Change in Bangladesh.* Lanham, MD: Lexington.

Sagan, S. D. 2009. *Inside Nuclear South Asia.* Palo Alto, CA: Stanford University Press.

Sanyal, S. 2012. *Land of the Seven Rivers: A Brief History of India's Geography.* New Dehli: Penguin.

Sethi, A. 2012. *A Free Man: A True Story of Life and Death in Delhi.* London: Jonathan Cape.

Sha, A. 2014. *The Army and Democracy: Military Politics in Pakistan.* Cambridge, MA: Harvard University Press.

Shewly, H. J. 2013. "Abandoned Spaces and Bare Life in the Enclaves of the India–Bangladesh Border." *Political Geography* 32:23–31.

Springate-Baginski, O., and Blaikie, P., eds. 2013. *Forests People and Power: The Political Ecology of Reform in South Asia.* London: Routledge.

Tanner, S. 2009. *Afghanistan: A Military History from Alexander the Great to the War Against the Taliban.* Philadelphia: Da Capo.

Thant, M.-U. 2011. *Where China Meets India: Burma and the Closing of the Great Asian Frontier.* London: Faber.

Learning Outcomes

An active seismic zone, responsible for the Himalayas, continues to cause problems.

Two major river systems flow through the region.

Monsoons are major climatic influences.

Much of the region was incorporated, directly and indirectly, into the British Empire and then partitioned after colonial rule ended into separate states, largely along religious lines.

Agriculture is an important activity, with the Green Revolution playing a significant role in improving yields in selected regions.

Industrialization occurs in selected urban sites. The region is an important textile manufacturer in the global economy.

Religious rivalries, especially between Muslim and Hindus, continue to mar social and political relations.

This is a region of massive population growth. Rates are beginning to fall as the region moves through the demographic transition with some countries experiencing a demographic dividend.

Urban growth has been dramatic with cities increasing in size and importance.

The conflict between India and Pakistan is a global geopolitical hotspot.

10

The Middle East and North Africa

The Middle East and North Africa (MENA) could just as easily and accurately be called South West Asia and North Africa (Figure 10.1). However, the term "Middle East," problematic and colonial as it is, has entered the popular lexicon and is still the most commonly used term. At first glance, it is an unlikely world region because it covers two separate continents: Africa and Asia. Popular views characterize it as an Arab Muslim world, but not all peoples in this region are Arab, not all Arabs are Muslims, and a majority of Muslims are not Arab. It does have a certain geographic coherence. It straddles a zone roughly 15 to 40 degrees latitude North with a similar geography of vast areas of aridity with occasional pockets of river valley and mountain greenness. It also has a shared experience of Islamic, Ottoman, and European imperialism and a similar trajectory of postcolonial independence. It is the hearth region for three of the world's great religions, and religious affiliations continue to shape conflicts and interactions. Oil and natural gas are important elements in the political economy of this region.

LEARNING OBJECTIVES

Identify the region's tectonic and climatic distinctiveness, and connect it to environmental problems of human settlement.

Define *palimpsest* and relate it to the region's diverse history of imperialist influences.

Survey the region's early agricultural activities and evaluate its lack of industry and the development of fossil fuels.

Recognize the intensively cultivated Nile delta and recall the three main stresses on this largely rural landscape.

Outline the diversity of religions, languages, and ethnicities in the region and discuss its demographic and gender features.

Describe the region's main drivers of urbanization and interpret its distinct forms of urbanism and settlement.

Examine the complex and unstable political geography of the region, and list the major geopolitical actors and influences.

The Environmental Context

Seismic Activity

This region sits between the boundaries of two massive tectonic plates, the African and Eurasian Plates, which are zones of intense tectonic activity. Earthquakes in Turkey and Iran are regular occurrences. Istanbul, formerly Constantinople, was almost razed to the ground in 557 CE. More recently, 17,000 people were killed in Turkey during the 1999 Izmit earthquake, and half a million were made homeless. In 2003, an earthquake registering 6.6 on the Richter scale caused the deaths of almost 30,000, injured 30,000, and destroyed much of the ancient city of Bam in Iran. Both Istanbul and Teheran are in active earthquake zones and, with their large dense populations, are particularly vulnerable to seismic disaster. Istanbul is currently undergoing a massive urban renewal program to make the city less vulnerable to earthquakes.

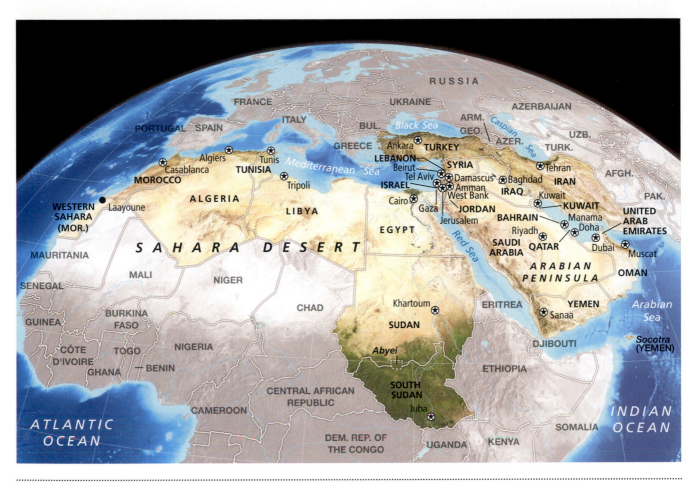

10.1 Map of region

An Arid Region

The most distinctive physical feature of this region is the long dry season. Rain falls mainly in the winter and rarely reaches more than 24 inches a year, with most of the rain falling in the higher mountainous regions.

Across the middle of the region, in the Sahara and Arabian Peninsula, rainfall rarely reaches more than 4 inches a year (Figure 10.2). Along the northern rim of this region and in coastal and more mountainous regions, rainfall can reach up to 30 inches a year. Because of the hot summers, two-thirds of all water is lost to **evapotranspiration**. Much of the region is characterized by desert with slivers of more fertile land along the coast and river valleys. North Africa is dominated by the Sahara, the world's largest desert that begins close to the Mediterranean and extends south to the **savannah** of central Africa. Satellite images show how the dense settlements of light along the fertile track of the Nile and beside the Mediterranean coast contrast with the relative emptiness of the surrounding desert.

In this arid region, the location of water was a determining factor in the location of cities. Cities, such as Aden,

Casablanca, Haifa, and Tunis, developed along the coast as the seawater provided means of transport as well as a moderating influence on temperatures. In other cases, fresh river water was the key. Cities grew up on the banks of the Euphrates, Nile, and Tigris. Damascus grew on the banks of the Barada River. In other cases, aquifers and underground wells were important sites. Marrakesh developed beside aquifers while Abu Dhabi, Doha, Jerusalem, and Medina were all originally built beside a freshwater well.

Water availability is declining especially in the Gulf States, where consumption is now nine times the renewable water resources. Demand is met through the importation of water and expensive desalinization plants. There are now over three hundred desalinization plants in the Gulf States. In 1962, Saudi Arabia had 556.6 cubic meters of water supply per person; by 2010 it had only 87.4. The comparable figures for UAE are 1376.0 and 20.0 and for Qatar, 1036.0 and 33.0. Extravagant water consumption is now seen as a mark of wealth. The average UAE resident uses 550 liters per day. The respective figure for Germany is 149. The Middle East is set to face an absolute water shortage by 2030.

10.2 The region is defined by its aridity. Sand dune landscape in Oman (photo: John Rennie Short)

Water is a source of political conflict. In many of the poorer countries, governments fail to provide accessible and cheap fresh water, giving traction to religious welfare groups. The source of many a political conflict lies less in ideology and more in the simple basic necessity of life, such as fresh water, adequate food, or employment opportunities. Israel's reluctance to give up its occupancy of the West Bank, despite international condemnation, is in part an unwillingness to lose control over West Bank **aquifers**.

There are conflicts around shared river systems in the Middle East. The management of the Jordan River system is made complicated by the fact that Jordan, Israel, Syria, and Lebanon all have rights and interests. The Tigris and Euphrates flow across the borders of Iraq, Turkey, and Syria. The Nile is another area of contention as Egypt draws more from the Nile than its neighbors. Ethiopia is a source of over 80 percent of the Nile's water supply but withdraws very little. Conflicts between Ethiopia, Egypt, and Sudan over Nile water are likely to emerge in the future. Water is such a scarce and valuable commodity in this region that when and where river systems cross international boundaries, they become a source of conflict more

often than platforms for cooperation. Thirsty neighbors often do not make the best of neighbors.

Desertification and Salinization

Two major environmental stressors on this region are also related to the lack of water: desertification and the salinization of soils (Figure 10.3).

Desertification was first noticed along the southern fringe of the Sahara in the 1970s. A series of drought years in the 1980s and 1990s increased soil erosion and activated fossil dunes that extended the desert southward into the savannah. It is not only drought that creates desertification but also the timing of rains throughout the year. Less rain, rain falling in fewer, heavier (more erosive) showers, and a shorter rainy season are all causes of desertification. The desert is extending its infertile grip across the northern and southern boundaries of the Sahara and across much of the Middle East. Predictions of drier and hotter weather over the next 50 years mean that desertification will become an increasing cause of concern. Current estimates suggest that desertification is impacting one-fifth of all land in MENA.

Desertification is not simply a result of drier conditions; it is exacerbated by overgrazing and population displacements. Population pressure puts more pressure on the land while social disturbances such as war and famine disrupt the traditional methods of husbandry and the regular movement of livestock. With displacements, people move into the more marginal lands leading to deforestation and more desertification. Desertification is not simply the result of climate change but also accelerated by political disruption.

The salinization of soils is a major environmental problem in MENA. The intense heat means that salts in the soils are raised closer to the surface and in drier conditions there is not enough rain to wash them away. The result is increasing salinity, land degradation, and reductions in fertility. In Oman, for example, salt-affected lands now constitute 70 percent of the agricultural areas of the country. Many aquifers in the region are also experiencing increased salinization as rainwater percolates through the more saline soil to the groundwater. The salinity of coastal aquifers in Israel has increased in the last 30 years.

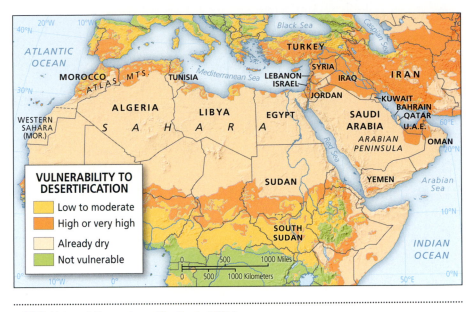

10.3 Vulnerability to desertification in MENA

Historical Geographies

Early Empires

The rise and fall of empires left imperial legacies that continue to shape the contemporary geography of this region. Around the Mediterranean coast, Greek and Roman Empires left a legacy of ancient cities and classical architecture. The rise of various Persian empires, for example, established the reach of Persian speakers across the region.

From the seventh century onward, an Islamic empire emerged from the desert of Arabia quickly extending control throughout the region. In 638 CE, a Muslim army defeated a Persian force, signaling the arrival of a new empire that soon extended its reach across MENA and into Spain, transforming animists into Muslims and making Arabic a dominant language and the language of the dominant. The empire spread Islam and Arabized much of the region. The new empire drew on existing Byzantine and Persian culture and traditions, so new cultural forms were grafted onto existing cultures and systems of organization. Muslim traders, sailing routes across the Indian Ocean, spread the faith as well as trading connections from the Middle East to China, India, South Asia, South East Asia, and East Africa and even into the coasts and islands of the Pacific.

The Rise and Fall of the Ottoman Empire

The Ottoman Empire began from a small territory in Anatolia around 1300 and lasted until 1924. The Turkish-speaking Ottomans were originally one of the nomadic tribes pushed west in the wake of Mongol expansion. In 1389, the Turks defeated the Serbs at Kosovo. In 1453, they captured Constantinople and overthrew the Byzantine Empire, inheriting its territory in Greece and southeast Europe. In 1517, they defeated the Arabs. The Ottomans advanced into Europe and displaced the commercial empires of Genoa and Venice in the Eastern Mediterranean and Black Sea. MENA and Europe became the center of a new Muslim caliphate centered in Constantinople—present-day Istanbul. Turkish Muslim migration into Europe and the conversion of Christians to Islam changed forever the religious make-up of the Balkans. During its period of greatest expansion, the Ottoman Empire extended across southeast Europe, North Africa, the Middle East, and into central Asia. By 1683, they had pushed all the way to Vienna, where they were defeated by a pan-European military force. Their defeat marks an inflexion point as expansion then turned into contraction.

But even by 1862 the empire still stretched across Europe, Africa, and the Middle East (Figure 10.4). Turkey's defeat in World War I marked the end of the centuries-old empire.

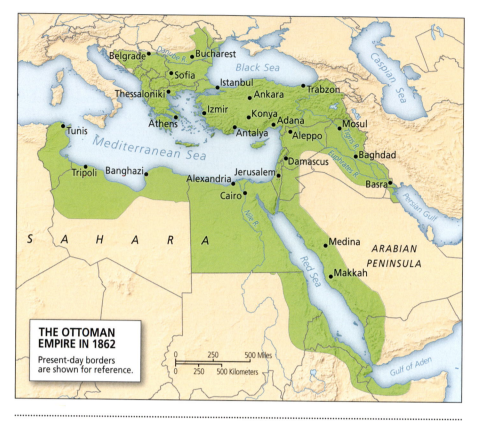

10.4 The Ottoman Empire in 1862

European Influence

After World War I, the winners—Britain and France—divided up the region into separate spheres of influence. Britain gained control of the territory now consisting of Iraq and Palestine, while France took Syria and Lebanon. Only the Arabian Desert, considered geopolitically insignificant and economically unimportant, was given to the Arabs. The Europeans were mainly concerned with a political stability in order to extract resources at cheap prices.

Places as Palimpsest: The Case of Tunis

Palimpsest originally refers to a manuscript with some of the original writing erased to make room for later writings. We can think of places as palimpsests with layers of influence. Consider the case of Tunis as a palimpsest of imperial writing.

The city of Tunis is the site of Carthage, a rival to Rome that was destroyed in 146 BCE in the Punic Wars fought between Rome and Carthage. It then became a Roman city linked to the wider empire throughout the Mediterranean, a legacy best revealed in ruins that are spread around the city and the magnificent Roman mosaics held in the city's Bardo Museum. The city was captured by a Muslim army in 698 CE as part of the rapid expansion of the Arab Muslim Empire across North Africa and into Spain and Europe.

The Arab influence is visible in the heart of the city, the **Medina**, a tight high-density area with a convoluted street pattern and five mosques. The oldest mosque was built in the ninth century ACE. This medieval Arab district is now a World Heritage site.

The city and the land along the coast came under the influence of the Ottoman Empire, and fine houses built for Ottoman officials can still be found in all their tiled richness.

Tunis then came under French rule in 1881, and a new town was laid out on a gridiron plan. Colonization involves a rewriting of urban space to show who is in control and who is being controlled. The garden suburbs of the French colonizers were in rich contrast to the spontaneous settlement at the edge of the city that housed the huge influx of migrants from the countryside. The contrast between the cellular Medina and the geometric colonial order is striking and intentional. The French built a central boulevard surrounded by roads and bordered by wide pavement that specifically echoed the Champs Elysées (Figure 10.5). It was built to make visible the connection between France and Tunisia, to show the rationality and technological superiority of France compared to a medieval religiosity. After independence in 1956, the street was renamed after a revolutionary hero and is now known as Avenue Habib Bourguiba, where it came to represent a modern independent Tunisia and was the place for celebrations of national identity and solidarity. In late 2010 it became the scene of demonstrations, part of the **Arab Spring**.

Tunis contains layers of this complex past: Roman ruins, Arab mosques, Turkish-inspired architecture in individual buildings and in the decoration on internal courtyards, French colonial architecture, and modern design. French is taught at school and is the language of the universities. Many Tunisians speak French and Arabic as well as other languages.

10.5 Avenue Habib Bourguiba in Tunis
(photo: John Rennie Short)

Economic Transformations

Agriculture

Before oil, there was agriculture. This region is one of the world's cradles of early agriculture. In Mesopotamia, the land between the Tigris and Euphrates Rivers, early peoples developed agriculture through irrigation and the careful exploitation of animals, seeds, and plants. In the drier areas, lacking water and with depleted soil fertility, nomads such as the **Bedouin** herded goats and camels.

This is a dry land with less than 10 percent of the land available for agriculture. The availability of water determines agricultural practices. There is dry farming, especially in the northern part of the region, where the main crops are wheat and barley. In the less arid areas alongside the Mediterranean coast, in the upland areas and along the northern rim of the region, such as upland Turkey and Iraq, there is sometimes enough rainfall for crops. In the more interior, lowland regions on the southern rim, the lack of water makes agriculture difficult without extensive and expensive irrigation. But where irrigation is possible, such as the Nile Valley, very high levels of productivity in the growing of rice and cotton are possible.

Much of the rural population is employed in small peasant holdings eking out a tough existence in an unforgiving land. We can get some sense of the relative economic weight of the agricultural sector by looking at the figures in Table 10.1, which lists, for four countries, the percentage of the labor force employed in agriculture and also agriculture as a percentage of the nation's GDP. The figures indicate a large agricultural sector in terms of employment but not as fully commodified or mechanized as in the United Kingdom and the United States.

There have been various schemes to modernize farming. However, the greater use of **cash crops** often has a socially regressive impact on peasant family farmers unable to compete with the larger more capitalized operations. Unable to afford the more expensive forms of modern agriculture and without the economies of scale to take on debt, there are now major differences within the rural economy between the capital-intensive sector of fully commodified agriculture and the low capital-intensive high labor sector of small-scale family farming. Small-scale farmers are particularly vulnerable to drought and desertification because they have less access to irrigated water. In Egypt, 93 percent of all farmers till patches of fewer than 5 acres. And in Tunisia, more than half of all farms are fewer than 12 acres. The modernization of agriculture that has increased in recent years has positively impacted those large landholdings with access to capital as they have developed larger acreage and more intensive irrigation for export-led agricultural development. Small peasant farmers have seen living standards decline, a major reason behind the rural-to-urban shift of the population in the region. Rural poverty is endemic in much of the region, one reason behind a rural Islamic conservatism. The Arab Spring is often thought to have started

TABLE 10.1	Agriculture in MENA	
	PERCENTAGE OF GDP	**PERCENTAGE OF LABOR FORCE**
Egypt	14.6	29.0
Tunisia	8.7	14.8
Turkey	8.2	25.5
Morocco	14.0	39.1

when a Tunisian by the name of Mohamed Bouazizi set himself alight on December 17, 2010. His family was unable to survive on their family plot of land and moved to a nearby village, where he found employment selling fruit from local farms. The corrupt police regularly humiliated him, and finally he set himself alight. The flames set off the Arab Spring, which is often remembered as a series of protests in large public spaces in big cities, but it began in the impoverished setting of rural Tunisia.

Lack of Industry

As a colonized region, MENA had few opportunities to industrialize. It acted as a source for raw materials shipped back to the colonial centers for manufacturing and export. Its main role under the colonial yoke was as a provider of raw materials such as cotton and oil. With independence, a number of countries such as Egypt soon introduced a form of **import substitution** whereby goods were made in the country rather than being imported from abroad. Import substitution of fabricated goods was seen as a platform to industrialize and modernize. Turkey, Syria, and Iraq all followed similar policies. While this did create some domestic industries, their protection from overseas competition and their connection with circuits of political power led to an inefficient sector that was dismantled during the 1980s and 1990s as a form of structural readjustment to the wider global economy. The large-scale privatization that occurred was similar to what happened in post-Soviet Russia; it was a process dominated by big capital and powerful domestic political operatives. In the short to medium term, it led to the enrichment of the well-connected few and loss of employment opportunities for many.

While there are pockets of industrial agglomeration, the economies of this region are marked by a low level of industrial development. This has hampered the emergence of a relatively large-scale affluent middle class along the US model.

The Age of Oil

It is perhaps a simplification to consider the economic geography of this region under the title Age of Oil. A simplification, but not a complete stretch because oil extraction, and increasingly natural gas extraction, plays an outsized role in the economy of the region, essentially creating a major division between those countries with and those without oil. And even between those with differing sizes of reserves.

This region has long been a source of oil for the global economy. Ensuring the supply of cheap oil was the major reason for colonial and neocolonial interventions by the British, French, and United States. Initially, the industry was owned and controlled by major multinational companies repatriating most of the profits back home to their shareholders. However, things began to change radically

in the 1960s and 1970s as the national governments gained control of the oil industry and collected more oil revenues. Newly independent states organized themselves into an oil-producing cartel, **OPEC (Organization of Petroleum Countries)** in Baghdad, in 1960 that came to wield enormous power as it controlled much of the global oil supply. OPEC members also began refining their own oil rather than simply shipping it raw (Figure 10.6). Today OPEC countries are responsible for 70 percent of the world's current oil supply. Current members are shown in Table 10.2; notice the importance of countries in MENA and especially the Gulf States.

In 1973, OPEC declared an oil embargo to punish the United States and Europe for supporting Israel during the **Yom Kippur War**. Oil prices quadrupled. The effect on the global economy was enormous. Oil was made more expensive for countries such as the United States that had long relied on cheap imported oil. The price increase stimulated the search for alternative oil sources even in hostile environments such as offshore and in the Arctic. For poor countries without oil reserves, the effect was a dramatic reduction in current accounts as money had to be spent on importing expensive oil. And in one of the quickest transfers of global wealth, oil-producing countries reaped a windfall of vast oil revenues. The sums are enormous: consider just one country, Libya, which during the peak of the oil boom, received $1 billion a week in oil revenues.

Formerly marginal countries now had enormous fiscal power. Massive oil revenues allowed the Saudis to become an important regional military power and fund their export

10.6 The world's largest refinery in Dahran, Saudia Arabia (HO/EPA/Shutterstock)

TABLE 10.2 OPEC Members	
COUNTRY	PERCENTAGE OF WORLD OIL RESERVES
Algeria	0.7
Angola	0.6
Ecuador	0.5
Iran	9.4
Iraq	8.6
Kuwait	6.2
Libya	2.9
Nigeria	2.3
Qatar	1.6
Saudi Arabia	16.2
UAE	6.0
Venezuela	18.2

of an austere Islam around the world. The rich Gulf States were able to construct mosques and madrassas and bankroll Islamic groups, some benign and others not so benign, around the world. The vast oil revenues allowed some countries, especially the small Gulf States with vast reserves and small populations, to build dazzling new city skylines and reward their citizenry with comfortable lifestyles and generous social welfare. In a country such as Qatar, for example, most of the work, skilled and unskilled, is done by foreign labor while Qataris sit comfortably on an ocean of oil. Under Mohammed Gadhafi, the oil money in Libya was used to support a lavish presidential lifestyle and to fund terrorist groups around the world from the IRA in Ireland to the PLO in the Middle East. In contrast, countries without oil, such as Egypt and Syria, had a harder time projecting power or rewarding citizens with generous welfare.

Rural Focus

Stress on the Nile Delta

The Nile delta is a low-lying area fanning out from Cairo for 100 miles to the sea (Figure 10.7). Every year for at least three millennia rains in the headwaters of the upper Nile bring billions of gallons of fresh water and millions of tons of nutrient to the delta and the narrow strip of fertile land just under 500 feet wide either side of the river. In Figure 10.7, the Nile looks like a green thin necklace across the barren desert. There are three seasons: the flooding season, the planting season, and the dry season. The nutrient-replenished soil supported intensive agriculture. Much of the land was rented out to peasant farmers on small lots. The delta is a patchwork of small fields connected by 25,000 miles of irrigation ditches.

But it is under stress. First, there were the dams built upstream. The High Aswan Dam was built between 1960 and 1970. It had a dual purpose, to reduce flooding that could threaten people and land downstream and to provide power for a growing economy. However, it also reduced the water supply and the nutrient load to the delta. Farmers now had to use chemicals and fertilizers to maintain fertility, which came with its own set of problems, especially for the cash-poor peasant farmers unable to pay for the chemicals. For even those who could afford them, their application raised pollution levels. The damming continued with the Merowe Dam in Sudan and the Grand Ethiopian Renaissance Dam in Ethiopia that opened in 2017. These further reduce the supply of water and nutrients; and the lack of nutrient-rich water to the delta makes it more vulnerable to coastline erosion.

Second, climate change poses special problems. The longer hotter summers reduce water through greater evapotranspiration and increased soil salinity. Saltwater intrusion at the mouth of the delta is reducing

10.7 The Nile Delta (Tsado / Alamy Stock Photo)

fertility. The delta is also threatened by sea level rise. Most of the Nile delta is only 10 feet above sea level. The 3-foot rise in this region predicted by 2100 will mean that almost a third of the delta will become part of the Mediterranean.

Third, rapid urban growth is reducing the supply of land and generating more pollution. The delta contains only 3 percent of the land surface of Egypt but now holds more than half of the country's population, close to 50 million people. In the burgeoning cities along the river, including Cairo and Khartoum, lack of proper sewage treatment and poor waste management are polluting the waterways that drain into the delta.

Social Geographies

The Geography of Religions

Today, MENA is predominantly but not exclusively Muslim. The other religions, many of which preceded Islam, laid down sediments of beliefs and religions that survive. The survival was, in part, due to the ideology of Islam that gave a special place to other people of the book, Jews and Christians, who were considered protected communities and allowed to hold property and practice their religion. They were heavily taxed and had an inferior position; they were not allowed to proselytize and were excluded from power. But when we compare the position of

Jews and other religious minorities in medieval Europe, this treatment was an exercise in tolerance.

JEWS

For millennia, Jews lived throughout the Middle East and North Africa, concentrated in the towns and cities. There were Jewish peddlers in Morocco and more assimilated Jews in Egypt and Iraq who were integrated into Arab society. Jews in MENA largely escaped the horrors of the Holocaust that blighted Europe Jewry. The position of Jews changed dramatically after the creation of Israel in 1948. Wars between Israel and the surrounding Arab countries made life difficult as Jews were now associated with **Zionism** and Israel. Anti Jewish legislation was introduced in Iraq, and when Nasser ruled Egypt from 1956 to 1970, he stated that Jews were not Egyptians. After the fall of the Shah in 1979, the position for Jews in Iran became even more difficult as they were portrayed as Zionists and supporters of the Shah. Throughout MENA, Jews left for Israel or the West, especially the United States. The region lost a Jewish presence that had survived for thousands of years.

CHRISTIANS

Christianity predates Islam. The Christian Church in Egypt is one of the oldest in the world (Figure 10.8). Its adherents are known as **Coptic Christians**, after the language that they used to speak. There are between 10 and 15 million Coptic Christians in Egypt and half a million in Sudan. Long a presence in towns and villages, they now suffer from persecution, especially during outbreaks of Muslim fundamentalism. There are now half a million Copts in North America.

Maronites, Arabic-speaking Christians who follow the teaching of a fifth-century hermit, Maron, were originally persecuted by fellow Christians. They hid in the mountains of Lebanon, where they survive today. Currently they number around 1 million and play an important role in the religious-based political structure of Lebanon.

There are 2 million Assyrian (or Syriac) Christians living in Iraq, Turkey, northeastern Syria, and northwestern Iraq. Facing persecution and marginalization, Christians are leaving these war-torn countries. The number of Assyrian Christians in Iraq fell from 1.4 million in the late 1980s to around 400,000 by 2010. In Syria they number less than 40,000.

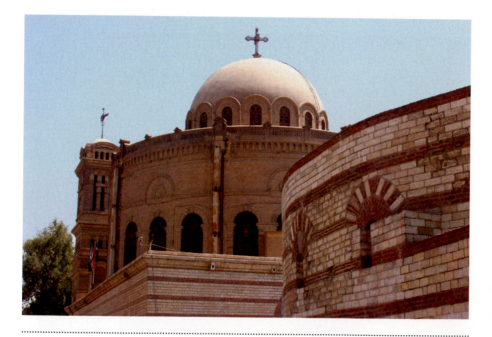

10.8 Christian Church in Cairo (Aude/Wikimedia Commons (CC BY-SA 3.0))

In 2015, the extremist Sunni group ISIS attacked thirty-five farming villages of Assyrian Christians in the Khabur Valley region, desecrating churches and kidnapping women and children. More than two-thirds of the total population of Assyrian Christians now lives outside the Middle East, and the exodus is likely to lead to the loss of Aramaic, one of the oldest continuously written and spoken languages in the region, the language of Jesus Christ. More than 150,000 live in Sweden.

YAZIDI AND DRUZE

There are many other religions groups. The Yazidi believe in a monotheistic universe under the care of seven angels. Their religion combines ancient Mesopotamian beliefs with elements of Sufi Islam and Christianity. They are concentrated in a region across Iraq, Syria, and Turkey. More than 650,000 live in northern Iraq. Since the 1990s, persistent persecution has forced more than 200,000 Yazidi to move to Europe. In 2014, the Sunni fundamentalist group ISIS targeted the Yazidi for extinction as an act of Islamic purification.

The Druze with an estimated population of 1.5 million have a distinctive and largely secretive religion with a belief system that combines elements of different religions and beliefs that first emerged in the eleventh century. The believers are tightly concentrated in a region now bisected by Syria (700,000), Lebanon (215,000), and Israel (140,000). The border between Israel and Syria separates Druze villages in the Golan Heights. I remember visiting the region in 1987 and seeing Druze villagers communicating with each other by shouting across the barbed wires and minefields. In all three countries, although a minority population, they play a significant role. One important element of their faith is an edict that they should show allegiance to the existing power, and so the Druze serve in the Syrian as well as the Israeli armies. In Syria they formed an important part of the officer class in Assad's army. In Lebanon they play an important role in the politics of the state. And in Israel they have citizenship and are active in the police and the military. They are Arabic speakers in the Israeli Defense Forces and are often security guards at Israeli airports. More recently, the Druze fear exclusion, as Israel is defined more as a Jewish state. Druze, because they have Arabic names and speak Arabic are often confused with and share similar discrimination faced by Palestinians. More than 310,000 Druze live outside this core region, mostly in South and North America.

MUSLIMS

The centuries-old religious diversity across the region is turning into religious homogeneity as Jews, Christians, and other religious minorities leave Muslim countries. MENA is increasingly Muslim as other religious minorities decline in significance and continue to leave the region due to persecution and marginalization.

While MENA is a core Muslim region, it is not numerically dominant. Out of the world's total of 1.6 billion Muslims, only 321 million live in MENA. A billion live in South and South East Asia, 242 million in sub-Saharan Africa, 44 million in Europe, and 5 million in the Americas. The largest Muslim counties in the world are not in MENA; they are in Pakistan, India, and Indonesia.

The main division within Islam is between **Sunni** and **Shiites**. Both share the basic tenets of Islamic teaching: a belief in Mohammed as the messenger of God, daily prayers, pilgrimage to Mecca, alms, and fasting during Ramadan. Their argument lies in their different interpretations of the correct succession after the death of the Prophet Mohammed. The Shiites believe that the succession should be hereditary and follow family bloodlines. The first caliphs after Mohammed were Sunni choices from outside his immediate family. The fourth was Ali, Mohammed's son-in-law, and thus more acceptable to the Shiites. When he was killed at a battle in 661, a rift developed between the two branches that was further widened when Ali's son, Hussein, was killed at the battle of Karbala 680 CE, or 61 in the Islamic counting. It is now a Shiite sacred site.

Sunnis are the dominant form of Islam with more than 1.3 billion followers. Shiites are a minority with less than 250 million, concentrated in Iran, southern Iraq, southern Lebanon, and eastern Saudi Arabia and neighboring Bahrain (Figure 10.9). The split has widened in recent years in part because it embodies two major and competing powers in the region, Saudi Arabia and Iran. Saudi Arabia is one of the richest Sunni Arab countries where the fundamentalist teaching portrays Shiites as non-Muslims. Since the revolution in 1979, Iran is the only official Shiite state. There is conflict between the two countries, and their proxies in Iraq, Syria, Lebanon, and Yemen, for geopolitical dominance, especially in weak, dysfunctional, or collapsing states that provide a space of opportunity for projecting national interests.

Events in Iraq after the fall of Saddam Hussein brought the conflict between Sunnis and Shiites into direct contact in one shared national space. In 2006, Sunni extremists bombed the al-Askari mosque in Samarra, Iraq, destroying the magnificent golden dome. It was a holy site for Shia Muslims. The bombs set off a spree of revenge killings and attacks on Sunni religious sites. A year later another bomb destroyed the remaining minarets in Samarra. When a car bomb killed at least sixty-eight people close to the al-Abbas mosque in Karbala, Iraq, in 2007—another sacred Shiite site—it renewed Shiite-Sunni conflict. Death squads, murders, and forced displacements solidified and reinforced the differences between the two communities. The conflict spiraled out from Iraq into Syria.

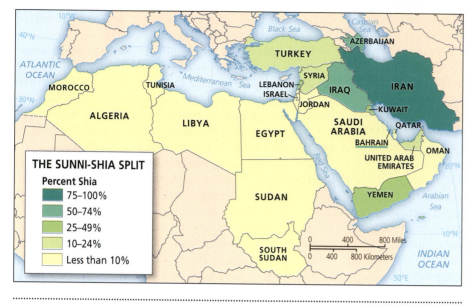

10.9 The Sunni-Shia split

The rise of a more fundamentalist Sunni Islam across MENA and in South and South East Asia has put the minority Shia communities into danger in countries such as Pakistan and Afghanistan.

The region remains a contested space between the two branches of Islam, their primary states, their proxies, and their terrorist offshoots. The divisions are exacerbated in weak or dysfunctional states.

The Geography of Language

There are four main language groups in the region—Arab, Hebrew, Turkish, and Persian—as well as a range of local languages. Arabic is by far the most dominant language, read and taught throughout the region. It is the liturgical language of Islam, the official language of twenty-seven states, and widely spoken in an arc of countries from the west coast of North Africa to the Middle East. There are different regional dialects; the largest and most dominant is Egyptian Arabic spoken by 54 million people. There is a standard Arabic, taught at schools and understood by most educated Arabs, as well as more localized dialects such as Yemeni Arabic, Gulf Arabic, Lebanese Arabic, and Tunisian Arabic, that may or may not be mutually intelligible. The Arabic spoken in the streets will vary by country and even region. More educated Arabs can speak both the local dialect and standard Arabic. Written Arabic is more formalized in what is now termed **Modern Standard Arabic**.

Hebrew as a language has an ancient history and a more recent emergence. It was the language of the early Israelites and then disappeared as an everyday language around 300 CE, surviving only as the liturgical language of the Jews. It was revived in the nineteenth century as part of early Zionism and is now one of the two recognized languages of Israel, the other being Arabic (see Figure 10.22). Almost 90 percent of Israeli Jews are proficient in Hebrew, including 70 percent of Israeli Arabs. In 2014, there were demands from some Israeli politicians to make Hebrew the sole official language, in effect hardening the notion of Israel as a Jewish state.

Not all Muslims speak Arabic. Turkish is the official language of Turkey and Cyprus and recognized as a minority language in Iraq and, reflecting the reach and legacy of the Ottoman Empire, also in Bosnia, Greece, Iraq, Kosovo, Macedonia, and Romania. A major Muslim holiday is Eid al-Adha, which marks the end of the **hajj** pilgrimage. Most Muslims in the Middle East and the West use the term "Eid" for the holidays. But in Turkey the holiday is "Bayram," a word that is also used in Central Asian countries such as Turkmenistan and Uzbekistan and by Muslims in the Balkans. The use of the word represents the limits of Turkish influence under the Ottoman Empire.

Persian was the language of the various Persian empires. It is the official language of Iran, Afghanistan (known as Dari), and Tajikistan but also spoken in parts of Azerbaijan, Iraq, Russia, and Uzbekistan. Its distribution maps the extent of the Persian empires. The Arabic term "Farsi" is sometimes also used to denote Persian. The language has retained much of its form and sounds over the centuries. A classic work, the epic poem *Shahnameh* (The Book of Kings) written around one thousand years ago is still accessible to modern Persians. Iran is the core state of the Persian-speaking world.

The Geography of Ethnicity

Language, religion, and ethnicity overlap; they split, blend, and reassemble in complex patterns.

KURDS

One of the largest ethnic groups is the Kurds. They have a distinctive language linguistically closer to Persian than Arabic. The population of around 30 million inhabits a core region that in the aftermath of World War 1 was divided between Turkey (15–20 million), Iran (8 million), Iraq (5 million), and Syria (2 million) (Figure 10.10). In terms of a percentage of a nation's total population they are, in order: Turkey, 20 percent; Iraq, 17 percent; Syria and Iran, 10 percent. These absolute and relative numbers

10.10 The Kurdish heritage

are estimates as they rely on people divulging information that they may not wish to share with national authorities for fear of reprisal or discrimination. They are most likely underestimates.

With a shared sense of identity, they are a nation without a state. The Kurdish nation is separated out into different states where they form a minority. In both Turkey and Iraq, they were marginalized by traditional power elites. A modern Kurdish resistance movement emerged in Turkey in the 1970s. There is an ongoing military campaign by the Turkish state in the east of the country where most Kurds live. In Iraq, the Kurds were oppressed by Saddam Hussein, but since 1992 and especially since 2003, they achieved a self-governing regional autonomy, initially under the protection of US air power. The Kurdish region has remained relatively peaceful and prosperous as the surrounding region collapsed. The Kurdish region is expected to export a million barrels of oil a day when the conflict dies down. In recent years they faced threats at the borders of their region from ISIS. The fighting at its greatest extent stretched over 650 miles. This conflict has also allowed the Kurds to claim control over a large swathe of territory that they have reclaimed from ISIS. Some of the support for ISIS in the region is from Sunnis fearing that Kurdish involvement was a prelude to the establishment of a Kurdish state. With oil revenues and control over a substantial part of territory, are we witnessing the early emergence of a distinct Kurdish state in their traditional homeland in Iraq and Syria?

AMAZIGH

Another distinct language group is the Amazigh linguistic-ethnic group in North Africa. Often called Berber, they prefer the term "Amazigh." There are around 30 million Amazigh across North Africa. They constitute half the population of Morocco and one-third of the population of Algeria, with minorities in Libya and Tunisia. In Morocco they are concentrated along the Mediterranean littoral in provinces such as Nador. They formed a short-lived republic in the 1920s and today constitute a periphery to the political core of the mainly Arabic-speaking elites living along the Atlantic coast and in the interior centered in Rabat. Their provinces often have lower infrastructural investment than the Arabic-speaking regions. They make up a significant element in the stream of 4 million Moroccan emigrants to Europe and especially France. Remittances from Europe are a significant element in the local economies. They were long considered second-class citizens compared to Arabs and then the European colonists. But in recent years their distinct culture is being recognized. Amazigh is now an official language in Morocco and a national language of Algeria. Notable Amazigh include the geographer Ibn Battuta (1304–1368/69) and contemporaries such as the great soccer player Zinedone Zidane, who embodies the diasporic nature of the community. He was born in 1972 in Marseilles to Amazigh parents and played in the World Cup–winning French national team. More than 2 million Amazigh live in Europe.

The Youth Bulge

One of the distinctive features of the demography of this region is the **youth bulge**. More than 60 percent of the total population is aged less than 24 years, and 30 percent of the population is aged between 15 and 29, an unprecedented ratio of the young and very young to the old (Figure 10.11). This youth cohort is the result of the high birth rates of the last two to three decades and the recent decline of birth rates. The median age in Libya is 24, and in Yemen it is 17. Compare this to the United States, where the median age is 37. In Egypt, 60 percent of the total population is aged under 30, and 18 percent are in the 17- to 25-year range, the dominant age for social protest and feelings of alienation.

In other regions of the world, this youth bulge is a demographic dividend as more people join the productive workforce. In MENA, by contrast, the bulge is associated with high unemployment and restricted employment opportunities. Youth across the countries of MENA have unemployment rates currently twice the world average. And in a twist from the situation in the United States, where the better educated have lower unemployment, it is often the opposite in MENA, where the educated youth have some of the highest unemployment rates because there are fewer opportunities for their skills and education compared to

10.11 The youth bulge: young women in Tunis (photo: John Rennie Short)

the unskilled and less well educated. The result is a large mass of discontented youth. Part of the demographics behind the Arab Spring, Islamic fundamentalism, and the recruitment to militant groups is the discontent of this youth bulge unable to find employment or job opportunities commensurate with their education.

Surveys reveal a high level of youth discontent as people respond that there is no space for them in the society as currently constituted. Young people across the region describe their state as one of "waithood" as they are unable to join the job market and make enough money to establish a home, career, and family of their own.

The Geography of Women's Lives

Traditional Islam is highly gendered with prescribed rules for men and women and codes of behavior in public spaces (see Figure 10.11). In Saudi Arabia, women must remain totally veiled in public and until recently were not allowed to drive a car. A conservative Islamic resurgence in the region since the 1970s replaced more liberal interpretations with increasing gender segregation and restriction of women to a domestic role, more rigid dress codes, and removing women from positions of legal and judicial authority.

There is also a strain of conservative Arab feminism, in contrast to a more liberal Arab feminism, that sees less

strict separation as a form of Western modernity to be resisted. Islamic fundamentalism today is more of an invented tradition of relatively recent vintage than a "pure" historical artifact. The fundamentalist and the secular represent not the past and the future, respectively, but alternative conceptions of the present.

Islamic law allows women full legal equality once they attain puberty, but the precise interpretation of how this equality is expressed varies across the region and through time. Tunisia and Turkey, for example, have more liberal traditions than say Saudi Arabia. We should be careful, however, in assuming a constant Islamic prescription for women. The role of women is also influenced by political changes. For example, in 1923, the newly independent republic of Turkey allowed women more freedom in public space, but recent years have seen the rise of more conservative interpretations of dress and behavior. After the fall of the Shah in 1979, more restrictions were placed on the behavior and dress of women in public space as the conservative clerics fashioned more austere interpretations as part of their creation of an Islamic republic.

Today the status of women varies widely. Where the government relies on conservative clerics to remain in power, much less equality is afforded to women. The status of women can also quickly change. Women were allowed to vote in Kuwait when voting was first introduced in 1985. The right was removed in 1999 and regranted in 2005. In 2015, Saudi women were allowed to vote and run for the first time in local elections. However, the legal system remains highly gendered. In court proceedings, females have to deputize male relatives to speak on their behalf, and the testimony of one man equals that of two women. The United States is a staunch ally of Saudi Arabia.

MENA has one of the lowest female participation rates in the formal economy. In 2016, while the ratio of female-to-male labor force participation was 0.89 in Norway, it was 0.57 in Kuwait, 0.42 in Turkey, and only 0.25 in Saudi Arabia. The global average is 0.51, while it is 0.81 for the United States.

Urban Trends

This was long a region of towns and cities as well as desert and oasis. Cities were centers of trade, commerce, and sites of religious devotion.

An Urban Explosion

The last 40 years have witnessed rapid urban expansion as people have moved to the town and cities (Table 10.3). Urban population growth is now twice the level of overall population growth. Cities now account for 70 percent of the total population, compared to less than 48 percent in 1980. Even in countries with a lower level of urbanization, such as Egypt (43 percent), Morocco (60), Sudan (34), and Yemen (34), urban growth rates are high and there is a distinct rural-to-urban shift of the population.

This rapid urbanization is not accompanied by a commensurate increase in employment opportunities. There is an urbanization of poverty and severe strains on service provision in cities. In much of the non-oil economies, there is a proliferation of informal settlements with limited property rights. More than 50 percent of Egypt's urban population lives with uncertain and insecure tenure rights and limited access to services such as fresh water and sanitation.

Gulf Urbanism

In the oil-rich countries such as Saudi Arabia and the Gulf States, the level of urbanization is much higher. More than 85 percent of people in UAE live in cities. In Qatar and Bahrain, it is 98 percent, both of them city-states where the urban conditions for citizens and high-skilled workers are very good. For the unskilled workers from South Asia, the conditions are less than ideal.

One distinctive feature of the urbanism in the oil-rich states of Saudi Arabia and the Gulf States is the large number of foreign born. In Qatar the 250,000 nationals make up only 12 percent of the city's total population. More than 80 percent of people living in Dubai are foreign born.

Consider the case of Abu Dhabi, the capital of UAE, where more than 90 percent of the city's 1.6 million population are expatriates, working people born outside of the country. Less than 10 percent are nationals, but they benefit most from the wealth from oil, and they are the only groups allowed to own property and are largely guaranteed jobs for life. Some are very wealthy, while the rest are affluent. Their lifestyle is paid for by oil.

Around 30–40 percent of the population is middle-class expatriate workers on short-term contracts whose living costs are paid for by their companies. They are the engineers, doctors, teachers, and technicians that come from Scotland, the United States, and Egypt. They live in villas and apartments, the size and quality dependent on their income and status. The low-wage service workers live at high densities in rental accommodations in the city. The construction workers are men recruited mainly from Pakistan, India, and Bangladesh but also some from Sri Lanka and Nepal. They are housed in labor camps, often in overcrowded conditions. One-half of the city's total population is men from South Asia. The typical South Asian construction laborer works for 12 hours day 6 days a week, for $354 a month. Most of them have not been home for 2 years.

Another feature of the new cities of the Gulf States is the city as urban spectacular, employing **starchitects** (highly prized architects brought in to design high-profile projects) for the big statement and the grand flourish (Figure 10.12). A combination of lots of money and a desperate need for world respect and global recognition leads to an architectural frenzy. Abu Dhabi has a Zaha Hadid–designed performing arts center and a Frank Gehry–designed Guggenheim museum. Its central business district is a concrete forest of new tall buildings. Currently Dubai boasts the world's tallest building, the 160-floor Burj Khalifa and the 77-floor Emirates Park Tower, the tallest hotel in the world. Tall buildings are spectacular, and as urban skyline they create a visual signature of a globalizing modernity, a visual spectacle intended to make us marvel and, above all, take notice in the overcrowded world of signature skylines. Almost a quarter of a million people labored to construct Dubai's aim of becoming a global city.

City-States and Primate Cities

The smaller countries such as Bahrain, Kuwait, and Qatar are essentially **city-states**, where the city and the state are essentially one and the same. The city is the state. National identity is thus associated with the city skyline and the hosting of urban spectaculars. One reason for Qatar's bid to host the 2022 World Cup was its ability to build all the stadia in a relatively small area.

In a number of states across MENA, one city dominates. Tunis's 2.7 million constitutes more than a third of the nation's 11 million. It is a primate city dominating the economic, cultural, and political life of Tunisia. There is a marked contrast between the more cosmopolitan, more secular cities and the small towns and more rural areas

TABLE 10.3 The Largest Metros in MENA	
CITY	**POPULATION (MILLIONS)**
Cairo	20.3
Tehran	15.2
Istanbul	14.6
Baghdad	8.7
Casablanca	6.8
Riyadh	6.5
Algiers	5.3
Ankara	5.2
Khartoum	5.1
Alexandria	4.9

10.12 Al Burj luxury hotel in Dubai is built to represent the sails of an Arab dhow. (Photo: John Rennie Short)

that are relatively homogenous and tend to be more conservative in outlook and religious observance.

In Egypt, Cairo is the undisputed center of religious, political, and economic life of the country; similarly with Tehran in Iran and Baghdad in Iraq. Elsewhere the dominant city competes with other cities. In Turkey, Istanbul is by far the largest city, but Ankara is the formal political capital.

Informal Settlements

Most cites in the region and especially the largest cities such as Baghdad, Cairo, Istanbul, Khartoum, and Tehran all have substantial numbers of informal settlements as poor migrants construct their own neighborhoods and find work in the informal sector. And in this fractious region of the world the large informal settlements are also the site of struggle as various groups compete for the hearts and minds of a population denied access to many services. The attraction of the Muslim Brotherhood in the Cairo slums was their ability to provide basic services that the official state was unable or unwilling to provide. Similarly across the region, pious charity often provides an opportunity for religious groups to spread their message to urban dwellers abandoned and ignored by the state.

City Focus: Istanbul

The city straddles both continents and embodies the varied and long contacts between East and West (Figure 10.13). Founded in 330 CE by the Emperor Constantine as Constantinople, it became the capital of the Byzantine Empire. As an important repository of Greek classical thought and a transmission point for the intellectual flowering of Arabic and Persian scholarship, Constantinople stayed lit when Western Europe sank into the **Dark Ages**. The Byzantines kept the sparks of culture alive that subsequently caught fire in the Renaissance of Western Europe. There were also more destructive exchanges. The city was pillaged in the Fourth Crusade, its treasures looted and taken back to Western Europe. The churches of Venice are filled with the theft.

When it became the capital of the Ottomans in 1453, it became the center of an empire that embraced the continents of Africa, Asia, and Europe. At its furthest extent, the city's power reached to the very gates of Venice. Under the longest reigning Sultan, Suleiman the Magnificent (r. 1520–1566), not only was the empire extended but also the city was adorned with new mosques. The city is still filled with the most beautiful of religious architecture, including the intimate Christian Church of St. Savior in Chora, the serene Hagia Sophia, and the stunning Blue Mosque. The city skyline is punctuated with **minarets** and Genoan towers, Ottoman mosques, and Byzantine churches. The Grand Bazaar built in 1455/56 continues to attract shoppers and bargain hunters (Figure 10.14).

10.13 Location of Istanbul

10.14 The Grand Bazaar, Istanbul (photo: John Rennie Short)

As the Ottoman Empire waned, there was a growing admiration for an ascending West. In 1856, the Sultan abandoned the Topkapi Palace for a new neoclassical palace at Dolmabahce, which would look at home on the banks of the Seine or the Thames as it does on the Bosporus.

With the fall of the Ottoman Empire and the creation of Ankara as the capital of modern Turkey, Istanbul was relegated to a more minor role in world affairs. It was only in the 1970s with the entry of foreign investment that Istanbul began to grow mainly through rural-to-urban migration as peasants moved off the land to search for opportunities in the big city. The population of the metro region has now passed 14 million. The city plays an important part in the economy of the country, with more than 27 percent of national GDP. It is an important tourist destination with the number of tourists from Europe as well as the Middle East now exceeding 11 million.

Istanbul sits on the border between east and west, a crossroads for cultural exchanges. To walk the streets of Istanbul is to see women dressed in the most contemporary fashions as well as in the *hijab*. Founded as a secular republic, modern Turkey is still situated between East and West, and at the heart of this unfolding drama is the city of Istanbul.

Geopolitics

The political geography of this region is complex. Today's patchworks of nation-states consist of fragments and constructs of postimperial collapse. There are states with more than one nation, such as Iraq, and nations without states, such as the Kurds. And increasingly, in selected parts of the region, states sometimes have to fight for sovereignty within their own boundaries with non-state actors.

Unstable States

States in this region emerged more from imperial boundaries and geopolitical considerations and deals between great powers than as organic units with shared national consciousness. Take the case of Iraq, for example. As an Ottoman Territory, it came under British control after World War I and then it became an independent country in 1932. It lacked internal coherence divided as it was between a Kurdish north, a Shia south, and a Sunni center (Figure 10.15). These divisions were sometimes masked but were revealed anew after the United States invaded in 2003.

Lebanon has three competing religious groups, and in Saudi Arabia there is a tension between the dominant Sunni Muslins and the Shia Muslims in the eastern part

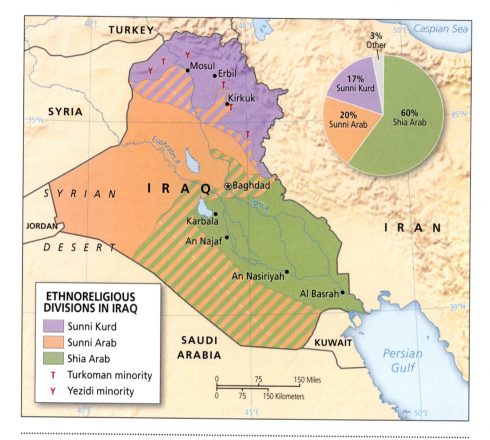

**ETHNORELIGIOUS
DIVISIONS IN IRAQ**
- Sunni Kurd
- Sunni Arab
- Shia Arab
- T Turkoman minority
- Y Yezidi minority

10.15 Ethnoreligious Divisions in Iraq

of the country. Bahrain has a Sunni monarch ruling over a Shia majority.

There is often tension within states as different groups compete for power, and the goal of nation building—taken for granted in much of Europe and North America—is still an ongoing, highly charged process.

Other Actors

A standard definition of the state is an organization that has monopoly control over a territory. However, in this region other actors play an important role because of the weakness of certain states and the competition between different communities in weak or dysfunctional states. It is important to note that the division between state and non-state actors is fluid as non-state actors can become part of a state and can also be linked to other states. There is a complicated relationship between state and non-state groups acting as proxies of other states. Iran supports Hezbollah and Hamas. Saudi Arabia was long a supporter of the Taliban and in 2015 sent jets into Yemen to bomb Sana'a as it considered the Houthis regime a proxy for their arch-rival, Iran. In some cases non-state actors slip and slide into formal state actors. It is complicated; so let us consider five examples: Hezbollah, Hamas, Houthis, Al Qaeda, and ISIS.

HEZBOLLAH

Hezbollah is a Shiite group in Lebanon formed in response to the Israeli occupation of southern Lebanon in 1982. Its resistance forced Israel to end its occupation of southern Lebanon and dealt the Israeli forces another heavy blow in 2006 during a thirty-four-day war. It began as an armed group, but it is also a charity, dispensing aid, and is also now a political party in Lebanon. With aid from Syria and Iran, it has as many soldiers as the Lebanese official army. It has been accused of terrorist bombings in Argentina, Israel, and Lebanon. It has achieved some legitimation as a mainstream political party, but its support of the Assad regime in Syria has undermined its broader appeal in the Arab world. In the war against ISIS, it forged alliances with Shiite and Christian villages in Syria against the Sunni rebel forces.

HAMAS

Similarly, Hamas has a multidimensional nature as both a military and political force. It was founded in 1987 to liberate Palestine from Israel and to establish an Islamic state in the region. Its military wings launch attacks against Israel. In 2006, it won a majority in the Palestinian elections.

HOUTHIS

Houthis are a Shiite group in Yemen long on the margins of political power. They were founded in 1992 and gained more power as the Yemeni state collapsed. In 2015, they took over control of the capital city of Sana'a and now rule over a large part of the country. In 2015, the Saudi air force, in collaboration with other Gulf States prosecuted a bombing campaign to overthrow the Houthis regime. Because they are Shiite, the Saudis consider them allies of their archrival, Iran.

AL QAEDA

Al Qaeda is a global terrorist organization with significant offshoots in MENA, including Al-Qaeda in Islamic Maghreb, Al-Qaeda in Syria, and Al-Qaeda in the Arabian Peninsula (AQAP). It promotes an Islamic fundamentalism that sanctions war against non-Muslims and non-Sunni Muslims. It is marked by violent terrorist acts.

AQAP is based in Yemen and Saudi Arabia but also organizes terrorist attacks in Europe and North America. It is also against the Saudi monarchical system. It is a shared enemy of both Saudi Arabia and Iran, one of the few things the two countries agree upon.

ISIS

What gives extra power to these non-state actors are political contexts of weak states, dysfunctional states, or states characterized by internal convulsions along ethnic and religious lines and an economic system that offers few opportunities for the majority of people, especially young people. This is the context for the rise of one of the most violent non-state actors in MENA. ISIS (Islamic State of Iraq), also known as Daesh, is a Sunni Muslim movement that emerged in postinvasion Iraqi from Sunni terrorist groups affiliated with Al Qaeda involved in the civil war between Sunni and Shia. It became a military force when it linked up with former **Baathist** military personnel displaced by the postinvasion Shiite dominance in Iraq. By 2013, it was a terrorist group motivated by Sunni Islamic fundamentalism, including violence against Shia Muslims and Christians, with a muscular military staffed by former Iraq Army personnel. It became self-funded by capturing oil depots, taxing residents and businesses under its control, extortion, and smuggling. In the failed states of Syria, undergoing civil war, and Iraq, beset by religious tensions, ISIS flourished, and by 2014 it had captured territory across Iraq and Syria. Figure 10.16 shows the extent of ISIS control in 2014, at its greatest extent. Over the subsequent years, the engagement by troops from Iraq and Syria as well as Kurdish and Iranian forces and Allied plane strikes shrunk the territorial hold of these extremists. It is now less of a territorial entity but remains a terrorist threat.

It gained some measure of support from the alienated and bored Muslim youth who feel marginalized in Europe, North America, or Australia. Of its 31,000 fighters in Iraq and Syria, almost 12,000 were foreign nationals from eighty-one countries. Fundamentalism appeals to a youthful population without jobs or hope or any sense of belonging. ISIS also has a sophisticated marketing using social messaging and the Internet. Its images of pornographic violence also attract the unglued and the violent. ISIS lost much of its appeal in the areas directly under its control as it could not provide jobs, clean water, and adequate sewage disposal. The actual experience of government often requires a different skill set than armed conflict or the espousal of religious creeds. There was also evidence of rising tension between local jihadist and foreign fighters. ISIS has extended its reach beyond Syria and Iraq, through the declaration of separate provinces in the Afghanistan-Pakistan border region, Libya, Egypt, Yemen, Saudi Arabia, Russia (North Caucasus), and Nigeria. In some cases this involved encroaching on Al Qaeda as in the Afghanistan-Pakistan border area, Russia, and Yemen. In other cases it involved the incorporation of existing insurgencies, such as Boko Haram in West Africa.

The notion of the **Caliphate**, the creation of a transnational authority guided by Islamic principles, is a long-held belief of many Sunni fundamentalists. Their dream is of creating a society firmly grounded in adherence to fundamentalist Islamic principles. In reality, the original caliphate was a more multicultural progressive and open society. But these current ideas are not based on historical reality but on contemporary interpretations. So it is not the recreation of an actual caliphate but the creation of a Caliphate imagined as a response to modernity.

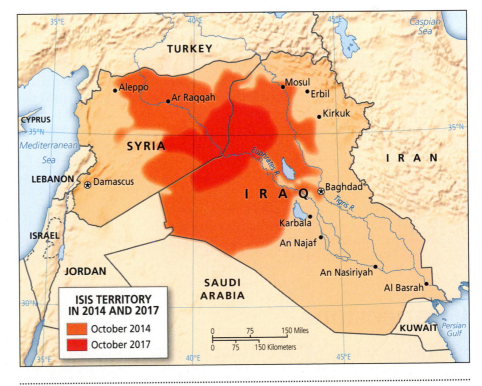

10.16 ISIS-held territory in 2014 and 2017

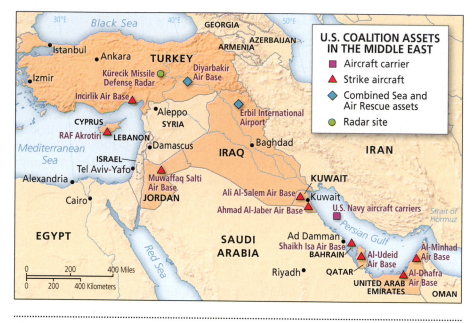

10.17 US Coalition assets in the Middle East

Alliances with the United States

The United States plays a huge role in this region. Saudi Arabia and the Gulf States are allies of the United States, and the United States has military bases in Bahrain, Kuwait, UAE, and Saudi Arabia (Figure 10.17). The United States fought the Desert Storm campaign in 1991 against Saddam Hussein to reinstate the Kuwait emir. In the minds of many in the Middle East, the United States props up the rule of unelected families and regimes. The United States is one of the main arms suppliers to Saudi Arabia and the Gulf States. The rulers, in turn, are criticized for their willingness to have US bases. Osama bin Laden was first motivated in his campaign against the United States after the Saudi royal family allowed US forces to enter the Kingdom in 1992 in the war against Saddam Hussein. The bonds between the United States and the UAE are particularly close. The UAE has participated in every US-led coalition since 1991, and the United States has both an air base and a deep water harbor facility. The UAE, unlike many of the other states, has actually sent military personnel in coalition-led campaigns to Afghanistan, Somali, and Libya.

Connections

MENA has for centuries been part of wider global connections. Here we will look at an intellectual connection of the past as well as the contemporary connection of airlines.

The Contribution of the Islamic World to the European Renaissance

The roots of the European **Renaissance** lie in North Africa, Muslim Spain, and the Middle East. The renewal of interest in classical thought that erupted in Renaissance Europe was based on the work of Arab and Persian scholars who had kept alive and extended the intellectual tradition of classical Greece.

The foundation of renaissance science and critical inquiry was laid in Muslim cities such as Baghdad under the Abbasid caliphs, and in Cordoba, Toledo, and Granada under the Moors. We have a linguistic legacy of their contribution in the Arabic origin of words such as *alchemy*, *algebra*, and *algorithm*. Arab, Jewish, and Persian scholars kept knowledge alive and moving forward while Europe was in intellectual retreat and decline.

In a variety of cities, Christian, Jewish, and Muslim scholars translated the work of the ancients and worked on new themes. Ibn Khaldun (1332–1406) was born in Tunis. His great work, *The Muqaddimah* (Ibn Khaldun), is perhaps one of the earliest social science works that combines both history and geography. It influences scholars to the present day. His economic work, for example, influenced John Maynard Keynes and Arthur Laffer. His statue sits in the middle of the main street in Tunis.

The scholars of Muslim Spain, North Africa, and the Middle East kept alive and improved upon the classical knowledge. The European Renaissance had its roots in the intellectual activity of the early Muslim world.

Airlines and Global Hubs

The Gulf States, flush with oil revenues, have invested in airlines and airports. Three Gulf States have spent substantial sums in establishing global airlines moving people and goods around the world through their home hubs. The major hub of the Emirates airline is Dubai, one of the world's largest airports, in terms of international passengers, and most certainly one of the plushest (Figure 10.18). It handles five times the capacity of Chicago's O'Hare airport, the busiest in the United States. The airport directly employs 90,000 people and contributes over $26 billion to the economy. The airport is now responsible for a third of

Dubai's GDP. Etihad is based in Abu Dhabi, and Qatar is based in Doha.

The three airlines present a challenge to US international airlines such as Delta and United. The US companies claim that these three airline companies receive an unfair advantage through interest-free loans from their respective governments who also give them tax and customs duty exemptions. The prohibition of unions makes labor costs cheaper.

10.18 Dubai airport (photo: John Rennie Short)

The Gulf State airlines are acutely aware of possible backlash and so spend in overseas market areas to promote good will and to promote their brand name, especially through ownership and sponsorship of sports franchises. In the United Kingdom, Emirates sponsors the English Premier League team, Arsenal. Etihad owns Manchester City in England and has signed an agreement to be the exclusive carrier of the Washington Capitals (ice hockey) and the Washington Wizards and Mystics (men and women basketball teams). Etihad also advertises in *The Washington Post*, where it notes prominently that it buys Boeing aircraft and so supports the US economy.

For the three Gulf States, investment in airplanes and airports is a way to rebalance the economy from a precarious reliance on oil revenues. The airlines also project the soft power of the Gulf States to a global audience, especially to the increasing number of people travelling on international routes.

Subregions

Saudi Arabia and the Gulf States

This group of countries is fortunate to be blessed with substantial oil reserves (Figure 10.19).

Saudi Arabia is one of the largest countries in the region, second only to Algeria and Sudan in areal extent, but much of it is inhospitable desert. It has the second largest oil reserves

in the world, after Venezuela. About a fifth of all known oil in the world lies beneath the Arabian sands. Almost 80 percent of government revenue comes from oil sales.

Saudi Arabia has almost 300 billion barrels of oil in shallow, easily accessed reserves that make drilling and extraction very cheap. The cost of extraction in Saudi Arabia and the Gulf States is thus much cheaper than in most of the rest of the world and especially compared to the more hostile environments of the Arctic or of deep-water sites.

10.19 Map of Saudi Arabia, Yemen, and the Gulf States

Saudi Arabia can still make a profit when oil is little over $10 a barrel. Venezuela, in contrast, requires over a $90 a barrel to break even. Saudi Arabia produces 30 million barrels a day. In 2014, Saudi Arabia had $750 billion in cash reserves after several years of oil at over $100 a barrel. It costs only 45 cents to purchase a gallon of oil in Saudi Arabia.

However, the heavy reliance on oil means the changing price of oil has a significant impact on the Saudi society. There is international competition, and traditional buyers of Saudi oil such as the United States are developing domestic energy supplies through fracking. The Saudi policy of maintaining production levels in order to destroy competitors also means that the price stays low. The result is declining government revenues. This has become problematic as the government seeks to play a bigger military role in the region—military hardware is expensive—and the majority of Saudi have become used to the generous social benefits provide by a cash-rich government. The majority of Saudis is aged under 30 years; the citizens are used to full employment and benefits. If the revenue crunch continues, Saudi Arabia may have to reduce these benefits and demand more work from Saudi citizens that, in turn, may provoke a populist backlash.

Oil revenues have allowed the country to build up its hard power of advanced military arms as well as the soft power of promoting an austere form of Islam around the world. Saudi Arabia spends around $2.5 billion a year promoting **Wahhabism**, an austere form of Sunni fundamentalism, in mosques and schools around the world.

The country is ruled by one family, the al-Sauds. The dynasty has held on to power through the ability to distribute some of the oil revenues and by carefully attending to austere religious leaders. There are religious police, and women are not allowed to vote, drive a car, or go out in public unless accompanied by a male family member. The country is mainly Sunni Islam but with a Shia minority in the eastern region close to Bahrain.

There is a cluster of small but rich Gulf States, including Oman, Kuwait, Bahrain, Qatar, and UAE (which includes seven emirates, the two largest of which are Abu Dhabi and Dubai). These are oil city-states with huge oil revenues and a heavy reliance on imported labor. A building boom of skyscrapers and urban spectaculars has transformed their city skylines. Generous welfare payments keep living standards for most citizens very high while the rulers are fabulously wealthy. The average wealth per adult in Qatar is $157,000 and $144,400 for the UAE. The condition of the unskilled workers is often more problematic. In Qatar, South Asian laborers regularly work 12- to 15-hour days in searing heat, and sleep ten people to a room in unsanitary and overcrowded labor camps. Wages are low and working conditions poor. In 2013 alone, 185 Nepalese workers died on construction sites, and from 2010 to 2014, over 700 Indians died.

There are also differences in terms of political stability. While some are very stable such as Oman, Bahrain is more unsettled because the country has a Sunni ruler with a mainly Shiite population. Saudi Arabia propped up this unpopular regime during the Arab Spring, and now a wave of repression has undercut the Shiite democratic movement.

Although they have a similar profile of oil-based revenues and a large amount of expatriate workers, there are some differences. Saudi Arabia has the most austere form of Sunni fundamentalism. There is a looser interpretation in the Gulf States such as Bahrain, Kuwait, and Dubai. Even within the UAE, for example, the consumption of alcohol is permitted in Dubai but not in Abu Dhabi.

Yemen

Yemen was formerly divided up by two empires: an Ottoman-ruled territory (North Yemen) and a British-ruled territory (South Yemen). The North, centered on Sana'a, was mainly Shia, and the South centered on Aden was mainly Sunni. North and South Yemen fought wars against each other in 1972 and 1979. The North was supported by the Eastern Bloc and the South by Western powers. There was also a bitter civil war in South Yemen in 1986.

Yemen only emerged as a country in 1990 when North and South merged. The new country was a patchwork of different tribal and religious groups. Around a third of the population is Shia, concentrated in the northern province of Saada, and the rest are Sunni.

The Shia Houthis, reacting against what they saw as discrimination from the Sunni majority, have mounted rebellion against the central government since 2004. In the wake of the turmoil after the Arab Spring, they took control over much of the country. The country is now locked into the Sunni–Shia conflict. The Saudi government funds the forces supporting the former government, and Saudi-UAE air strikes have shelled the old city of Sana'a since 2015 after Houthis gained control. In 2017, Saudi military blocked the ports, delaying delivery of much needed food and medical supplies. The indiscriminate bombing of Sana'a and Houthi strongholds by Saudi and UAE planes along with food shortages and lack of medicine resulted in a humanitarian disaster. More than 8 million Yemenis now live on the very edge of famine. The conflict has displaced 2 million people and caused a cholera epidemic that so far has infected over a million people. Bombings and unrest in the cities have displaced more than a million out into the rural areas, where there are few water wells.

The country is currently split between Houthi-held areas in the west and Al Qaeda–dominated territory in the east of the country.

Water shortages have reached critical levels as 20 million people, more than 80 percent of the total population, do not have enough fresh water. A combination of dry climate, rapid population growth, mismanagement, and conflict has created a humanitarian emergency.

Iran and Iraq

Iraq and Iran: with similar sounding names, they are often confused (Figure 10.20). Both have substantial oil reserves, are current members of OPEC, and are the cradles of ancient civilizations. But they differ.

Iran was a powerful cultural and political power before the rise of Islam and retains a pre-Islamic cultural basis and its distinctive language. It is the ancient core of the Persian empires, with a predominantly Persian-speaking Shiite population, although 26 percent of the 65 million population speaks Turkic languages and almost 10 percent speaks Kurdish. It sits in a strategic position, bordering at least seven different countries with a coastline on the Caspian Sea and the Gulf. It also has control over the narrow Hormuz Strait through which passes more than 80 percent of the world's oil tankers. It holds the key to warm weather seaports for landlocked countries in the region, and it actively fosters good relations with central Asian republics, especially those with Persian-speaking populations.

To understand Iran—at least for people in the United States, perhaps startled by its explicitly anti-US stance—it is important to recall recent history. Starting in 1917, the country came under the control of the United Kingdom, and the Anglo Persian oil company essentially ran the country. In 1926, an army officer, Reza Shah, became the official ruler, claiming the ancient title of Shah, though he came from a modest background. His son Mohammed Reza replaced him in 1941. When the Iranian government sought to nationalize the oil industry, it was bitterly opposed by British and US oil interests. The CIA saw it as part of a global communist conspiracy and organized an illegal coup against Prime Minister Mossadegh and reestablished the oppressive rule of the Shah. These actions laid the basis for an anti-Americanism. The Shah's rapid policy of top-down modernization provoked a religious backlash. In 1979, the Shah was overthrown, and a theocratic republic was established. The rise of the clerics meant the Islamization of the state and the society.

Anti-US sentiment was reinforced when Iran fought a bloody 8-year war against Iraq that, under Saddam Hussein, was supported by the United States. And when an Iran airplane was shot down in 1988 by a US fighter with all 290 passengers killed, anti-Americanism was baked into the political discourse of the country. However, there is a significant resistance to the dominant narrative of the fundamentalist clerics. With more than 60 percent of the total population aged under 30 years, young Iranians, especially the more educated, have a more nuanced view of the world.

Iraq emerged from the remnants of the Ottoman Empire, and its national boundaries were a result of imperial cartography rather than national coherence. There was always an uneasy alliance between the Kurdish north, the Shiite south, and the Sunni center. Under Saddam Hussein, the Sunnis held political power with the Kurds and Shiites on the margins. When Hussein invaded Kuwait in 1990/91, he was quickly defeated by the US-led coalition, but he remained in power. The Kurdish north was protected by US air power, but the Shiites in the south paid a heavy price for their attempted insurrection.

In 2003, Iraq was invaded by a US-led coalition that wrongly claimed that the country had weapons of mass destruction. The invasion toppled Hussein but led to the unraveling of the country as the Kurds effectively broke away to the safety of their core region, and Shia and Sunnis engaged in a bitter and bloody sectarian struggle. The country was devastated by the US invasion and subsequent civil unrest and has yet to fully reassemble, if it

10.20 Map of Iran and Iraq

ever will. Perhaps it will split into its three constituent elements. Tens of thousands of Iraqis died in the wake of the invasion, and more than 2 million were displaced. In 2015, a significant part of Iraq came under the control of ISIS.

The Cockpit of the Middle East: Israel, Jordan, Lebanon, and Syria

The cockpit of the Middle East includes Israel, Jordan, Lebanon, and Syria (Figure 10.21). The territory was part of the Ottoman Empire that was divided up by France and Britain at the end of World War I. Britain gained control over what are now Israel, Jordan, and Iraq, while France had control over Syria and Lebanon. The control was authorized by the League of Nations with the understanding that the territories should advance to independence rather than remain as colonies.

Jordan is in the middle of the region, an almost landlocked country surrounded by five countries, including Israel. Its central location has meant that it is impacted by population displacement: Palestinians displaced by the establishment of Israel and more recently by people fleeing war in Iraq and Syria. More than half of its 9.7 million population are nonnationals. With no oil and limited water, Jordan survives on agricultural produce and remittances. Its vulnerable position means it has to balance various geopolitical options: in 1991, it supported Iraq; in 1994, it signed a peace treaty with Israel. The country maintains relatively good relations with the United States and Israel.

Syria was a patchwork state carved out by the French. Although 90 percent Arab, it has various religious groups—Sunni (70 percent), Shia (15 percent), and Christian (10 percent)—and significant numbers of Druze and Kurds. For a short period from 1958 to 1961, it combined with Egypt to form the United Arab Republic, a short-lived experiment in pan-Arabism. It has few oil reserves and thus limited ability to project hard power.

Since 1970 the country has been ruled by the Assad family, who belong to a Shia sect known as the Alawites that number around 2 million, a tenth of the total population. Alawites were preferred for the more lucrative military and government jobs. Religious differences were exposed in the uprising associated with the Arab Spring and expressed in a civil war that erupted in 2010 and has multiple actors, including the Syrian Army, Sunni fundamentalist groups, ISIS, Kurdish forces, and a variety of others. More than a quarter of a million people were killed in the fighting and up to 10 million have been displaced as they seek to escape from the violence and retribution. Tracts of the country are now under the control of various factions. By 2015, the country had disintegrated with at least 1,000 different militia and insurgency groups, warlords, and criminal gangs. Syria is also the setting for wider conflicts: Saudi Arabia/Iran, Turkey/Kurds, Sunni/Shia, United States/Russia and Syria, which makes a settlement harder to come by and prolongs the agony for the Syrian people.

The French carved Lebanon out of Syria. It has a Mediterranean coastline and two parallel mountain ranges running roughly north to south that frame the Bekka Valley. It has a complicated social geography and a contentious political geography. It is a multifaith state

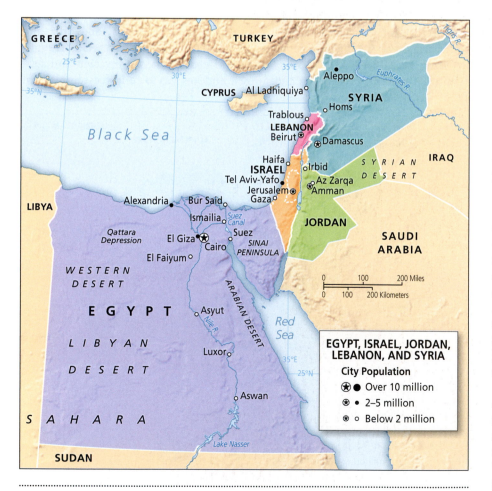

10.21 Map of Egypt, Israel, Jordan, Lebanon, and Syria

that officially recognizes five different Islamic groups and eleven different Christian groups. These religious differences formed the basis of the bitter civil war fought between 1975 and 1990. Today, in order to balance the religious faiths, the president is always a Christian, the prime minister is always a Sunni Muslim, and the speaker of the house has to be a Shiite Muslim. There are more than 400,000 Druze. No census has been taken since 1932 as the results may undermine the existing distribution of political spoils. Corruption is made worse by this guaranteed political power and patronage. Sharing a border with Israel is problematic, as it has provided a haven for anti-Israel forces as well as the setting for Israeli counteroffensives. Israel invaded south Lebanon in 1978 and again in 1982, occupying the territory until 2000. Syrian troops occupied part of the Bekka Valley until 2005. It is still a divided country. Hezbollah gained prominence and local support because it effectively defeated Israeli forces in 1986 and now effectively runs the southern part of the country.

Hinge States

There are three countries that play a significant role in the region but cannot be easily lumped together with other countries: Egypt, Israel, and Turkey. They are significant hinge states that link the region with different parts of the outside world. Israel is the most distinctly different country in MENA and the one most closely allied to the United States.

EGYPT

Egypt is the only Afro-Asian country with a national territory that reaches down the Nile Valley into Africa and across the Red Sea into Asia (see Figure 10.21). It plays an important role in the Arab world. It is the largest Arab country in population, with 99 million people. One in three Arabs in the world is Egyptian. It also plays a pivotal cultural and political role. It was one of the first independent Arab countries; in 1922, it broke away from British control, after centuries of Ottoman rule. It was and remains a center for pan-Arabism, both secular and religious, and it wields influence through its cultural importance with more than sixteen universities attracting Muslim students and scholars from all over the world. More than 2 million Egyptians work overseas.

It remains such an important US ally in the fight against Islamic terrorism and as an anchor of Arab support for Israel that the United States has supported a series of authoritarian and military-backed regimes. Even after grabbing power in an effective coup d'état in 2013, the United States soon rewarded the Egyptian military with $3 billion in military aid.

As with so many countries in MENA, Egypt has a very marked youth bulge with more than half of the population aged less than 25 years and almost one in five between 17 and 25 years, the prime years for youth alienation and attraction to militancy. Neither the private market nor the government is able to provide jobs for this bulging youth cohort.

The military plays an outsized role in the political and economic life of the country. The army has regularly intervened in political affairs. In 2013, it overthrew the elected, although increasingly unpopular, leader and installed an Army general as leader. In the resultant social conflict, 2,000 to 3,000 were killed, and there are now more than 42,000 political prisoners. The army-backed government directly and indirectly owns 50 percent of the economy, and the officer class forms an important economic and political elite.

Egypt faces an insurgency, especially in the Sinai region, where it is now linked to ISIS.

ISRAEL

An anomaly: Israel is a Jewish state in the heart of the Arab world, surrounded by Muslims (see Figure 10.21). It is a settler state built amid indigenous Palestinians. How did this happen?

Israel grew out of a nineteenth-century Zionist project to save European Jewry. In Eastern Europe, Jews were viciously persecuted and subject to pogroms, while in Western Europe assimilation was eroding a distinctly Jewish identity. Later, the Holocaust gave an existential need for a home for the Jews.

The Zionist project in the region began in the late nineteenth century when settlers looked to establish a Jewish homeland. They looked at different regions around the world, including East Africa and Palestine, then part of the Ottoman Empire. At the time, more than half a million Arabs, Bedouins, and Druze lived in the region. Pogroms forced many Jews to leave Eastern Europe. There were two main waves from Russia, one in 1881 and another in 1903, which brought more activists and socialists. At the end of World War I, Palestine came under British control. In the famous Balfour Declaration of 1917, Lord Balfour wrote in a letter to Lord Rothschild, "His Majesty's Government views with favour the establishment in Palestine of a national home for the Jewish people . . . it being clearly understood that nothing shall be done which may prejudice the civil and religious rights of existing non-Jewish communities in Palestine."

In 1921, Labor Brigade pioneers moved into the Valley of Harod to establish a **kibbutz**. Zionists such as the German-born Arthur Ruppin (1876–1943) believed that the control of land especially in concentrated blocks was important for defense and to provide the basis for a Jewish state. He bought land on the coastal plains from the absentee Arab landlords. Arab nationalism grew as the Arab peasants were displaced and the former Ottomans were becoming more self-consciously Arabs and Jews. There was a securitization of space as each group defended their territory and created paramilitary forces. There were Arab-Jewish confrontations that by 1947–1948 erupted into a full-scale civil war. The day after the state of Israel was founded, in May 14, 1948, the armies of Egypt, Jordan, Iraq, Syria, and Lebanon invaded. The Israeli Defense Forces (IDF) successfully repelled the attacks. In July of that same year, the Arab town of Lydda was forcibly emptied of Arabs.

Israel remained under constant threat. In 1967, the Egyptian army entered the Sinai and, in a preemptive strike, Israel destroyed the air forces of four Arab states and, within 6 days, giving the name to the war, captured the Sinai, East Jerusalem, the West Bank, and the Golan Heights. Israel now had "occupied" territory that was opened for settlements, an especially attractive proposition for ultra-orthodox Jewish settlers who saw their claim to the territory as sanctioned and approved by their God. The large number of settlers plays a major role in domestic Israeli politics.

Israel is now the recognized homeland for Jews. In 1897, only 50,000 Jews lived in Palestine, and they accounted for less than 1 percent of 0.4 percent of world Jewry. By 1950, it was 10.6 percent, and now with 6 million Jews in Israel, it is 45 percent. By 2050, the majority of all Jews in the world will live in Israel.

Israel is a Jewish homeland that has a substantial non-Jewish minority estimated at around 2 million people, or almost one-fifth of the population. Many have very little commitment to the state and if anything a deep resentment against the Israeli state. Many Palestinians refer to May 15 as Nakba Day, meaning Day of the Catastrophe, to refer to the displacement of around 700,000 Palestinians after the creation of Israel.

The demographics are interesting with the lowest birth rates among secular Jews and the highest birth rates recorded for two groups: the non-Jewish, mainly the Arab population, and the ultra-orthodox Jews who were long exempted from many taxes and military service. This creates a demographic time bomb for the state, because more non-Jews threaten the reality of Israel as a Jewish state while the ultra-orthodox do not serve in the military, compulsory for all other Israeli Jews. Since 1980 the percentage of children attending ultra-orthodox schools has risen from 4 to 20 percent, which leaves fewer people to pay taxes and serve in the IDF.

The urban geography of Israel is sometimes described as "urburb," referring to the meshing of urban and suburban in an ever-expanding frontier of new residential communities separated by open land. The two major cities of Jerusalem and Tel Aviv also embody the different characteristics of contemporary Israel. Jerusalem has a more theocratic atmosphere where religion hangs heavy in the air and conflict is a fact of everyday life, especially after 1967 when east Jerusalem with its substantial Arab population was annexed. Tel Aviv, situated on the coast, is more hedonistic and more explicitly modern, cosmopolitan, and secular.

Israel is one of the more successful states in the MENA in terms of economic growth. Its economic growth in recent years has created a more unequal society than the one envisioned by the early Jewish pioneers. However, the geopolitics remains vexing.

First, there is the issue of occupation. Israel has control over the West Bank, East Jerusalem, and the Golan Heights (Figure 10.22). These areas, especially East Jerusalem and the West Bank, contain considerable Arab populations. There are more than 2.5 million Arabs in the West Bank. Since 1967, there have been substantial Jewish settlements; there are now 390,000 Jewish settlers in the West Bank and 375,000 in East Jerusalem. There is considerable friction between the two communities. Israel is building a 25-foot, 400-mile wall along the West Bank that follows the 1949 Armistice line but also includes several illegal Israeli settlements. Proponents of the wall see it as a defense against terrorist attacks, while critics point to the dismemberment of the territory of a future Palestine. The occupation raises moral issues for a democratic country committed to personal freedoms.

Second, there is the issue of Israeli "policing" of Palestinian territory, especially Gaza. The Gaza Strip is a narrow coastal strip of land between Israel and Egypt. It is under Palestinian Authority, as is the West Bank, and houses 1.7 million people, almost all of whom are Muslims. There are now no Jewish settlers. Israel launched an attack against the region in order to remove Hamas. In 2014, after rockets were launched against Israel from Gaza, the IDF launched a major offensive, shelling the area. More than 10,000 homes were damaged or destroyed, and 1,500 people were killed, almost 70 percent of them innocent civilians. Israel lost sixty-five soldiers. More than 100,000 residents in Gaza were still displaced even a full year after

10.22 This sign in Jerusalem is in Hebrew, Arabic, and English. (robertharding / Alamy Stock Photo)

the invasion. Many locals refer to the Gaza Strip as an open-air prison locked up between Israel and Egypt.

Third, there is the issue of governing the occupied territories. The West Bank, for example, is nominally under the authority of the Palestine Authority, but Israel controls the water and the finances and polices security zones in the territory in order to secure the safety of settlers. There is a vigorous debate within Israel between settlers and their political representatives and more secular Jews worried about the costs—military and moral—of occupation.

There is a geopolitical paradox that involves discussing two viewpoints, often considered in isolation. Israel is one of the few states that is threatened with annihilation. It is also one of the few democratic states that is occupying territory to which it has no internationally recognized mandate. The Israeli author Ari Shavit asserts that those on the left address occupation and ignore intimidation, while those on the right address intimidation and dismiss occupation. "Only a third approach," he argues, "that internalizes both intimidation and occupation can be realistic and moral and get the Israel story right."

TURKEY

There are two major powers that do not fit the usual perception of this region as an Arabic-speaking area, the predominantly Persian-speaking Iran and Turkic-speaking Turkey.

Modern Turkey emerged from the ashes of the Empire after World War I. The boundaries of the modern state were only fixed in 1923. At the outset, Turkey was reimagined as a secular modern state. Western clothing was encouraged, the fez was banned, and the Gregorian calendar and the Roman alphabet were adopted. Turkey embraced a Western-influenced modernity as a national defining character. This establishment secularity is now confronted by the recent rise of more religious-based political parties and social movements. There is an Islamic revival: in 1945, there was one mosque for every 1,000 Turks; by 1985, there was one for every 700. The Justice and Development Party came to power in 2002 as a more overtly and explicitly religious-based political movement.

Turkey plays a hinge role. Just as Egypt links the Middle East with North Africa, Turkey stands at the crossroads of Asia and Europe, with territory on either side of the traditional continental division of the Bosporus Straits (Figure 10.23). During the Cold War, it was a strategic ally against the Soviet bloc with control over the narrow passage of the Bosporus Straits, where Soviet ships and submarines had to pass in order to enter the Mediterranean. Turkey's strategic location continues as it forms a haven and a bridge to Europe for refugees fleeing civil war and unrest in Iraq and Syria. Turkey houses almost 2 million refugees from the conflict in Syria. Towns in southern Turkey along the border with Syria are now sites of refugee camps, aid missions, and political intrigue.

The overwhelming majority of the population is Muslim, mainly Sunni. There is also a significant Kurdish population. Of the 81.9 million people, more than 14 million are Kurds. Their traditional region is in the east and southeast of the country, but more than half now live outside of this region in towns and cities throughout Turkey and many are fully integrated into Turkish society. From 1984 to 2000, the Kurdistan Workers' Party (PKK) escalated the struggle for cultural recognition into a bloody campaign. Much of the traditional Kurdish region was under repressive military rule and was closed to foreign travellers. More than 40,000 people died in the war between Turkish forces and the PKK. Since 2000 the violence and repression has lessened, and in 2002, the Turkish government allowed a more open expression of Kurdish identity with Kurdish now taught in schools. The rise of a strong Kurdish region in northern Iraq and Syria worries the Turkish government, who fears the emergence of a greater Kurdistan, including parts of Turkey.

There are large differences in Turkey between the more cosmopolitan and Western-orientated cities, especially

10.23 Map of Turkey

Istanbul, and the more conservative and Islamic small towns and rural areas of the Anatolian plateau.

North Africa

The North African region of MENA consists of Algeria, Morocco, Tunisia, and Libya (Figure 10.24). The population in all these countries is concentrated mainly along the coast, because moving inland toward the Sahara means less rainfall and declining soil fertility. All of these countries have a huge youth bulge and sluggish economies and have problems with non-state groups dedicated to creating Islamic republics.

The long-standing dictatorships in Libya and Tunisia were undermined during the Arab Spring as populist revolts overthrew the existing power structures. Post Arab Spring the two countries have diverged.

LIBYA

From 1969 until 2011, Libya was ruled by the dictator, Muammar Gaddafi. The steady oil revenues allowed him to promote insurgencies and terrorism around the world while maintaining a luxurious lifestyle for his family at home. He funded the war in Liberia in the 1990s along with terrorist groups such as the PLO and the IRA. He had continental pretensions and had himself declared King of Africa in 2011. When it was revealed that Libyan operatives were responsible for the downing of the Pan Am flight 103 over Lockerbie, Scotland, in 1988, trade sanctions were imposed that reduced revenues and clipped his power. It widened the gap between his family's extravagant lifestyle and the majority of the Libyan people. The sanctions were lifted in 2003. The regime collapsed in 2011 as the Arab Spring swept over the country and armed militia groups overthrew the dictatorship in a bloody civil war that cost the lives of 25,000 people. NATO air support helped the anti-Gaddafi forces prevail. Since the overthrow, the country has collapsed into chaos, the huge arms cache of the Gaddafi regime now circulating widely in the country. The dysfunctional state of affairs gives space and opportunity for human traffickers to ship migrants across the Mediterranean Sea to landing places in Europe. The

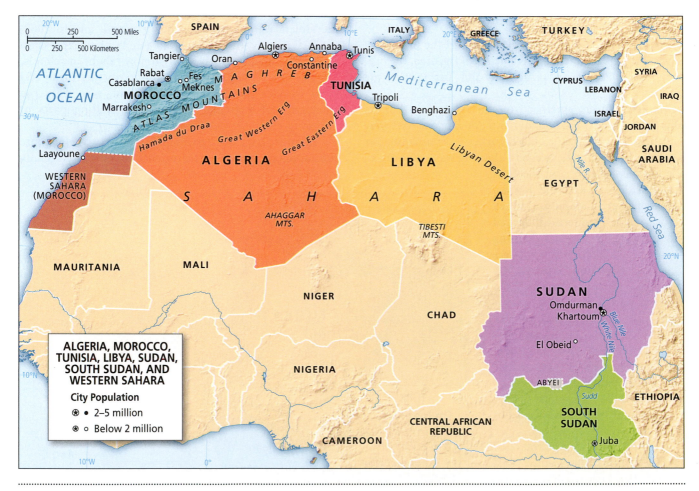

10.24 Map of Algeria, Morocco, Tunisia, Libya, Sudan, South Sudan, and Western Sahara

country has split into regional and sectarian rivalries, especially between the rival cities of Tripoli, Benghazi, and Misurata, providing opportunities for non-state actors such as ISIS.

TUNISIA

Before the Arab Spring, Tunisia was an authoritarian regime that promoted a secular state open to Western influence. Tunisia, unlike Libya, was more connected with Europe, with more non-oil investment and tourist flows, especially from France and the United Kingdom. Things changed in the Arab Spring of 2011 when the long-time dictator, Zine el-Abidine Ben Ali, was overthrown. When elections were held for the first time, the moderate Islamist party Ennahda won the election but later lost after voters felt it had done little to prevent extremist attacks. A secular party was returned to power in 2014 that was more concerned with stability than civil rights. The more moderate and secular nature of the country has made Tunisia a target for the extremists, eager to disrupt the country's tourist industry that is reliant on European visitors. More than 6.12 million tourists visited the country in 2014, most of them European. The industry supports over half a million jobs and is responsible for 16 percent of total GDP. When in March 2015, three terrorists killed twenty-one tourists at the Bardo Museum, popular with tourists for its wonderful collection of Roman mosaics, it was a calculated blow against the Tunisian tourist trade. Only 3 months later, a lone gunman killed thirty-eight tourists, mostly British, while they were sunning on a holiday beach. A tiny, but poisonous minority was trying to destroy the country's links with the outside world.

ALGERIA

Algeria is the largest country in Africa and the Arab world, though much of it is desert. The total population is almost 40 million, most of it concentrated in the narrow coast of the Mediterranean. Algeria was part of France's colonial empire, first annexed in 1830. More than a million people from France immigrated into the country: known as **pied-noirs**, they were a majority of the population living in the cities of Algiers and Oran. Muslims felt marginalized and organized for independence. After a bitter struggle with hundreds of thousands of casualties, the country gained independence in 1962. The oil industry was nationalized and industrialization was pursued. The economy is still based on the export of oil and natural gas.

In the first free elections, held in 1990, the Islamic Salvation Front (FIS) had a sweeping victory, but the army intervened so that they could not take power. An insurgency campaign against the authorities resulted in 100,000 deaths with the unrest only diminishing by 2000, when the government pursued a policy of national reconciliation.

However, protests in the streets during the Arab Spring prompted some reform, but there is still resentment against the entrenched power of the economic and political elites.

The dominant city is Algiers, situated on the coast. With a metro population of 5.4 million, it is by far the largest city and a classic example of a primate city. While the majority of the population is Sunni Muslim, there are ethnic differences. Around 30 percent speak Amazigh, and there is strong demand for greater autonomy, especially from the Amazigh core region of Kabylia in the north of Algeria in the Atlas Mountains. The French colonial legacy remains strongest in language; most educated Algerians speak French, and the language is widely used in mass media.

MOROCCO

The geography of Morocco is typical of the North African countries of MENA, with a coastal narrow plain along both the Atlantic and the Mediterranean, where most of the larger cites are located, a mountain range, and then as one moves southward, a desert region. Morocco has no major oil deposits and so is more reliant on agriculture and tourism. Phosphates are an important export.

In the south, Morocco has controlled Western Sahara since Spanish troops left in 1976. This dry territory with a population of less than half a million is disputed. It is under Moroccan control, but many of the local Sahrawi support the independence movement organized by the Polisario Front who waged an insurgency until a cease-fire was signed in 1991. They claim the territory should be referred to as the Sahrawi Arab Democratic Republic.

The largest city is Casablanca with a metro population of around 6.8 million; it is by far the largest city in the country. And although Rabat is the political capital, Casablanca is the chief port, major financial center, and location for multinational corporate offices.

Official languages are Arabic and Amazigh, though French is widely spoken among the elites and educated. Amazigh is spoken in the core region along the northern Mediterranean coasts and along the spine of the Atlas Mountains running from the southwest to the north. As with many of the unequal societies in MENA, Islamic fundamentalist groups provide an appeal to the many young people unemployed and underemployed.

The Southern Rim

The southern rim of MENA includes Sudan and South Sudan (see Figure 10.24).

SUDAN

Sudan came under British control in 1899. The British ruled the country as two separate regions, a Muslim north and a Christian-animist south. These two regions were the basis of the independent country that emerged in 1956. It was an unhappy political marriage. Power was concentrated with the Arab Muslims in the north centered in Khartoum. From 1983, and especially from 1989 onward, northern politicians followed a more strictly Islamic line. The southern non-Arab province always felt marginal, and there were two civil wars with the regime in Khartoum in 1955–1972 and an especially bitter and violent one from 1983 to 2005. The later unrest led to the deaths of over 2 million people, with more than 2.5 million displaced in the conflict. The government employed paramilitary force, the Janjaweed, to burn villages, destroy crops, and terrorize locals. The region broke away in 2011.

Oil has attracted significant Chinese investment in the country, and oil revenues are now close to 70 percent of Sudan's GDP.

There is continuing political conflict within the country. The Sudanese government followed a policy of effectively starving the civilian population of South Kordofan, a region bordering Sudan and South Sudan, to undercut their support of rebels.

SOUTH SUDAN

South Sudan is one of the world's newest countries; it became independent in 2011. Independence has not proved an end to the country's problems as the south Sudanese elites soak up most of the foreign aid and spend on military forces rather than on roads or schools. There is a conflict between two tribal groups, the Dinka (35 percent of the population) and the Nuer (15 percent of population), as well as conflict between rebel militias. A third of the country's population is displaced or living in foreign refugee camps. The country relies on foreign aid.

There are continuing territorial disputes with Sudan, over the region of Abyei, where there is a substantial Dinka population, and along the border region where there are major oil deposits.

One of the world's newest nations is spiraling into civil war, violence, famine, and massive population displacement.

Focus: Dubai's Race Against Time

Dubai is in a race against time. It has the smallest oil reserves of the Gulf States. Before the oil wells run dry, the emir is trying to develop a post-oil economy by branding Dubai in the global imagination and building a glittering new city in the desert sands.

Dubai is one of the seven emirates that make up the United Arab Emirates. In 1969, oil started to flow from offshore wells, inaugurating a major transformation of a dusty, dry backwater town into the site of key global importance. Foreign oil companies initially controlled the oil industry, but in 1971 the separate emirates achieved independence as a federation, the United Arab Emirates (UAE), and gained greater control over the oil revenues.

Dubai's oil reserves are dwarfed by those of Saudi Arabia and are only a twentieth of its fellow UAE emirate Abu Dhabi. Current estimates suggest that the oil will run out in 2035, so the race is on to develop a post-oil economy. It has achieved some success; almost half of income now comes from non-oil sources such as tourism and the airport. Dubai is a major airport, home of Emirates airlines, with flights from Europe and North America linking destinations in the Middle East with South and South East Asia.

The city attracts **flight capital**. For those who have substantial assets, especially cash, the world banking system is monitored, and as Switzerland has to open up its banking secrets to more scrutiny, Dubai is one of the centers where money can be parked, hidden, and invested with few questions asked or answered. A range of financial services has grown up to service the rich and superrich. Banking and financial services are a growth industry in the city.

The city is also embarking on a building frenzy of new hotels, apartments, and office buildings (Figure 10.25). They are built to meet the burgeoning tourist trade and commercial demand. Many of the residential units are speculative purchases by overseas buyers. Building in Dubai is used as a way to brand the city with signature buildings. Globally recognized architects are employed to design dazzling buildings meant to impress on the global imagination the image of Dubai as a modern city. The Burj Khalifa, currently the tallest building in the world, and the Al Burj hotel, one of the few six-star hotels in the world, are just two of the most significant and recognizable buildings (see Figure 10.12). A major construction project involved pouring a rock base into the sea in the shape of a palm to build luxury hotels and expensive villas.

Dubai is the scene of intense construction. As with all property cycles, there are slumps as well as booms. During the global financial crisis in 2008–2009, liquidity issues meant that Dubai had to rely on the wealthier Abu Dhabi to fund the Burj Khalifa, with the result that one of the most iconic buildings in Dubai is named after the ruler of Abu Dhabi, not the ruler of Dubai.

The building of the city and the development of the economy are very reliant on foreign labor. Only around 10 percent of the total workforce is local Emiratis. More than half are laborers from South Asia, especially Pakistan, India, and Nepal brought in to do construction work. Working and living conditions for these workers are often very poor. Female labor is brought in from South Asia and the Philippines as domestic workers. Again while wages may be high, conditions can be tough. Because there are few labor laws, foreign workers are subject to immediate deportation, and with limited legal rights, few can complain. Expatriates with higher job skills such as engineers and doctors make up a third of the workforce. This labor is globally sourced with Italian chefs, Egyptian doctors, Scottish engineers, and US oil workers adding to the cosmopolitan mix of the city. All foreign workers are on short-term contracts and are not allowed to retire in the city no matter how long they have worked there.

10.25 Luxury tower blocks in Dubai (photo: John Rennie Short)

Dubai tries to achieve global recognition by hosting signature events. Film and sports festivals, international conventions, and global conferences are just some of the many ways used to promote a positive image of the city to the wider world.

Unlike other Arab oil states with vast resources, Dubai had to connect more with the modern contemporary world. Its standards of public behaviors and tolerance are closer to the West than Saudi Arabia. Dubai has to integrate more into the ways of the external world in order to promote a post-oil economy. Can Dubai win the race against time? Can it construct a sustainable and competitive economy after the oil has run dry?

Select Bibliography

Abouzeid, R. 2018. *No Turning Back: Life, Loss and Hope in Wartime Syria*. New York: Norton.

Aitchison, C., ed. 2012. *Geographies of Muslim Identities: Diaspora, Gender and Belonging*. Chichester, UK: Ashgate.

Allen, T. 2001. *The Middle East Water Question*. London: I. B. Tauris.

Anderson, E. W. 2013. *Middle East: Geography and Geopolitics*. London: Routledge.

Anderson, E. W., and L. D. Anderson. 2010. *An Atlas of Middle Eastern Affairs*. London: Routledge.

Anderson, S. 2013. *Lawrence in Arabia: War, Deceit, Imperial Folly and the Making of the Modern Middle East*. New York: Doubleday.

Antrim, Z. 2018. *Mapping the Middle East*. London: Reaktion.

Bayat, A. 2013. *Life as Politics: How Ordinary People Change the Middle East*. Palo Alto, CA: Stanford University Press.

Ben-Dove, M. 2002. *Historical Atlas of Jerusalem*. London: Bloomsbury Academic.

Cockburn, P. 2015. *The Rise of Islamic State*. London: Verso.

Cole, J. 2002. *Sacred Space and Holy War: The Politics, Culture and History of Shi'ite Islam*. London: I. B. Tauris.

Cole, J. 2014. *The New Arabs: How the Millennial Generation Is Changing the Middle East*. New York: Simon and Schuster.

Cook, C., and K. Bakker. 2012. "Water Security: Debating an Emerging Paradigm." *Global Environmental Change* 22:94–102.

Davis, D. K., and E. Burke. 2011. *Environmental Imaginaries of the Middle East and North Africa*. Athens: Ohio University Press.

D'Odorico, P., A. Bhattachan, K. F. Davis, S. Ravi, and C. W. Runyan. 2013. "Global Desertification: Drivers and Feedbacks." *Advances in Water Resources* 51:326–344.

El Shaer, H. M. 2015. "Land Desertification and Restoration in Middle East and North Africa (MENA) Region." *Sciences in Cold and Arid Regions* 7:7–15.

Fromkin, D. 2009. *A Peace to End All Peace: The Fall of the Ottoman Empire and the Creation of the Modern Middle East.* New York: Holt.

Gelvin, J. G. 2017. *The New Middle East: What Everyone Needs to Know.* Oxford: Oxford University Press.

Gettleman, M. E., and S. Schaar, eds. 2012. *The Middle East and Islamic World Reader.* New York: Grove Press.

Ghannam, F. 2002. *Remaking the Modern: Space, Relocation, and the Politics of Identity in a Global Cairo.* Berkeley: University of California Press.

Glass, S. 2016. *Syria Burning: A Short History of a Catastrophe.* London: Verso.

Ibrahim, F. N. 2003. *Egypt: An Economic Geography.* London: I. B. Tauris.

Kanna, A., ed. 2013. *The Superlative City: Dubai and the Urban Condition in the Early 21st Century.* Cambridge, MA: Harvard University Graduate School of Design.

Keenan, J. 2009. *The Dark Sahara: America's War on Terror in Africa.* London: Pluto Press.

Kimenyi, M., and J. M. Mbake. 2015. *Governing the Nile River Basin: The Search for a New Legal Regime.* Washington, DC: Brookings Institute.

Kinzer, S. 2008. *All the Shah's Men: An American Coup and the Roots of Middle East Terror.* Hoboken, NJ: Wiley.

Lynch, M. 2012. *The Arab Uprising: The Unfinished Revolutions of the New Middle East.* New York: Public Affairs.

Mansfield, P. 2013. *A History of the Middle East.* London: Penguin.

Mikhail, A. 2012. *Water on Sand: Environmental Histories of the Middle East and North Africa.* New York: Oxford University Press.

Morton, M. Q. 2017. *Empires and Anarchies: A History of Oil in the Middle East.* London: Reaktion.

Quataert, D. 2005. *The Ottoman Empire 1700–1922.* Cambridge: Cambridge University Press.

Rabinowitz, D., and A. B. Khawla. 2005. *Coffins on Our Shoulders: The Experience of the Palestinian Citizens of Israel.* Berkeley: University of California Press.

Richards, A., and J. Waterbury. 2009. *A Political Economy of the Middle East.* 3rd ed. New York: Westview Press.

Rogan, E. 2009. *The Arabs: A History.* New York: Perseus.

Rosovsky, N., ed. 1996. *City of the Great King: Jerusalem from David to the Present.* Cambridge, MA: Harvard University Press.

Shafir, G. 2017. *A Half Century of Occupation: Israel, Palestine and the World's Most Intractable Conflict.* Berkeley: University of California Press.

Shavit, A. 2013. *My Promised Land: The Triumph and Tragedy of Israel.* New York: Spiegel and Grau.

Smith, D. 2016. *The Penguin State of the Middle East Atlas.* 3rd ed. New York: Penguin.

Stewart, D. J. 2012. *The Middle East Today: Political, Geographical and Cultural Perspectives.* 2nd ed. London: Routledge.

Weiss, M., and H. Hassan. 2016. *ISIS: Inside the Army of Terror.* New York: Simon and Schuster.

Wilson, R. 2012. *Economic Development in the Middle East.* London: Routledge.

World Bank. 2011. *Poor Places, Thriving People: How the Middle East and North Africa Can Rise above Spatial Disparities.* Washington, DC: World Bank.

Learning Outcomes

The Middle East and North Africa (MENA) have a shared history of Islamic, Ottoman, and European imperialism.

The most distinctive physical feature of this region is the long dry season. Water is a scarce and valuable resource, and where river systems cross international boundaries they become a source of conflict more often than platforms for cooperation.

Two major environmental stressors in this region are desertification and salinization of soils.

The region relies on the extraction of oil and gas.

Religion has had an immense influence on the geography of the region as MENA is predominantly Muslim today. Minority religious groups, such as the Jews, Christians, and Yazidi, have been persecuted and ostracized forcing their numbers to greatly reduce.

Islam is divided between Sunni and Shiite followers. Their differences have led to political strife in the region.

Despite the popular conception of MENA as a region composed of Arabs, there are many other different groups distinguished by language, including Hebrew spoken mostly in Israel; Turkish in Turkey, Cyprus, and other local states; and Farsi spoken in Iran, as well as in several other nearby neighbors.

One of the largest ethnic groups in the region is the Kurds, who have a distinctive language and a population of around 30 million inhabitants across several countries.

The Amazigh population is also approximately 30 million with a distinct language and inhabitants that are spread across several countries in North Africa.

One of the distinctive features of the demography of this region is the youth bulge. More than 60 percent of the total population is aged less than 24 years, and 30 percent of the population is aged between 15 and 29.

The last 40 years have witnessed rapid urban expansion. Cities now account for 70 percent of the population.

The smaller countries such as Bahrain, Kuwait, and Qatar are essentially city-states where the city and the state are essentially one and the same.

Most cities in the region also contain large areas of informal settlements.

Non-state actors, such as Hezbollah, Hamas, the Houthis, Al Qaeda, and ISIS, have gained influence over the geopolitics of the region due to weak and dysfunctional states and an economic system that offers few opportunities for the majority of people, especially young people.

The United States has long played a huge role due to its military bases scattered throughout the region.

The region can be broken down into the smaller subregions of Saudi Arabia and the Gulf States blessed with oil reserves; Iran and Iraq; the cockpit of the Middle East composed of Jordan, Lebanon, and Syria; several countries with more distinctive backgrounds, including Egypt, Turkey, and Israel; North African countries; and the poor countries of Sudan and South Sudan that occupy the "Southern Rim."

Sub-Saharan Africa

Our earliest human ancestors lived in Africa and then spread out from the continent to inhabit the world. Sub-Saharan Africa remains one of the poorest regions of the world. Fifty percent of Africans south of the Sahara live on less than $2 a day. It is particularly vulnerable to climate change and consequent environmental challenges. The colonial appropriation of resources has left a legacy of economies precariously dependent on a narrow range of primary products. Driven in large measure by China's demand, the commodities boom of the first decades of the twenty-first century that led to the notion of Africa rising is now replaced by a slump in commodities prices. Africa plays an important role in the global imagination. For many outside of Africa, it remains a dark continent, too often associated with images of famine and civil war. In recent years, however, things have improved. While the border zone with North Africa is a zone of instability, much of the rest of the region has seen some modest growth and improvements in democratic accountability. Sub-Saharan Africa is a region of rapid population growth and quickening urbanization (Figure 11.1). It has a rich diversity of cultures with eight hundred different language groups and over one thousand distinct ethnic groups.

The remains of the first known human, and our shared ancestor, called Eve, who lived roughly 150,000 years ago, were found on the border between present-day Tanzania and Ethiopia. Around 100,000 years ago, people started leaving the continent to populate the world. Africa is our ancestral home.

The Environmental Context

Africa is one of the largest remnants of the ancient continent named Gondwanaland that subsequently spilt into Australia, Antarctica, and South America. The connection is still visible in the jigsaw-like fit between the northeastern coast of South America and the indented coastline of West Africa. Not all of Africa is so old. Between 25 million and 5 million years

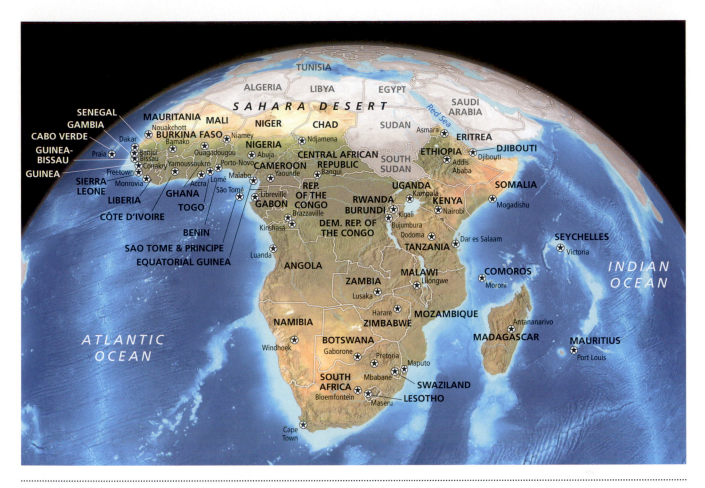

11.1 Map of region

Climate and Vegetation

Climate and vegetation belts run parallel and mirror each side of the equator. In the north there is a vast swathe of desert, the Sahara Desert, which runs from Mauritania to Somalia. The process of desertification is shifting the desert frontier southward, generating an environmental stress on local peoples and is one of the major reasons, along with civil unrest and political instability, behind the frequent food shortages and periodic famines. Chad, Mali, Niger, and the countries in the Horn of Africa have all experienced food shortages. Moving south toward the equator, the rainfall increases and the vegetation turns into semiarid scrub, grassland, and then equatorial rainforest (Figure 11.2). In the semiarid region, the grasses and

ago, volcanic activity produced a rift valley and mountain ranges. The Rift Valley, part of a 6,200-mile rupture, and Mount Kilimanjaro at 19,341 feet, are two of the more dramatic expressions of this more recent tectonic activity.

shrubs are watered during the short wet season. Climate change is creating drier conditions that, combined with overgrazing, lead to soil erosion.

Along the equator, the abundant sun and high rainfall create the ideal growing conditions for lush tropical rainforest. The intense heat causes air to rise and creates the **Intertropical Convergence Zone (ITCZ)**, a mixing of northerly and southerly air masses. The narrow ITCZ moves across the equator in sympathy with the overhead sun. In January, for example, the sun is highest south of the equator, and the ITCZ is located in a more southern position, bringing more rain. The city of Mongu in Zambia, for example, receives around 8 inches of rain in January and December, but hardly any in June and July. In midsummer, the ITCZ moves north, bringing more rain to regions just north of the equator. The city of Abeche in Chad, for example, gets almost of all its rainfall from June to September, with 8 inches on average each August.

The effects of the ITCZ and climate and vegetation distribution are also shaped by proximity to the coast and altitude. The Luhya people of the highland region of

11.2 Savannah in Tanzania (photo: Lisa Benton-Short)

western Kenya divide the year according to the rhythms of the agricultural season. *Simyu* is the hot, dry season of midwinter. *Wafula*, from March to May, is the start of rainy season.

South of the equator the climate and vegetation mirror each other, with forest giving way to grassland, semidesert, and then desert (the Kalahari Desert). At the southern tip of the continent, a Mediterranean climate of hot, dry summers and warm, wet winters is perfect for South Africa's flourishing wine industry.

Environmental Challenges

Africa faces a number of environmental challenges that include poor soils (a feature it shares with other parts of the ancient continent of **Gondwanaland**, including Australia), desertification, deforestation, soil erosion, and loss of ecosystems. In semiarid regions, desertification is caused by drier climate, overgrazing, population pressure, and the clearance of trees for timber and fuel, which exposes delicate soils to erosion. A combination of environmental and human factors is at work as a warmer, drier climate in association with political instability puts more pressure on an already strained ecosystem. The tropical rainforests are also being cleared for fuel, wood, grazing, and plantations. The rainforest has steadily shrunk in size, leaving fragmented ecosystems that make rare species more vulnerable to extinction.

Madagascar is a prime example of the loss of tropical forest. The island has lost around 50 percent of its forest since 1950 through logging, legal and illegal fire grazing and cultivation. Between 1990 and 2005, the island lost on average 37,000 hectares per year. This deforestation is particularly damaging because the island contains around 200,000 species, many of them found nowhere else in the world. The deforestation of Madagascar is reducing global biodiversity.

CLIMATE CHANGE

The most major environmental challenge is climate change. It is particularly acute in Africa because of the widespread poverty and the fiscal inability of many governments to respond effectively and quickly. According to the IPCC, the temperature across the continent is likely to rise between 1°C and 3°C by 2050. As the climate becomes drier and more desertification occurs, Africa will become especially vulnerable to water shortages. This change will impact agricultural yields and further exacerbate food shortages. Sea level rises will threaten lowlying coastal areas.

Flooding, changing rain patterns, and extreme heat waves may leave most African countries poorer in 2100 than they are today. The cost of climate change adaption could lead to a 10 percent loss in GDP. This is an especially cruel turn of events: Africa is one of the most impacted regions yet has one of the lowest per-capita carbon footprints of any populated region in the world.

Historical Geographies

Africa has a rich history before the coming of the Europeans. Empires rose and fell. The Aksum Empire emerged in the first century CE in Eritrea and Ethiopia, while in West Africa the Kingdom of Ghana lasted from 700 to 1000 CE. The sophisticated cities of the Mali Empire were important trading centers linking routes across the Sahara connecting North Africa with sub-Saharan Africa. And the Zulu Empire extended across South Africa until defeated by the British in the late nineteenth century. But to understand contemporary Africa, it is important to consider the deep and lasting legacy of Arab and European colonialism.

The Arab Muslim Empire at its greatest extent reached North Africa as far west as Spain. Its territorial control and cultural influence extended the Horn of Africa and along the coast of East Africa. The lasting legacy was the installation of Arab elites and the diffusion of Islam. Arab merchants trading across the Indian Ocean spread the Muslim faith. The area between North Africa and sub-Saharan Africa became a zone of interaction, including cooperation as well as conflict between the Muslim world and the world of the animists. This liminal zone is still a setting for geopolitical struggle.

A series of city-states emerged along the east coast of Africa as hubs linking maritime trading routes with the interior. Arab traders made up a significant element of the population of Zanzibar. After the conflict between the local Arab traders and the Portuguese in 1698, Zanzibar came under the control of the Sultanate of Oman, until 1890 when it became a protectorate of the United Kingdom. The island city-state thus reflects the complex history of Swahili locals, Arab and Portuguese merchants, and British imperialists.

European Colonialism

The Europeans extended their influence into Africa. First, the Portuguese established bases along the coast as they extended their trading network along the west coast of Africa, around the Cape of Good Hope, up the coast of east Africa, and then on to South and South East Asia. Later, the Dutch and then the British established coastal footholds in South Africa.

Even as late as 1880, European colonialism was limited to British, French, and Portuguese coastal footholds. The European scramble for Africa began in 1884 when Europeans drew up a system for claims and territorial appropriations. By 1914, the whole of Africa, apart from Ethiopia and Liberia, was partitioned among the European powers (Figure 11.3). The French had a huge swathe of north and West Africa. The British gained territory all over the continent, the Spanish had small territories along the west coast, Portugal secured control inland from their coastal holdings on the east and west coast, and Germany had three chunks of land on the east and west coast. Even the king of the tiny nation of Belgium laid claim to a vast territory in the heart of tropical Africa.

European rule lasted for over 50 years before the independence movement gained the necessary strength to throw off colonial rule.

Anticolonial independence movements strengthened after Sudan achieved independence in 1956. Other countries followed—Ghana in 1957, and then Kenya in 1963. The last colonial power left Africa in 1975 when a revolution in Portugal led to independence for Angola, Guinea-Bissau, and Mozambique. Zimbabwe achieved independence in 1979, and in 1990 Namibia escaped from the control of South Africa.

The Legacy of Empire

While not all problems of contemporary Africa are the result of European colonialism, the historical experience proved disastrous. First, there was an effective stripping of the resource base. The economic history of the Belgian Congo highlights the dramatic exploitation. The vast territory was the personal property of the King of Belgium; it was a colony seventy-six times the size of Belgium organized to enrich the royal family. Almost 20 million people were brought under a brutal regime to provide slave labor to work on rubber plantations. Through murder, starvation, exhaustion, and consequent

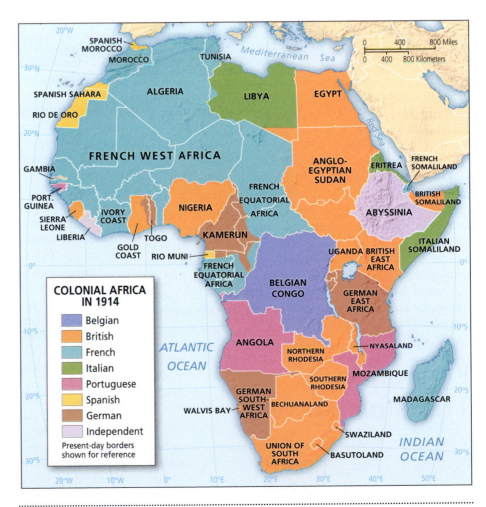

11.3 Colonial Africa in 1914

disease, more than 10 million people, out of 20 million, were eradicated by a brutal system of resource exploitation. Civil society was ripped apart, and local economies were destroyed.

Second, economies were orientated toward the exploitation and export of raw materials. Africa became a primary commodity producer for the global market. The emphasis was on exporting wealth out of Africa, not on local production or creating trade links between different parts of Africa. The economic geography was orientated around the export of raw materials, rather than internal linkages. Transport routes linked the interior to coasts—a characteristic that still hampers Africa's development and makes national economic coherence very difficult.

Third, many of the colonial powers used a form of indirect rule to save on costs and military manpower. The colonial authorities reinforced tribal differences as a way to keep order, favoring some chiefs over others and some tribes over rivals. The result: tribalism was reinforced under the colonial regime. When Belgium took control of Rwanda in 1914, after Germany's defeat in World War I, they favored the minority Tutsis over the majority Hutus. The resentment between the groups was part of the basis of the **Rwanda genocide** of the 1990s.

Fourth, the boundaries of the newly independent countries often reflected colonial boundaries rather than coherent territorial units. A national legacy of different and often competing ethnicities often made postindependent nation-building very difficult: a task that was made all the more difficult because European colonialism never envisaged a governing role for Africans. The Europeans, by and large, did not produce a cadre of educated Africans.

The postcolonial legacy included arbitrary national boundaries, landlocked countries, poor transport connections outside of primary commodity extraction, and distorted economies. Civil society was never encouraged, and political participation was suppressed. After the quick and hasty exit of many European powers, the new nations often lacked a national civil society or a national culture of political participation. The legacies of colonialism made it difficult for postindependent sub-Saharan Africa. Given the enormous difficulties, the advances are considerable.

Economic Transformations

In the last two decades, much of sub-Saharan Africa has seen economic growth. Foreign direct investment increased from $2.8 billion in 1990 to $57 billion in 2013. Gross domestic product also increased with 5 percent annual increases from 2000 to 2014. More than 300 million people became part of a more affluent middle class.

Economic growth has occurred, but full development is hampered by a continued heavy reliance on a few primary commodities; entrenched poverty, especially in rural areas; and a crippling lack of good infrastructure. There are only 3 kilometers of road per 100 square kilometers of territory. For the richest countries in the world, it is 134. In Denmark, the figure is 168, and in the United States it is 68.

The Importance of Primary Commodities

The colonial legacy means that African economies still have a heavy dependence on a small number of primary products. One-half of all export revenue comes from the export of commodities, with a consequential vulnerability to global price changes.

There is only one train line in the entire country of Mauritania. It runs for 437 miles from the interior to the port city of Nouadhibou. Each day the 1.5-mile-long train pulls 220 hoppers filled with 22,000 tons of iron ore. Passengers can travel on the train either as stowaways squatting in the open air among the iron ore or pay the $3 to sit on uncomfortable benches sharing the 16- to 21-hour journey with goats and donkeys. The resources of the country are mined and then shipped overseas with little value added or economic multipliers to the national economy. The processing and manufacturing take place elsewhere.

This railway in Mauritania is an apt metaphor for the economic geography of sub-Saharan Africa, as it highlights the reliance on primary products, the export of wealth, and the lack of value-added work.

In Angola and Nigeria, more than 90 percent of foreign revenue comes from oil, in Zambia 90 percent comes from copper, and in Ghana 90 percent comes from oil, cocoa, and gold. When commodity prices are high, revenue flows in, but when prices fall, government revenue is slashed. The volatile nature of global commodity prices makes the economies subject to rapid and violent change.

Primary production can be a **resource curse** as the reliance on just a few commodities can reduce economic diversification and economic efficiencies, create wide swings in government revenue that make it difficult to follow through on long-term social investment such as education and health, and can more easily lubricate a slide into corruption and cronyism. When massive revenues come from the licensing of just one or two commodities, then there are vast opportunities for corruption to become endemic.

Between 1990 and 2013, China's rapid growth sucked in primary products from around the world. China was a huge and growing market for primary products from Africa. There was a commodity boom. China's trade with Africa, only $4 billion in 1996, increased to $200 billion by 2010. Africa's exports to China increased 40 percent each year from 2000 to 2006. The commodities boom filled

11.4 Chinese investment in road construction in Nairobi, Kenya (photo: Lisa Benton-Short)

government coffers, and some of the newfound wealth percolated down to the population.

China invests heavily in Africa's infrastructure by building high-speed train links, highways, and a regional air traffic system (Figure 11.4). These are much-needed developments as high transport costs and poor linkages characterize much of the transport geography of sub-Saharan Africa. The Chinese gained something in return. They invested money, employed Chinese labor, and reaped profits both directly in the commercial undertaking of building these projects but also indirectly as the better transport links make it easier and cheaper to export commodities and import manufactured Chinese goods.

The commodity boom in the 1990s and 2000s was based on growing Chinese demand. However, Chinese economic deceleration reduced demand and led to a decline in prices. In 2015, Africa's exports to China declined by over a third from the previous year.

income. More people now move to the cities either permanently or seasonally in search of a livelihood, fueling urban growth. People also try to move overseas. By 2010, 65,000 Gambians lived abroad, almost 4 percent of the population. They send back money from their earnings to buy things like houses and tractors. Today, 20 percent of Gambia's GDP consists of remittances.

Women have a particularly difficult time, often shouldering burdens as wage laborers and family farm laborers, as well as being responsible for housework and childrearing (Figure 11.5). Women do the bulk of agricultural labor. Women's contribution to the household income has also tended to shake loose the traditional patriarchy of African rural households.

A recent status report on African agriculture points to productivity increases in the past 15 years, especially in countries that invest in agricultural development such as Burkina Faso, Ethiopia, Ghana, and Rwanda. There is a new green revolution involving new technologies, scaling up of agricultural practices, more private sector investment, and co-production organizations such as grower associations. There is also an increasing commodification of land and more agricultural production is moving up the value-added chain. Still some farmers are growing to meet their immediate needs, but there are also more capitalized, commodified forms of farming.

Problems continue to exist in the rural sector. Two-thirds of all farmers in Africa lack the necessary capital to improve soil fertility and that means that productivity gains of 28 percent in yields over the past 15 years lag well behind similar farmers in Asia with an 88 percent increase. Only 10 percent of farmers in Africa are able to access formal lending institutions.

Changing Rural Society

Sub-Saharan Africa is still a predominantly rural society. But the nature of this rural society is changing.

There is a downward trend in the value of agricultural products on the world market, and African farmers, struggling with low social fertility, lack of access to credit, and poor transport links, have extra difficulties. There is also casualization of labor in many rural sectors. People formerly employed on a regular basis are now hired on a daily, weekly, or seasonal basis only. Across sub-Saharan Africa, small-scale commercial farmers struggle to make a decent living. This has led people to look for alternative sources of

11.5 Woman carrying yams and a baby in Rwanda (photo: Lisa Benton-Short)

Government and nongovernmental organizations (NGOs) have tended to favor large-scale agriculture at the expense of small-scale farmers, though some aid programs and government policies are now focusing more on women and small-scale farmers.

Small-scale mining, involving individual or collective labor-intensive mineral extraction using little capital, is another element in the trend toward rural nonagricultural income diversification.

A LAND GRAB

There is a shift across the continent toward larger scale plantation-style agriculture in which much of the investment comes from, and most of the profits go to, overseas markets. Between 2000 and 2009, there was an increase in the amount of arable land of over 26 percent while there was an overall decline in the rest of the world. This was almost entirely due to plantations. Over 450 plantation projects in eighteen countries now grow cereals, palm oil, biofuels, rice, and sugar cane. The trend is also associated in places with a land grab. Land ownership is ambiguous over millions of square miles, making it easy for well-connected and well-financed corporations to appropriate land.

Manufacturing and Services

Things are changing. Manufacturing employment is increasing. Kenya, Ethiopia, and Rwanda have promoted manufacturing development. We will see how dairy products in Rwanda are producing value-added products such as cheeses and yogurt from milk. The processing of local raw materials is also the basis for manufacturing in Senegal. Trade deals with the United States have encouraged textile manufacturing in Kenya, Lesotho, Madagascar, and South Africa. There are some success stories. Take the case of Mauritius. Since independence in 1969, the Mauritius government has diversified from its traditional reliance on sugar plantations with the creation of an export-processing zone, a free port in Port Louis, tax incentives to encourage foreign direct investment, and the establishment of upmarket tourism.

Tourism is a vital source of foreign exchange. Safaris in east Africa, urban tourism to South Africa, and ecological and cultural tourism around the continent are all sources of revenue. International tourist receipts account for 33 percent of exports in Ethiopia, 23 percent in Tanzania, and 18 percent in Senegal. The industry can grow even more provided that political stability exists. Foreign tourists are dissuaded with negative press, especially those making long flights from Western Europe and North America.

Tourism provides not only foreign exchange and employment opportunities but also a way to marketize and hence give financial incentive to protect the biological diversity and ecological uniqueness of Africa. This could be destroyed or degraded under traditional models of economic development.

Rural Focus

Agriculture in Rwanda

Rwanda is home to 11 million people; by 2050, that number could be 20 million. Feeding a population expected to double in 30 years is a difficult task for any country. In Rwanda it is a formidable challenge because 63 percent of the population is poor, living on less than $1.25 per day. Compounding the problem is Rwanda's geography: the country is small, landlocked, the most densely populated in Africa, and defined by a green mountainous landscape that earned it the nickname "Land of a Thousand Hills." There are challenges to farming hillsides. Rwandan agriculture is largely rain fed, and irrigation systems are few and far between. Farming the steep slopes is difficult work, and severe rains can wash away the hand-built terraces. The average farm size in Rwanda is small—about 0.6 hectare. Compare that with the average family farm in the United States of 89 hectares. There is pressure to move even higher onto the hillsides. Hillside terracing has encouraged the planting of multiple crops on a given farm. This lowers the risk and also increases the diversity of food sources.

Many farmers in Rwanda have two, three, or four cows. One challenge has been how to collect milk from so many individual farmers. Milk is now collected in stainless-steel containers and then bicycled to collection centers (Figure 11.6). Getting milk to the collection center is no easy task. In Rwanda, as in many other countries in Africa, smallholder dairy farmers struggle to gain access to the market and to establish credit worthiness in order to expand their businesses.

Fifty jobs are created for every 1,000 liters of milk that go through the dairy value chain, such as making cheese or yogurt, between on-farm jobs and milk transportation. One example is Blessed Dairies, which has established a rapidly growing yogurt and cheese processing business; it supplies both of these products to RwandAir. When the company first started processing in 2012, they processed 1,000 liters of milk each day. As the market demand has increased over the years, this has grown to 3,000 liters processed per day, and today the company employs forty-four people. Rwanda has done an impressive job of increasing milk productivity and reaching out to rural farmers, but consumption has lagged behind production. A major challenge is to increase milk consumption. One strategy has focused on improving the distribution and increasing visibility of milk outlets. The city of Kigali boasts dozens of MilkZone outlets where people can refill containers of milk. They are becoming as ubiquitous as 7-11s in the United States and have helped to improve the distribution of clean and healthy milk.

11.6 Milk collection center in Rwanda (photo: Lisa Benton-Short)

Social Geographies

Demographics

Currently only 16 percent of the world's population lives in Africa, but this is likely to grow to around 40 percent, 4.5 billion, by the end of this century. It is estimated that by 2050, Nigeria will replace the United States as the world's third most populous nation, after India and China.

In much of sub-Saharan Africa, except South Africa, birth rates remain high at between 4 to 6 children per woman. All of the top-ten countries and all but one of the top-twenty countries with the highest birth rates are located in sub-Saharan Africa (see Table 11.1). With double the world's birth rates, sub-Saharan Africa is the global center for rapid population growth.

Much of Africa, especially rural Africa, is still at the early stage of the demographic transition, with relatively high birth rates and with death rates just beginning their downward slide. The high birth rates predominate in rural areas where people need children to help them work the fields. Children become a way to insure against the vagaries of growing old in a society with limited social welfare for the elderly. The high birth rates also reflect the lack of birth control measures, which have been more successfully adopted in much of South America and Asia.

The rapid population growth strains the ability of the government or the market to respond with enough jobs.

TABLE 11.1	Top Ten Countries in Birth Rates*
COUNTRY	**BIRTH RATE**
Niger	7.46
Somalia	6.12
Mali	5.92
Chad	5.79
Angola	5.79
Burundi	5.66
DR Congo	5.66
Gambia	5.53
Uganda	5.46
Nigeria	5.41
United States	1.90

*Births per woman

Half of the population in the region is under 17 years old (Figure 11.7). It will be several decades before the demographic dividend will come into play; for the moment the rapid growth creates problems for provision of child care, education, and employment for young people.

Consider the case of Nigeria where crude death rates fell from 29.6 per 1,000 in 1950–1955 to 14.9 in 2005–2010. Crude birth rates in the same time periods fell only slightly from 46.1 to 42.2. In the longer term, birth rates will come down, but in the meantime the rapid fall in death rates and the slight fall in birth rates will lead to a rapid increase in population. In 1955, the population of Nigeria was 41 million; in 2019 it is estimated to be just over 200 million. The median age in the country is only 17.9 years.

The very young population by itself is not the cause of social instability and political turmoil, but it is a contributory factor, along with limited educational and employment opportunities. The lack of jobs for these young people is a significant factor in the growth of the informal economy, the persistence of organized crime, and the political instability as young people find little space in the existing society. Explosive growth with limited economic opportunities will slow economic growth until the demographic dividend kicks in.

HIV/Aids

More than 2 out of every 3 of the 34 million people in the world who are HIV-positive live in sub-Saharan Africa. The prevalence of HIV/AIDS in sub-Saharan Africa plays a significant role in depressing life expectancy. More than 1 million people die every year from the disease. It is estimated that AIDS reduces average life expectancy to 47 years; without the HIV/AIDS epidemic, it would be closer to 62 years.

There are 50 million AIDS orphans, children who have lost a parent to the disease. Nine out of every ten HIV-positive children in the world live in sub-Saharan Africa, where 25 million people are living with AIDS. While it is less of a death sentence in the rich world, where people and public health programs can afford drugs, the widespread poverty in sub-Saharan Africa makes the disease that much more impactful on life expectancy and quality of life. South Africa is the epicenter of the disease, with slightly lower rates of infection in East Africa and West Africa. Populations most affected are young women, children, sex workers, male homosexuals, and people who inject drugs.

In some countries such as Rwanda, aggressive public health programs are reducing the spread of the disease, the mortality rate, and the impact of the disease. Awareness and monitoring programs since the late 1990s have made it a major public health issue in the country, though the HIV prevalence rate among adult ages 15 to 49 years is still high at 3.1 percent.

Socioeconomic Geographies

We can identify different types of countries based on fertility rates, urbanization levels, and economic trajectories. Five categories are noted in Table 11.2. *Diversifiers* are coming to the end of rapid birth rates and have more robust economies. The largest country in this category is South Africa. *Early urbanizers* have between a third and half of their people living in cities. Fertility rates are still high. This category is mainly found in West Africa. *Late urbanizers* mainly found in East Africa are more rural. *Agrarians* are still predominantly rural with high birth rates. Many of them are landlocked, which heightens their

11.7 Young children in Africa (photo: John Rennie Short)

TABLE 11.2	Typology of African Countries			
CATEGORY	**URBANIZATION (PERCENT)**	**NO. OF CHILDREN PER WOMAN**	**GROSS NATIONAL INCOME ($US)**	**EXAMPLES**
Diversifiers	40–65	3 or fewer	+ $10,000	Mauritius, South Africa
Early urbanizers	35–50	5	$1,000–$4,000	Ghana, Senegal
Late urbanizers	Less than 30	4–6	$1,000–$2,200	Ethiopia, Kenya, Rwanda, Tanzania
Agrarians	Less than 30	6	Less than $1,600	Chad, Malawi, Niger
Resource-based	40–78		$500–$20,000	Congo, Nigeria, Zimbabwe

difficulties in expanding their economies. *Natural resource-based countries* rely on the export of a few commodities. They are marked by higher than average urbanization rates.

Urban Issues

Rapid Urbanization

Sub-Saharan Africa is experiencing a very rapid urbanization. Urban residents only constituted 14 percent of the population in 1950. Today it is closer to 40 percent and will reach 50 percent by 2030. There is a wide variety; the *early urbanizers* identified in Table 11.2 have much higher levels of urbanization than the *later urbanizers* and very much more than the *agrarians*.

Urbanization in the region is occurring not just in the very large cities (see Table 11.3) but also in cities of fewer than 500,000 and along the urban–rural interface. African urbanization occurs both in the growth of very big cities as well as in the expansion of urban villages.

Informal Urbanism

Rapid urban growth has overwhelmed the ability of the state or the private sector to adequately meet the demand for housing, employment, and infrastructure. Urbanization across much of sub-Saharan Africa is associated with informal settlements, congestion, and inadequate infrastructure (Figure 11.8). More than half of all urban dwellers live in slums. African urbanism has an informal improvised feel. In a remarkable case of resilience, residents have built their own accommodation, made employment, and created their own cities. Urban farming is an example of the resilience of urban residents. Almost 60 to 80 percent of Brazzaville's vegetables are grown by slum dwellers. In Brazzaville the vegetable farmers earn on average five times the national average income. Urban farming was long restricted, but now in cities such as Harare more than half of all urban residents grow most of their own food.

There is a relative and absolute large number of people in informal housing, commonly referred to as slums. Almost 2 out of every 3 urban dwellings are slums. The figure is higher in certain countries such as Angola, Central

TABLE 11.3	Ten Largest Metros in Sub-Saharan Africa	
CITY	**COUNTRY**	**POPULATION (MILLIONS)**
Lagos	Nigeria	18.8
Johannesburg	South Africa	13.7
Kinshasa	Congo/DRC	12.0
Luanda	Angola	7.9
Dar es Salaam	Tanzania	6.1
Nairobi	Kenya	5.9
Abidjan	Ivory Coast	5.4
Kano	Nigeria	4.5
Cape Town	South Africa	4.1
Dakar	Senegal	3.6

Data from http://www.citypopulation.de/world/Agglomerations.html

11.8 Kibera informal settlement in Kenya (photo: Lisa Benton-Short)

African Republic, and Mozambique, where it is 4 out of every 5. And even in those countries where 40 percent of the housing stock is not considered slums—a significant proportion—between a quarter and a third lack access to toilets or reliable public water. The urban poor face myriad challenges, such as health risks due to poor living conditions and overcrowding, livelihood risks from vulnerable employment, external shocks from events such as natural disasters that disproportionately affect them, and governance risks because they do not receive adequate policy attention.

The formal economy is simply not large enough to provide employment for all urban residents, and so the majority operates in the informal economy. Two out of every three urban residents in the region earn less than $4 a day.

Because city governments have such limited funds, infrastructure provision remains a recurring issue. City governments lack the resource to adequately find transport, education, and clean water provision. Poor transport hits the poor particularly hard, limiting their job searches and employment possibilities. The urban poor have to spend almost 40 percent of their income on transport.

Cities are the sites where the very rich as well as the very poor and rising middle class compete for resources and space. A majority of the middle class in sub-Saharan Africa lives in cities. A third of the urban population of South Africa is considered middle class.

Cities have high rates of poverty and inequality. In some cases, slum renewal and gentrification mark the power of the rich over the poor. There is an economic apartheid, with the wealthiest inhabiting separate cities, living in compounds, and having a very different urban experience than the majority of urban poor.

City Focus: Lagos

Lagos, Nigeria, is a prime example of a megacity in the developing world. In 1950, when it was the capital of the British colony of Nigeria, the population was less than 300,000. By 2018, the population of the metro area was estimated between 16 million and 21 million, making it one of the largest in the world. The wide estimate is a function of the difficulty in counting informal settlements as well as the struggle between the local and central states. Migration from throughout West Africa keeps the city's average annual growth rate at 6 percent, a wave of rural-to-urban migration that seems unlikely to subside. By 2025, the city's population will be close to 35 million.

This level of immigration outweighs the ability of a formal economy or housing market to accommodate. The result is a city dominated by the informal and illegal economy and squatter settlements. The city is a stunning example of the ability of ordinary people to cope and generate seemingly endless entrepreneurial activity.

The 2010 BBC documentary *Welcome to Lagos* tells the story of Makoko, a slum neighborhood of around 200,000. Many of the dwellings rest on stilts over a lagoon. The documentary was a celebration of the ingenuity and vibrancy of slum life in the city. While it noted problems of flooding, irregular power supply, and poor environmental conditions, it was lyrical in its invocation of the resiliency and vitality of marginal groups living in the slums. Rather than a depiction of gloom and darkness, it highlighted the incredible dynamism and energetic entrepreneurship of the slum dwellers.

The rate of growth has, in a way, overwhelmed the city. Unlike many cities where slums are distinct from elite areas, illegal settlements and informal markets have engulfed most of this city. The problems are exacerbated by its political system. A series of regimes have operated kleptocracies that have impoverished the entire nation. Wealth is unevenly distributed, while the incomes of those in the middle shrink as costs outpace wages. Corruption is rife, and civil society is dominated by economic relations where the powerful and rich control the weak and poor. Most people work for someone else, paying this person a cut from their income.

The provision of basic services is problematic. Many of the poor lack access to safe drinking water. Only the wealthy have access to the government-provided pipe-borne water. Many neighborhoods use water from boreholes drilled into the ground. A vibrant industry in

11.9 Map of Lagos

the city is the making and selling of sachets, small pouches of water in sealed plastic bags, for the equivalent of around $0.06. The industry is not regulated, and there have been cases of contaminated water being sold. Tankers also sell water that is obtained from groundwater. The tankers are often contaminated. Water is expensive, and the poor often spend up to 30 percent of their household income on water from sachets or tankers.

The city continues to change, with gentrification and slum removals leading to population displacement and growing inequalities. Conflicting interpretations abound. Lagos as a new urban order: people making their way without the benefit of a functioning urban government. Or Lagos as dystopia: massive population growth paired with a shrinking economy and an inefficient public sector, leading to urban destitution.

City Focus: Cape Town

Cape Town sits at the southern tip of Africa close to the Cape of Good Hope (Figure 11.10). The first Portuguese explorers landed in the region in 1487, and the Cape was named Good Hope because of the promise of the riches of the East Indies. The Dutch East Indies Company established the port of Cape Town around 1652. The British took over effective control of Cape Town in 1795.

The metro population is now just over 4 million. The city plays an important role in the government of the region and the country. The different branches of government are spread around the country. Pretoria is the seat of the executive, Bloemfontein the judicial capital, and Cape Town the legislative seat. Johannesburg is the largest city with a metro population of 4.5 million.

The system known as *apartheid*, the Afrikaans word for "separation," lasted from 1948 to the early 1990s. It was system that identified three racial categories of black, white, and colored (a mixed-race category that constitutes a majority category in Cape Town, at the time often known as Cape coloreds) and forcibly segregated people into different areas and effectively different lives, with living standards much higher for whites compared to blacks. Cape Town has a significant colored population. In the system, blacks were forced to live on the edge of the cities in townships and were restricted from staying in the city after dark. They had to make the long trek to the city for work but had to leave again in the evening. The demarcation was unequal; the best places were reserved for the whites.

District 6 was a multicultural area in Cape Town, housing blacks, coloreds, Muslims, and Christians. The neighborhood embodied the complex racial history of the city, a heterogeneous area with people of mixed race and immigrants from East and South East Asia. Under the 1966 Group Areas Act, it was declared a white-only area and designated to house 3,500 **Afrikaan** white railway workers. Over the next 15 years, the original inhabitants, many of whom had lived for decades in the neighborhood, were forced to leave, many of them to Cape Flats over 55 kilometers away. Most of the existing houses were demolished and over 60,000 people were removed. The clearance only finished in 1984. In 1997, the area was finally opened to blacks. There is now an active process of seeking restitution and adjudicating land claims in the neighborhood.

Today the racial breakdown of the city is 42 percent colored, 15 percent white, and 39 percent black. While the apartheid policy is now abolished, there are still marked inequalities by race. Although there is now a small, wealthy black elite, the majority of blacks are poor. A large black and colored middle class has emerged, suppressed during the apartheid regime, yet neighborhoods are still racialized. There is even distinct residential differentiation within the different language groups of the more affluent white community. The upmarket suburbs in the north of the city are mainly Afrikaans speaking, while the affluent southern suburbs are predominantly English speaking.

Cape Town is one of the more cosmopolitan cities in Africa; it is situated on a beautiful site and blessed with a mild Mediterranean climate and a long tradition of liberal attitudes as well as the legacy of apartheid.

The city has major water supply issues. A long drought that began in 2015 across the region limited the supply of water. Water consumption was cut in half and very severe restrictions on water usage lasted until December 2018.

11.10 Cape Town from Tabletop Mountain (photo: John Rennie Short)

Geopolitics

Across the continent there has been a trend toward greater democratization. In 1991, only three countries were considered free according to the indices prepared by the organization Freedom House, but by 2014 this number had increased to twenty-six. Problems remain, as elsewhere in the world.

Tribalism and National Cohesion

The vast majority of sub-Saharan countries only achieved independence within the last 60 years. National identity often had to be created from a wide variety of ethnic and tribal groups in national boundaries that reflected imperial division rather than national communities. One result is the instability of national communities as they break into constituent parts. A classic example is the Biafra-Nigerian War fought in 1967 to 1970. The region of Biafra, dominated by the Igbo people, broke away from the northern Nigerian–dominated federal government. The federal government dominated by other tribal groups blocked the Biafra region and caused mass starvation. Hunger was employed as a political weapon. Almost 3 million people died mostly from starvation and disease. It was one of the first televised civil wars, bringing the images of starving children onto television screens all over the world. International media coverage drew attention to the plight of Biafrans, and it was the first postcolonial conflict that generated international concern and relief organizations. These indelible images of war and starvation and their association with Africa were burned into the Western imagination. Sadly, the tragedy was repeated in the wars in Liberia and Sudan and others throughout the region.

A more recent example is the 2007–2008 Kenyan crises that occurred after a presidential election result was widely condemned as rigged. Underlying the conflict was tribal conflict between the Kikuyu on the one side and the Luhya, Luo, and Kalenjins on the other. Almost 1,500 people were killed and half a million displaced during the 2-month unrest. It ended with the creation of a coalition government, but tribal differences continue to undermine national cohesion.

Tribal differences continue to play a significant role in African national politics because they are often used by politicians to gain support, identify enemies, and create scapegoats. Over time it seems the tribal differences have less ability to generate political violence, but they are regularly used to generate political support.

Area of Unrest: Eastern Congo

While it would be inaccurate to present a picture of all of Africa as a place of perennial war, there are specific regions that have been very badly affected by long-term wars and endemic civil unrest. The eastern Congo region, including the provinces of Ituri, Maniema, North and South Kiva, and Tanganyika, is the site of decades of violence as various insurgency groups seek to gain control over the valuable mineral deposits of gold, tungsten, tantalum, and tin (Figure 11.11). After the genocide in Rwanda, Hutu soldiers fled to eastern Congo. The Tutsi-dominated Rwanda government invaded the eastern Congo and installed a new government in 1996. Two years later, the Rwanda-Congo agreement dissolved, and rebel groups backed by the different governments fought for supremacy along with various ethnically based local militias.

Since 1994 more than 5 million people have been killed, 3 million displaced, and 1 million raped. There are more than 2.7 million people living as internally displaced refugees.

A Geopolitical Fracture Line

There is no clear-cut division line between sub-Saharan Africa and North Africa. In some classifications, Sudan is included in North Africa, while in others it is considered part of sub-Saharan Africa; similarly with Mauritania. There is no simple solution to the confusion. We can think of the division as arbitrary and constantly shifting according to the criteria we employ. The CIA World Fact Book, for example, does not even make a distinction, simply using the regional category of Africa. In other words, there is no simple or conclusive division.

Yet there is a difference as we move south from the coast of the Mediterranean to the equator. The Arab Muslim

11.11 Map of Eastern Congo

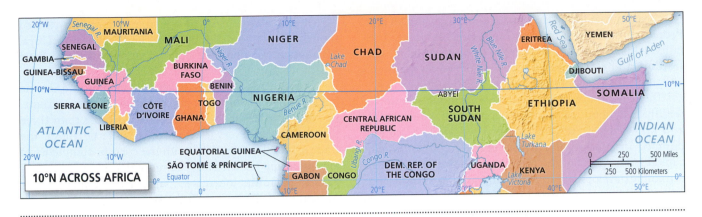

11.12 Line of 10 degree North across Africa

influence slowly peters out toward the south. Arabic is an official script used with other scripts in Mauritania, Chad, Djibouti, Sudan, Eritrea, and Somalia, but not in Mali or Niger. Islam is the official state religion of Mauritania and Somalia, but not Mali, Niger, Chad, or Sudan. Mauritania is the only member of the Arab Maghreb Union that also includes Algeria, Libya, Morocco, and Tunisia.

The division between Arab and non-Arab Africa runs through roughly 10 degrees latitude (Figure 11.12). The line is not straight because the Muslim influence in the east "bends" southward, the result of Arab traders in the Red Sea and Indian Ocean, to include most of Somalia and the coast of Kenya and Tanzania. In the case of Sudan, the 10-degree line now approximates the division between Sudan and South Sudan, although there are still non-Muslim areas north of this line. The people of the Nuba Mountains in the region of Kordofan are in Sudan rather than South Sudan. They follow Islam as well as Christianity and a form of **animism** and are unwilling to live under the sharia laws dictated from Khartoum. The Sudanese government is unwilling to give up the region because of its rich oil reserves.

The division between a Muslim north and a non-Muslim south is most obvious in nations that straddle this line such as the Central African Republic and Nigeria. The Central African Republic sits on the edge of the division with Chad and Sudan to the north and Congo to the south. Almost 80 percent of the 4.7 million people are Christian and the remainder Muslims. Christians are mainly sedentary farmers and live in the south, while the more nomadic Muslims live in the north of the country. A civil war in 2013 was fought along this religious division, overlain with struggles over land and political power. Muslims in the south of the country moved to the north of the country or overseas to escape violence. The Muslim population of Bangui fell from 138,000 to 900. More than 1 million people were displaced by the civil unrest and violence.

In Nigeria, the Muslim population is concentrated in regions in the north of the country. The religious differences are reinforced by different affiliations with the Hausa and Fulani tribes, which are mainly Muslim, and the Igbo and Yoruba, which are mostly animist and Christian. A new capital was established in the middle of the country at Abuja to link the Muslim and non-Muslim areas. The fissure is now starker because of the operation of Boko Haram, a militant and violent Islamic fundamentalist guerrilla group which since 2009 has terrorized the northern part of Nigeria as well as the border areas of neighboring Niger, Chad, and Cameroon with military attacks, kidnappings, and extortion as part of their goal of establishing sharia law. A rough translation of Boko Haram is "western education is sinful." By 2014, more than 10,000 were killed and 1 million made refugees as Boko Haram extended their influence into across Niger, Chad, and Cameroon—kidnapping, recruiting, and bombing.

If anything, the line that was traditionally porous, liminal, and fluid is now becoming more fixed and harder, the division reinforced by civil war, insurgence, and population displacement.

Connections

The Slave Trade

Africa was connected to the global economy in a most brutal fashion. For four hundred years, human beings were shipped from Africa to the New World. Estimates suggest that almost 12.5 million people were transported against their will across the Atlantic. The slaves built the economy of the New World as domestic servants and unpaid labor in plantations, mines, and farms. The vast majority were taken from their homes in west and west central Africa and transported by ship to be sold at slave markets in present-day Brazil, Cuba, and the United States. Close to 1.8 million people died at the collection points and on the overcrowded ships.

Slavery provided cheap labor at enormous human cost. People were ripped from their homes and subject to brutality, while being subjected to the basic negation of their freedom and very humanity. The sin of slavery continues to haunt the racial politics of North and South America.

The slave trade constituted an enormous transfer of wealth as economically active people with unique skills and knowledge were forcibly lifted from one place and dumped in another. It was the forced export of human capital from Africa to the New World. Slave trade limited economic development in Africa and provided cheap labor and a vital knowledge base for the New World colonizers. In west and west central Africa, populations stagnated during the slave era. Estimates suggest that the mid-nineteenth-century population in the region would have been double what it was without the pernicious effects of the slave trade. The forced loss of population impacted the local societies, destroying and undermining cultures and their future trajectories. As the loss of population transfer stunted economic growth in large parts of Africa, the African presence in the New World transformed the societies as well as the economies. Africans played a huge role in building the New World and the legacy of Africa lives on. The African presence radically changed the New World. Cuisine, music, language, social relations, and religion across North, Central, and South America were all shaped by the forced removal of Africans to the Americas and their subsequent ability to survive and resist this inhumane treatment.

Making Their Way to Europe: The Contemporary Migrant Tide

Sub-Saharan Africans make up an increasing proportion of the now almost 500,000 people per year who try to reach Europe from the Middle East and Africa (Figure 11.13). The migrant stream has increased because of political instability, economic distress, and the opportunities afforded by the collapse of civil society in Libya that makes it easier for smugglers to ship people across the Mediterranean. More than 43,000 people made the journey across the central Mediterranean in 2013. The numbers increased to 170,000 in 2014; by 2018 it had fallen to an estimated 40,000.

There are two main routes from Africa. An eastern route connects Somalia, Ethiopia, and Sudan into Libya. Migrants move along the other western route from West Africa across the Sahara to the coastline of North Africa.

Cities such as Agadez in Niger, Gao in Mali, and Shabha in Libya are transshipment points along this western route. In 2015, Agadez smugglers transported more than three thousand people every week across the desert to Libya. The trade is illegal but operates with the corrupt complicity of local officials. The migrants pay a total of $2,000 to smugglers to make the journey to North Africa, often paying in stages to different operators. It costs almost $300 to be taken by smugglers from Agadez to southern Libya. The weekly revenue for the Agadez smugglers is estimated at around $1 million. While profitable for the smugglers, it is hazardous for the migrants. It is a dangerous and perilous journey with dead bodies regularly turning up in the desert or washed up on the Mediterranean shorelines. Almost 2,000 people drowned trying to make their way across the sea in the first 6 months of 2015. Smugglers put too many people on unsafe and overcrowded boats.

The pull of economic opportunities across the Mediterranean Sea draws migrants. Estimates suggest that at least 600,000 people are waiting on the southern shore of the Mediterranean to make the trip across the sea to Europe. People are even willing to make the trip all the way from Gambia, a country with no industry, few natural resources, and an oppressive government. Poverty is endemic and hunger is common. Tiny Gambia's biggest export is people seeking to find opportunities abroad. Gambians are willing to travel the 2,300 miles to Agadez,

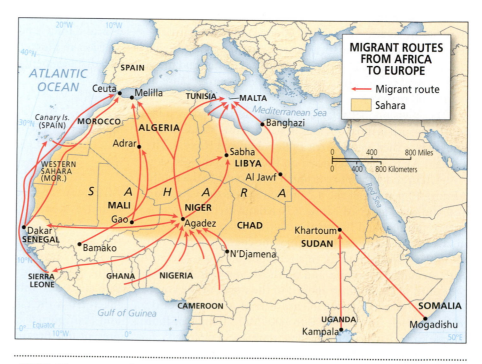

11.13 Migrant routes from Africa to Europe

the 850 miles across the desert to Sebha, and then another 415 miles to the coast, where they will encounter a dangerous sea journey in an overcrowded and leaking boat.

Recycling Electronic Waste

One consequence of the pervasive use, constant updating, and discarding of computers, handheld devices, and smartphones is a rising tide of e-waste. In the United States, the average person disposes of 66 pounds of electronic waste each year. Therefore 46 million tons of e-waste is generated annually. Only around 2.7 million tons are disposed of in a legal and safe manner. The rest is illegally and surreptitiously dumped and most is shipped overseas and ends up in poor countries. Richer countries have higher environmental standards, making it expensive to dispose of these products and their toxic materials. It is cheaper to load them onto a container and ship them to another country. In many poorer countries, environmental standards either do not exist or, if they do, are routinely ignored. And despite attempts to curb illegal and unlicensed recycling of e-waste, the stream continues to grow. The e-waste industry is now a $7 billion industry that connects the rich world and increasingly China to poorer countries. The United States and China combined are responsible for about a third of the world's total e-waste stream.

Electronic recycling and e-waste sites are now common in African cities, especially in West Africa. In places such as Ikeja in Lagos, discarded electronics arrive by container ships from Britain, South Korea, Germany, and the United States, where they are repaired, recycled, and sold with some ending up in the city dumps and landfills. Agbobloshie is a suburb of Accra on Ghana, where tons of e-waste are burned to retrieve valuable metals.

This informal economy provides jobs and income but at a high environmental cost and a public health danger. The burning, burial, and processing of e-waste at sites like Agbobloshie is a dangerous business because the process releases include lead, mercury, arsenic, and dioxins that pollute the air, poison the soil, and profoundly impact the health of the workers and residents.

Subregions

The Sahel

The Sahel is a term used to describe the ecological zone between the Sahara Desert and savannah that is up to 620 miles wide and 3,360 miles long. The belt runs across the continent from the Atlantic Ocean to the Red Sea (Figure 11.14).

11.14 Countries of the Sahel

It is a tough region with a semiarid climate. The infrequent and irregular rainfall can only support annual grasses and small trees, which survive by losing their leaves in the dry season. To cope with the tough conditions, the local people traditionally practice a seminomadism in which they move north during the wet season and south during the dry season. When traditional husbandry and agricultural practices are disturbed, the population is susceptible to food shortages.

Desertification along the desert's southern edge is a major problem in this zone and especially along its border with the Sahara, where the vulnerability is made worse by soil loss and erosion. Environmental conditions have worsened with desertification, drought, and severe water shortages. A severe drought lasted from 1968 to 1974, and another occurred in 2010. The drought in association with civil unrest, which breaks the normal lines of supply and chains of commerce, creates the preconditions for famine and starvation.

The countries of the Sahel include Burkina Faso, Chad, Mali, Mauritania, and Senegal. They are linked by their experience of French colonialism that has left a legacy of a French-speaking black elite and continuing political, economic, and military connections with France.

The countries are poor with marked inequalities of income between a tiny wealthy elite and the majority of poor farmers and nomads.

Social division is often overlain by ethnic differences. In Mauritania, for example, Berbers Arabs known as Beydanes took black slaves. Slavery was widespread and only abolished in 1981. Finally in 2007 a law was passed that allowed the prosecution of slaveholders. Beydanes continue to be the political and economic elite, and the darker skinned people of the south form a permanent underclass.

Most of the countries of the Sahel are poor with only mining producing export revenue. The majority of the rural population is nomadic, raising sheep, goats, and cows and moving with the seasons. Drought and civil unrest are pushing people off the land and into the cities.

Islam is the dominant religion. Chad at the edge of the region has a slightly higher proportion of non-Muslims who are mainly Christians. Only Mauritania is a specifically Islamic Republic where Arabic is the official language. French is a common language in all of the countries for the educated elite, with most of the population speaking local languages.

The region is also subject to geopolitical struggles between state and non-state actors. There is an active Al Qaeda affiliate known as al-Qaeda in the Islamic Magreb (AQIM). It considers the governments of the region apostate for not enforcing a strict interpretation of sharia law. The group aligned with Berber separatists and took control over large parts of northern Mali in 2012 before being defeated by French troops. France, because of its long connections and colonial legacy, plays a major military role in the area, propping up hard-pressed government troops. More than four thousand French troops were deployed in 2013 across five countries in the Sahel. AQIM had attacked Malian soldiers and taken hostages, especially in the towns of Timbuktu and Gao, which they briefly held in 2012–2013. Timbuktu has for centuries been a repository of books and manuscripts. In the fourteenth century, Timbuktu had emerged as an important trading center. By the fifteenth century the city was a center for learning and knowledge. Thousands of manuscripts were produced, originals as well as copies of texts brought along the trade routes. The volumes, in exquisite script, touched upon a variety of subjects, including poetry, medicine, legal scholarship, history, and geography. One golden age ended in 1591 with the invasion of Moroccan forces. This was a recurrent pattern of open scholarship alternating with closed theocracies. Almost 377,000 manuscripts were held by families in the city and in the surrounding towns and villages who hid them from outsiders, including the colonial French. When the city and region came under the control of Al Qaeda, in 2012, the manuscripts were again at risk. Local librarians managed to ship almost 350,000 manuscripts to safety in southern Mali by jeep and boat.

Horn of Africa

The Horn of Africa consists of Djibouti, Eritrea, Ethiopia, and Somalia (Figure 11.15).

Ethiopia is by far the largest country in the region with over 102 million people. It has a distinctive history with a long Christian tradition and is one of the few countries in the continent that resisted European colonialism apart from a brief occupation by Fascist Italy from 1936 to 1941. The country was long ruled by Emperor Haile Selassie from 1930 to 1974. He was deposed by a revolutionary movement that turned into a military dictatorship that ruled from 1974 to 1991. During this time the country was racked by conflict, invaded by Somalia, and experienced a major drought. Harvest failure and social unrest created a famine from 1983 to 1985 that killed almost 8 million people. The brutal regime ended in 1991, but a full democracy is only now beginning to appear. In recent years economic growth has flourished. Between 2005 and 2015, the economy grew at an average rate of 10 percent. In 2016, the Chinese built and financed a railway between the capital city, Addis Ababa, with the port of Djibouti creating a vital link for the landlocked nation. There are plans for 1,500 more miles of railway to crisscross the nation. The country has cheap labor and abundant electricity but needs infrastructure improvement. It is attracting investment from Belgium, Italy, Netherlands, China, and India for manufacturing and commercial farming. Four out of every 5 jobs are still in agriculture, and unemployment hovers around 20 percent. It is still difficult to do business because of government bureaucracy. When the government confiscated land for factories and farms, it lit sparks of protest and resistance in the Oromia and Amhara regions. The country has overcome civil war and famine, but full development is hindered by corruption, ethnic and religious divisions, and recurring environment problems.

There are two dominant religious traditions in the region: a long-established Christianity, especially strong in Ethiopia and Eritrea, and Islam brought by the traders from the Middle East. Somalia is a predominantly Muslim country.

The colonial influence in the region has contemporary geopolitical consequences. The colonial spheres of Italy, France, and Britain created three distinct areas known as British, French, and Italian Somaliland. These regions became the basis for postcolonial nations whose boundaries cut across religious, ethnic, and clan lines.

In 1960, on the eve of independence, French Somaliland voted to become an autonomous part of France, gaining independence in 1977 as Djibouti, a small country around the Gulf of Tadjoura in the Gulf of Aden. Djibouti retains a French legacy. Arabic and French are the two official languages of the close to 1 million population. It is a predominantly Muslim society strategically located at the mouth of the Red Sea and the Gulf of Aden and a principal port for the entire region. It has a continuing border conflict with Eritrea over jurisdiction of the Ras Doumeira peninsula. The United States maintains a large base, Camp Lemonier, used as a strategic air base and for drone strikes throughout the Middle East. It was recently extended to 500 acres. The United States will spend $1.4 billion over the next 20 years to expand and upgrade the base. It is the only permanent US base on the continent of Africa, and one with effective range over the Sahel and the Middle East.

Somali is principally a Muslim country that was previously known as British Somaliland. The national boundaries of the new nation did not align with ethnic identities. In 1977, war broke out when Somali forces invaded Ethiopian territory to annex the Ogaden, a region populated by ethnic Somali Muslims. A military dictatorship ran out of popular support as various secessionist movements resisted central government control. Since the mid-1990s, the country has lacked a strong central state, and powerful militias and Islamic groups such as **Al Shabaab** have created a weak and dysfunctional state. The weakness of the states allows not only religious movements and clan militias to flourish but also criminal enterprises. Somali pirates terrorized the busy shipping lanes of the Red Sea and Gulf of Aden by boarding and taking over ships and claiming ransoms from the ship owners. The piracy was only reduced when an international coalition of armed vessels provided an effective show of force.

Eritrea, like Djibouti and Somalia, is made of various ethnic groups; the largest is the Tigrinya, who constitute 55 percent of the total population. They are Christian. After the Italian defeat in World War II, the territory was federated with Ethiopia in 1950. The Eritrean Liberation movement was established in 1961 and waged a military campaign against the Ethiopian authorities. Finally, the country won independence in 1993. The boundary between Eritrea and Ethiopia is still in dispute, and it was the basis for the Eritrean-Ethiopian conflict from 1998 to 2001. Since independence, the country has been run by an authoritarian state in a one-party system in which

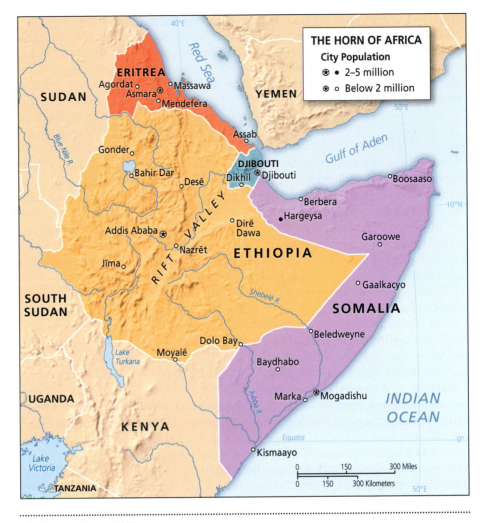

11.15 The Horn of Africa

military conscription is compulsory and long. The harsh conditions have led to a significant outflow of people, and remittances now form 32 percent of GDP. Somalis and Eritreans constitute a significant part of the stream of migrants seeking economic opportunities and political freedoms in Europe.

West Africa

West Africa is an arc of countries from Senegal to Nigeria (Figure 11.16). A defining physical feature is the shift from the semiarid north to the wetter, more tropical south. In one of the large countries such as Nigeria, with a population of close to 200 million, the north-south division in physical geography is also overlain with an ethnic and religious divide between a Muslim north and a non-Muslim south of the country.

Inside the region there is a wide variety of tribal and ethnic affiliations, including the Hausa (50 million), Yoruba (40 million), Igbo (35 million), Mande (30 million), Akan (20 million), and Fulani (20 million), as well as smaller groupings, including Abong, Ashanti, Fulani, Ga, and Wolof. Many nation-states in the region contain a mixture of different groups. When these differences are compounded by differences in economic and political power, national cohesion is often stretched thin. Ethnic, religious, and tribal tensions remain. They tend to be submerged by economic growth and then resurface with economic recessions and political conflict. Civil wars that once racked Sierra Leone and Liberia have dissipated.

The political boundaries of present-day West Africa reflect European colonial influence. European powers, starting with the Portuguese and later the English and French, established slave sites all along the coast, transshipment ports for people forcibly taken into slavery and then shipped across the Atlantic. The transatlantic slave trade was centered in West Africa. In the subsequent scramble for Africa, the European powers grabbed territory along the coast and moved inland.

The Portuguese laid claim to Guinea Bissau and Cape Verde. Togo and Cameroon came under German control; Britain claimed Gambia, Sierra Leone, Ghana, and Nigeria; while French locked up Guinea, Côte d'Ivoire, Burkina Faso, and Benin. When these countries achieved independence, their national boundaries reflected imperial arrangements. Gambia, for example, is only a sliver of territory, the result of Britain's territorial claim to control the mouth of a river rather than any national cohesion or identity.

The one exception to European colonialism was Liberia; as a country, it was founded by black American settlers escaping the crushing slavery of the United States. The capital, Monrovia, is named after the US president who encouraged schemes to settle blacks from the United States in Africa. The descendants of the US settlers became known as Americo-Liberians. They form a politically powerful economic elite. The country's main social-ethnic fracture line runs between the Americo-Liberians and the indigenous peoples.

A major colonial legacy was the reorientation of the commercial economies to primary production for the foreign markets. This led to a reliance on a few primary commodities for foreign exchange, such as diamonds, groundnuts, oil, and rubber. The Firestone rubber plantation in Liberia is the country's largest employer. It supplies

11.16 West Africa

rubber to the United States for manufacturing into rubber tires and rubber goods. Almost 40 percent of all latex in the United States comes from this plantation, a 220 square mile enclave in the country that contains between 8 million and 15 million rubber trees. No manufacturing from rubber takes place in Liberia.

Oil is found in West Africa. However, in Nigeria it has led to a resource curse, when a valuable commodity undermines rather than promotes the country's economic development, reinforces trends toward corruption and cronyism, and makes the economy and government revenues very dependent on the rapidly fluctuating price of just one commodity. Nigeria's economy follows a roller coaster with the price of oil. In more recent years, the economy is diversifying away from the heavy reliance on oil.

Across the region the dominant trends are high population growth rates, a very youthful population, rapid urban growth, and the emergence of an urban middle class.

11.17 Central Africa

Central Africa

This region consists of Cameroon, Congo, Democratic Republic of Congo (DRC), and Equatorial Guinea (Figure 11.17). It is a region of high rainfall and lush vegetation. The tropical rainforest long resisted European colonialism. The DRC, for example, was one of the last territorial regions of Africa annexed by a European power. Originally the private domain of the King of Belgium, it became a Belgian colony and finally achieved independence in 1960.

The countries range in size from tiny Equatorial Guinea with a population of close to 1.3 million, to DRC, formerly known as Zaire, with 84 million. All of the countries experienced colonial rule.

Most of the people work the land. Population growth rates are high with limited economic development beyond primary commodity production. There are some cash crops such as bananas, coffee, and cocoa, but most foreign revenue mainly comes from the sale of minerals and especially oil, now exploited in all of the countries of the region. The oil industry has proved more of a resource curse than a platform for widespread economic development. There is marked income inequality and few democratic institutions.

East Africa

The division between this region and Central Africa has a clear physical dimension (Figure 11.18). The Rift Valley is the world's greatest geological depression that runs north

11.18 Part of the 3,500-mile Rift Valley (Emad Aljumah/Getty Images)

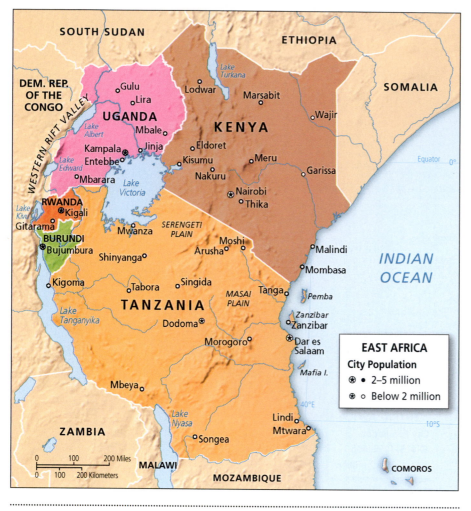

11.19 East Africa

There is a difference between the low-lying, hot, and humid coastal areas and the higher, less humid highland areas inland from the coast (Figure 11.19). These climatic differences are reflected in soil fertilities and resultant population densities. The highland areas of Rwanda, Burundi, and Kenya have much higher population densities than the rest of the region.

Along the coast of East Africa, there is a strong Islamic influence resulting from centuries of trade with the Arab world. Dar es Salaam—the name means "port/place of peace" in Arabic—is still one of the largest cities in East Africa.

The region has complex a colonial history. Germany annexed what is now Burundi, Rwanda, and Tanzania. After Germany's defeat in World War I, Burundi and Rwanda came under Belgian control while the British gained Tanzania. Britain was the single colonial power in Kenya and Uganda.

Ethnic conflict continues to plague areas of this region; the most well-known is the Rwandan genocide, where the conflict between Hutus and Tutsi resulted in the death of almost 800,000 people. The region that borders Rwanda, Burundi, and the DRC continues to be a zone of political instability and violent confrontations between state and non-state actors. But not everywhere is riven by ethnic tension. Tanzania, for example, perhaps because it has such a rich diversity with few large groups, is relatively conflict-free. Conflict is more marked where there are just a few groups jockeying for power, as in Kenya.

The bulk of the population is still involved in agriculture, and although cities are growing rapidly, the majority of people make their living from the land. Foreign tourists seeking the safari experience bring much needed foreign revenue into Kenya and Tanzania (Figure 11.20).

Southern Africa

This region of Africa includes the economic powerhouse of South Africa, one of the more developed countries on the continent (Figure 11.21). Its strategic location attracted the interests of European traders. Conflict between the British and early Dutch settlers led to the Boer Wars that finally ended in 1901. *Boer* is the Dutch word for "farmer."

to south separating out East Africa from west-central Africa. A product of volcanic activity between 35 and 25 million years ago, the tear in the earth's surface is still a visible presence in the landscape.

11.20 Wildlife in Tanzania (photo: Lisa Benton-Short)

From 1948 to 1994, the country was run on an apartheid system with a rigid separation between blacks and whites. Blacks could not live in the central cities and were forced to live in townships on the edge of the city. The apartheid system finally ended in 1994, though there is still a legacy of white affluence contrasting with black poverty (Figures 11.22 and 11.23). South Africa's economy is buoyed with agricultural exports, minerals, and manufacturing.

Enclosed in the territory of South Africa are two small landlocked countries: Lesotho and Swaziland, with populations, respectively, of approximately 2 million and 1.5 million. Both countries are poor, predominantly agricultural, with few mineral resources and a pervasive incidence of HIV/AIDS.

Export revenues gained from minerals—copper in the case of Zambia—dominate the economies of the former British colonies of Zambia, Zimbabwe, and Malawi. All the countries have received significant amounts of Chinese investments in recent years, especially in infrastructure such as railways that make it easier to export primary products. Thousands of Chinese workers are building railways across the region.

Zimbabwe is only just emerging from decades of authoritarian rule by one man. Robert Mugabe led the country after its independence from the United Kingdom in 1980 until 2017. He and his political allies and cronies appropriated the land of white settler farmers and oppressed the Ndebele tribal group while promoting his own Shona tribal group. Decades of disastrous economic policies and the collapse of the export agricultural sector led to hyperinflation and an $11 billion debt. By 2017,

11.21 Map of southern Africa

11.22 Affluent area in Cape Town, South Africa (photo: John Rennie Short)

11.23 The "black township" of Soweto, South Africa (photo: Lisa Benton-Short)

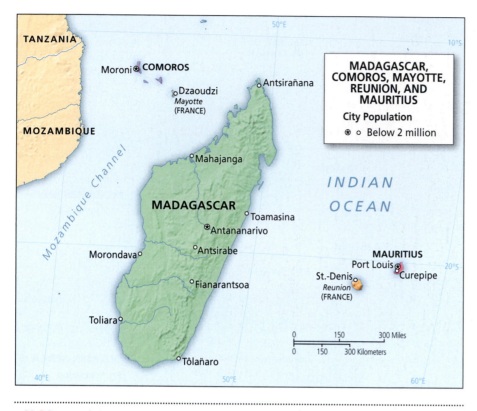

TANZANIA

MOZAMBIQUE

Mozambique Channel

Moroni ⊗ COMOROS

Dzaoudzi
Mayotte
(FRANCE)

Antsiranana

Mahajanga

MADAGASCAR

INDIAN

OCEAN

Toamasina

⊗ Antananarivo

Morondava

Antsirabe

MAURITIUS
Port Louis ⊗
Curepipe

Fianarantsoa

St.-Denis
Reunion
(FRANCE)

Toliara

Tôlañaro

**MADAGASCAR,
COMOROS, MAYOTTE,
REUNION, AND
MAURITIUS**
City Population
⊗ ○ Below 2 million

0   150   300 Miles
0   150   300 Kilometers

11.24 Map of Chagos Islands, Madagascar, Mauritius, and Reunion

social unrest and military intervention marked the end of Mugabe's power.

Namibia and Botswana are very dry, semidesert countries blessed with valuable mineral deposits. Their wild desert beauty and political stability make them an attractive tourist destination for long-haul tourists from Europe and North America as well as an investment opportunity from around the world. In countries with political stability, such as Botswana and Namibia, tourism is a growing industry.

Further north are the wetter, more tropical countries of Angola and Mozambique. They were long part of Portugal's empire in Africa and only achieved independence in 1974. Most people are poor farmers. Government revenues derive from oil and minerals. The resource curse ensures that very little of the oil revenues percolate down to the vast majority of the population.

Islands

There are a number of island territories off the southeast coast of Africa, including the Chagos Islands, Madagascar, Mauritius, and Reunion (Figure 11.24). The first settlers to arrive on Madagascar came from South East Asia around 2,500 years ago. They made their way across the ocean in the tides that flow counterclockwise around the Indian Ocean.

These islands were strategically important for global trade. Arab traders landed here as they explored the Indian Ocean, and from the sixteenth century onward, European traders used them as stopovers as they made their way around the Cape of Good Hope sailing to India and Asia. Before the opening of the Suez Canal in 1969, these islands were on the main trade route from Europe to Asia, important and strategic sites for the merchant and military navies of Europe. The Portuguese, Dutch, French, and English/British all fought to gain control. The changing colonial history of Mauritius tells the story; first it was Dutch, from 1638 to 1710, then French from 1715 to 1810, and then came under formal British control from 1810 to 1968. The French held control of Madagascar until 1960. Reunion remains French, and the Chagos Islands are still controlled by Britain and home to an important US military base.

With colonial control the islands became sugar plantation export economies. Slaves were brought in from Africa, and later, when slavery was abolished, over half a million indentured laborers were shipped from India.

Mauritius achieved independence in 1969 and became a republic in 1992. It bears the imprint of French rule, in the names of places across the island. The main city is Port Louis while the smaller towns include Chemin Grenier and Souillac. The British legacy is evident in traffic patterns as cars drive on the left. The shared legacy is retained in the continued use of English and French and the everyday use of **Mauritius Creole** that combines French and English with Asian and African influences. Since independence, the Mauritius government has sought to diversify from its reliance on sugar plantations with the creation of an export-processing zone, a free port in Port Louis, tax incentives to encourage foreign direct investment, and the establishment of upmarket tourism (Figure 11.25).

Mauritius has an ongoing dispute with the United Kingdom over the Chagos Islands, where the British government evicted three thousand islanders in order to construct a military base on one of the islands, Diego Garcia, which it leased to the United States. The base played an important part in the Cold War and is still used to extend US global military reach.

The small island of Reunion, with a population of less than 865,000, remains a department of France. Colonization began in 1645, and the local economy was based on slave labor and then indentured labor to work the plantations.

11.25 Luxury resort on Mauritius (photo: John Rennie Short)

Focus: Mobile Phones and Banking in Kenya

The mobile phone has changed lives in Africa. It has replaced networks of fixed line communication that were often inadequate or unreliable. It has allowed millions of Africans to leap-frog the landline and become connected to services in a different way.

Mobile phones have opened up financial services for millions of people who previously lacked bank accounts or access to credit. One example from Kenya is M-PESA (*Pesa* is a Swahili word for "money"), which allows users to transfer small amounts of money. In short, it's a digital wallet. Kenyans use M-PESA and its PIN-secured text messaging to transfer money from one account to another, pay utility bills, or deposit money into an account stored on their cell phones. It's been credited with reducing crime by substituting cash for PIN-secured virtual accounts and eliminating the need for people to take bundles of cash with them in person. It has freed up people to spend time doing other things instead of standing in long lines at the bank or traveling long distances to get to the bank, both of which can be problematic in infrastructure-constrained countries such as Kenya.

M-PESA began as a joint partnership between the Kenyan telecom company Safaricom and a $1.5 million grant from the United Kingdom's Department of International Development. Originally it was conceived as a system to allow microfinance-loan repayments to be made by phone. But it has become a larger financial platform. Almost 83 percent of the population 15 years and older has access to mobile phone technology. As a result more than 20 million Kenyans use M-PESA, including 70 percent of the country's poor (people living on less than $1.25 a day). Some $2 billion are exchanged each month, which represents about 25 percent of Kenya's GDP.

M-PESA is not a traditional bank—it does not pay interest on deposits or make loans. And it does not have traditional bank offices. Instead, it relies on thousands of towers for network coverage. Kiosks like these are found in every village and town, allowing rural populations access to financial services that would have otherwise been impossible. The platform development has created tens of thousands of jobs.

Inadequate and inaccessible financial services help keep the poor trapped in poverty. Without access to finance, the poor can't invest in tools to increase productivity, start a microenterprise, invest in education or health, or even take time to search for better opportunities. Many

In 2005–2006, the crippling disease of chikungunya affected close to 30 percent of the entire population.

France annexed Madagascar in 1897 and French rule continued until 1960. The newly independent country initially had a great reliance on France, and then flirted with the Eastern bloc, but since the 1990s, it has shifted to a more market-based economy. Political instability is constant with military coups and periodic widespread protests. The island nation with a population close to 26 million relies on tourism, agriculture, and extractive industries. Deforestation and illegal logging pose major environmental challenges made all the more pressing because of the unique ecology of the island. Madagascar is an area of biological diversity and uniqueness. Animal and plant species evolved independently over the last 60 million years ago, when the island broke off from the neighboring continent. Madagascar's ecology is unique; the long isolation meant that distinct plant and animal species evolved here and nowhere else. It is a biological hotspot of diversity and uniqueness. More than one hundred different species of lemur are only found on the island. While the French government is financially able to bear the cost of declaring national parks in the upland tropical forest areas of Reunion, Madagascar, with less money and a much larger population, struggles to balance saving the forests and perhaps generating over the longer term a green tourist economy with the immediate economic gains of exploiting the forest.

We can see an example of species loss. The dodo was a flightless bird found only in Mauritius. When the first Dutch sailors arrived on the island in 1598, they hunted the bird, as it was an easy-to-catch source of food. Within a century the bird was hunted to extinction. The last recorded sighting of the 3-foot-tall adult bird was in 1662. It was particularly vulnerable as it was unable to escape human predators through flight. The dodo bird is not only a reminder of the vulnerability of unique ecosystems in small islands but also of the human capacity to cause extinction.

of the world's poor do not have formal bank accounts. While M-PESA cannot replace all the functions of a bank, it has revolutionized access to financial services for the poor, empowering them in new ways, and supporting entrepreneurial creativity.

M-PESA is one of the most successful mobile phone–based financial services in the developing world, making paying for a taxi ride easier in Nairobi than in New York City. M-PESA is a good example of how a relatively small investment can transform millions of lives.

Select Bibliography

Africa Economic Outlook. 2016. http://www.oecd.org/dev/african-economic-outlook-19991029.htm.

African Agriculture Status Report. 2016. http://agrinatura-eu.eu/2016/09/2016-african-agriculture-status-report-aasr/.

Aryeetey-Attoh, S. 2010. *Geography of Sub-Saharan Africa.* New York: Prentice Hall.

Balehegn, M. 2015. "Unintended Consequences: The Ecological Repercussions of Land Grabbing in Sub-Saharan Africa." *Environment: Science and Policy for Sustainable Development* 57:4–21.

Bhorat, H., A. Hirsch, S. Kanbur, and M. Ncube. 2014. *The Oxford Companion to the Economics of South Africa.* Oxford: Oxford University Press.

Bosker, M., and H. Garretsen. 2012. "Economic Geography and Economic Development in Sub-Saharan Africa." *The World Bank Economic Review* 26:443–485.

Bryceson, D. F. 2002. "The Scramble in Africa: Reorienting Rural Livelihoods." *World Development* 30:725–739.

Bryceson, D. F., and J. B. Jønsson. 2010. "Gold Digging Careers in Rural East Africa: Small-Scale Miners' Livelihood Choices." *World Development* 38:379–392.

Ciochetto, L. 2017. *Globalization and Sustainability in Sub-Saharan Africa.* London: Imperial College Press.

Collins, R. O., and J. M. D. Burns. 2013. *A History of Sub-Saharan Africa.* Cambridge: Cambridge University Press.

Eastwood, R., and M. Lipton. 2011. "The Demographic Transition in Sub-Saharan Africa: How Big Will the Economic Dividend Be?" *Population Studies* 65:9–35.

Estache, A., and Q. Wodon. 2014. *Infrastructure and Poverty in Sub-Saharan Africa.* New York: Palgrave Macmillan.

French, H. W. 2014. *China's Second Continent: How a Million Migrants Are Building a New Empire in Africa.* New York: Alfred A. Knopf.

Goebel, A. 2015. *On Their Own: Women, Urbanization, and the Right to the City in South Africa.* London: Queen's University Press.

Grant, R. 2015. *Africa: Geographies of Change.* Oxford: Oxford University Press.

Grant, R., and M. Oteng-Ababio. 2012. "Mapping the Invisible and Real 'African' Economy: Urban E-waste Circuitry." *Urban Geography* 33:1–21.

Hamer, J. 2016. *The Bad-Ass Librarians of Timbuktu.* New York: Simon and Schuster.

Hochschild, A. 1998. *King Leopold's Ghost.* New York: Houghton Mifflin.

Hugon, P. 2008. *African Geopolitics.* Princeton, NJ: Markus Wiener.

Keltie, J. S. 2013. *The Partition of Africa.* Cambridge: Cambridge University Press.

Lee, M. C. 2014. *Africa's World Trade: Informal Economies and Globalization from Below.* London: Zed Books.

Lundgren, C., A. Thomas, R. York, and C. Lundgren. n.d. *Boom, Bust or Prosperity? Managing Sub-Saharan Africa's Natural Resource Wealth.* Washington, DC: International Monetary Fund.

Maddox, G. 2006. *Sub-Saharan Africa: An Environmental History.* Santa Barbara, CA: ABC-CLIO.

Mandela, N. 1994. *Long Walk to Freedom: The Autobiography of Nelson Mandela.* Boston: Little, Brown.

Melese, G., and M. M. Mulinge. 2013. *Impacts of Climate Change and Variability on Pastoralist Women in Sub-Saharan Africa.* Kampala, Uganda: Fountain.

Meredith, M. 2011. *The Fate of Africa: A History of the Continent Since Independence.* New York: Public Affairs.

Nunn, N., and D. Puga. 2012. "Ruggedness: The Blessing of Bad Geography in Africa." *Review of Economics and Statistics* 94:20–36.

Packenham, T. 1991. *The Scramble for Africa.* New York: Avon.

Richmond, Y., and P. Gestrin. 2009. *Into Africa: A Guide to Sub-Saharan Culture and Diversity.* Boston: Intercultural Press.

Scholvin, S., and S. Andreasson. 2015. *A New Scramble for Africa.* Burlington, VT: Ashgate.

Schulz, N. 2015. "Dangerous Demographics? The Effect of Urbanisation and Metropolisation on African Civil Wars, 1961–2010." *Civil Wars* 17:291–317.

Silva, C. N. 2015. *Urban Planning in Sub-Saharan Africa: Colonial and Post-colonial Planning Cultures.* New York: Routledge.

Stock, R. 2013. *Africa South of the Sahara: A Geographical Interpretation.* 3rd ed. New York: Guilford Press.

Vreyer, P., and F. Roubaud. 2013. *Urban Labor Markets in Sub-Saharan Africa.* Washington, DC: World Bank.

Learning Outcomes

Similar belts of climate and vegetation straddle either side of the equator—from hot, wet rainforest through savannah to desert.

Weather patterns are impacted by the Intertropical Convergence Zone.

Desertification and soil erosion are major problems.

Climate change has a particularly heavy impact in this region because of population vulnerabilities and governments' lack of financial resources.

While the eastern coastal areas witnessed Arab traders and empires, the whole of this region was impacted by European colonialism.

The European colonial powers used Africa to extract raw materials and produce; this left an imperial legacy of artificial national boundaries, poorly integrated economic systems, lack of mass education, and increased tribalism.

The resource curse is felt with full effect in countries dominated by the export of primacy produce, especially oil.

China is undertaking major investments in the region, especially in transport.

Predominantly agrarian, many countries are diversifying.

The rapid urban growth is marked by the creation of a new middle class as well as informal economies and slum housing.

Most of the region is experiencing very rapid population growth despite the scourge of HIV/AIDS.

Tribal politics continue to undermine state legitimacy and national cohesion in some countries.

While much of sub-Saharan Africa has seen democratization and reductions in political violence, the Eastern Congo remains a violent zone of conflict.

There is a major fracture line between the mainly Muslim and predominantly Christian areas of the Sahel.

The following subregions can be identified: Sahel, Horn of Africa, West Africa, Central Africa, East Africa, Southern Africa, and a range of island nations off the east coast.

12

Australasia and Oceania

LEARNING OBJECTIVES

Contrast the geologic origins of the region's physical landscapes and evaluate its major environmental vulnerabilities.

Distinguish the region's diverse histories and experiences of early human settlement.

Outline the region's distinct experiences of geopolitical incorporation from European exploration to independence.

Survey the predominant economic activities of the region's major countries and of the islands across Oceania.

Examine the diversity of cultures across the region and discuss the variety of challenges facing indigenous peoples.

Recognize Australia's rural Murray-Darling River Basin and explain its economic importance and environmental crisis.

Describe Australian urbanization and its associated issues and interpret its cities' high quality of life.

This region covers a vast territory, including much of the Pacific Ocean (Figure 12.1). It has the smallest population total of the world regions discussed in this book. Despite the overall low density, there are intense concentrations of people in cities large and small. The region contains tiny vulnerable island states as well as the island continent of Australia. Distanced from the dominant centers of economic and political power for centuries, making it a setting for the imaginative geographies of island paradises, it is now an integral part of the modern world and is in the forefront of major environmental issues stemming from global climate change, including warming temperatures and sea level rise.

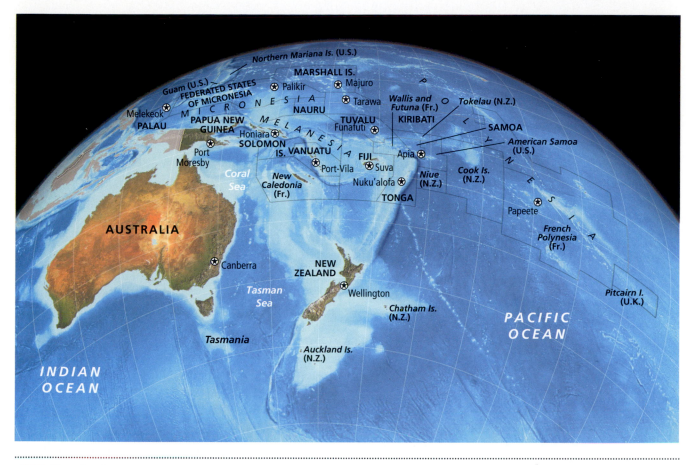

12.1 Map of region

A Landscape of the Very Old and the Very New

The physical landscape of Australasia and Oceania is a mixture of the very old and the very new. The island continent of Australia is a remnant of the giant landmass of Gondwanaland that began to break up around 180 million years ago. New Zealand and other nearby islands are more recent, part of the **Pacific Ring of Fire**, the arc of tectonic activity at the edges of plates that ring the Pacific Ocean. While Australia is a zone of limited tectonic activity, the islands of the ring are scenes of intense contemporary activity because they are located along the fault line where the Australia Plate collides with the Pacific Plate. Not all the volcanic activity occurs at the edges of plates. Hawaii's islands are not on the edge of a plate but located above a **volcanic hotspot** where molten magma breaks through the surface of a plate. Lava flows continue to build land into the sea. The big island of Hawaii is made up of five separate

volcanoes. The most active is Kilauea. The island is still under active construction by the earth's forces.

There are also coral islands made from the skeletons of polyps. These very low-lying coral islands can take a circular shape known as an atoll. The largest in the world, Kwajalein in the Marshall Islands, is 170 miles long. Most of the coral atolls took their present form only in the past 5,000 years when sea levels stabilized after the melting of the ice sheets.

The largest coral reef system is the Great Barrier Reef that stretches through 1,400 miles parallel with the northeast coast of Australia (Figure 12.2). Much of it is protected by a Marine Park designation that limits fishing and building developments. It is one of the beautiful natural wonders of the world. I have visited it on at least seven occasions and never failed to be captivated by its wondrous beauty. It is a huge though delicate organism requiring clear water and stable temperatures. It is threatened by runoff from the land that darkens and pollutes the water, global climate change that is increasing sea temperatures, and the invasive crown of thorns

12.2 Great Barrier Reef off the northeast coast of Australia

starfish. In 2016 alone, the reef lost almost 30 percent of all its coral because of ocean heat. Half of the reef is now destroyed and most of the remaining reef will shift to a new ecological state.

Environmental Vulnerabilities

Much of the region is subject to violent storms. Hurricanes in this part of the world are known as cyclones. In the South Pacific, the cyclone season runs from November to April. There are numerous examples of devastating storm events. On Christmas Eve of 1974, Cyclone Tracy destroyed most of Darwin, a city on Australia's north coast. The storm leveled three-quarters of all buildings in the city. Fatalities were relatively few—seventy-one people—given the intensity of the storm and the damage it caused, because of the early warning system that allowed most people to evacuate the city before the storm hit. In 2015, Cyclone Pam, with winds of over 168 miles per hour, devastated many communities in the island nation of Vanuatu. The country has a population of 276,000 spread over sixty-five islands. More than 47,000 live in the capital city of Port Vila. The violent storms caused flooding, landslides, and powerful sea surges that overwhelmed many coastal settlements. Nine out of every ten buildings in the country were impacted, and there were severe fresh water and food shortages. The impact of big storms on small and relatively impoverished countries is particularly devastating compared to the impact

on larger more affluent countries like Australia. In Vanuatu, the whole country was affected, the government's capacity for effective response was very limited, and foreign aid was desperately needed. This was much like the earthquake in Haiti in 2010.

Many of the Pacific Island nations have mounting environmental issues from dying coral, damaged coastal ecosystems, marine pollution, declining fish stocks (especially the overfishing of bigeye and yellowfin tuna), and ocean acidification. We need to recalibrate the ideal of a South Pacific paradise.

Climate Change

Climate change is causing a warming of the ocean that is increasing cyclone activity and the acidification of the ocean. Increasing acidification of the oceans places stress on the coral reef, a vital resource for fishing and tourism.

The most serious threat is sea level rise. The small, low-lying island nations of the Pacific are very vulnerable to sea level rise as well as the increased cyclonic activity associated with climate change. The highest point in the Marshall Islands is only 33 feet above sea level, and the majority of people live close to the shore. Across many of the South Pacific islands, altitude rarely exceeds 10 feet. Kiribati has an average elevation of only 7 feet. Almost 60 percent of all infrastructures on the island chains are within a quarter of a mile of the coastline with a total replacement value of around $22 billion. Kiribati, the Marshall Islands, and Tuvalu have almost 95 percent of their buildings within this limit. The predicted 13-inch rise in sea level by 2050 and the 3.5-foot rise predicted by 2100 will displace thousands of people and cost billions of dollars.

As the waves become higher, more of the land is inundated with salt water, even during normal tides and especially during storms, killing off vegetation, plants, and crops and polluting fresh water sources.

For Kiribati, Samoa, Vanuatu, and many of the other islands in the South Pacific, even the most conservative of predictions constitutes an existential threat. The people of this region could become climate change refugees.

The government of Kiribati has purchased 12 square miles of land on Fiji as a possible future settlement in the event of forced evacuation. The islands of the South Pacific are on the front line of climate change catastrophe. The plight is all that more desperate because of the limited budgets for adaptation and the limited impact of their own mitigation. The island nations of the South Pacific are small in population, resources, and wider political influence. They are so small and so distant from centers of political power that they have little voice in climate change discussion and negotiations. With some of the lowest carbon footprints on the planet, they are paying the heaviest price. The threat of territorial extinction is caused not by their own activities but by the heavy carbon footprint of others. It is a footprint so heavy it is likely to sink them.

Maintaining Biological Heritage

After the break-up of Gondwanaland, Australia and Oceania remained in evolutionary isolation. The result is flora and fauna found nowhere else in the world. The most obvious examples are the kangaroos, emus, wombats, and platypuses.

In some cases endemic species have dispersed. Eucalyptus trees that evolved in the dry, hot, fire-tempered regime of Australia were successfully exported across the world. You can smell the eucalyptus tree in California as well as in New South Wales.

However, the very distinctiveness of local ecosystems has made them vulnerable to the invasion of foreign species. The rabbit was introduced by the British in 1788. With prolific reproduction rates and few natural predators, their numbers soon exploded, causing major damage to the grasses and flora that maintained the beef and sheep industries. The damage was so considerable that in 1893 a rabbit-proof fence was started in Queensland to keep them out of farming areas. Between 1901 and 1907, a rabbit-proof fence (three in fact) was built across the continent to halt the westward move of the rabbits into the rich pasture lands of Western Australia (Figure 12.3). The rabbit population declined dramatically in the 1950s due to **myxomatosis**, and the rabbit fences were no longer needed.

The introduction of rats and predators, such as stoats and domestic cats, has wreaked havoc on the bird life of Australia and New Zealand. Before the coming of humans, flightless birds such as kiwis and cassowaries could evolve and flourish because there were no natural predators. But once the rats arrived with the first Maori settlers and the domestic cat was introduced by the Europeans, the flightless birds were an easy prey. A quarter of New Zealand's bird species are now extinct, with many just hanging on. The kiwi, the national bird and emblem of the country, is on the edge of extinction. The threat to the native wildlife is so severe in New Zealand that a national campaign was launched to rid the country of imports such as rats and stoats. The country's Department of Conservation regularly conducts massive aerial drops of deadly toxins to kill rats. Saving native wildlife, such as the kiwi, is now seen as a patriotic act.

The problems of wildlife in Australia and New Zealand are made worse by climate change. Since many of the species are endemic and found nowhere else, they have nowhere else to go. Endemic species have a higher risk of extinction. Most of the predicted climate change

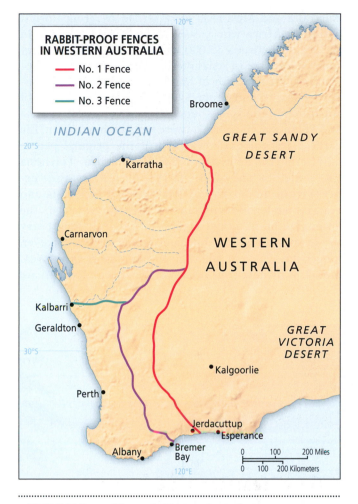

12.3 Rabbit-proof fences in Western Australia

variability takes climate into regimes that the species have never experienced before. So with no prior experience and nowhere to go, they are at greater risk of extinction. It is estimated that a third of all rainforest mammals, birds, reptiles, and frogs in the Australian wet tropics will either become endangered or extinct by 2085. And more than 9 out of 10 species will experience population decline of more than 50 percent, putting them at risk of eventual extinction.

The Peopling of the Region

The distinction between Australia, Melanesia, Micronesia, and Polynesia results from the different timing of the peopling of this region. Genetic evidence shows that

Australia and Melanesia were settled by groups originating in Africa who passed through Asia, arriving in the region between 35,000 and 65,000 years ago. They spent some time in Asia interbreeding with the Denisovans, an extinct form of hominid similar to the Neanderthals, who constitute 3 to 5 percent of the DNA of Australian Aborigines and Melanesians. Micronesia and Polynesia, in contrast, were peopled much later. The earliest archeological evidence of the original people of Micronesia dates from 3,500 years ago. Polynesia was settled by one of three ways. According to the **express train model**, they passed through Melanesia and reached western Polynesia around 2,000 years ago. The **entangled bank model** argues for a slightly longer stay in Melanesia, and the **slow boat model** assumes an even longer stay. Current scientific opinion favors the express model. Whichever one is correct, what is undisputed are the amazing skills, bravery, and determination of these early people to set out on a vast watery wilderness in the hopes of finding new land and new opportunity. It is a recurring story, but rarely has the journey been so vast or so difficult. The peopling of Micronesia and Polynesia is a testament to the navigation and sailing skills of these early peoples.

The settling of Polynesia occurred in stages. In the first wave, from around 1200 BCE, people sailed from near Papua New Guinea to settle Fiji, Tonga, and Samoa. Around 600–800 CE in the second wave, people ventured further into the ocean and landed on the Cook, Society, and Marquesas Islands. After a long pause, because of immense distances and greater difficulties, the maritime explorers then sailed further east to reach Hawaii, New Zealand, and finally Easter Island by around 1200 CE.

In the past 200 years, the peopling of this region was impacted by the colonial incorporation of the lands into foreign control. Australia and New Zealand attracted settlers from Britain and Ireland (Figure 12.4); then from the 1950s onward, the rest of Europe; and more recently, from all over the world. One of the most interesting population movements occurred in Fiji, where labor was imported from India starting in 1879. The Indo-Fijian population constituted just over half of the total population at its peak and still constitutes a third of Fiji's total population.

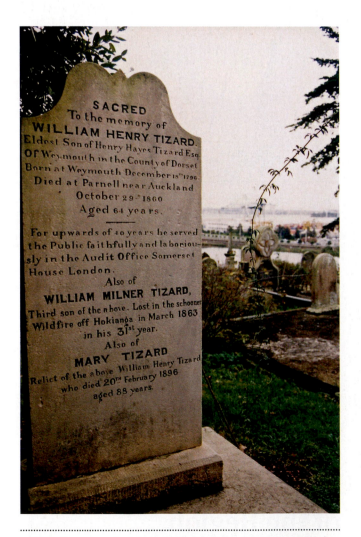

12.4 Gravestone in Auckland tells one part of the story of the European settling of New Zealand (photo: John Rennie Short)

The Geopolitics of Incorporation

Australia and Oceania were on the other side of the world from the global colonial powers of Europe, which meant their incorporation occurred later than regions such as South America or even South and South East Asia.

There were some early explorers and early annexations. There is some evidence that Portuguese ships explored the northern coast of Australia in the late sixteenth century. A Spanish ship, with a Portuguese navigator, probably reached Vanuatu in 1606. In the same year a Dutch ship reached the west coast of Australia. Under the promotion of the Dutch East India Company, which had a base in Batavia (now known as Jakarta), a succession of Dutch ships then sailed to Australia, extending European knowledge of the coastline. In 1642, Abel Tasman sailed to Tasmania (he named it Van Diemen's Land after the head of the Dutch East India Company), New Zealand, Tonga, and the Fiji Islands. The sea between Australia and New Zealand is still known as the Tasman Sea. European contact was limited because the region was so far away with no obvious commercial opportunities, unlike the Spice Islands of South East Asia.

The Spanish expanded from their base in the Philippines to claim Guam, the Marianas, and the Caroline Islands. By the late seventeenth century, the English were sending ships to the region and the ex-pirate and early geographer William Dampier was the first Englishman to set foot on Australia in 1688. It took over 100 years, however, before a more formalized exploration was launched. Captain James Cook was instructed by the British government to sail to the Pacific as part of astronomical observations, but with secret orders to find and claim territory in the southern hemisphere. He sailed from Plymouth in 1768 and by 1770 had discovered Tahiti and New Zealand. In April of that year he landed at what he called Botany Bay, close to present-day Sydney. He unknowingly sailed past the narrow harbor entrance to what is now Sydney. He then sailed up the coast of Australia, naming places as he went along. Cape Tribulation off the coast of Queensland marks a trying time when the ship *Endeavour* struck a reef, now known as Endeavour Reef, part of the Great Barrier Reef. The British claimed Australia and New Zealand as part of the British Empire.

Australia was now British, although no one asked the Aborigines; the possession of the country was validated by other European powers and not by any treaty with the indigenous people. But what to do with a distant place, literally on the other side of the world? The region was subject to what one Australian historian referred to as the tyranny of distance: far from the center of political and economic powers which, before jet travel, meant weeks and often months for the transfer of information, decisions, goods, and services.

At the end of the eighteenth century, with a mounting prison population, and the closure of the now United States as a dumping ground for criminals, the British authorities decided to ship convicts to this faraway land. In January 1788, the first convict fleet sailed into Sydney Harbor, after a brief stay at Botany Bay.

Other maritime nations made claims on the region. The French were keen to expand their empire. In 1842, they laid claim to a number of island chains now referred to as French Polynesia, including the Marquesas and the Windward Islands that include the fabled island of Tahiti. In 1853, they took possession of New Caledonia, initially as a penal colony. Later, it became a valuable source of nickel. France continues to have significant presence in the region in three territories—French Polynesia, New Caledonia, and Wallis and Futuna—with a military presence and extensive investments.

In the second half of the nineteenth century, the remaining island chains were annexed by colonial powers.

In 1876, Britain and Germany agreed to divide up the southern Pacific into separate spheres of influence. The British gained control over Kiribati, Cook Island, Fiji, Pitcairn Island (still British), Tonga, Tuvalu, the Solomon Islands, and Vanuatu. The German Empire claimed part of New Guiana and purchased islands from Spain in 1899. The German presence is memorialized in the naming of the Bismarck Archipelago northeast of Papua New Guinea. After Spain's defeat in the Spanish-American War, the United States gained possession of the Philippines and Guam. Spain also sold the rest of their island possessions to the German Empire. The United States laid claim to Hawaii, and the full annexation was formalized in 1959 when it became a US state. The United States still has significant presence in the region with formal control over American Samoa, Guam, the Mariana Islands, and Wake Island and informal control over the Federated States of Micronesia, the Marshall Islands, and Palau. The United States maintains military bases throughout the island chains.

New Zealand has an association with the Cook Islands, Niue, and Tokelau. Norfolk Island is an external territory of Australia, and the Pitcairn Island, populated by descendants of mutineers from HMS *Bounty*, is an overseas territory of Britain.

Achieving Independence

In the past 50 years, some of the islands have achieved independence. Political independence is easier to achieve than economic independence. For many the road from colony to postindependence was winding. Nauru, with a population of less than 14,000, was annexed by Germany in 1889, along with the Marshall Islands. After Germany's defeat in World War I, it was ruled by Britain under a League of Nations mandate. Then after World War II it was administered under a UN mandate by the United Kingdom, Australia, and New Zealand. In reality, Australia was the main force. The tiny island nation only became independent in 1968.

Fully independent nations now include Fiji, Samoa, Tonga, Tuvalu, Kiribati, and Nauru. Some rely on remittances and use innovative strategies. Tuvalu, total population of approximately 11,000, relies for foreign currency on Tuvaluan men working on merchant ships, sales of fishing licenses, and the lease of its Internet domain name of .tv.

The small island nations may be politically independent, but their precarious economic situation makes economic independence very difficult, especially for the more remote and smaller island nations. Regular political and fiscal crises raise the possibility of failed Pacific states.

Connections

Nuclear Testing in the South Pacific

This region is literally on the other side of the world to some of the more populated areas of the world. And with vast swathes of island chains owned or controlled by foreign powers, it made the region a favored site for nuclear bomb testing.

France tested nuclear bombs in the desert of Algeria, but when that country achieved independence in 1962, that was no longer an option. France turned its attention to its distant island territories of the South Pacific, suitably far enough away from France. A total of 193 atmospheric and underground tests were conducted in the region through the mid-1990s. Starting in 1975, the tests were carried out underground. In 1987, when the region was declared a nuclear free zone, France ignored the call and continued to explode devices at Moruroa until 1996. In 2009, the French government offered $13.5 million to compensate victims of the nuclear weapons testing program.

The British also conducted testing in the region. They exploded bombs in Maralinga in South Australia in 1956–1957. Plutonium contamination remains at the site. Nuclear bomb tests were also conducted in the South Australian desert at Emu field, at Monte Bekko Island off the coast of Western Australia, and a 1.8-megaton bomb was exploded at Christmas Island in 1957.

The United States had an extensive test program in the region. There was so many that the term "Pacific Proving Ground" was used. Between 1946 and 1962, the United States detonated sixty-seven nuclear weapons in the Marshall Islands. Bikini Atoll, a series of islands, was a favored site. In 1946, the first nuclear explosion in peacetime was detonated, destroying one of the coral atolls (Figure 12.5).

12.5 The 1945 nuclear explosion on Bikini Atoll (Photo courtesy of the Library of Congress)

In 1954, a 15-megaton bomb was exploded on Bikini Atoll. And in a bizarre adoption, the island gave the name to the swimsuit. The explosion created a 20-mile-high mushroom cloud and radioactive fallout. The explosion was 1,000 times more powerful than the bomb dropped on Hiroshima. In 1962, a missile warhead was detonated at a high altitude above Christmas Island.

The nuclear weapons testing destroyed beautiful coral atolls, displaced thousands of local people, showered large areas of Polynesia with plutonium, poisoned military personnel and civilians close to the test sites, and contaminated the world's atmosphere.

Australia

Australia is the largest and most populous country in the region (Figure 12.6). It has one of the longest continuous human occupancies in the world, outside of Africa.

Indigenous People and Land Rights

The first peoples to settle Australia came across the narrow land divide from South East Asia somewhere between 60,000 and 50,000 years ago when more of the world's oceans were locked up in giant ice sheets and more land was exposed. They quickly dispersed throughout the entire continent, and by 40,000 years ago had settled as far south as Tasmania off the south coast. They had a major impact and were a contributing factor in the extinction of the mega fauna and the transformation of the forest and grassland through sophisticated fire management. The total population before contact with Europeans is estimated between 750,000 and 1 million.

The subtle adaptation and transformation of the physical environment created different types of cultural ecological regions in precontact Australia, including temperate areas with medium to high population density, arid zones with low densities, and semiarid areas with medium to low densities. There were also coastal and estuarine areas and ecological cultural areas with more monsoonal-type weather. The ecologies and cultures differed. We can consider one case study from the central desert region inhabited by the Arrernte tribe, where, because of the poor soils and low and erratic rainfall, population densities were low, approximately one person per 80–200 square kilometers. The total Arrernte precontact population, across central interior Australia, is estimated between 8,000 and 10,000. Food supply was scarce and tool development was limited by the need for portability. The landscape was a complex system of communal land titles, based on lineage, family, and place of conception. Responsibility for "managing" specific sites often lay in the hands of senior elders, both men and women. Most sites were gender specific in their ownership and responsibility patterns.

12.6 Map of Australia

asymmetry in power between the whites and the Aborigines. The encounter was one of possession and dispossession. The Aborigines were marginalized.

When the British arrived in 1788 to establish a permanent settlement, the land was summarily annexed. The indigenous people were dispossessed of their land. There was resistance but the superior firepower of the white settlers eventually led to the great dispossession and population crash of the Aboriginal population. Only in the more remote and less attractive areas of the country did Aboriginal communities survive. The dry interior region was one of the last areas to be colonized. A regime of effective government regulation and surveillance tightly controlled the Aborigines. From 1911 to 1953, Aborigines were moved and relocated to reserves and missions with no heed given to their concerns or wishes. From 1953 on, they were classified as wards of the state and required official permission to marry, leave reserves, dispose of property, drink alcohol, or open a bank account. These restrictions were only lifted in 1964. Even until the late 1960s, Aboriginal children were often removed from their families and placed in orphanages or white foster homes.

In the 1970s, there was a change in attitudes as more people recognized the injustice committed against the indigenous people. Protests by the Aboriginal community also highlighted their plight and the claims. The most significant change occurred with land rights. The indigenous people of Australia now have recognition of their land rights, culture, and citizenship. However, there is still a chasm between their living standards and employment opportunities compared to the nonindigenous. Unemployment is three times higher for indigenous peoples compared to nonindigenous, while median household income is only half. Life expectancy for indigenous men and women is 59 and 65 years; the respective figures for the nonindigenous are 77 and 82 years.

Environmental Challenges

There are a number of environmental challenges facing Australia. It is a large, dry continent where rainfall is only prevalent along some of the coast and in the tropical

The complex interlocking titles were less monopoly controls and more relational and totemic, allowing individuals and groups a broad range of claims, responsibilities, and bargaining options that enabled long-term occupancy of a harsh environment with irregular water and limited food sources. A cosmology connected people to the land: It was dominated by an emphasis on ancestral traces in the country, the flattening of time between the time of the ancestors and the present, and the importance of specific land features, such as waterholes, watercourses, trees, hills, and gorges as both vital resources and totemic sites. Space rather than time was the predominant axis of cosmological meaning. The cosmology bound people to the land in intricate webs of meaning that also sustained long-term economic usage. According to traditional stories that have been recorded, the landscape was shaped by ancestral figures. Early anthropologists referred to this as the Dreamtime.

The initial encounter between the Aborigines and white settlers was one of mutual unintelligibility. There were fundamental differences in how they viewed land and what land ownership involved. There was a basic conflict over land and resources. There was also a huge north (Figure 12.7). A famous poem by Dorothea Mackellar sums it up:

> I love a sunburnt country,
> A land of sweeping plains,
> Of ragged mountain ranges
> Of droughts and flooding rains.

Later in the poem "My Country," first published in 1908, she referred to the "wide brown land."

After Antarctica, Australia is the driest continent on the planet, where rainfall in the interior trails off into insignificance. Paradoxically, it is also a land of flooding, as flash floods quickly overwhelm river channels.

The country is subject to drought. The worst occurred from 1997 to 2009, and some scientists suggest that global climate change may be a contributing factor leading to longer, drier, hotter summers. There are also regional processes at work. In El Niño years, such as 2015, less rain falls in Australia. The longer term trend is toward drier conditions. The dry conditions affect both rural and urban areas. The long drought reduced water supply in the farming districts along the Murray-Darling river basin, making water more expensive. The small-scale farmers were especially vulnerable to financial pressures. The drought also impacted the cities. Perth's severe water shortage prompted the construction of a desalinization plant that now provides 17 percent of the city's water supply. The other major cities also look to desalinization as a way forward in a drier future.

The dry climate over much of the year provides the ideal conditions for wildfire, known in Australia as bushfire. Bushfires are especially common in the summer months from November through March. Dry vegetation catches fire, sometimes set deliberately, and hot, dry winds drive them across the sunburnt landscape. Black Saturday 2009 in Victoria was one of the worst fire events with loss of life, property, livestock, and wildlife. One hundred and seventy-three people died, and thousands lost their homes.

A Resource-Based Economy

Australia's wealth is based on a resource-based economy that was originally part of the British imperial systems. Wheat, wool, and gold were important, but now the dominant primary exports are coal and minerals. Since the 1990s, the economy was turbo-charged with mining-led growth and a resource-driven economy. Iron ore is the country's most valuable export.

This long commodity boom meant that Australia did not experience the global financial crisis of 2008–2009 and the great recession to the same extent as other rich countries. It meant that Australia got richer, both relatively and absolutely. In 1990, Australians' living standards were on average roughly around 90 percent of US levels; by 2012, it was closer to 115 percent. However, there are signs that the long commodity book is coming to an end as China's breakneck growth falters. In 2014, there was a 30 percent fall in iron ore prices, 15 percent fall in coal, and 20 percent fall in wheat. Weakening demand from China and the drop in commodity prices meant an increase in the federal deficit. The federal government plans to cut spending over the next 4 years by half a billion dollars.

Multicultural

For decades there were two main social groups: Aborigines, who were marginalized from the politics and culture of the nation, and the dominant Anglo-Celtic settler group. Immediately after World War II, the immigrant catchment area widened to include southern and Eastern Europe. However, a whites-only immigration policy lasted until the 1970s. In the past four decades, there were two

12.7 The dry interior of Australia (photo: John Rennie Short)

main changes. The first is that Aborigines have achieved greater cultural prominence and political rights. In 2008, a Sorry Day was declared as an act of national apology for what white Australia did to Aborigines.

The second change is that Australia is now a more multicultural country. The country has one of the highest immigration rates for a developed country, with 25 percent foreign born. At least 46 percent of the population has at least one foreign-born parent. The waves of immigrants have changed in composition from British and Irish to more Greeks, Italians, and Slavs in the 1950s to more Vietnamese in the 1970s and Indians and Chinese from the 1980s. Today migrants come from China, India, and the Middle East.

An Urban Nation

Australia is one of the most urbanized countries in the world. Close to 9 out of every 10 Australians live in cities. Australians tend to live in big cities with close to 15.6 million of the almost 25 million living in just five cities—Melbourne (5 million), Sydney (5 million), Brisbane (2.3 million), Perth (2 million), and Adelaide (1.3 million) (Figure 12.8).

Australian urbanization is best understood as a collection of urban primates. Each of the cities is the main urban center and capital of the five most populous states, New South Wales (Sydney), South Australia (Adelaide), Victoria (Melbourne), Queensland (Brisbane), and West Australia (Perth). Before the federation of Australia in 1901, each of the states was a separate colony organized around a major coastal city. They were the cultural, economic, and political centers of their respective states.

The heavy urban concentration comes with heavy congestion costs, traffic jams, and lack of affordable housing. The cost of congestion is estimated to be around $6 billion in 2020 for Sydney and $5 billion for Melbourne. There are strong agglomeration economies at work in these cities with central and inner-city locations at a premium. In Sydney the dwelling prices can regularly reach to over five times the average income, a figure that leads to a huge debt encumbrance. Existing property owners receive windfall gains while those trying to move into the housing market for the first time and without housing equity have a much harder time. The urban housing markets are also subject to severe cycles. Cheap interest rates and high demand encourage developers to build new apartments and houses;

12.8 Brisbane (photo: John Rennie Short)

this leads to a bubble, with prices collapsing and then rising again in the next cycle.

For those in Sydney's inner suburbs, 40 percent of available jobs in the city are within 30 minutes travel distance: in the outer suburbs, less than 5 percent of jobs are within that travel range. I was reminded of this in 2015 when I spoke to a young woman who had migrated from the Punjab in India. She lived in the outer suburbs of Bankstown, and it took her over 2 hours to work at a central city hotel on the 8 a.m.-to-2:30 p.m. shift. She worked a second job in a local restaurant until 10 p.m. Amazingly, she found time to take courses at a business college.

Although very unequal, the overall quality of life in Australian cities is still high compared to many other cities in the world. According to a 2015 ranking by *The Economist*, the ten most livable cities include Melbourne (1), Adelaide (5), Sydney (6), and Perth (7). Also mentioned was Auckland (8) in New Zealand and, in order, Vienna, Vancouver, Toronto, Calgary, Helsinki, and Zurich. The least livable cities were Tripoli, Lagos, Port Moresby, Dhaka, and Damascus.

Cities in Australia, as well as New Zealand, are some of most livable cities in the world.

City Focus: Sydney

Sydney, with a metropolitan population close to 5 million, is the largest city in Australia and the wider region. The area around the coast was the home of indigenous people for thousands of years before Captain Cook landed at Botany Bay in April 1770. The British claimed the territory. After the Declaration of Independence by the United States, the British had nowhere to dump their rising number of prisoners. Why not send them to the other side of the world, to establish a distant colony? In January 1798, eight ships with 850 convicts landed at Botany Bay, guided by Cook's reports. In less than 3 weeks, they moved from the coastal settlement to the sheltered harbor of Sydney Cove. More than four thousand convicts were shipped to the fledgling community over the next 4 years. There was no need to build a prison; it was an open-air gulag.

The Antipodean gulag soon became a thriving port as the interior of Australia was opened up to mining and farming. Wool and coal, and timber and wheat were shipped through the city, and it emerged as the political and economic center of the state of New South Wales. It was the primate city of the state. By 1900, the population reached close to half a million. The population grew with immigration from Britain and Ireland.

The city grew after World War II as migrants were attracted to the economic opportunities. The lack of labor meant that workers' wages and conditions were some of the highest in the world. The predominantly Anglo-Celtic migrant stream was widened as migrants from southern Europe, especially Greece and Italy, and the Balkans came to the city.

Always in competition with Melbourne, the capital city of Victoria, Sydney gradually became the preeminent global city for Australia as multinational companies and foreign banks headquartered in the city, with the beautiful beaches and warm sunny climate part of the attraction. The opening of the Sydney Opera House gave the city an internally recognized iconic building, something that Melbourne lacks. When Sydney hosted the Summer Olympics in 2000—Melbourne hosted the much less publicized Summer Games in 1956—it stamped the city's iconography on global consciousness.

Recent years have seen significant changes. The Olympics site was held in a polluted inner-city area, Homebush Bay. There is substantial gentrification of the inner suburbs. Suburbs such as Paddington are now an important part of the city's gay scene. The most expensive housing is closer to the beach with lower income groups living in the outer suburbs. The closeness to the beach is an indicator of wealth and status in the city. The old dock areas have also been refurbished as entertainment complexes. More than 1.5 million people were born overseas with increasing numbers of immigrants from Asian countries. More than a third of Sydneysiders can speak a language other than English.

The Sydney Opera House opened in 1973 (Figure 12.9). It replaced an old tram shed. It sits on a spit of land protruding into Sydney Harbor. The Danish architect Jorn Utzon won the design competition in 1957. Although there are significant changes from Utzon's original design, especially in the internal layout—one of the many difficulties leading the architect to resign from the project—his basic idea for a building that references its maritime location remains wonderfully intact. The roofs, which echo sails as well as ocean waves, are works of art as well as building forms. I have visited the Opera House many times, and it never fails to evoke a sense of wonder and awe at its form. Sitting in a prominent position, surrounded on three sides by water, it is one of the great works of art of the twentieth century, a building so beautiful and emblematic that its roof outline now signifies the whole city, if not the entire country.

Rural Focus: The Murray-Darling River Basin

The Murray-Darling River basin is a large area in southeast Australia that covers over 400,000 square miles and sprawls across five states (Figure 12.10). It is an important rural area responsible for almost 40 percent of Australia's agricultural output. Farming is dependent on irrigation by the river waters under the control of a commission. Over a hundred years, 240 dams were built and water

12.9 Sydney Opera House (photo: John Rennie Short)

12.10 The Murray-Darling River Basin (photo: John Rennie Short)

was distributed without much concern for environmental impacts. In the twenty-first century, a long drought created problems of increasing salinity and threatened the long-term health of the river system. Under a management plan, announced in 2010 and passed in 2012, a cap was placed on water withdrawals. Many residents of the small towns in the region, farmers, and wine growers argue that the new allocations could undermine farming in the region. Since the plan's implementation, there was also conflict between two upstream states, New South Wales and Victoria, who wanted less water reduction than downstream South Australia, where the problems of salinity were more severe. The plan has also been criticized for poor enforcement, especially in the crucial upper reaches, and lack of transparency about the allocation of water-drawing rights.

Despite the problems in implementation, the plan at least represents a response to a growing environmental crisis. Much of rural Australia is getting hotter and drier,

creating major environmental problems such as water availability and water quality.

New Zealand

New Zealand is often linked with Australia. And there are similarities: both are settler societies dominated by Anglo-Celtic populations, both have English-speaking parliamentary democracies, and both are still linked to the Commonwealth with shared cultural values and sporting traditions, such as rugby and cricket.

But there are differences. New Zealand was settled by Polynesians, called Maoris (whereas Australia was originally settled much earlier by Melanesians), whose chief-based, militarized society provided very stiff resistance to British domination. The Treaty of Waitangi signed in 1840 recognized Maori land claims, in principle if not always

NEW ZEALAND
City Population
⊛ ○ Below 2 million

North Island

South Island

SOUTHERN ALPS

Tasman Sea

PACIFIC OCEAN

Cook Strait

Stewart Island

Whangarei
Takapuna
Auckland
Manukau
Hamilton
COROMANDEL PENINSULA
Tauranga
Rotorua
Lake Taupo
Taupo
Gisborne
New Plymouth
Napier
Wanganui
Hastings
Palmerston North
Nelson
Wellington
Christchurch
Timaru
Queenstown
Oamaru
Dunedin
Invercargill

12.11 Map of New Zealand

colony. It was part of the British imperial trading system, providing wool and dairy produce to the UK home market. When the United Kingdom joined the European Economic Community in 1973, New Zealand had to find new markets. New Zealand became an example of the shock doctrine of neoliberalism in the 1990s with major cuts in social spending and a greater reliance on the private market. The economy has become less dependent on agriculture and on the British market. It now sells goods and produce throughout the world.

As with Australia, the indigenous people, the Maori, resisted marginalization and in recent years there has been a more vigorous expression of Maori identity and culture. And also similar to Australia, the ethnic mix has changed with more migration from Asia.

Auckland is the largest city in the country with close to 1.5 million people.

As part of a discussion about national identity in a changing world, New Zealand held a national referendum about the national flag. The current flag has the Union Jack. The debates have been ongoing since the end of World War II with many critics arguing for a new flag to better represent an independent sovereign nation. In a referendum in 2015/2016, a small majority of 56 percent voted to keep the present flag.

in practice. There is no comparable founding document in Australia, where the indigenous people were not considered a sovereign nation.

Also, geologically, New Zealand is younger than Australia and is largely of recent volcanic origin.

The country has two main islands, conveniently named after their relative locations, North Island and South Island (Figure 12.11). The spine of the South Island is formed by the Southern Alps, providing a majestic glaciated scenery. Most of New Zealand is precariously located on the Pacific Ring of Fire and subject to earthquake activity. An earthquake struck Christchurch in 2011, measured at 6.3 on the Richter scale, with a death toll of 180.

New Zealand was colonized by the British and formally controlled from 1840. In 1907, it became a self-governing

Oceania

The total population of the island nations in Oceania, excluding Australia and New Zealand, is only around 2.3 million, with people living on remote islands far from the main centers of the world. This distance has allowed the creation in the popular imagination of the South Seas as a tropical paradise far from the cares of the modern world. This vision has haunted explorers, visitors, and tourists since the seventeenth century. The reality is, of course, very different.

Islanders face problems and opportunities different in detail because of a small population with limited resources living on remote islands, but similar in broad character to the rest of the world.

Subsistence agriculture still plays an important role in many of the islands. Economic integration into the wide global economy has occurred through exploitation of natural resources and plantation economies, both of them emerging with colonial annexation, and more recently by tourism.

Oceania can be divided up into the three different subregions of Melanesia, Micronesia, and Polynesia, a division based on the initial settling of the region and associated genetic and language differences.

Melanesia

Melanesia consists of Vanuatu, the Solomon Islands, Fiji, and Papua New Guinea, which we considered in the chapter on South East Asia (Chapter 8) because it shares an island with Indonesia (Figure 12.12).

Vanuatu has a population of 288,000 spread over sixty-five islands. More than 47,000 people live in the capital city of Port Vila. Climate change leading to rising sea levels threatens this island nation. Coastal settlements are moving to higher ground to escape being washed away. The island is very vulnerable to cyclones. Cyclone Pan in 2015 caused tremendous devastation. Poor and isolated island states, such as Vanuatu are especially at risk from these events.

The Solomon Islands is an island chain located on an active tectonic fault subject to volcanic eruption and earthquakes. It was a British possession, previously called the British Solomon Islands to reinforce the colonial point. The name was changed just before the country became independent in 1978. The islands were first drawn into a wider commercial orbit, when sailors would land on the island to capture men and ship them off to work in the sugar plantations of Fiji and Queensland. Coconut plantations were established on the islands by foreign investors. A chemical company exported them to its factory in Sydney. There were few benefits to the locals, a case typical of early plantation agriculture in the South Pacific. Most locals still carry on with traditional farming enabled by the explicit reservations of land for Solomon Islanders. Three-quarters of the population is engaged in subsistence farming and fishing. The tropical forest is dangerously exploited with all accessible forest likely to be cut down by 2030. The forests are clear felled and whole logs are shipped to Japan. Mineral resources have yet to be fully developed, but most economic development is hindered by the lack of infrastructure and associated transport difficulties. Inward foreign investment was much reduced during the civil war that lasted from 1998 to 2003.

The original settlers of Fiji were Melanesian. They occupied a third of the 332 islands that make up Fiji, although now 9 out of every 10 people live in either one of the two main islands, Viti Levu and Vanua Levu. The capital city of Suva has a total population of 77,000. In 1874, the islands came under the control of the British Empire. Sugar plantations were established, but the Fijians still retained most of their land, unlike the dispossession in Hawaii. The British imported laborers from India; the first arrived in 1879. Working conditions were deadly but eventually some escaped from the bonds of indentured servitude. Without access to the land available to native Fijians, they became shopkeepers, traders,

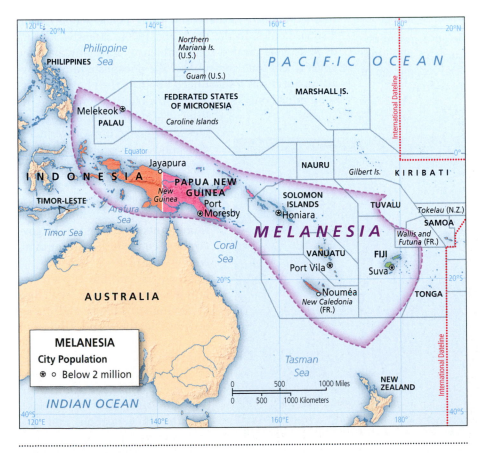

12.12 Map of Melanesia

and business people and a central part of the commercial economy of the country. By 1996, just over half the population of Fiji was Indo-Fijian. The city of Nadi, population of close to 43,000 and major hub of the tourist area as it is close to the international airport, has the largest Hindi temple in the southern hemisphere. In 1970, the island nation became independent from the United Kingdom. With a total population close to 900,000, Fiji is one of the more affluent sovereign states in the South Pacific though poor infrastructure limits development. In 2013, the Constitution guaranteed access to safe and clean water.

It is well endowed with primary resources, including fish, forest, and minerals. Sugar and coconut plantations dot the islands (Figure 12.13). The main exports and sources of foreign exchange are sugar, mineral water, timber, textiles, and food products, such as coconuts and processed fish.

Postindependence politics has been a series of democratic governments disrupted by military coups. The Army is traditionally staffed by Melanesian Fijians who try to limit the power of the Indo-Fijians and enhance the power and status of native Fijians. Since 2000, more Indo-Fijians have left the country and now they constitute just over a third of the total population. The constant civil unrest undercuts the tourist economy as it is a long-haul destination and so easily substituted for closer and safer destinations if there is civil unrest.

In 2015, in order to heal the rifts, the government declared a new national motto "Togetherness in Harmony for a Prosperous Fiji." It was both an official recognition of ethnic conflict and an attempt to move beyond it.

12.13 Plantation workers harvesting sugar cane in Fiji (photo: John Rennie Short)

Micronesia

Micronesia consists of over 2,100 islands studded across the South Pacific (Figure 12.14). There are coral islands, atolls, volcanic islands, and low-lying islands. The Caroline Islands consist of approximately 500 small coral islands; the Gilbert Islands are sixteen atolls and coral islands; and the Mariana Islands are volcanic islands. The Marshall Islands are atolls and low-lying islands. All the islands are vulnerable to cyclonic storms and sea level rise. They all share the similar problems of remoteness and hence high transport costs for basic goods.

There are two different forms of political territory. The sovereign states include the Federated State of Micronesia, Kiribati, the Marshall Islands, Nauru, and Palau. Then there are the three US territories of Guam, the Northern Marianna Islands, and Wake Island. The United States also exerts major influence across the region. The Marshall Islands and Palau, for example, have liminal status between complete independence and totally controlled by the United States.

The total population of Micronesia is 536,000, and the population ranges from 167,000 in Guam to 11,000 in Nauru.

The problem for the small sovereign states, lacking US support, is highlighted by the example of Nauru, the third smallest country in the world, where 11,000 people live on 8 square miles. In the 1960s and 1970s, extensive mining of phosphate created some affluence, especially after the government took control over the phosphate mining royalties, but it was a short-lived boom as the phosphate was soon mined out. The country now requires foreign aid. Its difficult economic position generates a variety of innovative financial policies. It operates as a tax haven and more surreptitiously as a center for money laundering. It runs detention centers for the Australian government to house illegal immigrants to Australia. The government employs 95 percent of all employed people.

There is low life expectancy, only 60 years for men, with one of the highest obesity and diabetes rates in the world. The poor health outcomes are a result of the shift away from the traditional diet of fish and coconut to imported processed foods that are high in salt and sugar, lack of exercise, poor health education, and cultural norms that associate girth with wealth. This is an issue prevalent across the South Pacific.

Nauru also sells political recognition, swapping loyalties between the Republic of China, also known as Taiwan, and the People's Republic of China in return for aid money and for recognizing Abkhazia, the breakaway region of Georgia in return for Russian aid.

Nauru is surrounded by ocean, and the supply of fresh water is a major problem, forcing the government to construct three desalinization plants.

Polynesia

Imaginary lines that connect the triangular points of New Zealand, Hawaii, and Easter Island define the territory of Polynesia. It consists of over 1,000 islands; most of them were built by the volcanic activity of hotspots rather than tectonic plates, as in the case of Samoa and the Hawaiian Islands (see Figure 12.14).

The colonial legacy lives on with French control over French Polynesia, Chile's possession of Easter Island, and US control over Hawaii, now a state, and American Samoa, effectively a US colony. Other island nations, including the Cook Islands, Niue, Norfolk Island, Pitcairn Island, and Tokelau, are "in association" or are formal territories of Australia, Britain, or New Zealand. Only Samoa, Tonga, and Tuvalu are sovereign states.

Samoa, with a total population now close to 199,000, was fought over among the United States, Britain, and Germany. In a treaty signed in 1929, the eastern island came under the control of the United States, today present-day American Samoa, while the island in the west came under German, British, and then New Zealand control. It became an independent country is 1962. So now we have an independent Samoa and the unincorporated US territory of American Samoa.

Like other Pacific islands, the people have to be creative in order to survive in a global economy. In 2011, Samoa "jumped" a day when it moved west of the International Date Line, thus making it 3 hours ahead of Sydney time rather than 21 hours behind. Despite attempts to diversify, the economy is still dependent on foreign aid; remittances from Samoans working overseas, especially in New Zealand; and agricultural exports, including coconut products and copra.

EASTER ISLAND

Three extinct volcanoes form the structure of Easter Island, a remote island in the South Pacific at the farthest edges of this world region; it is so far east that it is a territory of Chile.

Easter Island is an ecological morality tale. It is the very opposite of sustainable development: destructive

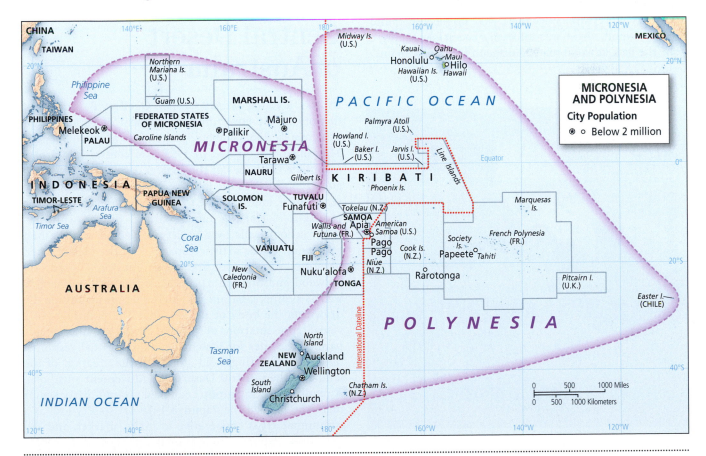

12.14 Map of Micronesia and Polynesia

development made worse by the inability to leave or find other resource alternatives and a social system that encouraged conflict rather than cooperation. It is a lesson for the global society.

It was first settled by Polynesians between 1,000 and 1,500 years ago, relatively recent by global standards, because people had to travel so far to land on a tiny prick of earth in a vast ocean.

The islanders built 887 moai monumental head statues that sit enigmatically in the landscape (Figure 12.15). They are massive, the largest 33 feet high and weighing 82 tons. Most of them faced away from the ocean, looking toward clan lands and perhaps guarding the villages. A recent study suggests that the location of freshwater was the key to the location of the statues. The construction, transportation, and erection of these massive monuments required sophisticated social organization and vast resources of food and wood.

The island was forested when the first people arrived, but between 1400 and 1600 the island was laid bare as the people cut down all the trees. Deforestation was exacerbated by rats that had no predators and ate the palm trees which were the habitats of birds whose guano had fertilized the soil. No trees meant no birds and hence declining soil fertility. A slash-and-burn form of agriculture in association with the limited space and declining fertility added extra pressure. Since there were no reefs surrounding the island, there was an easily accessible supply of fish. The lack of wood made it difficult for people to build boats to collect fish, a valuable form of protein. The lack of food created more social conflict. It became a society in ecological distress. The islanders had committed a form of ecological suicide by destroying their resource base. This self-inflicted environmental damage was facilitated by the environmental conditions of a dry small remote island without soil nutrient replacement. As food became scarce, the remoteness of the island made escape impossible; clan tensions were heightened and led to even more statue building and more destruction of the forest. When Europeans visited the island in the eighteenth century, they found fewer than 2,000 people where once there were up to 15,000.

Subsequent history was not kind to the islanders. Slave raiders took away much of the population in the 1860s, and visitors brought small pox, devastating those left on the island. Diseases from visiting whalers further reduced the population. By 1877, there were only 111 people living on the island. The territory was annexed by Chile in 1888. A local movement continues to lobby the Chilean government for indigenous land rights.

Focus: Art of the Central Desert of Australia

In 1971, a young Sydney teacher, Geoffrey Bardon, went to work in the government settlement of Papunya, approximately 250 kilometers northwest of Alice Springs/Mparntwe, established in 1959 as a central location to assimilate the tribes of the central desert. He developed good rapport with the local people, some influence with senior men, and was able to encourage and promote the production of art. A mural painted on the white exterior school walls by a variety of people was the first expression of the artistic capacities of the locals. Later, paintings were made, first on composition board and plywood and then on canvas, using natural pigment as well as synthetic paints, employing designs traditionally drawn on the body or in the sand. The paintings were visual representations of spiritual geography since people painted the sites for which they had spiritual responsibility. The paintings told stories of the creation of the land and the **Central Desert Art Movement** was born. A regular painting group coalesced around thirty senior men who formed the Papunya Tula Artists Pty Ltd. as a communal artist cooperative to make, market, and sell paintings. Papunya Tula annual art sales now average close to $A5 million. The success of Papunya was a model for other communities.

12.15 Moai statues in Easter Island (Aurbina/Wikimedia Commons)

Federal and state governments also stimulated the Aboriginal cultural economy as a way to develop the tourist industry and generate employment opportunities for indigenous communities. The Aboriginal Arts Board was established in 1973 to encourage and foster indigenous arts and crafts. Arts centers were established in many remote indigenous communities, and art advisors were employed to help in promoting, selling, and marketing art produced in the local community.

Arts centers have expanded, from sixteen in 1980 to just over one hundred, and there are now between six thousand and seven thousand indigenous people making art and artifacts for sale. Art sales now form a significant source of revenue for Aboriginal communities throughout Australia and especially in central Australia. Some of the art is sold directly from the art centers, but much of it is funneled through dealers and galleries (Figure 12.16).

The emergence of this Central Desert Art Movement influences the character and form of a new national imaginary and shapes a new global representation of Australia as a country with an indigenous presence. It is now possible to buy "Aboriginal" motifs in a range of objects from scarves to screen savers, from carpets to dishcloths, and from t-shirts to sweaters. Qantas cabin crews have ties, skirts, and dresses with "Aboriginal" designs drawn from the motifs of indigenous artists from the central desert region. This is part of a wider appropriation of Aboriginal art motifs in the material and cultural representation of Australia. Major banks cover their ATM machines in desert dot painting motifs; the Australia Post issues stamps with the image of paintings from Papunya Tula. Corporate headquarters in Melbourne and Sydney, Parliament buildings in all the states, and federal government offices in Canberra display prominently the work of Aboriginal artists, and Australian embassies around the world show the work of indigenous artists in permanent installations

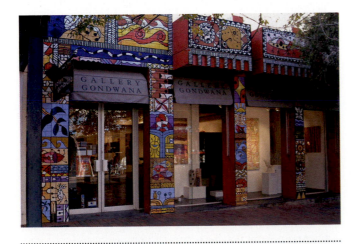

....
12.16 Aboriginal Art Gallery in Alice Springs/Mparntwe (photo: John Rennie Short)

and occasional exhibitions. Aboriginal aesthetics inform the national identity through its appearance in contemporary design, public space, and national representations in the wider world.

Focus: Obesity in the South Pacific Islands

This region holds a dubious record: eight of the top-ten countries in the world with the highest obesity rates are in the South Pacific. In order, they are as follows: Nauru, the Federated State of Micronesia, the Cook Islands, Tonga, Niue, Samoa, Palau, and Kiribati. Over 94 percent of Nauru's population is considered overweight and 73 percent in Kiribati. The United States is ninth on the list at 74 percent.

What are the reasons? Diet is the main culprit. The traditional diet of fresh fish and fruit was replaced by imported food that is high in sugar and salt. Some researchers suggested that colonial settlers who tried to teach "proper" food habits as a form of civilizing the "natives" first started the shift. Traditional meals were considered old fashioned. Purchasing and eating imported food became a sign of social status. In small island communities, the adoption of new food norms by influential figures had an immediate effect. Traditional knowledge of food collection and preparation was soon lost while the old lifestyles dominated by physical labor and high calorific expenditure were replaced by more sedentary lifestyles and greater net calorific intake. There is also cultural bias toward the heavy, with traditional cultural emphasis on girth as a sure sign and symbol of wealth and prestige.

The food supply in the remote Pacific islands is limited in choice, with emphasis on the prepackaged and processed. Typical diets across the island nations are now high in salt and sugar with limited nutritional value. It is often difficult to get fresh fruit on tropical islands in the South Pacific! Nearly all food in Nauru is imported from Australia, Japan, and New Zealand, and the average daily caloric intake for males aged 20–39 years is a staggering 8,700 calories. The recommended daily intake for men of this age is around 2,400.

One argument is that body types differ and so a standardized measurement of obesity works against bigger and heavier bodies that are more common in the South Pacific. However, the recent rates of growth of body mass suggest that obesity is real and not just a function of applying a one-size-fits-all measurement gauge. On both Nauru and the Cook Islands between 1980 and 2008, the increase in average body mass index was four times higher than the global average. The effects are becoming more obvious, with marked increases in diabetes and heart disease and mounting pressure on health services.

Select Bibliography

Armitrage, D., and B. Ashford. 2014. *Pacific Histories: Ocean, Land, People.* New York: Palgrave Macmillan.

Arthur, W., and F. Morphy. 2005. *Macquarie Atlas of Indigenous Australia: Culture and Society Through Space and Time.* New South Wales: Macquarie Library.

Barnett, J., and J. Campbell. 2010. *Climate Change and Small Island States: Power, Knowledge and the South Pacific.* London: Earthscan.

Beer, A. 2012. "The Economic Geography of Australia and Its Analysis: From Industrial to Postindustrial Regions." *Geographical Research* 50:269–281.

Blainey, G. 1975. *The Tyranny of Distance: How Distance Shaped Australia's History.* Sydney: Macmillan.

Broome, R., and R. Broome. 2010. *Aboriginal Australians.* Crows Nest, NSW: Allen & Unwin.

Callaghan, P., and S. Hendy. 2013. *Get Off the Grass: Kickstarting New Zealand's Innovation Economy.* Auckland: Auckland University Press.

Cronkite, E. P., R. A. Conard, and V. P. Bond. 1997. "Historical Events Associated with Fallout from Bravo Shot-Operation Castle and 25 Years of Medical Findings." *Health Physics* 73:176–186.

Dibblin, J. 1990. *Day of Two Suns: US Nuclear Testing and the Pacific Islanders.* Lanham, MD: Rowman and Littlefield.

DiNapoli, R. J., C. P. Lipo, T. Brosnan, T. L. Hunt, S. Hixon, and A .E. Morrison. 2019. "Rap Nui (Easter Island) Monument (*ahu*) Locations Explained by Freshwater Sources." *PLoS ONE* 14 (1): e0210409. https://doi.org/10.1371/journal.pone.0210409.

Enright, M. J., and R. Petty. 2013. *Australia's Competitiveness: From Lucky Country to Competitive Country.* Hoboken, NJ: Wiley.

Fisher, D. 2013. *France in the South Pacific.* Canberra: ANU Press.

Flood, J. 2006. *The Original Australians: Story of the Aboriginal People.* Crows Nest, N.S.W.: Allen and Unwin.

Gammage, W. 2011. *The Biggest Estate on Earth: How Aborigines Made Australia.* Sydney: Allen & Unwin.

Garden, D. S. 2005. *Australia, New Zealand, and the Pacific: An Environmental History.* Santa Barbara, CA: ABC-CLIO.

Hanlon, D. L. 1998. *Remaking Micronesia: Discourses over Development in a Pacific Territory, 1944–1982.* Honolulu: University of Hawai'i Press.

Hawley, N. L., and S. T. McGarvey. 2015. "Obesity and Diabetes in Pacific Islanders: The Current Burden and the Need for Urgent Action." *Current Diabetes Reports* 15:1–10.

Hughes, T., et al. 2018. "Global Warming Transforms Coral Reef Assemblages." *Nature* 556:492–496.

Keen, I. 2004. *Aboriginal Economy and Society: Australia at the Threshold of Colonisation.* Melbourne: Oxford University Press.

Keneally, T. 2006. *A Commonwealth of Thieves: The Improbable Birth of Australia.* New York: Doubleday.

Keppel, G., C. Morrison, J. Y. Meyer, and H. J. Boehmer. 2014. "Isolated and Vulnerable: The History and Future of Pacific Island Terrestrial Biodiversity." *Pacific Conservation Biology* 20:136–145.

Kumar, L., and S. Taylor. 2015. "Exposure of Coastal Built Assets in the South Pacific to Climate Risks." *Nature Climate Change* 5:992–996. http://www.nature.com/articles/nclimate2702.

Luthy, R. G., and D. L. Sedlak. 2015. "Urban Water Supply Reinvention." *Daedalus* 144:72–82.

McKnight, T. L. 1995. *Oceania: The Geography of Australia, New Zealand, and the Pacific Islands.* Englewood Cliffs, NJ: Prentice Hall.

Molloy, M., and W. Larner. 2013. *Fashioning Globalisation: New Zealand Design, Working Women and the Cultural Economy.* Chichester, UK: Wiley-Blackwell.

Moore, C. 2004. *Happy Isles in Crisis: The Historical Causes for a Failing State in the Solomon Islands, 1998–2004.* Canberra: Asia Pacific Press.

Nuclear issues in the South Pacific: Hearing before the Subcommittee on Asia and the Pacific. 1995. Committee on International Relations, House of Representatives, One Hundred Fourth Congress, first session, November 15, 1995, vol. 4.

Rapaport, M. 2013. *The Pacific Islands: Environment and Society.* Honolulu: University of Hawaii Press.

Schwarz, A.-M., et al. 2011. "Vulnerability and Resilience of Remote Rural Communities to Shocks and Global Changes: Empirical Analysis from Solomon Islands." *Global Environmental Change* 21:1128–1140.

Suarez, T. 2004. *Early Mapping of the Pacific: The Epic Story of Seafarers, Adventurers and Cartographers.* Hong Kong: Periplus Editions.

Weller, R., and J. Bolleter. 2012. *Made in Australia: The Future of Australian Cities.* Crawley: UWA Publishing.

Zhang, Y. 2003. *Pacific Asia: The Politics of Development.* London: Routledge.

Learning Outcomes

This region consists of Australia, New Zealand, and a number of island chains that stretch across the vastness of the Pacific Ocean.

There is a range of landscape—from the weather-eroded shield of central Australia, the active volcanic landscape of New Zealand, to the beautiful coral islands in the tropical waters.

The region is particularly vulnerable to climate change because of the large number of endemic species, low-lying coastlines, and the track of cyclones.

The region was peopled in the following order: Melanesia, Micronesia, and Polynesia.

Much of the region came under the dominance of the European powers and the United States. While many countries have achieved independence, many are still under informal control.

The islands of the South Pacific were sites for the testing of nuclear explosions with devastating results.

Australia is by far the largest country by both area and population in the region. A long resource boom has raised standards of living beyond levels in the United States. It is one of the most urbanized nations in the world, with the majority living in just five large cities.

Indigenous people in Australia and New Zealand are becoming more potent political forces, changing debates about national identity, history, and land rights.

New Zealand has shifted its export bias from the United Kingdom to a more global market.

The small island nations in the Pacific are especially vulnerable to climate change because of their location, coastal orientation, and lack of finance for adaptation.

They are also economically vulnerable because of their size, location, and lack of economic alternatives in a global economy.

13

North America

North America consists of two large countries of continental proportions: Canada and the United States (Figure 13.1). They both extend from the Atlantic to the Pacific. While Canada has a large territorial extent, its relatively small population is largely restricted to a 100-mile zone north of the border with the United States. The United States has the largest economy in the world and a huge military with global reach. Across the region there are marked regional differences from the frozen tundra of northern Canada to the subtropical metropolis of Miami and from dynamic metro economies to declining industrial regions.

LEARNING OBJECTIVES

Describe the region's varied physical landscapes and distinguish environmental challenges as geologic or anthropogenic hazards.

Explain the region's pre-Colombian settlement and impacts of the Colombian Encounter, and then discuss development of European settler societies.

Survey the region's shift from primary production, to industrialization, and then to a postindustrial economy.

List the particular dangers associated with commercial agriculture in the rural United States.

Examine the region's mobile, aging population and its immigrant experience and relate these to its dynamic gender, race/ethnicity, and class relations.

Recognize the region's metropolitan and suburban forms and identify evident social differences and unhealthy environments.

Outline the region's three border areas and their geopolitical tensions and evaluate the role of the United States as a global superpower.

The Environmental Context

This region consists of landscapes from across the continuum of geologic time. The Canadian Shield, a plateau of eroded granite rocks, is the oldest part of the continent created over 600 million years ago. Then there are more recent tectonic activities. The Appalachian chain was formed between 400 million and 225 million years ago. The Rockies, a mountain chain over 500 miles long, was formed 80 to 40 million years ago by the Pacific Ocean Plate, slowly moving east, pushing under the western-moving North American Plate. These plates are still on the move, their active boundary the scene of earthquakes and volcanic activity.

During the ice ages of the past 2 million years, glaciers, both in advance and in retreat, sculpted the landscape, especially in the northern and more mountainous parts of the continent, leaving behind lakes and moraines. Since the end of the last Ice Age, rivers have deposited sediment in the great central basin of the Mississippi and in other riverine basins and along the coasts. Land is still being created by sediment depositing at the mouth of the Mississippi.

In the center of the continent the interior lowlands raise slowly west to the Great Plains and stretch right up to the foothills of the Eastern Cordillera.

Cardinal Physical Geographies

To move from north to south across this vast continent is to move from the very cold to the warm, from the frigid tundra to the balmy semitropical. The average high January temperature in Echo Bay in the Northwest

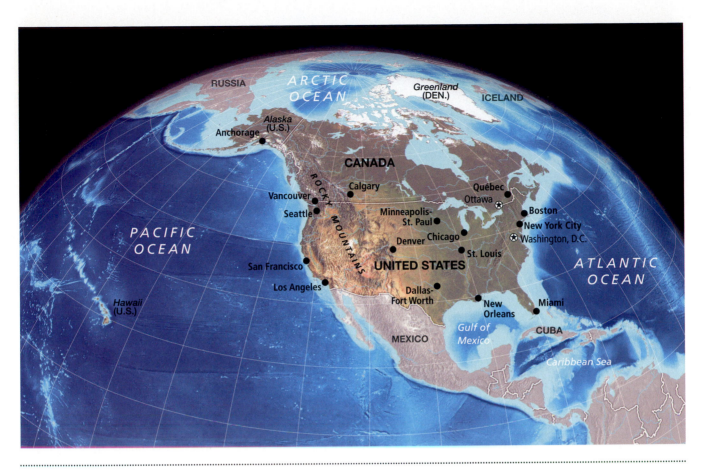

13.1 Map of region

Territories in Canada is –5.8°F. In the same month, Miami Beach in southern Florida reaches an average high of 73°F. No surprise then that **snowbirds** from Canada and the northern United States constitute a significant part of the winter migration to Florida.

There are corresponding shifts in vegetation cover moving from north to south (Figure 13.2), from the barren tundra through the temperate rainforest to the lushness of the subtropical (Figures 13.3 and 13.4).

To move from east to west is to move from a continental climate, with hot, humid summers and cold winters, especially marked in the interior, to the more temperate climes in the west. This broad east-west geographic categorization is influenced by altitude. Winter temperatures in the high Rockies can plummet. And by latitude: Southern California has a hot, dry Mediterranean climate while the cool and wet conditions in coastal Alaska are ideal growing conditions for temperate rainforests.

There are significant desert areas that receive less than 10 inches of rain each year. There are cold deserts such as the Columbian Basin and hot deserts such as the Arizona-New Mexico Plateau and, along the US southwest border with Mexico, the Sonoran and Chihuahua Deserts (see Figure 13.2).

Weather patterns are influenced by the trajectory of the **jet stream**, a fast-moving, narrow band of high-altitude wind at 30,000 feet above sea level. A more southerly jet stream in winter brings the harsh cold from the tundra to the Midwest and Eastern Seaboard while the same trajectory in summer brings much needed coolness to hot and humid parts. During El Nino events, the jet stream takes a more southerly track, bringing more snow to the Rockies. In a La Niña year, the jet stream is pushed further north, bringing more snowfall to the Midwest during winter and hot, dry weather during the summer.

Environmental Challenges

The continent is shaken and torn by seismic activity, especially along the west coast, where the Pacific Plate is moving east and sliding over the North American Plate. There are volcanic eruptions and earthquakes along this plate boundary. The **San Andreas Fault** in California is a major fault line. Further north is the **Cascadia subduction zone** that runs 700 miles off the coast of the Pacific Northwest. A rupture along this zone will cause a major natural disaster affecting more than 7 million people. Schools along the coast are already building extra floors

NATURAL VEGETATION REGIONS OF NORTH AMERICA

- Tundra
- Boreal forest
- Temperate forest
- Grassland
- Arid
- Tropical forest

13.2 Natural vegetation regions of North America

to provide some form of shelter in the wake of a tsunami wave, and evacuation routes are now posted (Figure 13.5).

Seismic activity is found throughout the continent. An earthquake in 2011, registering 5.8 on the Richter scale, shook the Washington Monument at the symbolic heart of downtown Washington, DC. Across much of the east and central part of the United States and Canada an earthquake is always a possibility.

At times the distinction between natural and **anthropogenic hazards** becomes less certain. The rapid uptick in earthquake activity in parts of the Great Plains and especially in Oklahoma is a byproduct of **fracking** as excess water is pushed into rock formations at very high pressure, causing structural instability in the bedrock.

A major environmental challenge is posed by over a century of urban-industrial growth that has created a toxic environmental legacy of pollution and degradation. Resource exploitation and industrial production in the late nineteenth and first three-quarters of the twentieth century

were often undertaken with few or no pollution controls and no sense or appreciation for the longer term environmental consequence. Effluent from factories, mines, and furnaces found their way into the air, the water, and the earth. The most polluted areas are categorized in the United States as **Superfund sites**. There are 1,317 that require major environmental remediation due to hazardous material contaminations. They are concentrated in the older industrial areas of the country as well as regions of mining and resource exploitation. Canada's polluted sites are recorded in the Federal Contaminated Sites Inventory. Currently 23,078 sites are listed that contain a range of contaminants, including polychlorinated biphenyls.

As well as highly polluted sites, there are also **brownfield** sites of environmental contamination that are abandoned, unused, or underused. Brownfields are particularly common in the older industrial areas. To bring them into more productive and ecologically healthy spaces requires significant investment in environmental clean-up.

Another major challenge is the product of both affluence and technological sophistication of the society in North America that combines to produce a very heavy **ecological footprint**. This footprint measures how much land and water area a city requires to produce the resources it consumes and to absorb its wastes. Ecological footprint is measured in global hectares per capita (gha). The global average is around 2.6. In San Francisco the value was calculated at 7.1 gha, while in Calgary, Canada, the value was 9.8 gha. Winters are cold in Calgary, and most people use cars to get around.

Another measure is a city's **carbon footprint**, which is the total amount of greenhouse gases it produces. The basic unit is tons of CO_2. The global average is 1.19 metric tons per person. One study measured the carbon footprint of twelve cities: Beijing, Jakarta, London, Los Angeles, Manila, Mexico City, New Delhi, New York, São Paulo, Seoul, Singapore, and Tokyo. The number included direct emissions from the metropolitan area, as well as emissions produced in the metro area but consumed elsewhere, such as goods manufactured in cities. Four of the cities have footprints smaller than the global average: Delhi, Manila, São Paulo, and Beijing, in part because of relatively high usage of public transportation. London was close to the global average, while Los Angeles had by far the largest, followed by Singapore, New York, and Mexico City.

The United States and Canada are ranked fifth and sixth, respectively, in terms of their ecological footprint in per-capita terms. The sheer size of the US population, combined with their affluent lifestyles, makes it the most environmentally impactful society in the world.

The heavy footprint is expressed in the long-term deforestation of much of the eastern woodlands since the coming of Europeans, the loss of biodiversity, and overall in the increasing stress on fragile ecosystems. Although the general story is one of loss of habitat and declining biodiversity, there are a few signs of ecological hope. Take the case of the gray seals. These creatures used to be much more common off the coast of New England, but their numbers had declined so much by the 1990s that there were only 2,000 seals reported near Cape Cod. But protection helped them recover. The 1972 Maritime Mammal Protection Act

13.4 Miami in Florida has a subtropical climate and lush vegetation (photo: John Rennie Short)

made it illegal to kill them. An aerial survey by drones that looked at seal populations using Google Earth found that there are now between 30,000 and 50,000 seals. And with more seals, there are more sharks. Sightings of great whites increased from 80 in 2014 to 147 in 2016. Cue the music for *Jaws*.

Environmental Hazards and Climate Change

North America has a range of environmental hazards from dangerous ice storms to destructive **hurricanes** and punishing **tornadoes** that develop in the hot interior of the continent. All these hazards are becoming more pronounced due to climate change. Let us illustrate with five examples from across the continent.

13.3 Boreal forest in Alaska (photo: John Rennie Short)

13.5 Evacuation tsumani posting in Ketchikan, Alaska (photo: John Rennie Short)

First, there is the warming of the Arctic areas. A rapid rate of loss of Arctic sea ice and snow cover is due to increasing temperatures. There is a positive feedback mechanism as less snow cover exposes more rock surface that then warms in the long summer and retains rather than reflects heat. The Arctic areas of the world including North America are experiencing the fastest rates of increasing temperatures, an average of 3.5°C since the beginning of the twentieth century. The persistent warming is creating significant and extensive Arctic changes, including the thawing of permafrost, which in turn releases more carbon and methane into the atmosphere, changing the habitat range of many animal species and increasing the vulnerability of others such as the polar bears. There are also cultural impacts, as indigenous people have to adapt to markedly changing environmental conditions. Melting permafrost and rising sea levels threaten many indigenous communities across the northern part of North America. Around four hundred Inupiat people inhabit the town of Kivalina in northwest Alaska above the Arctic Circle. They were originally a nomadic people but were persuaded by the US government to live in permanent settlements such as Kivalina. Warming is reducing the length of the winter, destroying the icy sea that protects the town and

13.6 Fires in Pacific Northwest (photo: John Rennie Short)

exposing the settlement to a higher sea level and punishing waves. The town is in imminent danger of being washed away. Scientists estimate that the town will be underwater by 2025. Plans are now being developed to abandon the site and relocate the people to a less vulnerable place.

Second, there is the increasing prevalence of wildfires in the broad sweep of the West, ranging from California through the Canadian Rockies into the boreal forest of interior Alaska (Figure 13.6). Over the last few decades, and especially since 2000, the wildfire season is getting longer with more fires, bigger fires, and more damaging fires. More than 5 million acres were destroyed in the boreal forests of Alaska in one recent season. More than 8 million acres have burned since 2000. In 2017 and 2018, fires in California had devastating effects. In 2017, two huge fires, one north of San Francisco and another north of Los Angeles, killed forty-five people, burned more than a half million acres, and destroyed over two thousand structures. In 2018, fires devastated large areas of California.

There are more people now living beside forests in the **wilderness–urban interface (WUI)**. In an uninhabited forest, regular fires are a part of the natural ecological cycle, and they clear the brush. There are certain types of vegetation in this region that regenerate through fire. But when more people in expensive homes live close by, then forest fires are suppressed, leading to the build-up of the flammable brush and vegetation. Fires, when they do break out, then tend to be larger, more destructive, and more difficult to control. Having more people close to forest also raises the risk of accidental fires. Climate change is making the West hotter, drier, and more vulnerable to fire. The higher temperatures wick away moisture from the trees, making them more vulnerable to fire. The combination of the build-up of combustible material and hotter, drier climate leads to more fires.

The cost of protecting people and property in the WUI is so expensive it has shifted the priorities of the US Forest Service. So much money is spent on protecting property from megafires that programs for preventing wildfires and protecting habitats and wildlife are much reduced. In 1990, firefighting accounted for only 13 percent of the Forest Service budget; now it eats up more than half.

Third, there is the increasing probability of more extreme events. Consider the case of the **derecho** that knocked down power lines in the suburbs of the Washington-Baltimore metro area.

In the late evening of Friday, June 29, 2012, electric power went out all over the Washington metro area. As the power failed, television screens went blank, email links were lost, air conditioning units simply stopped, and traffic lights ceased to function. More than 1.4 million people lost power, some of them for almost a full week, in the oppressive humidity of midsummer. Without power the area was no longer a functioning modern city. The anonymous reliability of modernity was replaced by the improvisation and inventiveness of individuals, family support systems, neighborhood alliances, and community connections. The vital necessity of readily available power that defines contemporary urban living was made acute by its absence. The lack of power highlighted the dependence on it. The taken-for-granted was revealed as the absolutely essential, and the urban fragility of modernity was brutally exposed. Critical infrastructure is only experienced as vital when it fails.

The storm that knocked out metropolitan Washington developed quickly. Unlike hurricanes that can brew for days in the Atlantic and Caribbean before hitting the shore, providing much needed advance warning of the storm's intensity and track, this damaging storm began as a small storm cell close to Chicago barely 10 hours previously. A series of thunderstorms developed in the early afternoon of Friday, June 29, 2012. The storm turned into a derecho, a long-lived, violent, straight-line, convective windstorm. This derecho, with fierce winds of over 80 mph, moved rapidly east and south at around 60 mph. By 10:30 that evening the strong winds knocked down trees throughout the region, short-circuiting much of the metro's power system.

Put simply, the basic fuel of derechos is warm air. Initial thunderstorms activate a hot, humid air mass that is then lifted along its leading edge. The very hot air in front of the air mass generates the strong updrafts that keep the derecho moving. Derechos emerge during long spells of hot weather. The year 2012 was, until then, the hottest in over 100 years with temperatures in the continental United States at 3.2°F degrees higher than the average for the twentieth century. Temperatures were above normal for every month in the summer of 2012, part of a national increase in the incidence of hotter summers in the continental United States and part of an even wider global trend of a warming of the planet. Previously, largely isolated to the hot, humid area of the Great Plains and Midwest in the high summer, derechos are becoming more common across a wider spread of territory, as land surface temperatures increase (Figure 13.7). The June 29, 2012, derecho pushed all the way to the Atlantic coast of Maryland and impacted part of the densely populated **Megalopolis**. Previously, few derechos had made their way past western Maryland.

The increasing summer temperature, which is now redirecting derechos from the largely rural interior sections of the country to the urbanized seaboard of Megalopolis, is increasing the vulnerability of the densest urban region in the United States to quick-moving damaging winds. The vulnerability is heightened because derechos are particularly devastating; unlike hurricanes, a more common problem for the eastern seaboard, they develop very quickly and unleash their full impact with very little warning. Derechos are difficult to predict; they emerge quickly and move rapidly across the landscape. It is difficult to plan for winds of more than 80 mph with no prior

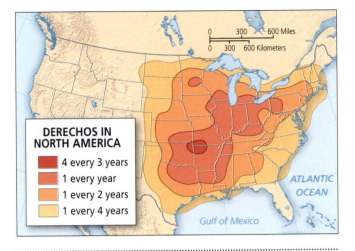

13.7 Derechos in North America

warning or, at best, only a few hours notice. Derechos are one of the least predictable and least known types of the now increasingly common "extreme" events produced by global climate change.

Most people think of increasing land temperatures when they think of global climate change. But the sea is also warming. There is a steady increase in sea surface temperature (SST), which provides the background to our fourth and fifth case studies.

Fourth, the increase in SST has an impact on coral reefs around the world. The 113-mile-long Ocean Highway links the mainland of South Florida with Key West and runs roughly parallel to a magnificent 360-mile coral reef system, the third largest barrier reef in the world. The coral reef cover was halved in size from 1996 to 2016. High SST in 1997–1998 and again in 2014 and 2015 caused heat stress that bleached the coral. Bleached coral is weakened and thus more prone to disease. It is not only an environmental problem—important though that is—it is also an economic crisis. The $36 billion local fishing and tourism industries dependent on people visiting the coral reefs are under threat as well as the precious sea life of the coral reefs.

Finally, increasing SST is also the backdrop to the increasing intensity of hurricanes along the eastern seaboard. Average SST in the Caribbean is now averaging 2°F degrees above historic monthly maximums. Increasing SST is making hurricanes, and especially major hurricanes, more likely. The water is warmer and stays warmer for longer, thus effectively increasing the amount of fuel for hurricanes and the time available for hurricane formation. Warm seawater is the energy source of hurricanes. As the ocean water stays warmer longer, the hurricane season is extended and more large hurricanes are created. Increasing SST is making hurricanes, and especially major hurricanes, more likely and shifting the hurricane zone farther north.

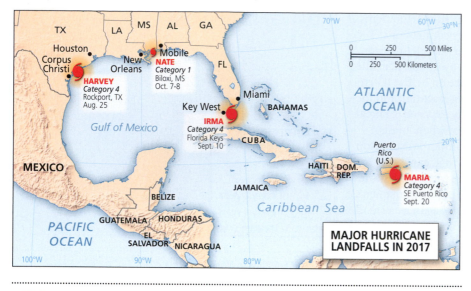

13.8 Major hurricane landfalls in 2017

On average, there are 8–10 storms in the Atlantic in the typical season. Since the mid-1990s, there has been an average of 15–16 storms a year. They are growing in intensity. In 2004 alone, four big hurricanes, Charley, Frances, Ivan, and Jeanne, caused $42 billion worth of damage.

Hurricanes can also maintain their power as they move farther north. Hurricane Sandy's destructive power in October–November 2012 as it moved through the Caribbean and along the Atlantic seaboard was in part due to warmer SST, which meant it could retain its power even as it moved, causing storm damage to New Jersey and New York. What was once only a one-in-100-years event is now turning into a one-in-20-years or even one-in-10-years event.

The 2017 hurricane season was one of the most active (Figure 13.8). Hurricane Harvey, a category 4 storm, hit Texas and Louisiana in August. Over 51 inches of rain deluged the city of Houston, flooding 100,000 homes and causing $200 billion worth of damage.

We should be careful to use the term "natural disaster" in this case. Flood damage was exacerbated by a lack of land use planning that allowed a steady conversion of permeable surfaces into the impermeable surface of roads, cement, and concrete. Floodwater had nowhere to go as the city's unplanned growth transformed the urban landscape from green to gray. Flood damage was caused by heavy rainfall on square miles of impermeable surfaces.

Hurricane Irma slammed South Florida in early September with another category 4 storm that had rampaged through the Caribbean. Later that same month, Hurricane Maria reached category 5 wind speeds of 160 mph, devastating Puerto Rico. A month later much of the island still lacked electricity with limited access to fresh water. The final death toll was nearly 5,000 people .

Climate change is impacting the entire region from increasing risk of drought and fire in the West, dryness as the new normal in the Southwest, the worsening severity of flooding along the coast, to the shrinking of the Colorado River.

In North America, provinces, states, cities, and counties are responding to the challenge of climate change with policies of adaption and mitigation. Adaption involves responses such as raising the sidewalks in Miami Beach. The city is vulnerable to sea level rise. It is estimated that by 2030 the sea level along the immediate coastline will rise between 6 and 12 inches. During heavy rain, more than 6 inches can fall in the city in 45 minutes. But there is also "sunny day flooding," when high tides create floods even when the skies are clear. The city of Miami Beach is devoting $400 million over the next 5 years to adapt the city to this flood threat by raising sidewalks and improving water and sewage engineering.

The year 2017 was the most expensive one for weather disasters in the United States. Three major hurricanes, wildfires, flooding, tornadoes, and drought brought the total bill to $306 billion. The previous high was $215 billion in 2005. These costs estimated in a 2018 report are adjusted for inflation, so they represent a real and substantial increase. Wildfires alone in 2017 cost $18 billion, triple the previous record. Weather is not the same as climate, but climate change increases the risk of extreme weather events. Perhaps 2017 may be a harbinger of years to come.

Climate change has different effects in different parts of the region. The impact also varies by socioeconomic status, with health effects such as deaths from extreme heat, longer allergy seasons, more polluted air and water having a proportionately larger impact on lower income groups. Climate change is not only an environmental issue, it is a social concern.

Historical Geographies

Pre-Columbian

When the first European settlers came to North America, they entered a land long inhabited. It was not empty wilderness but an occupied and populated land.

The very first humans came from Asia at the end of the last Ice Age. It is estimated that around 14,500 years ago people from North East Asia travelled along the Pacific coast, living off the bounty of the sea. An ice-free land corridor emerged around 12,600 years ago that made movement even easier. After 2,000 years the corridor was made less accessible by the growth of dense forest. But by then humans had dispersed throughout the New World.

The first settlers were foragers, gatherers, and hunters who followed the large megafauna of the era such as the musk ox and the woolly mammoth. Many of the large animals were hunted to extinction. Climate change probably played a role but so did human actions. The ecology of the New World was shaped by these early human settlers.

New tools were developed around 10,000 years ago. Well-made scrapers and knives with fluted bifacial points made hunting more efficient. A cache was found in Clovis, New Mexico. Another big change came with the domestication of plants. Around 3,000 years ago, squash and maize were first grown. A variety of different cultures developed as the local environments became the basis for material culture and spiritual beliefs. A number of broad categories can be identified (Figure 13.9).

13.9 Pre-Columbian culture areas in North America

Along the Arctic margins, later arrivals from Eurasia, such as the Inuit, developed hunting techniques to survive the harsh winter climates.

Along the Northwest coast, the abundant supplies of game and fish provided a rich protein base that enabled a sophistical and rich material culture that included monumental wooden sculptures, beautiful wooden tools, and intricate carvings.

Woodland culture developed in the Northeast and Southeast. People in settled villages developed techniques to grow beans, maize, and corn. The town of Cahokia near St. Louis had a population of around 30,000 in 1300 CE.

In the Southwest and California, people lived by hunting, farming, and foraging. In the Plains the hunting of large animals such as the bison and buffalo augmented farming. The coming of the horse in the Columbian Encounter dramatically shifted this emphasis from farming to hunting.

The Columbian Encounter

North America attracted European powers eager to establish territorial claims and exploit the resource base. The New World represented an enormous economic opportunity with its natural wealth of land, minerals, timber, and fur. Spain was more interested in the mineral wealth of Central and South America, and so the North American region was always more on the periphery of Spanish interests. The French, in contrast, established trading bases throughout the St. Lawrence and spread their influence via the inland waterway system. The French legacy lives on in the names of US cities such as Detroit, Baton Rouge, and New Orleans. Early English settlements were established along the Atlantic seaboard and New England.

We have already discussed some of the consequences of the Columbian encounter in previous chapters. To summarize: the Europeans brought diseases that proved fatal. It is estimated that 90 percent of the population of the New World died within a hundred years of the first European contact. This demographic holocaust undermined many indigenous societies.

Consider the case of the Mandan, a settled agricultural people who lived on the northern plains of what is now North Dakota. They appear in the notebooks of Lewis and Clark as their expedition explored westward. The expedition wintered with the Mandan in 1804–1805. Clark described them as "brave, humane and hospitable, the most friendly of all disposed Indians inhabiting the Missouri." The Mandan, numbering at their peak perhaps 12,000, had been living in the region since at least 1500. However, in 1778, a smallpox epidemic ravaged the community. Plagues of rats in 1825 ate their main source of wealth and food: underground caches of corn. A weakened community was further devastated by epidemics in 1836. By 1838, most villages were abandoned as the population shrunk to little more than 300. The remnants took to foraging and hunting, a pale shadow of a vibrant settled civilization. This is a distinctive story, but it is also part of the larger narrative of indigenous demographic collapse, cultural disintegration, and loss of land.

Although it varies in timing and extent, the big picture of the Columbian Encounter is one of indigenous people losing control over their land. The colonists and the governments bought, appropriated, and stole their land, erasing much of the indigenous presence. We still have a legacy of place names across the region from Penobscot Bay in Maine to the town of Ketchikan in Alaska.

Settler Societies

From the seventeenth to the twentieth century, North America became a **settler society** populated by European migrants. Settler societies are formed by imperial powers in the peopling and colonization of territory. The emigrants settle and construct their own states that in turn develop into distinct nations. Other settler societies include Argentina, Australia, New Zealand, South Africa, and Uruguay.

In North America by the eighteenth century the two dominant imperial powers were Britain and France. After the French defeat in the global Seven Year War (1756–1763), known in North America as the French and Indian War, Britain became the dominant power. French speakers in the Maritime Provinces, known as Acadians, were forcibly removed during this war and deported to what is now the United States. Others were deported to France and Britain before making their way to Louisiana, where Acadia became Cajun. New Orleans and the region around the mouth of the Mississippi to this day have a strong French legacy in language and cuisine. The descendants of early French settlers still maintain a presence in Canada, especially strong in Quebec. The French legacy continues in the French-speaking communities across Canada and in the official national policy of French and English bilingualism.

Soon after Britain achieved dominance in the continent, the settler society fractured into two distinct forms of national sovereignty. In 1763, Britain drew a Proclamation Line along the crest of the Appalachian Mountains to restrict settler settlement (Figure 13.10). The British were eager to maintain good relations with Native Americans, especially those such as the Iroquois that had fought with them in the war against the French. The British government also felt that the settlers should help defray the costs of the recent war and introduced various taxes. The American settlers, in contrast, saw the land beyond the line as a rich opportunity being kept from them by a distant sovereign and the new taxes as unfair. The resistance to the sealing off of land speculation and taxation without representation led to the War of Independence and eventually to the creation of a new republic. Divided loyalties cut across family bonds. Benjamin Franklin was long a supporter of the British Crown until his conversion to republicanism. He was a reluctant revolutionary. His illegitimate son, friend, and partner, William Franklin, refused to give up his post as royal governor of New Jersey, was jailed for 2 years for opposing the revolution, and eventually moved back to England. He remained a staunch loyalist.

While a revolutionary rupture inaugurated the United States, Canada remained under the British Crown, becoming a federal dominion in 1867. In 1931, it became fully independent from the United Kingdom.

The differences remain. The United States is a republic, whereas Canada is a federal parliamentary democracy and a constitutional monarchy. There are also differences in political culture with Canada sharing a similar ideology to the United Kingdom and Europe in terms of the more prominent role of government, while many in the United

THE PROCLAMATION LINE OF 1763
- Proclamation Line
- Thirteen colonies
- Other British land

13.10 The Proclamation Line of 1763

States still have a strong distrust of central government. Canada is more liberal, while the United States is more neoliberal. Two countries so similar and close together also have differences that distance them from each other.

Economic Transformations

The economic history and geography of North America can be summarized as a gradual relative shift from primary producer for colonial markets, to more global markets, to more industrialization, and then, more recently, to a postindustrial economy. This trend varies over both time and space. While the economy as a whole has made this transition, there are still important spatial variations, with some regions still dependent on primary production while in selected metropolitan areas the advanced service sector is the most dominant.

The Primary Sector

After the territory of North America was brought under European control, the region was then connected to global routes and transcontinental migration paths. The land was a source of cheap and available land, once cleared of indigenous peoples, and the vast resource base of furs, timber, and minerals was exploited to sell on global markets. For most of the eighteenth and early nineteenth century, North America's main global economic role was as a provider of primary products. Timber, wheat, minerals, and staples such as cotton were exported to overseas markets, especially to the United Kingdom and Europe. North America was also linked to Europe through trading patterns that involved the import of slaves from Africa. The **triangular trade route** linked Europe, West Africa, and the eastern seaboard of North America in a trade of manufactured goods, human flesh, and primary commodities.

From the end of the nineteenth century, the national economies of Canada and the United States have shifted more toward manufacturing and services. However, primary production still plays an important role. There are **resource-based economies** such as oil and gas in Alberta, timber in British Columbia, or minerals in Wyoming. Take the case of Wyoming, where 20 percent of the state's economy is still based on mining. The next highest in the United States is Alaska with 15 percent and West Virginia with 11.5 percent. The Powder River Basin in Wyoming is one of the single biggest coal-producing areas in the country. However, the heavy reliance on this primary product also creates a resource curse as the price of coal rises and falls. The neighboring state of Idaho, in contrast with less than 1 percent of the state's economy from mining, has a

more balanced economic base with more manufacturing, technology and services and consequently higher growth rates over the medium to long term. While a resource endowment provides a favorable start to the region's growth, more sustainable growth depends on shifting away from a narrow reliance on primary production.

The economy of Texas was once based on primary production, especially oil. The peak oil production was in 1992, but the sector still plays a significant role in the state. The fluctuating price is a problem. In 2008, a barrel of oil was $145, but it plummeted to $30 in 2016 and in March 2019 was $63. Oil was both a gift and a trap. It provided revenues for the state, but large fluctuations make long-term investments, such as education, subject to change. There are economic benefits but also environmental despoliation and health issues. The oil extraction economy has negative impacts on water, air, and health.

More recently across the continent, fracking allows the extraction of oil from tight rock formations. But once again the increased risk of earthquakes and the environmental damage to water supplies and air quality mean that it is not an unalloyed benefit.

Across the continent, primary production continues to play an important role in regional and local economies, whether in the oil fields of Alaska, Alberta, and Texas or the fisheries of New England and the Maritimes. But because commodity prices fluctuate and the demand for labor declines, after the initial burst of early resource exploitation, they rarely provide long-term growth trajectories. The abandoned gold rush towns of the old West remind us of this. Today's equivalent of the gold rush may be the rapid exploitation of the Bakken Shale site and the oil in the Williston Basin that traverse the US-Canada border (Figure 13.11).

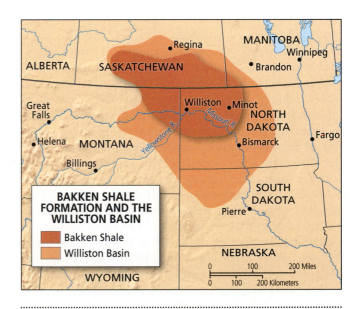

13.11 Bakken Shale Formation and the Williston Basin

FARMING

Farming is an important sector in the economy of both the United States and Canada. It is marked by heavy capitalization and mechanization. While the family farm continues to be an important and enduring foundational myth, the reality is that agriculture in North America is big business.

Agriculture, especially in the fertile plains and lowlands, has achieved high yields and great efficiencies to produce cheap and reliable food. Fewer people work on the land as machines and chemicals have replaced human labor. In 1850, famers made up around 64 percent of the nation's workforce, but by 2016 that figure had fallen to less than 2 percent.

There is still a substantial agricultural sector. In the United States, more than 3.2 million farmers operate 2.1 million farms. But while the average size of a farm is increasing, so is the average age of farmers. It was 50.5 years in 1982, but 58.3 years by 2012.

Farming and farming-related industries, such as food processing, continue to dominate many of the rural areas of the continent. But in these predominantly farming counties and provinces, the population is declining and aging. Farming is still important in both economic and cultural terms, but it is less of a dynamic boost to regional economies as it settles into a mature production phase with mechanization, capitalization, and declining labor needs.

There are costs to the efficiencies of this sector. Impressive corn production, for example, is dependent on large applications of pesticides and fertilizers. The result is polluted waterways. As nitrogen and phosphorous run off into the water supply, they supercharge the growth of algae. Toxic **algae blooms** in the Gulf of Mexico, the Chesapeake Bay, and Lake Erie are in large part due to nutrient pollutions from highly intensive agricultural production. Cheap food is being purchased at the expense of our polluted waterways and negative health outcomes. Consider the case of the region of Delaware, Maryland, and Virginia, known as Delmarva, where poultry farming is important. Hens caged in large indoor sheds are the source of cheap chicken. But it comes at a price. There is the underlying ethical question of animal cruelty but also environmental issues. In the Chesapeake Bay watershed, spring rains are loaded with fertilizers from farms and suburban lawns, and the nitrogen and phosphorous find their way into the bay. The mix of chemicals and animal waste triggers algae blooms, which soak up oxygen in the water and create dead zones in the bay. Similarly in the West, the giant feed lots where cattle are penned in produce cheap beef but at the cost of massive pollution runoff.

Efficient and industrial farming produces large quantities of cheap food. But there are costs associated with this great benefit.

The Rise and Fall of Manufacturing

From the late nineteenth century to the late twentieth century, North America shifted from a resource-based economy to a manufacturing economy. New factories produced goods that replaced imports and in some cases provided exports to the rest of the world. The result was the growth of industrial towns and regions in the Northeast, stretching east and west from Boston to Minneapolis and north and south from Toronto to St. Louis. In the earliest days there was often a very distinct specialization. The town of Gloversville in upstate New York, for example, was an early site for the making of gloves. Detroit became known for the making of cars, while Springfield, Massachusetts, made guns.

The exemplar city was Detroit. The city was a manufacturing center by 1900 with almost a thousand machine shops making ships, stoves, engines, and mining machinery. There was a pool of skilled labor and a network of local financiers. Detroit was soon producing two-fifths of the nation's car output. Henry Ford founded the Ford Motor Company in 1903 with twelve shareholders, all from Detroit. In 1908, the first Model T appeared on the market. It was a simple yet robust model that went through numerous design improvements to become one of the first mass-produced cars. Assembly lines and mass production had been developed over the previous century in a range of manufacturing industries, including bicycles and watches as well as the Chicago meat processing and packing industry. Ford did not invent mass production, but he refined and improved it. Car production became standardized, precise, and continuous. Cars became less a luxury item and more a regular purchase. In order to reduce labor turnover, Ford also paid high wages for the time. A well-paid industrial workforce was essential to decreasing labor turnover and maintaining worker allegiance. Ford was instrumental in creating the high-paid, blue-collar sector of the new industrial city.

Detroit was the forerunner of the modern industrial complex. Production was controlled by a small number of very large companies who operated under oligopolistic conditions. Production was mechanized and repetitive with long production lines. An organized working class was both relatively well paid and well cared for. There were outbreaks of labor unrest when the business cycle softened demand and management implemented layoffs and wage cuts. But, by and large, a stable system of capital–labor relations was established, and a relatively affluent working class was created. Thereafter this particular form of the industrial city was transformed by manufacturing decline, regional shift, and global shift.

The decline of manufacturing employment was in part a function of success. As industrial processes became ever more mechanized and routinized and eventually automated, then less labor was needed. Machines replaced laborers. Most cities witnessed industrial decline, but this was offset by the rise in service employment. Take the case of New York. In 1958, it had one of the largest concentrations of manufacturing employment in the country with more than 300,000 production workers. By 1997, the figure had fallen to less than 100,000. The numbers have steadily tumbled. However, the city's shift to a greater reliance on producer services, especially in the financial and banking sector, generated employment that offset the manufacturing decline. Cities and regions with more diverse economies weathered the deindustrialization much better than cities with a narrow reliance on manufacturing,

There was also a metro and a regional shift in manufacturing. At the city level, manufacturing plants became more suburban than centralized in the cities as older plants closed and new plants opened in the cheaper land of the suburbs. There was also a regional shift. Figure 13.12 highlights this shift away from the old industrial heart of the Northeast. Major industrial areas were established in the West Coast, South, and Southwest. Take car making, for example, long concentrated in Detroit. In 2018, it was announced by Toyota-Mazda that their newest plant in the United States would be located in Huntsville, Alabama, where almost 4,000 workers will produce 300,000 vehicles yearly. Cheap land, low wages, and restriction on unionization all make places like Alabama more profitable for new manufacturing than the traditional heart regions.

It is a complex pattern with marked deindustrialization of traditional urban manufacturing economies and more recent industrialization in selected sites. The division of **Rustbelt** and **Sunbelt** is often used to identify areas of economic decline and growth (Figure 13.12).

The Rise of a Service Economy

The United States and Canada have shifted toward more service economies, with higher absolute and relative numbers in the service sector.

This shift has a number of consequences. Let us consider just two. First, it allows a higher participation rate of women since traditional manufacturing was male dominated. Women have an easier time entering the job market.

Second, the advanced producer services, such as banking, insurance, and creative industries, are now the markers of metro success. The more dynamic metro regions such as San Francisco, Vancouver, or New York City tend to have a higher proportion of the advanced producer services while city regions that lack these tend to be less dynamic.

We can see the difference when we consider Schenectady in New York and San Jose in California. These two cities embody the differences between the Rustbelt and the Sunbelt.

In 1950, Schenectady and San Jose had similar populations: 92,061 and 95,280, respectively. At that time Schenectady's economy seemed secured. Over 27,000 people worked for General Electric. The city was flourishing with

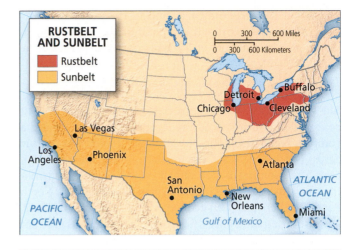

13.12 Rustbelt and Sunbelt

a vibrant downtown and a buoyant job market. Corporations such as American Locomotive Company provided employment for thousands of workers as well as a way of life. Social clubs and softball teams grew up around the connections workers made in the factory. Seventy years later most of those jobs are gone. The Locomotive Company closed in 1968, and General Electric shed over 90 percent of its jobs in the city. By 2017, the city's population shrank to just over 65,705, the houses lost value, and the credit rating of the city, always a sharp-eyed fiscal view of a city's economic health, was downgraded by Moody's Investors Service to the lowest in the state. The median household income in 2017 was $43,174.

By 2017, San Jose, in deep contrast, had a population of almost 1 million and was one of the larger cities in the United States. Decades of spectacular growth fueled in particular by the Silicon Valley boom in high technology and computer-related industries make San Jose one of the most prosperous and economically dynamic cities in the country. In 2017, median household income was $96,662. Among the companies headquartered in the city are Adobe, Cisco, and eBay.

Some cities such as Pittsburgh made a more successful transition. The city was long known as Steeltown, and the football team is still referred to as the Steelers. Over the past 30 years, the city has moved away from its early reliance on heavy industry, from Steeltown to "eds, meds, and feds," a shorthand for higher education, medical research, and the encouragement of federal government–related jobs and employment opportunities.

In the past 50 years, economic growth has become more uneven with dynamic metros such as San Jose but also areas of decline such as Schenectady. More than 52 million Americans live in economically distressed areas plagued by problems of poverty, poor education, and declining work opportunities. These areas are concentrated in the southern half of the country—mainly small and rural—but also large industrial hubs in the North and Midwest. Minorities represent over half of the population in distressed zip codes but only a quarter in prosperous ones.

Regional Differentiation

The twin processes of deindustrialization and metropolitan concentration of advanced producer services have created a complex regional mosaic of economic growth and decline across North America. There are declining former industrial areas and booming metros, affluent new suburbs and shrinking small towns. Across much of small town and rural North America, there has been economic decline, population loss, and rising despair. Policies to boost struggling areas have not been particularly successful. In Empowerment Zones, businesses receive tax credits for hiring residents of distressed communities and local governments get block grants to fund business assistance, infrastructure, training, and so on. Most studies show little or no impact.

North America is now more regionally differentiated because while some regions have chronic joblessness, others have problems of growth such as congestion, rising house prices, and lack of affordable housing.

Rural Focus

The Dangers of Farm Labor

Agricultural work in the United States can be dangerous. In 2015, the year of the latest statistics, workers in the agricultural sector had a work fatality rate higher than police officers and twice the rate of construction workers.

There are many reasons. More farms use unskilled immigrant labor because it is cheap, but that also means employing people with little or no experience dealing with agricultural equipment or large animals. On dairy farms, for example, workers have to deal with 1,500-pound animals. There were 6,700 injuries on dairy farms in 2015, and forty-three workers died. On large dairy farms, significant amounts of excrement are produced, most of it stored in pits. A typical size for a large dairy farm is a 10-foot-deep pond that stretches across 15 acres. On occasion, workers on tractors have tipped over into these pits and drowned.

There is little federal oversight of working conditions on farms. Farming across the United States is dominated by large commercial operations where at least half are unskilled immigrant workers and there is inadequate oversight of worker safety. The most dangerous job in the United States, it turns out, is not police work or construction work, but work on a large commercial farm.

Social Geographies

Center of Population

The mean center of the US mainland population is plotted for each Census decade since 1790 (Figure 13.13). The point marks the central fulcrum of the national population. In 1790, the mean center was located in Maryland, and over the years it has steadily moved westward in line with the westward shift of population. Between 1970 and 1980, the mean center crossed the Mississippi River, and by 2000 it was located in Phelps County, Missouri. By 2010, it shifted farther westward and southward to Texas County in Missouri. It is

13.13 The mean center of population of the United States

projected to be in Wright County, Missouri, by 2020. The slow, steady shift of the mean center marks the redistribution of the US population to the expanding metro areas of the South and West. Its slow progress, however, reminds us of the continuing population weight of the Northeast.

Demographic Trends

The most significant demographic trend in both countries is the aging of the population as population growth has declined. In 2018, the annual rate in the United States was 0.7 percent and significantly larger at 1.2 percent in Canada. The respective rates for 1960 were 1.7 and 2.3. In other words, over the past 60 years the growth has declined dramatically. There are more non-child households, more single-person households, and fewer traditional households of mother, father, and two children. This demographic change varies across space and is more prevalent in the larger metro areas. Almost half of households in Washington, DC, for example, consist of single-person households.

The population is aging as the birth rate declines and life expectancy increases. In 1960, Canada and the United States had life expectancies of 72.1 and 69.7, respectively. By 2018, the figures had increased to 82.8 and 78.6, respectively. Canadians live longer, on average, than their counterparts in the United States, in part because of a better health care system.

While the overall trend is toward an increase in life expectancy, there is also the disturbing decline in life expectancy in the United States in the last few years and especially among white middle-aged women in small towns. Many are suffering from the diseases of despair linked to alcohol and drug abuse, poor diet, and lack of hope. Women often have onerous domestic burdens of keeping families together, especially when they face economic trying times and employment distress. The weight of responsibility in difficult circumstances can have negative health outcomes.

The population of both countries is aging. In Canada, for example, the proportion of people aged 65 years and over in 1960 was 8 percent. It is now closer to 15 percent, similar to the United States, and likely to grow to 25 percent by 2036. The median age in 2018 in the United States is now 38.1 years. There are regional differences. Whereas Maine has a median age of 44.6 years, the median age is only 30.8 years in Utah, the result of persistently high birth rates and large families, and 33.9 years in Washington, DC, as more young workers move into the city to work.

There are consequences to the aging, especially in the area of entitlement. In the United States, for example, Social Security and Medicare are largely designed to cushion the lives of the elderly. But these programs were established where there were more workers per recipient. As the number of workers paying for these provisions declines relatively and the number of recipients increases as the population quickly ages, then the stage is set for a fiscal crisis as expenditures outrun the ability to pay. Canada and the United States, like many affluent European countries, face a major problem as the population ages and the social welfare programs become more expensive. The elderly tend to vote more than the young, so the interests of the elderly tend to receive more welcome political response. There is a growing intergenerational inequality as younger workers pay for benefits that they themselves may never receive.

Nations of Immigrants

As settler societies, the national identities of the United States and Canada were shaped by immigrant waves. In the early years, immigrants came from Europe, but in the past four decades the immigrant source areas have become more global.

Canada had distinct waves of immigration. The first wave consisted of British and French settlers at relatively modest levels. There was a larger second wave from 1890 to 1920 of people from continental Europe. This wave made Canada a more multicultural society. The 1976 Immigration Act widened access to the country. Canada is now a multicultural society with roughly a quarter million immigrants per annum. By 2016, more than one-fifth of Canada's population was born outside the country. And one in three of all children aged 15 years and under was born overseas or had at least one parent born overseas. The percentage of foreign-born varies, with high levels in major metro areas. In Toronto, 46 percent of people were born overseas, and the city is filled with different nationalities and ethnicities, giving it a very cosmopolitan feel. Small towns and rural areas on the Maritimes, in contrast, have a much lower percentage of foreign-born. In 2016, the foreign-born population of Prince Edward Island was only 5.8 percent.

It is a similar story for the United States, with two distinct recent waves. The first was from 1880 to 1930 when immigrants came from Europe, and the second was in the wake of the 1965 Immigration Act, which abolished a tight quota system and opened up the possibilities to immigration from regions of the world other than Europe (Figure 13.14).

The source of migrants has also changed over time from predominantly Europe even as late as 1980, and then widening to include Africa, Asia, Central America, and South America.

There is a distinct clustering with the percentage of foreign-born high in certain metro areas compared to less economically dynamic small towns and rural areas. Almost half of the foreign-born live in just three states: California, Texas, and New York, and most live in twenty major metro areas, including, in order of numbers, New York City, Los Angeles, Miami, San Francisco, Chicago, and Houston. Even within the metro areas there is distinct clustering. In the borough of Queens in New York City, the foreign-born constitute 47.5 percent of the total population. Compare

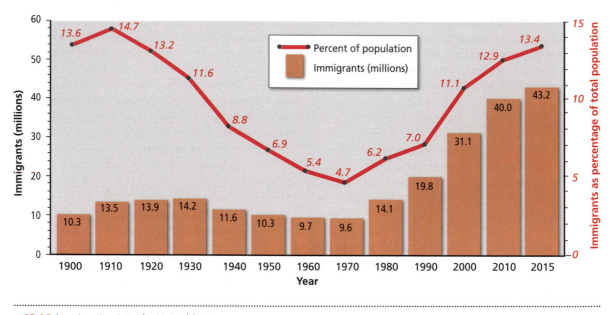

13.14 Immigration into the United States

this with Iowa County in Wisconsin, where the foreign-born constitute only 1.2 percent of the population. There is enormous diversity in some parts of the country and homogeneity in others.

A number of different types of immigrant gateways can be noted (see Table 13.1) Continuous gateways for immigration such as New York City and Chicago attracted immigrants in the past as well as the present because of family ties and employment opportunities. There are also emerging gateway cities such as Atlanta, Charlotte, Dallas, Miami, and Washington, DC, that had little history of foreign-born immigration but whose present economic vitality has attracted more. In bypassed traditional gateways, such as Buffalo and Detroit, poor economic performance fails to attract immigrants. There are also gateways that have emerged since 1945 and some that are reemerging. In the early twentieth century, immigrants were mainly attracted to central cities. In the most recent waves, however, suburban areas have emerged as immigrant gateways. Annandale in Virginia, for example, is known as Little Korea, while Wheaton in Maryland has a significant Hispanic population. Immigrant suburbs are more readily evidenced.

TABLE 13.1	Six Types of Immigrant Gateways in the United States
TYPE	**EXAMPLES**
Continuous	Boston, Chicago, New York, San Francisco
Former	Baltimore, Buffalo, Detroit
Post–World War II	Fort Lauderdale, Los Angeles, Miami
Emerging	Atlanta, Dallas, Las Vegas, Washington, DC
Reemerging	Denver, Philadelphia, Phoenix, Seattle
Suburban	Annandale, VA; Wheaton, MD

Of the roughly 45 million foreign-born in the United States, approximately 11 million, a quarter of all foreign-born are undocumented immigrants who have no formal right of stay. Undocumented immigration (some use the term "illegal immigration") is a major political issue, especially for lower income US citizens who feel that their position in the labor market is undermined by cheaper undocumented workers. The middle and upper income groups in the United States benefit most from high levels of immigration and especially undocumented workers because they provide cheap labor as nannies, gardeners, and service workers. The biggest burdens are carried by unskilled native workers whose wages tend to be suppressed if there are large numbers of workers willing to work in more difficult conditions for less money.

Gender

Social and economic changes have a significant impact on gender relations. The role of women has changed over the years. While lower status women have always worked, middle and upper income women were often formally and informally restricted to the home rather than the workplace. Since the end of World War II, there has been a steady rise in female participation rates in the workforce. Gender relations have also changed with the decline in the size of families. There are more single-person and childless households. Women are no longer as restricted to the home.

In part, the growing participation has a positive side: more women are able to enter the workplace. Although as the #metoo discussions revealed, this sometimes involved harassment. There is also a basic economic necessity because in order to sustain household incomes, more women need to work. For many working women, there is the extra burden of working outside while also still being primarily responsible for household labor.

There are also changing gender relations. Between 1975 and 2018, while, on average, men's income declined, women's income increased. One trend was for men to move from higher to lower incomes in the wake of deindustrialization and the decline of well-paying jobs in manufacturing, while women moved from low to higher incomes in the expanding service areas. The differences in income growth many be one reason behind the marked misogyny in much of US popular culture. As more men feel marginalized by the economy and cultural norms, it is easier for anti-feminism to emerge.

Female participation rates in both the United States and Canada are relatively high. In Canada, 78 percent of women aged 16 to 64 years are in the workforce; in the United States, 73 percent.

Female participation rates are also influenced by child care arrangements. The United States is one of the very few countries that does not mandate paid maternity leave, joining a group that includes Lesotho, Liberia, Papua New Guinea, and Swaziland. In contrast, Canada mandates 52 weeks. The United States simply does not invest in child care and maternity leave, making it more difficult for women to remain in the workforce and have a family.

Race, Ethnicity, and Class

Race is a slippery idea that has no real basis in human biology. It is a social construction rather than a biological fact. The concept of race has no genetic or scientific basis. All humans, regardless of race, are more than 99.5 percent the same at the DNA level. Racial categories are created and imposed, adopted and celebrated, rather than emerging from some blood heritage. In the rest of this section, imagine that the term "race" always has quotation marks around it to signify its social rather than biological reality.

The most distinctive racial categorization in the United States is between black and white. While Canada has its own racial issues, the black-white divide is less toxic because of the lack of slavery in the history of Canada.

In the United States, young black males are 21 times more likely to be killed by police than their white counterparts, and the median wealth of white households is 13 times greater than the wealth of black households. How do we explain these disparities? In large part, it represents the legacy of slavery and the operation of racist policies. While explicit racial policies of exclusion and marginalization are now considered illegal, there is still a burden of being black in the United States.

Figure 13.15 highlights the difference in median income between different groups. Notice how Asians have a median household income more than twice that of blacks. The difference between white and black median income is also substantial. While the position of blacks in the United States has improved, there are still substantial differences

in wealth and income and health outcomes between whites and blacks.

The geography of race in the United States was originally largely a function of slave economy with a significant number of blacks in the southeast and south. In 1900, the black population of the United States was largely southern and rural. In the past 100 years there has been a gradual shift from rural to urban of the black population as people left the confinement of the South and its explicitly racist policies to seek greater freedom in the cities of the North and West. Between 1916 and 1970, more than 6 million blacks in the South left because of declining employment opportunities and an explicit system of racial segregation and economic marginalization; this was referred to as the **Great Migration**. And it permanently changed the geographical distribution, with blacks becoming more urban and then more suburban.

It is at one and the same time both easy to talk about race and yet difficult. Easy because the black-white difference is still so clear and obvious that race talk is a permanent obsession in the United States. However, it is often difficult to talk about how race connects and disconnects with other dimensions of social difference such as social class, gender, and sexual orientation. The black community is as diverse as the white community with differences in class and gender often ignored when race becomes the only defining characteristic. More complex pictures emerge if we compare and contrast the social geographies of rich black men, poor black women, young and old, gay and straight. Race and ethnicity are important elements in social geographies but more often than not in association with other social characteristics.

There are also differences in socioeconomic status with resultant differences across a range of phenomena, including

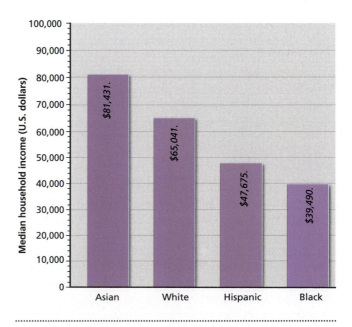

13.15 Median household income by race/ethnicity in the United States

health outcomes. Babies born in the poor areas of the country such as on Native American reservations, in eastern Kentucky, and in the Mississippi Delta will on average die 15 years earlier than children born in affluent coastal California. There are areas of persistent and entrenched poverty such as the lower Mississippi Delta. But there are also some emerging areas such as southern Oklahoma, where life expectancy has declined due to job loss and rising despair.

Urban Trends

Continuing Metropolitanization

The drift of population to large cities continues. The US Census employs the term "metropolitan statistical area" (MSA) to refer to urban areas with a core area of at least 50,000 and economic links to surrounding counties. Using this statistical, rather than political division of municipal boundaries, it is possible to measure the metropolitanization of the US population. In 1950, the metropolitan population was just over half at 56.1 percent of the total US population. By 2018, the figure was 85 percent. The US population is increasingly and overwhelmingly concentrated in metropolitan areas (Table 13.2). More than 90 percent of the country's entire population growth in the last decade occurred within MSAs. North America continues to become a more urban and metropolitan society. Metro areas in the United States now house 83 percent of the population and are the main sites for innovation and job growth. The 100 largest metro areas hold 69 percent of all jobs and are responsible for three-quarters of the nation's GDP. The most dynamic parts of the economy, with more job opportunities and higher wage rates, are in the cultural creative economies of finance, advertising, and all those sectors that require symbolic analysts. Data and narrative have replaced metal shaping and car manufacturing. This cognitive capitalism has a heavy urban bias, as it requires the close proximity of talented and creative people. There are strong economies of urban agglomeration because people living and working closely together generate the necessary increase in knowledge, creativity, and innovation.

Central Cities

The story of the central city of MSAs is complex. The first type of city is *steady decline*. A typical city in this category is Detroit, which experienced a peak of 1.84 million around the mid-twentieth century and then continuous decline; its 2017 population was 673,104, down from 951,270 in 2000. The city embodies the rise and rapid fall of the older, unbounded, industrial city. Other cities in this category include Akron, Baltimore, Birmingham, Buffalo, Cincinnati, Cleveland, New Orleans, Rochester, Toledo, and Pittsburgh. In these cities the loss of employment caused by the long,

TABLE 13.2 The Twenty Largest Metro Areas in USA	
METRO REGION	**POPULATION (MILLIONS)**
New York	20.3
Los Angeles	13.5
Chicago	9.3
Dallas–Fort Worth	7.3
Houston	6.8
Washington	6.2
Miami	6.1
Philadelphia	6.0
Atlanta	5.8
Boston	4.8
Phoenix	4.7
San Francisco–Oakland	4.7
Riverside-San Bernardino	4.5
Detroit	4.3
Seattle	3.8
Minneapolis	3.6
San Diego	3.3
Tampa	3.0
Denver	2.8
Baltimore	2.8

Source: https://en.wikipedia.org/wiki/List_of_metropolitan_statistical_areas

slow decline of manufacturing is yet to be replaced completely by new forms of economic growth. These cities also bear the brunt of an urban fiscal crisis as the steady loss of population and tax base erodes the revenues of the city. The second type of city is *continuous increase*. Here the story is of rising economic and population growth and ease of annexation. A typical case is San Jose, California, a Sunbelt city with an expanding economy based largely on information technology. Other examples of this type include San Diego, Las Vegas, and Orlando. The third model type is *growth interrupted*. Here examples include New York City, Atlanta, San Francisco, and Seattle. These cities' upward population trajectory saw some decline before returning to growth and eventually surpassing their previous population peak. In the case of Seattle, population peaked in 1960 and then declined before growth returned by the 1990 census. Then there is the *slowly resurgent* city, where previous peaks are not reached, but there is a slow, steady return of population. Examples include Boston, Philadelphia, and Washington, DC. In all three cases, the areal size of the city remained roughly the same. Washington, DC's population peaked in 1950 and then saw continuous decline until 2000, at which point the city's population began a decade-long resurgence.

Suburbs

Since 1950, North America has become more suburban (Figure 13.6). There is considerable variation. At one extreme are boomburbs characterized as municipalities of more than 100,000 that were not the major city of their MSA and experienced double-digit growth for three consecutive decades. At the other extreme there are "suburbs in crisis," defined as suburbs that witnessed population loss and economic retrenchment. Suburbs at risk are those built between 1950 and 1970. The inner-ring suburbs often have an older stock and in terms of desirability fall between the attractions of resurgent central cities or the new housing on the suburban edge. The older industrial suburbs are especially at risk. Dundalk, an industrial suburb in the Baltimore metro region, experienced increased poverty and declining income. The poverty rate for individuals increased from 9.2 percent in 2000 to 13.8 percent in 2017. The older industrial suburbs continue to decline.

Urban Differences

Across urban America we have the dynamic cities, such as San Jose, a Sunbelt city dominated by the booming information technology sector. On the other hand, we have the old industrial cities. Let's take a closer look at one of them.

City Focus: Baltimore

In 1950, Baltimore had a population of 950,000 and, like may cities in the United States, a vibrant manufacturing base providing jobs and economic security. The magnet of jobs attracted black migrants from the South. Since the mid-1970s, a steady loss of manufacturing jobs, due to off-shoring, relocation to suburbs and nonunion areas of the United States, and increased productivity, has eroded this economic base. This is a trend across the United States and across the world, but in Baltimore as in so many industrial cities in the United States, there were few employment alternatives, so the result was rising unemployment, especially for the unskilled. In neighborhoods across the country where there are concentrations of the unskilled and limited opportunities for retraining older workers or education for younger people, this major economic shift results in pockets of poverty.

The city lost population along with its tax base while housing in the poorer inner-city neighborhoods was abandoned (Figure 13.17). Things came to a head with riots in Baltimore in 2015. Any event has multiple causes, but there are at least three background factors that we should bear in mind. The first was the momentum of the police brutality narrative starting with events in Ferguson in 2014. The images of police violence and community perceptions of a cover-up were increasingly common, with each case reinforcing the sense of injustice.

The second is the lack of trust between police and minority black populations. Despite more black officers and more blacks in senior positions, there is still a gulf between blacks and police departments that community policing measures have failed to bridge. This turns into a chasm between poor blacks and the police because of the active policing of low-income areas. The policing of the cities in the United States is dominated by what amounts to a war against low-income minority neighborhoods. In 1980, the United States had a prison population of 500,000, but by 2018, this increased to 2.3 million as more young men, especially young men of color, were caught up in an expanding web of criminal incarceration. The narratives of tough on crime, **broken windows theory**, war on drugs, and militarization have all escalated into aggressive policing and a fractured trust between residents and police.

13.16 Affluent suburbs in Northern Virginia. Part of Megalopolis (photo: John Rennie Short)

13.17 Abandoned housing in Baltimore (photo: John Rennie Short)

Young black men are stopped more frequently and jailed more often and longer than white counterparts for similar activities. In Baltimore, one in three black men can expect to spend some time in jail during his lifetime.

The third element is the stifled economic opportunities and limited social mobility of many inner-city residents. Rising inequality in the United States has meant a small minority has done well, the middle class is squeezed, and those of lower income are trapped in funnels of failure. For young people caught in a web of multiple deprivations, street violence is commonplace.

One of the sites of the rioting in Baltimore was the blighted neighborhood of Sandtown-Winchester, which was also the scene of rioting in 1968 (see Figure 13.17). Over 37 years later, little progress has been made in a community that is 96 percent black and where almost half are children, more than double the national average. Some have moved out and some have moved on, but for those left, Martin Luther King's Dream is still just a dream.

And while the Baltimore riots focus attention on race, we also need to consider the issue of class. It is so much easier to talk about race in the United States than class, and so the wider issue of restricted opportunities for the semiskilled and unskilled, black as well as white and brown, is ignored. There is a squeeze on the semiskilled and unskilled, with the squeeze all that much tighter on the minority groups. The events in Baltimore, often seen through only the prism of race, are also freighted with concerns of class.

There is also the balkanization of metropolitan America by which declining central cities are cut off from the economic benefits of suburban growth. Baltimore's population declined from almost a million in 1950 to just over 611,648 in 2017. The wider Baltimore metro area, which includes Baltimore and surrounding suburban counties, has grown from 1.1 to 2.8 million in 2017, with the fourth largest median income in the United States.

County governments, not the city, reap all the benefits of the increased property and income taxes. There is a central city suburban fiscal disparity problem with the city hard-pressed to meet the mounting social needs of increasingly impoverished populations with a diminishing tax base. This fiscal squeeze promotes, in Baltimore as in other similar cities, an emphasis on flagship downtown developments such as football stadia, ballparks, Grand Prix events, and convention centers. These benefit downtown business interests but fail to do much for the inner city. Cities concentrate on attracting middle and upper income groups because they provide revenue. And across urban America we see pockets of gentrification and gleaming downtown towers beside stubborn pockets of poverty. Hamstrung by job loss, declining revenue, and population loss, many cities across the United States have the heavy lift of making up for decades of federal neglect and lack of a coherent and well-funded urban policy program.

To compound the problems, these neighborhoods also suffer from multiple deprivations that include unhealthy environments. Elevated lead levels in inner-city Baltimore make it difficult for children to learn and concentrate. So it is not just limited employment and educational opportunities but also a complex web of multiple deprivation that effectively traps people.

There are many Baltimores. Within the city boundaries there are old established elite areas such as Roland Park and more recently gentrified areas such as Federal Hill. The Baltimore of the riots was only part of the city, a swathe of inner-city neighborhoods impacted by job loss, poor education, and aggressive policing. But there are other Baltimores outside of Maryland. They include Akron, Birmingham, Cincinnati, Cleveland, Detroit, Pittsburgh, and Toledo. It is not just an inner-city problem. There are also the bleak areas in the cracks of the metropolis: the trailer parks and suburban rental units that house those pushed out of the city by gentrification and redevelopment. Baltimores of economic neglect, aggressive policing, and multiple deprivations are found throughout metro regions across the country. They are the places of despair that house the voiceless of the US political system, the marginalized of the US economy, and those left behind in the commodification of US society. The remarks of Martin Luther King Jr. made in 1966 still have resonance: "A riot is the language of the unheard."

City Focus: Vancouver

The west-coast Canadian city of Vancouver has witnessed tremendous population growth in the last decades. The metro population is now around 2.5 million with 16 percent of Chinese origin. For the city, it is 28 percent—and some locals now refer to it as Hongcouver. The city has changed from a British outpost to a Pacific Rim city with a very buoyant housing market bordering on overheated.

Chinese immigration restarted in 1984 when the United Kingdom agreed to hand over Hong Kong to mainland China in 1997. Canada was the first rich country to go after the wealthy. Under its Immigrant Investor Program, first introduced in 1989, foreign nationals could gain residency in Canada by loaning interest-free $800,000 (Canadian) to any of the provinces for 5 years. The program was very attractive for first Hong Kong and then mainland wealthy Chinese for whom it was a relatively cheap method to gain residence in a secure, safe country with generous social benefits. More than 130,000 individuals entered Canada through this program. Vancouver became a popular destination hub for linking families in Hong Kong and mainland China. There are so-called astronaut families with fathers working in China and mothers and children living in Vancouver.

While the program was a boon to the Chinese wealthy, it was seen increasingly in Canada as a too-cheap selling of their citizenship with negative effects on property markets. When the program was cancelled in February 2014, it had 59,000 pending applicants, 45,000 from mainland China. At its peak in 2005, the program was responsible for almost 11 percent of the roughly 250,000 immigrants allowed into the country each year.

Vancouver was a popular destination for affluent Chinese seeking a stable banking system, the rule of law, and the ability to retain wealth. The city was only a 10-hour flight from Hong Kong. Asian wealth flooded into the city and spurred city development (Figure 13.18). Property prices soared; a $20,000 house in East Vancouver in early 1979 now costs $1.3 million Canadian. Locals reap windfall benefits, but later entrants have to pay higher prices.

City Focus: Megalopolis

The largest city region in North America is a contiguous area of metropolitan counties formed by the consolidated metropolitan areas of Baltimore-Washington, Boston, Philadelphia, and New York (Figure 13.19). The region, Megalopolis, stretches 600 miles from north of Richmond in Virginia to just north of Portland in Maine and from the shores of the Northern Atlantic to the Appalachians. It covers 52,000 square miles, with a population close to 50 million. It contains just over 14 percent of the entire US population. Megalopolis is overwhelmingly suburban, with three out of every four persons living in the suburbs.

Megalopolis, like other large city regions across the country, is a place of increasing racial diversity. In the four combined MSAs that make up this extended region, the percentage of blacks, Asians, and Hispanics are, respectively, 20.2, 8.9, and 18.6 percent. The respective figures for the United States are 12.6, 4.8, and 16.3 percent. Megalopolis is one of the most racially diverse regions of the country. The suburban spread of racial minorities is also becoming more marked. The number of Hispanics living in the suburbs of New York MSA increased from 1.69 million in 2000 to 2.39 million in 2010; as a percentage of the total suburban population, they increased from 12.8 to 17.0 percent. In some metro areas, there are more minorities living in the suburbs than in the central city. In the Washington, DC MSA, for example, there were 857.380 Hispanics living in the suburbs compared to only 54,749 living in the city. The Hispanic experience in Washington MSA is predominantly a suburban one. Across the entire region the suburbs are becoming more racially diverse.

Megalopolis is home to significant numbers of immigrants from overseas. The foreign-born population has increased from 10 percent in 1960 to 23.8 percent in 2010—almost double the national average of 12.4 percent. Migrants are found in both central cities and in suburban areas; particular concentrations can be identified as immigrant gateways. One example is Tyson's Corner, Virginia, an archetypal edge city located off the Washington, DC, Beltway. The population of Tyson's Corner is around 20,000, with almost 35 percent foreign-born.

13.18 High-rise development in central Vancouver (photo: John Rennie Short)

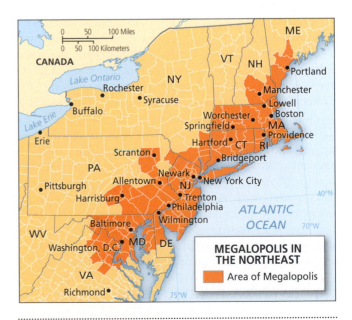

13.19 Megalopolis in the Northeast

Geopolitics

Borders

There are three borders in North America. From north to south, they are the northernmost border in the Arctic, the border between the United States and Canada, and between the United States and Mexico.

THE ARCTIC

A warming Arctic is now more accessible for resource exploitation. Rough estimates of the resource base suggest something in the order of 90 billion barrels of oil and 1.6 trillion cubic feet of natural gas, more than 10 percent of the world's untapped reserves. However, there are costs of drilling at such high and cold latitudes.

Russia, with a long presence in the Arctic region, is keen to exploit the possibilities. In 2001, in an act of scientific endeavor, political posturing, and national pride, the Russian ice breaker *Yamal* celebrated the new millennium with a trip to the North Pole, and in 2007–2008 Russia sent two miniature submarines to the floor of the Arctic Ocean. Russia now has twenty-two state-owned ice breakers and is currently building the world's largest. There are also plans to develop the vast resources. The state-owned oil company Rosneft signed a deal with ExxonMobil to exploit the oil reserves in the Kara Sea, off the coast of Siberia.

The US Coast Guard, in contrast, only has one heavy-duty ice breaker, the *Polar Star*, and the US Navy has none. While the US Navy invests in very expensive aircraft carriers to fight on the open oceans, it is woefully lacking in the ability to project power in a warming Arctic. US military strategy, it often seems, is geared up to fight the last war rather than adapting to changing conditions.

The rich resources and huge commercial opportunities raise issues of sovereignty in a region of contested sovereignty. Who owns the mining rights around the North Pole? The Arctic Council members include Canada, Denmark, Finland, Iceland, Norway, Russia, Sweden, and the United States. It provides a forum for negotiations while the United Nations Convention of the Law of the Sea (UNCLOS) provides a more binding legal framework. But as the ice melts and the seas become more open, issues of national sovereignty will arise in a region long ignored as an icy wilderness. Russia, like all the other nations involved, seeks to maximize its territorial claims and enforce exclusive economic zones. The case is still before the UNCLOS, and final demarcation of economic exclusive zones may take decades. In the meantime, Russia, with one of the largest presences in the region, is eager to assert its claims.

CANADA AND THE UNITED STATES

At 5,525 miles, the border between Canada and the United States is the longest international border between two countries. It stretches from the North Atlantic Ocean to the Pacific. Today's border is the end result of treaties and confrontation. When the United States emerged as a spate country, a border had to be drawn between the still loyal Canada and the new upstart nation. At the Treaty of Paris in 1783, a boundary line was drawn through the Connecticut River along the 45th parallel and through the St. Lawrence River and the Great Lakes. As settlers in both countries moved west, the border was fixed at the 49th parallel.

The creation of an International Boundary Commission in 1796 provided a forum for the two countries to sort out disputes. From 1872 to 1876, it established the exact boundary line through the Rockies. In 1925, the Commission was made a permanent organization responsible for surveying the border, maintaining boundary markers and a 10-foot clearance on either side of the line.

While the vast majority of the border is now agreed upon by the two countries, there are still some small areas of disagreement, including Machias Sea Island (Maine and New Brunswick), Dixon Entrance (Alaska/British Columbia), Beaufort Sea (Alaska/Yukon), and the Strait of Juan de Fuca (Washington/British Columbia).

Until 9/11, the border was only lightly policed. Since then, security has tightened, yet 300,000 people still cross it every day.

THE UNITED STATES AND MEXICO

The border between the United States and Mexico stretches across almost 2,000 miles from the Pacific to the Gulf of Mexico. Up to a million people cross each day along the forty-five official crossing points.

The border was defined by the 1848 Treaty of Guadalupe Hidalgo signed at the end of the Mexican-American War, which began in 1846 after the United States invaded Mexico. Under this treaty, Mexico lost almost half of its landmass, including what are now the states of Arizona, California, Nevada, New Mexico, and Utah as well as parts of Colorado, Kansas, Oklahoma, and Wyoming. Five years later the United States purchased a piece of land, the Gadsden Purchase, to ensure a railway right of way. The treaty and the Purchase define today's border (Figure 13.20).

For many years, the border was scarcely visible as people moved back and forth between the two national territories. From 1942 to 1964, the US government actively encouraged the immigration of seasonal agricultural workers from Mexico. In recent years, there has been a strengthening of border security, the construction of a 700-mile security fence, and the expression of a desire to build a wall along the entire border. There are at least half a million illegal crossings each year.

Half of the border is the Rio Grande. Under a deal signed between the two countries in 1945, the United States is obliged to give water from the Colorado River to Mexico while Mexico is supposed to give water from the Rio Grande to the United States. The problem is that there are water shortages in the entire region as historic drought conditions have lowered the water table and shrunk rivers and lakes. In some years, water in the Rio Grande can barely make it all the way to the Gulf of Mexico. Mexico "owes" the United States 350,000 acre-feet (the amount of water that will cover that many acres to a 1 foot depth). Increasing water demand on both sides of the border as fresh water is in diminishing supply compounds the problem. Water volume in the river is decreasing, and water quality is suffering due to pollution from fertilizers and sewage.

One problem is using a river as an international boundary; it moves as the channels snake and bend this way and then another. In 1864, heavy rains caused the river to flood over its existing banks and move the channels southward. Texas gained 700 acres from Mexico. It was only returned to Mexico in 1964, and the five thousand people who had made it their home as part of the United States were evicted. In order to tame the river's erratic ways, it was largely encased in a permanent artificial channel that fixed the boundary.

While there are many instances of conflict and strife along the border, there are also instances of communal connections. Each year at the Fandango Fronerizio in San Diego/Tijuana, dance platforms are erected beside each other across the border. The music and dancing transcend the border. The festival connects and heals, at least for a day, the scar that divides.

The United States as a Global Power

The United States is the world's superpower with the ability to extend its power across the globe. The rise to globalism is relatively recent. While the United States was expansionist from the outset, appropriating the land of the indigenous peoples, its military power was relatively weak compared to Britain, France, and Russia. After the Civil War, US foreign intervention increased. There was the westward view toward Asia. Alaska was acquired in 1867 from Russia, and a series of Pacific islands were annexed, including Midway (1867), Hawaii (1888), and Guam (1898). Then there were the opportunities provided by the decline of the Spanish Empire. The United States acquired Cuba, the Philippines, and Puerto Rico either as territories or protectorates. Between 1899 and 1902, the United States fought a bitter campaign against Filipino nationalist forces. The United

13.20 The Mexico-US border

TABLE 13.3 US Military/security interventions in South America

Ecuador 1962–63	Chile 1964–1973
Brazil 1961–64	Bolivia 1964–1975
Peru mid-1960s	Argentina 1970s
Uruguay 1964–1970	

States was also a serial interventionist in the affairs of postwar Central and South America (see Table 3.2 in Chapter 3 and Table 13.3).

The United States is now a global power with a sprawling archipelago of military bases around the world, a vast military-industrial-security complex backing a pervasive worldview dominated by visions of global hegemony. To justify the huge costs, an equally large counterforce needs to be imagined, created, or invented. This was easy during the Cold War when an "evil" Soviet empire fit the bill. With the USSR's demise came a brief period of uncertainty before global terrorism became the enemy.

Connections

Trade Associations

Trade associations are an integral element of economic globalization. The **North American Free Trade Association (NAFTA)** was established in 1994 between Canada,

Mexico, and the United States. It is one of the largest trade blocs in the world. The agreement sought to reduce barriers to trade and investment flows between the three countries. It is a coherent geographic entity with varying levels of economic development. Canada and the United States are considered rich countries; Canada's gross national income per capita is four times that of Mexico. There were fears that the disparity would lead to a downward pressure on wages and environmental regulations in a "rush to the bottom." In Mexico, the main effect was the increased growth along Mexico's northern border of **maquiladoras**, factories that import raw materials and export them northward. US agricultural exports to Mexico increased, especially beef exports. The cheap imports of rice and corn depressed the local agricultural sectors in Mexico. One consequence of the NAFTA was the establishment of the Commission for Environmental Cooperation (CEC) that in turn created the North American Environmental Atlas.

NAFTA provided a template for subsequent trade deals. The Central America Free Trade Agreement (CAFTA) was passed in 2004 and includes the United States and Costa Rica, Dominican Republic, El Salvador, Guatemala, Honduras, and Nicaragua. It aims to create a free trade zone that eventually will link up with NAFTA to create an even larger North-Central American free trade zone. Critics complain that it costs US jobs as companies move production to the low labor costs of Central America. But it also allows US companies even better access to Central America, and CAFTA is the tenth largest US export market in the world. In Central America, urban residents did better than rural residents. Richer urban residents now have access to cheaper goods. The reduction of import tariffs reduced household income in rural areas due to the import of cheap corn and rice. The clothing sector in all the non-US countries also benefited from tariff-free access to the huge US market. Both NAFTA and CAFTA are trade deals that reduce tariffs and increase the mobility of goods, services, and capital. They are not designed specifically to increase wages, increase environmental protection, or ensure social justice.

Miami as Capital of Central America and the Caribbean

Florida is now part of the United States, but it has a Caribbean connection. It shares a similar colonial history; it was Spanish from 1513 to 1821, with a brief period of British control from 1763 to 1783. It also shares a similar climate, able to grow tropical and subtropical plants such as sugar. Its weather system is more Caribbean than North American. It is no accident that the nickname of the University of Miami football team is the Hurricanes. Miami is also the regional capital of Central and South America, second and sometimes first home to many of the political and economic elites of Central and South America, with many airline connections to cities across Central and South America and throughout the Caribbean (Figure 13.21).

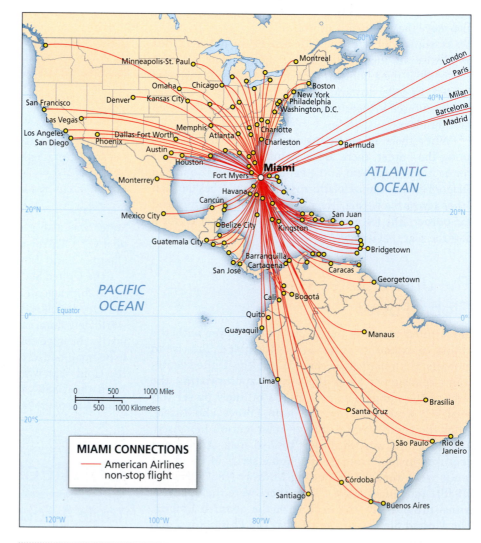

13.21 Miami connections

Subregions

This region consists of two countries, the United States and Canada. There are many similarities between the two: both are large continental territories and have a shared history as settler societies; and they are both immigrant nations and share a degree of affluence that makes both of them some of the richest countries in the world. There are also differences. The United States has a much bigger population and has the largest economy and military power in the world.

Canada is distinct from the United States in other ways. It has two official languages, English and French (Figure 13.22). The population is not spread evenly across the vast land. More than 90 percent of Canada's 35 million people live within 100 miles of the US border. While the United States broke away from the United Kingdom in the late eighteenth century, Canada stayed with the United States, only achieving full independence in the twentieth century, although the Queen remains the head of state.

Canada is the second largest country in the world after Russia. It is also more of an immigrant society than the United States with more than 6 million people born overseas—almost 20 percent of the population—compared to 12.5 for the United States. In a UN Survey, it ranked seventh out of 155 countries in terms of happiness. The United States ranked nineteenth.

Canada

From a regional perspective, Canada can be divided up in a number of ways. We can use the provinces as the basis for seven distinct regions.

Quebec is one of Canada's largest provinces, much of it part of the Canadian Shield, which limits the agricultural potential but provides a rich mineral resource base. It is a French-speaking province, the core area of early French settlement. The relationship between Quebec and the remaining mostly Anglophone Canada has varied. Calls for separation that were especially strong in the 1970s remain strong but now largely more muted after referendums on the issue. The controversy did lead multinationals and some banks to shift their head offices from Montreal to Toronto.

Ontario is the economic powerhouse of the Canadian economy. It is similar to Quebec in that the north part of the province is rich in minerals. The southern part, especially the triangular-shaped region from Windsor to Kingston, is rich farming land and a major economic center for manufacturing, now declining and expanding services. At its heart is the largest city in the nation, Toronto.

The Maritime Provinces include New Brunswick, Nova Scotia, and Prince Edward Island. Fishing, coal mining, and agriculture were the basis of the local economies. But this region lacking an industrial base became more marginalized by the late twentieth and early twenty-first centuries. The population declined as young people moved to the employment opportunities of cities such as Toronto. The population is aging. Tourism has emerged as an important industry.

Newfoundland and Labrador are on the margins of the country, poorly integrated in the national transport system. They are resource based, relying on minerals, forestry, and hydroelectricity. The heavy reliance on fishing was undercut by the rapid decline of cod stocks in the late twentieth century because of overfishing. In 1992, a moratorium on cod fishing tolled the death knell for a once thriving industry.

13.22 A bilingual nation (photo: John Rennie Short)

13.23 Farm building in Canadian prairie (photo: John Rennie Short)

The Prairie Provinces include Alberta, Manitoba, and Saskatchewan. Their name derives from the flat treeless grassland across their southern reaches (Figure 13.23). After they were linked to the Canadian core by the railway system in the late nineteenth century, immigration and agricultural settlements soon followed. Wheat was a dominant crop. Mineral exploitation, especially the production of oil and gas, transformed the economies, especially Alberta. The Athabasca oil sands, also known as the Alberta tar sands, are large deposits of heavy crude oil in northeastern Alberta. Oil and gas are now shipped by pipelines to other parts of Canada and the United States.

British Columbia entered the Canadian Federation in 1871, but it was not until the completion of the Canadian Pacific Railways in the mid-1880s that the province was integrated into the Canadian space economy. The railway stimulated trade, investment, and immigration and helped in the creation of a resource-based economy of fishing, forestry, minerals, and farming. With the rise of the Pacific Rim economies such as Japan and especially China, the province has moved to a more central position in the global economy. Vancouver is now a major city with a well-developed advanced services economy. The interior of the province is still based in declining resource based economy.

The North consists of North West Territories (NWT), Yukon, and Nunavut. These are the most sparsely populated regions of the country with a combined population of little more than 125,000. Minerals play an important part of local economies, as does federal employment. Compared to the rest of Canada, this region has the highest percentage of indigenous peoples. While the NWT has 36 percent, Nunavut, which was created in 1999, is 83 percent, mainly Inuit.

Canada was settled long before the first French and British settlers. The indigenous peoples lost much of their land, although there is now a strong movement for land claims and land compensation. Nunavut was created in part as a response to indigenous demands. The indigenous people of Canada, south of the Arctic Circle, are known as **First Nations**. Dealing with First Nations in terms of reconciliation and land claims is an important issue in contemporary Canada.

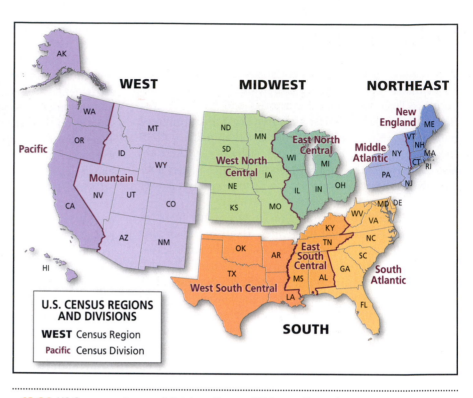

13.24 US Census regions and divisions (Source: US Census Bureau)

The United States

The regional diversity of the United States is a fascinating topic. While there is no settled agreement on what regional divisions to use, the debate is often intriguing. There is the division into states that provide a useful political geography of difference. Then there are the four Census Bureau metaregions of Northeast, South, Midwest, and West (Figure 13.24). These units provide the basis for comparing census data.

It is even more interesting when popular writers try to come up with regions with catchy titles. Joel Garreau identified nine nations (Table 13.4). Can you identify them on a map? More recently, Colin Woodward went for a more historical perspective and identified eleven regional cultures shaped by history but with a contemporary legacy. Figure 13.25, for example, shows how these regions differed in their support for Donald Trump in the 2016 presidential elections.

TABLE 13.4	Nine Nations of North America
New England	Quebec
The Foundry	Empty Quarter
Dixie	Mexamerica
The Islands	Ecotopia
Breadbasket	

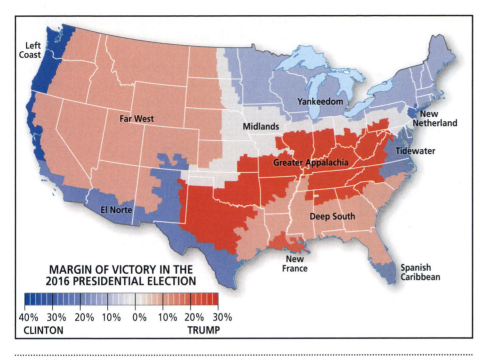

MARGIN OF VICTORY IN THE 2016 PRESIDENTIAL ELECTION

40% 30% 20% 10% 0% 10% 20% 30%
CLINTON   TRUMP

13.25 Regional differences in the 2016 Presidential election

the ecosystems and led to very marked decline in the salmon population. The river also passes through the nuclear reactors at Hanford, and nuclear waste has leaked out from the storage tanks into the rivers.

Dams are a core element of American infrastructure and provide many important services. However, aging infrastructure has led to many dams becoming obsolete, costly, and unsafe, threatening human life if they fail. By 2020, more than 65 percent of dams will be past their designated life span. These structures put a strain on American rivers and wildlife by blocking an estimated 600,000 miles of US rivers. Without a comprehensive plan for this failing infrastructure, the problem will continue to grow.

Focus: Engineering the Columbia River

The large rivers in North America are often highly engineered in order to generate electricity and or to minimize flooding.

The Columbia River rises in the Rockies of British Columbia and flows for more than 1,243 miles. In 1964, the United States and Canada signed a treaty to build and operate dams in the upper Columbia River. Four dams were built, three in Canada and one in the United States. Today there are close to four hundred dams for hydroelectrical power and for irrigation. Half of US hydroelectric power now comes from power stations on the Columbia River. The electricity currently is provided at cost, a boon to consumers and power-hungry industries such as aluminum smelting. The many dams have changed the ecosystem. While salmon ladders allow some fish to move upstream to spawn, the vast engineering has disrupted

Focus: Hawaii

The island chain was created by a volcanic hotspot located just to the east of Hawaii (Figure 13.26). The islands then drift in a northwesterly direction. Islands such as

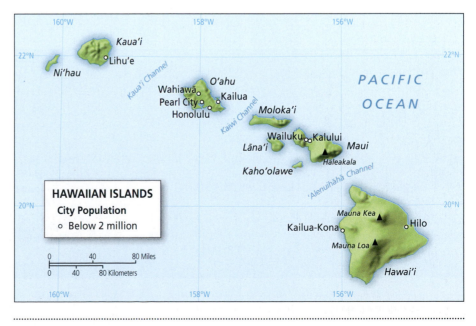

HAWAIIAN ISLANDS
City Population
○ Below 2 million

13.26 Map of Hawaiian Islands

13.27 Recent lava flows are evident in this rock formation on a beach in Hawaii. (Photo: John Rennie Short)

Kauai and Oahu are thus older and more eroded than the younger Maui and Hawaii. Volcanic activity is still a presence on the island (Figure 13.27).

The islands were settled relatively late in human history, about a thousand years ago, by Polynesians who created a sophisticated culture. Recognized by France and the United Kingdom as an independent state on November 28, 1843, they were annexed by the United States in 1893 when the US government forced the queen to abdicate. It became a state in 1959. In recent years there has been a revival of Hawaiian identity with some asking for November 28 to be recognized as La Kuokoa—Independence Day. Hawaii County Council now registers it in a list of state holidays as a way to overcome erasure of *kanaka* history. *Kanaka* is a word for native Hawaiian. Native Hawaiians constitute almost 80,000 of the 1.4 million population, with a further 210,000 putting Hawaiian as part of a mixed-race categorization.

As a US colony, the island chain became an important military base and plantation economy, especially for the growing of pineapples and sugar. Large plantations employed cheap imported labor housed in company towns. With the cheaper jet travel, the islands became a tourist attraction, and the high sea transport costs made the plantation products less economically viable. Large estates were sold off and developed as tourist resorts, replacing sugar and pineapples as the fulcrum of the Hawaiian economy.

The tourist economy was built around images of faraway islands with a languid tropicality cooled by balmy trade winds. Resorts were constructed around the images of a tropical paradise in hotels with ocean views and landscapes of palms and "tropical" vegetation (Figure 13.28).

Global warming is impacting the islands by the warming of oceans, especially during El Niño years, and the incidence of more storms and bigger storms. **Coral bleaching** occurs as the hotter ocean effectively kills the coral. Major coral bleaching events include 1997–1998 during a super El Nino, 2009–2010, and 2015 when 10 to 20 percent of the coral reef was lost. The tropical paradise is not immune to global climate change.

13.28 Luxury resort on Hawaii (photo: John Rennie Short)

Select Bibliography

Airriess, C. A., ed. 2016. *Contemporary Ethnic Geographies in America*. Lanham, MD: Rowman and Littlefield.

Baker, E. W. 1994. *American Beginnings: Exploration, Culture, and Cartography in the Land of Norumbega*. Lincoln: University of Nebraska Press.

Baptist, E. 2014. *The Half Has Never Been Told: Slavery and the Making of American Capitalism*. New York: Basic Books.

Barr, J., and E. Countryman. 2014. *Contested Spaces of Early America*. Philadelphia: University of Pennsylvania Press.

Benton-Short, L., ed. 2014. *Cities of North America*. Lanham, MD: Rowman & Littlefield.

Benton-Short, L. 2016. *The National Mall: No Ordinary Public Space*. Toronto: Toronto University Press.

Benton-Short, L., J. R. Short, and C. Mayda. 2018. *A Regional Geography of the United States and Canada*. Lanham, MD: Rowman & Littlefield.

Black, J. 2011. *Fighting for America: The Struggle for Mastery in North America, 1519–1871*. Bloomington: Indiana University Press.

Bone, R. M. 2013. *The Regional Geography of Canada*. Don Mills, Ontario: Oxford University Press.

Bunting, T., P. Filion, and R. Walker. 2010. *Canadian Cities in Transition: New Directions in the Twenty-first Century*. Don Mills, Ontario: Oxford University Press.

Case, A., and A. Deaton. 2015. "Rising Morbidity and Mortality in Midlife among White Non-Hispanic Americans in the 21st Century." *Proceedings of the National Academy of Sciences* 112:15078–15083.

Colten, C. E., and G. L. Buckley. 2014. *North American Odyssey: Historical Geographies of the Twenty-first Century*. Lanham, MD: Rowman and Littlefield.

Conzen, M., ed. 2010. *The Making of the American Landscape*. 2nd ed. New York: Routledge.

Edmonston, B., and E. Fong, eds. 2010. *The Changing Canadian Population*. Montreal: McGill-Queen's University Press.

Fenn, E. A. 2014. *Encounters at the Heart of the World: A History of the Mandan People*. New York: Hill and Wang.

Ganster, P. 2015. *The U.S.-Mexican Border Today*. Lanham, MD: Rowman and Littlefield.

Garreau, J. 1982. *The Nine Nations of North America*. New York: Avon.

Gordon, R. 2016. *The Rise and Fall of American Growth*. Princeton, NJ: Princeton University Press.

Harden, B. 2012. *A River Lost: The Life and Death of the Columbia*. New York: Norton.

Johnson, W. 2013. *River of Dark Dreams: Slavery and Empire in the Cotton Kingdom*. Boston: Belknap Press.

McGillivray, B. 2006. *Canada: A Nation of Regions*. Don Mills, Ontario: Oxford University Press.

Meinig, D. 1998–2004. *The Shaping of America: A Geographical Perspective on 500 Years of History, Volumes 1–4*. New Haven, CT: Yale University Press.

Natural Resources Canada. 2017. *The Atlas of Canada*. https://www.nrcan.gc.ca/earth-sciences/geography/atlas-canada.

NOAA. 2018. *Billion Dollar Weather and Climate Disasters*. https://www.ncdc.noaa.gov/billions/.

Short, J. R. 2006. *Alabaster Cities: Urban US since 1950*. Syracuse, NY: Syracuse University Press.

Short, J. R. 2013. *Stress Testing the United States*. New York: Palgrave Macmillan.

Steege, A. L., S. L. Baron, S. M. Marsh, C. C. Menéndez, and J. R. Myers. 2014. "Examining Occupational Health and Safety Disparities Using National Data: A Cause for Continuing Concern." *American Journal of Industrial Medicine* 57:527–538.

United States Geological Survey. 2018. *The National Map*. https://nationalmap.gov/small_scale/printable.html.

White, R. 1966. *The Organic Machine: The Remaking of the Columbia River*. New York: Hill and Wang.

Woodward, C. 2011. *American Nations*. New York: Viking.

Wyckoff. W. 2014. *How to Read the American West: A Field Guide*. Seattle: University of Washington Press.

Learning Outcomes

This region consists of Canada and the United States.

The continental extent creates a rich variety of climatic and vegetation zones.

The region has a number of environmental challenges, including the legacy of its industrial past.

The Columbian Encounter involved the loss of land for indigenous people and the creation of settler societies.

The region was globally connected initially through the export of primate product, but manufacturing and services replaced this early reliance, although there are still a number of regional and local resource-based economies.

The loss of manufacturing jobs is keenly felt in the older industrial regions and cities.

There is marked regional differentiation between declining and booming economies.

The center of population had shifted westward.

The population is aging, which creates problems for social welfare programs.

Both Canada and the United States are immigrant nations with a more recent widening of the source region of immigration from Europe to more global locales.

Race still plays a significant role in social differentiation, especially the black-white division in the United States.

Large metropolitan areas surrounded by suburban and in some cases resurgent central cities dominate the region.

While the Arctic and the Mexico-US border are sources of tension, the border between the United States and Canada is quieter.

The United States is the world's only superpower involved in interventions around the world.

Glossary

Afrikaans White Dutch settlers in South Africa. Also the name of their language.

Al Shabaab A Somali-based militant Islamic group.

algae blooms Rapid growth of algae in lakes often due to the amount of runoff high in phosphates and nitrogen from farms and lawns.

Andean Indigenous Belt A region of South America that runs from Columbia, through Ecuador and Peru, into Bolivia and parts of the Amazon rainforest where a significant indigenous population is primarily located.

animism The belief that nonhuman entities have spirits, prominent among hunter-gatherers yet evolving into subsequent cultures.

Anthropocene era The current geological period in which human activity is deemed responsible for the profound and fundamental restructuring of the environment.

anthropogenic hazards Hazards caused by human action.

aquifer An underground layer of rock that bears water.

Arab Spring A series of protests in countries in the Middle East and North Africa that began in 2010 and called for the removal of authoritarian regimes.

Arctic Council An organization whose members include Canada, Denmark, Finland, Iceland, Norway, Russia, Sweden, and the United States, which provides a forum for negotiations regarding the rights of sovereignty over areas of the North Pole.

Asian Infrastructure Investment Bank (AIIB) A China-led initiative to provide funds for development projects.

Azores-Bermuda High-Pressure Zone A subtropical, semi-permanent, high-pressure zone in the Atlantic.

Baathist A political party that ruled Iraq from 1968 to 2003.

Bali Hindu A religion in Bali that combines Hinduism and Buddhism as well as ancestor worship and animism.

Bamboo Network The name given to the extensive connections between Chinese family businesses in South East Asia.

banana republics The name, most often used as a derogatory term, given to El Salvador, Guatemala, Honduras, and Nicaragua because their economies were based on tropical primary commodities such as bananas, but were also highly corrupt and unstable.

basin An area where water drains from and flows into a river or a number of tributaries.

Bedouin A nomadic people living in the desert regions of the Middle East and North Africa.

biodiversity An indicator used to describe and measure the number of distinct plant and animal species in a given geographical area.

Bolivarism A term used to describe anti-imperialism rhetoric that promotes income redistribution, national sovereignty, and a distancing from the orbit of power of the United States.

Bollywood Name given to the globally influential movie industry that dominates Indian culture, as well as the moviemaking complex in Mumbai where many productions originate.

Brazilian Miracle A period of rapid economic growth in Brazil that lasted from 1968 to 1980.

breakaway Republics Refers to those countries that were formerly part of the old Soviet Union, but decided to separate from Russia after its fall.

BRICS An acronym for Brazil, Russia, India, China, and (sometimes) South Africa, which are grouped together due to the fact that they are all large countries with developing economies.

broken windows theory The idea that small issues such as broken windows need to be addressed in order to instill a sense of civic pride. Also used to refer to a policing strategy that addresses small issues such as panhandling in order to prevent more serious crimes.

brownfields A former industrial site affected by environmental contamination.

Byzantine Empire An empire that developed from the Roman Empire, centered on Constantinople that lasted from 285 until 1483.

cadmium A toxic chemical often created as a byproduct of zinc production.

caldera A volcanic crater caused by the destruction of the cone during violent explosions.

Caliphate The re-creation of a transnational authority guided by Islamic principles; it is a long-held belief of many Sunni fundamentalists.

carbon dioxide A gas produced from the burning of carbon. The increasing amount in the atmosphere is responsible for climate change.

carbon footprint The amount of greenhouse gases produced for any given person, entity, or practice.

Cascadia subduction zone A tectonic plate boundary that runs from Vancouver Island to Northern California.

cash crop A crop that is easy to bring to market and is also considered a primary good, such as wheat.

caste system A social system created by Hinduism and reinforced by the British in India, which is made up of social categories that determine not only position in the social hierarchy but also proportionate occupation, behaviors, and mores such as eligible marriage partners.

Central Desert Art Movement Beginning in the early 1970s Aboriginal artists in the Central Desert region of Australia began to make commercial art for a wider audience. The movement used traditional motifs in an abstract rendering of landscapes and sacred sites. Art sales are now a significant source of income for indigenous groups in the region and the art is now a significant source of national identity for the wider nation.

cerrado Name given to the forest and woodland area of Brazil.

chaebol   A system of large family-controlled firms that dominates the South Korean economy.

China's Sorrow   Another name for China's Huang He River due to the damage caused by its constant flooding.

circular migration   The movement of people from their rural homes to urban employment centers and back on a regular basis.

city-states   A country in which the city and the state are essentially one and the same.

civilization-state   A state that embodies a particular civilization.

climate change   A change in climate that takes different forms across the globe.

climate change adaption   Adapting to the challenges of climate change.

climate change mitigation   Attempts to reduce climate change.

cocaine capitalism   A name given to the large drug trade in Central and South America.

Cold War   A form of hostility between countries that is marked by threat, military posturing, and other acts short of actual warfare. Also a name given to the US–USSR conflict between 1945 and 1989.

Columbian Encounter   A term denoting the arrival of Europeans to the Americas, which resulted in vast demographic changes to indigenous populations.

Columbian Exchange   The global exchange of plants and animals unique to either the Old World or the New World.

command economy   A type of economic system that is controlled and planned by the central government.

commodification   A term describing how material becomes a resource when it is valued and traded, has a price equivalent, and enters the arena of things bought and sold.

Common Agricultural Policy (CAP)   A policy implemented in Europe after recent wartime food shortages in the mid-twentieth century that fixed food prices and limited supply from outside of a select group of European countries.

communism   A political belief system in which a socialist economic system is implemented to allow the state to maintain control over the means of production and to provide for the people according to their perceived needs.

Confucianism   A philosophy that stresses loyalty to family, upholding of social relations, and commitment to just social activity.

Coptic Christian   The Christian Church in Egypt, one of the oldest in the world.

coral bleaching   When water is too warm, coral will eject the algae in their tissues, making them turn white. It is a sign of stress and an indicator of poor health and even the death of the coral.

core-periphery model   A dynamic economic model which classifies countries as either core, periphery, or semiperiphery by the economic interactions between them.

coup d'état   An attempt by a group or individual to overthrow an existing government in order to implement their own regime.

cultural globalization   The transmission of cultural ideas and practices around the world.

Cultural Revolution   A movement in China prompted by Mao Zedong that lasted from 1966 to 1976. Officials and were removed from office, many were persecuted, and some were executed. Urban youth were removed to the countryside. It prompted widespread disruption and destruction.

Cyrillic alphabet   The alphabet used by Slavic languages, such as Russian, Bulgarian, and Serbian.

Dalai Lama   The spiritual and political leader of Tibet, who left the region in 1950 after the country was annexed by China.

Dalit   A member of the lowest caste in the traditional Indian caste system.

Darien Scheme   A plan in the late seventeenth and early eighteenth century by Scottish investors to establish a colony of Scottish settlers in the isthmus of Central America.

Dark Ages   The period in Europe from roughly 500 to 1100 associated with limited urban growth, economic development, or knowledge production.

deindustrialization   Refers to the declining size of the manufacturing sector.

delta   An area created by sediments carried by a river.

demographic dividend   Occurs when birth rates fall to a point that requires less investment in the very young, but before more investment is required of the elderly, and results in the relative and absolute increase of younger, more productive workers.

demographic holocaust   Occurred when European colonists came to the New World and brought diseases that killed millions of indigenous people.

demographic transition   A global change in mortality and fertility that occurred around 1800, but also refers to the four phases of transition.

demographic transition model   A model of different stages that begins with high birth and death rates, moves through increasing birth rates and declining death rates, to low birth rates and long life expectancy.

dependency ratio   A measure of the number of individuals below 14 and over 65 years when compared to the total working-age population.

derecho   A straight-lined, heavy wind storm that can develop and move quickly.

Desert Art movement   The development of Aboriginal art from the 1970s onward in the western desert of Australia.

desertification   A phenomenon in which areas that typically received greater amounts of precipitation become drier and take on characteristics more prevalent in desert climates.

developing country   A term used to describe countries that are progressing economically but are still not developed enough to be considered First World or developed countries.

development state   A state which takes a leading role in macro-economic planning, strategic investment, and in guiding economic growth.

diaspora   A group of people living outside of their homeland.

drumlins   A low mound of compacted earth shaped by glacial action.

Durand Line   A boundary implemented after the second Afghan War redefining the border between Pakistan and Afghanistan.

Dutch Golden Age   A period of strong economic development that centered on Amsterdam beginning in 1578 as the Dutch broke away from Spanish control.

easterlies   Prevailing winds that blow from east to west, named for their origin, not their destination.

ecological footprint   The amount of land required to support human activities.

economic globalization   The growing interdependence across national boundaries of economic flows and transactions.

economic liberalization   Refers to the movement from a more centralized form of economy to a freer, market-based economy.

El Niño   The systemic variation in sea temperature in the Pacific Ocean, off the coast of South America, typically in late December, that has an impact on the weather of the region and across other regions of the world.

Enlightenment   Intellectual belief system developed in seventeenth- and eighteenth-century Europe that stressed reason over belief and rationality over religion.

entangled bank model   Suggests a long history of cultural interaction between South East Asia, the Melanesians, and the Polynesians in the settling of the Pacific.

eskers   A long ridge of sand and gravel deposited by glacier meltwater.

ethnic cleansing   The targeting and forced removal of ethnic minorities by other groups and/or the nation-state.

Eurometro   A term used to describe the vast metropolitan area that encompasses most of Europe.

European Union (EU)   The economic and political alliance of most European countries.

evapotranspiration   The process by which water is transferred from land to the atmosphere by evaporation from soil and other surfaces and by transpiration from plants.

exclaves   Political fragments of a state not physically connected to that state and surrounded by the territory of one or other nations.

Exclusive Economic Zone (EEZ)   A 200-mile buffer zone that extends into the territorial waters surrounding a country and is prescribed by the United Nations Convention on the Law of the Sea over which a country has certain rights.

express train model   Suggests a recent and rapid expansion of Polynesian ancestors from Asia/Taiwan via coastal and island Melanesia, in the settling of the Pacific.

favelas   The name for marginal settlements or slums in Brazil.

feudalism   The dominant social system in medieval Europe, in which the nobility held lands that the peasantry were obliged to work.

finger lakes   Narrow finger-like lakes formed by glacial action. Good examples are found in the Finger Lakes region in New York State.

First Nations   The indigenous peoples of Canada who lived south of the Arctic Circle.

First World   A term used to describe countries that are considered to be rich, affluent, and have growing economies.

fjords   A long narrow sea inlet with steep sides.

flash urbanism   Urban development that uses more spectacular displays in order to demonstrate its prowess and attract tourists.

flight capital   Money that leaves other areas of the world, seeking a location with strong secrecy laws where money can be hidden and invested without scrutiny.

food insecurity   Used to describe a state in which there is not enough reliable, affordable, and nutritious food available.

foreign exchange   Currency from another country.

Fourth World   Refers to the indigenous peoples deemed "marginalized" and "dispossessed."

fracking   Injecting liquids at high pressure into underground rock formations to extract oil and gas.

free trade zones (FTZs)   Locations in which tariffs and other trade barriers are reduced or eliminated and goods, services, and capital are allowed to flow more freely between countries.

geographical imagination   A way of thinking about the world that considers the relationship between one's location and another and is sensitive to people–environment relations.

geologic hotspot   Volcanic regions fed by mantle from below the earth's surface. Not caused by tectonic plate movement.

global North   A term replacing the previously used "developed" designation for countries, named because of the location of most developed countries, which sits above the 30 degrees North latitude line.

global shift   The movement of manufacturing industries from the developed to the developing world.

global South   A term replacing the previously used "developing" designation for countries, referring to the dominance of such countries below the 30 degrees North latitude line.

Gondwanaland   A giant continent that broke up around 180 million years ago into Africa, Australia, Antarctica, India, and South America.

Great Game   The conflict between the British and Russian Empires in the nineteenth century for dominance in Afghanistan and South Central Asia.

Great Leap Forward   The campaign for rapid industrialization and collectivization in China. It lasted from 1958 to 1962 and created famine and at least 25 million people starved to death.

Great Migration   The movement that occurred between 1916 and 1970 of almost 6 million Africa Americans from the Deep South to cities in the North and West of the United States.

Great Patriotic War   Russian name for World War II.

Greek fiscal crisis   A crisis in Greece after the 2007/2008 financial meltdown caused by too much government debt.

Green Revolution   Name given to a series of innovations that were especially prominent from the late 1960s to the 1970s, which increased agriculture production through new high-yielding varieties of crops, irrigation, and the use of pesticides and fertilizers.

ground truthing   A method for mapping which references the relevant complex social geographies.

gulag   A system of labor camps maintained in the Soviet Union in the early to mid-twentieth century.

hajj   The pilgrimage to Mecca required of all devout Muslims at least once in their lives.

Han Chinese   The dominant ethnicity in China, composing more than 91 percent of China's total population.

Hanseatic League   A German-dominated alliance of merchant cities in the Baltic and northern Europe, especially strong from 1300 to 1500.

hijab   A veil typically worn by Muslim women.

Hukou   The household registration system in China that has been used to regulate and control rural citizens' movements.

hurricanes   Storms that develop over warm seas and oceans and then move across the surface of the globe that can generate high winds, tidal surges, and heavy rain.

Ice Age   Generally refers to any sustained period of cooler weather, but more specifically often used for the period from 2.5 million to 10,000 years ago when permanent ice sheets were created and the weather cooled dramatically.

imperial overstretch The tendency for empires to become involved in more foreign interventions that they can afford or manage successfully.

import substitution An economic development strategy marked by protectionist policies in order to bolster local growth.

Indigenismo A movement that seeks the reaffirmation of indigenous people and their rights.

Indochina The region that includes Cambodia, Laos, and Vietnam.

Industrial Revolution The period around 1800, centered in Britain, in which manufacturers rapidly discovered and implemented new ways of creating products.

infant mortality rate A measurement of the number of deaths of children under 1 year of age.

informal sector/economy Sector of the economy that is not recorded in government and official statistics, where few, if any, taxes are paid.

interglacial warming The period between ice ages.

Intergovernmental Panel on Climate Change (IPCC) A scientific organization established by the United Nations to provide objective information on climate change and its possible impacts.

International Date Line A line that runs from the North Pole to the South Pole that designates the point at which one calendar day turns to the next.

Intertropical Convergence Zone (ITCZ) A belt of low pressure around the equator.

invasions Name given to informal settlements in Lima, Peru.

jet stream A strong westerly wind that blows several miles above the earth's surface. The shifting nature of jet streams is responsible for changing weather patterns.

Joseon Dynasty The ruling system in Korea with one of the longest continuous political systems from 1392 to 1910.

just-in-time production An inventory strategy used by companies to meet immediate demand.

katchia badis Temporary unauthorized settlements on government land in Karachi, Pakistan.

kettle lakes Small shallow lakes formed by retreating glaciers.

Keynesianism An economic theory, named after famed economist John Maynard Keynes, in which government investments into economic activity are seen as viable, and sometimes necessary, contributors to economic growth.

Khmer Rouge A communist guerilla movement in Cambodia that held power from 1975 to 1979. Responsible for mass killings and mass starvation.

kibbutz A communal settlement in Israel.

Kobe beef Meat from cattle in Kobe, Japan. Considered a delicacy.

La Niña Often follows an El Niño event and is characterized by a decrease in sea temperature across the Eastern Central Pacific of up to 5°C, which also has an impact on the weather of the region and across other regions of the world.

latifundia An economic system in Latin America in which productive and accessible land was parceled out, often through Royal Charter, into large private estates.

Line of Control (LOC) The highly disputed, militarized region that separates Southeast Pakistan and Northern India.

loch A name used in Scotland for a lake or sea inlet.

loess Fine-grained windblown sediment.

madrassa Islamic religious school.

Mapuche An indigenous group inhabiting lands in central Chile and southwest Argentina.

maquiladoras Factories in Mexico, often close to the US-Mexico border, that export goods northward to the United States.

market economy A type of economic system that is not controlled by a central authority but rather left to the free and open collaboration and decisions of the participants in the market.

Mauritius Creole A French-based creole language spoken in Mauritius.

Mayan Collapse The abandonment of cities and towns of the Mayan Empire in the eighth and ninth centuries.

medina A high-density area found in the city center in the Muslim world.

Megalopolis Generally, the name for a very large city region. More specifically, it is the name given to the extended urban region that extends along the eastern seaboard of the United States from Boston to Washington, DC.

mestizo Refers to people of mixed European-indigenous origin, including the vast majority of the people of Central America.

middle-income trap Occurs when countries, after very rapid growth from low- to middle-income status, falter due to poor infrastructure—poor roads and inadequate sanitation, education, and health facilities—and low productivity.

minaret A tall tower in a mosque initially used to call worshippers to prayer.

Modern Standard Arabic A more formalized written standard form of Arabic that is commonly used across the Arab-speaking world, in contrast to the multiple forms of spoken Arabic.

Mongol Empire The empire founded by Genghis Khan that extended in the thirteenth and fourteenth century across Asia into Europe.

Monroe Doctrine A policy implemented by the United States that laid claim to geopolitical influence in the Central American and Caribbean region.

monsoon A seasonal wind in South East Asia that blows from land to sea in October to May (dry monsoon) and from sea to land in May to September bringing rain (wet monsoon).

monsoonal climate A climate marked by seasonal changing winds that bring in heavy rain.

moraines Glacial deposits of rock and sediment.

Moscow Plan of 1935 A blueprint for Soviet city structures, which was influential in creating the urbanized structure of modern Russia.

mulatto A term used to describe people with both black and white ancestry.

myxomatosis A highly infectious and fatal disease that affects rabbits. Introduced into Australia in 1959 to control the rapidly growing rabbit population.

narco economy A national or regional economy that relies heavily on the profits from the trade in illegal drugs.

nation Community of people with a common identity, shared cultural values, and a commitment and attachment to a particular area.

negritude A movement that values a renewed appreciation of African culture.

Neo-Keynesianism Based on the theories of John Maynard Keynes that posit government can influence economic growth through monetary policies to overcome price rigidity and imperfect labor markets.

neoliberalism An economic ideology that promotes deregulation, minimal or small government, low taxation, and free trade.

newly industrialized countries A term used more frequently since the end of the Cold War referring to countries that have since industrialized and matured into First World countries.

Nine dash line Name given to China's claimed territory in the South China Sea due to its presentation on Chinese maps with nine dashes.

nongovernment organizations Organizations that are not sanctioned or affiliated with any particular government or state but may be funded by particular countries. Examples include the International Monetary Fund, the World Bank, and the World Trade Organization.

North American Free Trade Association (NAFTA) A trade union between Canada, Mexico, and the United States.

North Atlantic Treaty Organization (NATO) A military alliance formed in 1949 between Canada, the United States, and European countries as a bulwark against the Soviet Union.

oligarchs A very rich business person who might also maintain a great deal of political authority. Oligarchs proliferated in Russia after the fall of the USSR as many previously state-owned assets were acquired below market value.

Opium War Conflict between China and the United Kingdom in 1839–1842 over the opening of China's eastern ports to European traders and especially the import of opium into China by the British.

organic farming A form of agriculture that does not include the use of synthetic chemicals and fertilizers in agriculture in order to protect the environment and provide for more sustainable long-term agricultural opportunities.

Organization for Economic Co-operation (OECD) A grouping of the richest thirty-five countries of the world in North America, Western Europe, East Asia, and South America.

Organization of Petroleum Countries (OPEC) A trade cartel of oil-producing states, mostly Middle Eastern, which uses its power to influence global oil markets.

Pacific Ring of Fire An arc of tectonic activity, including volcanoes and earthquakes, caused by the movement of tectonic plates that surround the Pacific Ocean.

palimpsest Originally refers to a manuscript in which the original script was overlain by subsequent scripts but some still remain visible.

pampas The grasslands of South America.

Pan Americanism The principle of political or commercial cooperation between the United States and the countries of South America, as well as those in Central America and the Caribbean.

paramilitaries An amalgam of gangs, ex-military, and military backed irregulars within a country, which were often involved in appropriations of peasant lands in South America.

Partition The division of British India into India and Pakistan in 1947.

Pearl River Delta A polycentric city of eleven compact cities, including Hong Kong, Macau, Huizhou, and Guangzhou, that is projected to soon increase its population to 65 million and is one of China's industrial growth regions.

peat bog A wetland filled with dead plant material.

permafrost A layer of soil that remains frozen all the year round.

pied-noir A French expatriate living in Algeria.

Pipelineistan A vast network of pipelines used to transport oil and gas deposits from Russia and central Asia to markets in Europe and China.

pivot to Asia Refers to the US commitment to building up military power in Asia in response to Chinese political and military shows of force.

pogrom An organized killing of a particular ethnic group.

polders Low-lying land that is drained of water and turned into productive land.

political globalization The growing political cooperation between states and their greater use of transnational organizations and global nongovernment organizations.

polycentric city A metropolitan area with more than one major city.

predatory elites Elite groups that use their political power to enrich themselves rather than the national welfare and at the expense of general social welfare.

primacy The dominance of a country by one city or region.

primary commodities Goods that come from agriculture, forestry, mining, and fishing.

primate city A city that is the largest in the country and is the center of economic and political life.

prime meridian The line of longitude designated as the zero degree line.

protectionism Any form of economic policy that seeks to support local companies at the expense of foreign corporations, usually through quotas, tariffs, or some combination thereof.

Qing Dynasty One of China's most dominant dynasties. China became China, as we know it today, under the Qing Dynasty (1644–1912) when the approximate boundaries of contemporary China were created.

remittances The money that temporary and permanent migrants send back to their home country.

Renaissance The cultural development in fourteenth-century Europe, centered in Italy, that drew upon classical learning.

resource-based economies Economies that are heavily dependent on the exploitation of primary commodities such as oil and gas.

resource curse The downside of a heavy reliance on one or more primary commodities such as oil, which includes corruption, wildly fluctuating government revenues, and stunted economic development.

Richter scale A logarithmic scale that measure seismic disturbances from 1 to 10.

Rose Revolution The democratic movement in Ukraine after the collapse of the Soviet Union.

Russification The ideological movement by Russian leaders to promote the active creation of a single Russian empire with one religion and a single language.

Rustbelt The region of the Northeast and Midwest associated with a declining manufacturing sector. Also sometimes known as the Frostbelt.

Rwanda genocide The slaughter of half a million Tutsis by rival Hutu in Rwanda in 1994.

Saffron Revolution Popular protest against the military junta of Myanmar in 2007. Buddhist monks played an important role.

San Andreas Fault The 750-mile active zone between the Pacific and North American Plates that runs from San Francisco to southern California.

savannah A grassy plain that contains few trees.

Schengen Area A single territorial unit created in 1985 that allows for the freedom of movement and capital mobility in Europe.

Scientific Revolution Rapid advances in scientific understanding in Europe in the seventeenth century based on observation, measurement, and experimentation.

sea surface temperature (SST) The surface temperature of seawater.

Second World A term popularly used during or before the Cold War, for describing communist countries.

service economy A segment of the economy that is made up of jobs and services performed as opposed to goods produced.

settler societies In general, refers to any society dominated by settlers from elsewhere. More specifically refers to societies such as Australia, Canada, and United States that were populated by settlers form Europe.

sharia law A legal system based on the precepts of Islam.

shifting cultivation Involves utilizing fields and gardens over a wide area through rotation in time, as opposed to the fixed cultivation that is necessarily dominant in the densely populated South Asian region, for example.

Shiite A sect of Islam distinct from Sunni whose adherents believe that the succession of the Prophet Mohammed should be hereditary and follow family bloodlines.

Silk Road A significant and highly influential ancient trade route that linked Europe with China.

slow boat model The theory that the settling of Polynesia started around 8000 BCE.

slum A term for the unplanned, often illegal, informal housing in cities that arises due to the inability of formal markets and public authorities to provide enough affordable and accessible housing. Slums are also referred to as "shanty-towns," "informal housing," and "squatter housing."

snowbirds Seasonal migrants who leave cold regions in the winter months for warmer climes.

South Aegean Volcanic Arc of Islands A chain of volcanic islands formed by plate tectonic activity in the Aegean Sea.

Soviet Realism A realist idealistic art style imposed by central authorities in the USSR from 1932 to 1988.

Soviet Socialist Republics (SSRs) Countries that were part of the Soviet Union, some of which maintained some relative autonomy.

Soviet Union (USSR) The world's first communist state established in 1917 that encompassed modern-day Russia along with a number of countries that surround it. The USSR collapsed in late 1991.

space-time compression The process by which improvements in transport make it quicker to cover the same distance and thus bring places closer together.

starchitects The name given to big-name, well-known architects who participate in construction to market or brand a city.

starchitecture See *starchitects*.

state A separate and distinct unit of political authority.

subak A self-organized collective of farmers who share irrigation water from a common source, found throughout Bali.

subduction When a tectonic plate is pushed under another plate.

Sunbelt The region stretching from California to Florida associated with warm weather, bright sunshine, and economic growth.

Sunni A sect of Islam distinct from Shia in that its adherents do not believe that the succession of the Prophet Mohammed should be hereditary and follow family bloodlines.

Superfund sites Areas of environmental contamination. Some of the most polluted sites in the United States designated for clean-up.

sustainability The ability to meet the needs of the present without destroying the needs of the future.

swidden agriculture An agricultural system in which temporary clearings are cropped for fewer years than they are allowed to remain fallow and involve "slash-and-burn" clearing practices.

syncretic religions Religions with beliefs and practices created from multiple religious traditions.

taiga Region found between the tundra and steppe regions of northern latitudes that is generally composed of coniferous forests.

theocracy A system of government that is ruled by the leaders of a religion or based on the beliefs of a single religion.

Theravada Buddhism A form of Buddhism based on the oldest known records of Buddha's teaching. Dominant form of Buddhism in Cambodia, Laos, Myanmar, Sri Lanka, and Thailand.

Third World A term used to describe countries that are poor with high population growth and numerous economic and societal problems.

T-O maps Early maps first described in the eighth century and printed in the fifteenth century, which depict a view from the northern hemisphere of the world as circular and surrounded by an ocean.

tornadoes A storm of violently rotating wind.

trade associations An agreement between countries that typically results in reduced barriers to trade and investment flows between the countries (e.g., NAFTA).

Trans-Pacific Partnership (TPP) An economic trade agreement signed in 2016 by twelve countries along the Pacific Rim.

triangular trade route A major sailing route that linked Europe, Africa, and the New World in the transport of goods and slaves.

tropical rainforest A forested area in the tropics marked by substantial rainfall.

tropical zone The region of the world that falls between the Tropic of Cancer (23.43 degrees North) and the Tropic of Capricorn (23.43 degrees South).

tsunami A large sea wave resulting from a tectonic shift.

typhoon A storm arising over warm sea water that produces heavy rainfall and punishing winds. The terms "hurricane" and "cyclone" are also used.

underdeveloped country See *Third World*.

urban primacy   Refers to a situation in which the concentration of a nation's population is in just one city.

urbicide   The use of violence to destroy a city.

U-shaped valleys   Eroded by glaciation, these valleys have very steep sides.

value-added chain   A series of vertically linked activities each of which add value. Examples include transforming milk into yogurt, steel into motor cars or financial statistics into financial knowledge.

Venetian Empire   A rich maritime trading empire centered in Venice that had port cities throughout the Adriatic, Eastern Mediterranean, and Black Sea. Lasted from the eighth to the eighteenth century with its greatest power around 1500.

Virgin Lands Campaign   A plan initiated in 1953 to open up vast areas of Kazakhstan steppe to grain production.

vodou   Also referred to as *voodoo*. Emerged in the New World from the mixing of African and Christian religious beliefs and practices.

volcanic hotspot   An area of volcanic activity. Not caused by plate tectonics but by plumes from the mantle.

Wahhabism   An austere form of Sunni fundamentalism in the Islam religion that is heavily promoted by Saudi Arabia.

westerlies   Prevailing winds that blow from west to east; named for their origin, not their destination.

Wilderness–urban interface (WUI)   The zone between protected wild land and land being developed.

Yom Kippur War   The war resulting from the attack in 1973 of Israel by Egypt, Iraq, and Syria.

youth bulge   A rapid increase in the number of people between the ages of 15 and 24 years.

zambos   A term used to describe those with black and indigenous ancestry.

Zionism   A belief and adherence to the idea that Jewish people should be allowed to establish and maintain their own state in Israel.

Index

Note: Page references followed by a "*t*" indicate table; "*f*" indicate figure.